Manpower
  #12
P

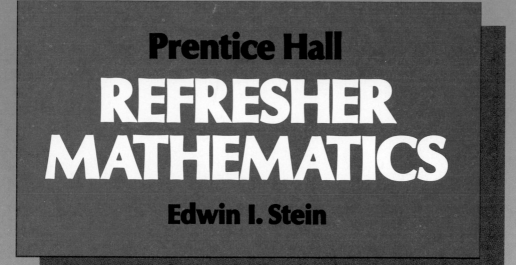

Prentice Hall

REFRESHER MATHEMATICS

Edwin I. Stein

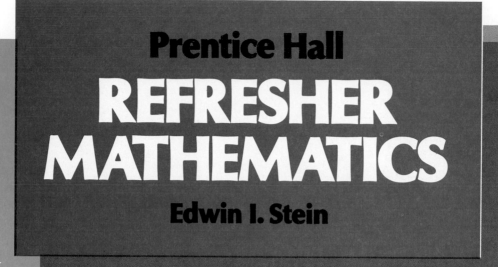

Prentice Hall

REFRESHER MATHEMATICS

Edwin I. Stein

Prentice Hall
Needham, Massachusetts   Englewood Cliffs, New Jersey

Credits

SYLVIA W. CLARK, MARY A. COSTICH
Editorial Development

L. CHRISTOPHER VALENTE
Art Director

RICHARD DALTON
Design Production

MARTHA E. BALLENTINE
Production/Manufacturing

LINDA DANA WILLIS, PETER BROOKS,
CAROLYNN DeCILLO, SUSAN GEROULD
Production/Design Services

KIRCHOFF/WOHLBERG, INC.
Design

BOSTON GRAPHICS INC., VANTAGE ART, INC.
Technical Art

SUSAN VAN ETTEN
Photo Research

JOHN MARTUCCI/L. CHRISTOPHER VALENTE
Cover Design

The author acknowledges the assistance of his wife Elaine, and the contributions made by Marilyn Lieberman, Mathematics Coordinator, Meadowbrook School, Meadowbrook, PA; Charlotte Jaffe, Director of Gifted Education, Clementon School, Clementon NJ, and Peter J. Jaffe.

ISBN 0-13-771122-0

Printed in the United States of America

    4 5 6 7 8 9   96 95 94 93 92 91 90

A SIMON & SCHUSTER COMPANY

Table of Contents

Chapter 4 Percent **223**

Unit 2 Consumer Applications 268

Unit 3

Measurement, Graphs, Statistics, and Probability 348

Chapter 8 Metric System 351

Chapter 9 Customary System, Time 371

Unit 4

Geometry 422

Chapter 11 Angles and Triangles 427

Algebra

528

Chapter 17 Graphing in the Number Plane   593

Unit 1

Basic Skills in Arithmetic

3

Inventory Test

7   **1.** Are enough facts given? If not, which fact is missing?

Mario ran 5 miles. How many minutes did it take him to run a mile?

19 **3.** Read or write in words: 95,237.

26 **5.** Round 6,849 to the nearest hundred.

37 **7.** Subtract: 9,856 − 3,682

61 **9.** Divide: $34\overline{)918}$

77 **11.** Read or write in words: 0.375

83 **13.** Round to the nearest tenth: 6.79

88 **15.** Add: 0.39 + 2.5

102 **17.** Multiply: 5.4 × 0.13

120 **19.** What decimal part of 25 is 16?

126 **21.** Multiply: 100 × 35

135 **23.** Write the standard form for 2.7 *million*.

149 **25.** Express $\frac{10}{25}$ in lowest terms.

153 **27.** Complete: $\frac{3}{5} = \frac{?}{20}$

157 **29.** Find the least common multiple of 12 and 16.

162 **31.** Find the least common denominator of the fractions: $\frac{1}{2}$ and $\frac{2}{3}$

170 **33.** Subtract: $5\frac{3}{4} - 2\frac{1}{3}$

179 **35.** Multiply: $\frac{2}{3} \times \frac{4}{5}$

195 **37.** 8 is what part of 20?

200 **39.** Write $\frac{3}{4}$ as a decimal.

210 **41.** Find the ratio of 25 to 75.

225 **43.** Write *seventeen-hundredths* as a percent, as a decimal, and as a fraction.

229 **45.** Write 0.12 as a percent.

233 **47.** Write $\frac{9}{10}$ as a percent.

242 **49.** What percent of 60 is 45?

12 **2.** Write an equation that represents your plan of solution in the following.

An auditorium has 400 seats in 25 rows. How many seats are in each row?

23 **4.** Write in standard form: Eight hundred seven thousand forty-eight

29 **6.** Add: 578 + 946 + 825

49 **8.** Multiply: 59 × 83

68 **10.** Name the greatest common factor of 54 and 72.

80 **12.** Write ninety-six hundredths.

86 **14.** Which is less: 0.7 or 0.65?

95 **16.** Subtract: 9 − 4.6

113 **18.** Divide: $0.2\overline{)6.8}$

123 **20.** 0.3 of what number is 12?

129 **22.** Divide: 4.9 ÷ 10

138 **24.** Write the short name for 16,300,000 in millions.

151 **26.** Rewrite $\frac{11}{8}$ as a mixed number.

155 **28.** Are $\frac{12}{15}$ and $\frac{28}{35}$ equivalent fractions?

159 **30.** Which is greater: $\frac{5}{12}$ or $\frac{5}{8}$?

164 **32.** Add: $4\frac{2}{5} + 7\frac{3}{10}$

177 **34.** Rewrite $4\frac{1}{2}$ as an improper fraction.

188 **36.** Divide: $2\frac{1}{4} \div \frac{1}{2}$

195 **38.** $\frac{1}{4}$ of what number is 36?

204 **40.** Write 0.8 as a fraction.

213 **42.** Solve and check: $\frac{n}{12} = \frac{27}{36}$

227 **44.** Write 9% as a decimal.

231 **46.** Write 40% as a fraction.

237 **48.** Find 25% of 93.

250 **50.** 80% of what number is 72?

Introduction To Problem Solving

Below is an easy step-by-step approach to help you master the techniques of problem solving.

1. **READ** the problem carefully to determine:

    a. The question asked.
    b. The given facts.
       • Some information may be hidden
       • Some information may be missing
       • Some information may not be needed

2. **PLAN** how to solve the problem.

    a. Choose an appropriate strategy.
       • Solve a simpler problem
       • Guess and test
       • Solve another way
       • Work backward
       • Make a drawing or model
       • Look for patterns
       • Make a table or graph
    b. Develop your plan.
       • Think or write out your plan in words.
       • Sometimes you choose an operation.
       • Sometimes you can write an equation or use a formula.

3. **SOLVE** the problem.

    a. Implement your plan.
       • Sometimes you make a graph, model, drawing, or table.
       • Sometimes you choose an appropriate method of computation
          —estimation
          —mental arithmetic
          —pencil and paper (estimate and compute)
          —calculator
    b. Find the solution.
    c. Answer the question being asked.

4. **CHECK** your solution.

    a. Check the accuracy. Check the units.
    b. See if your answer is reasonable, given:
       • your estimate
       • the information in the problem

Throughout this text, you will learn the specific strategies listed in the PLAN step that will help you in applying these problem solving techniques.

Reading to Understand a Problem

Problem Solving Step 1

1. **READ** the problem carefully to find
   a. the question asked
   b. the given facts

> To help identify the question asked and the given facts, rewrite the problem in your own words.

Problems

1. The highest point in Asia is Mount Everest, with an elevation of 8,848 meters. Mount McKinley, in Alaska, with an elevation of 6,194 meters, is the highest point in North America. What is the difference in their elevations?

   REWRITE
   Mt. Everest is 8,848 meters high.
   Mt. McKinley is 6,194 meters high.
   How much higher is Mt. Everest?

   QUESTION ASKED
   What is the difference in elevation?

   GIVEN FACTS
   Elevation of Mt. Everest is 8,848 meters.
   Elevation of Mt. McKinley is 6,194 meters.

> To find if all the given facts are needed or if a needed fact is missing, ask yourself: Do I have all the information to answer the question?

2. An airplane took off from Boston at 8:45 A.M. and arrived in Atlanta in 2 hours. What is the average speed in miles per hour?

   MISSING FACT
   The distance in miles
   from Boston to Atlanta.

   FACT NOT NEEDED
   Took off at 8:45 A.M.

When facts are given in the form of a table, graph, or map, read the title and labels carefully. The title states what is being showm, and the labels identify the specific facts being presented.

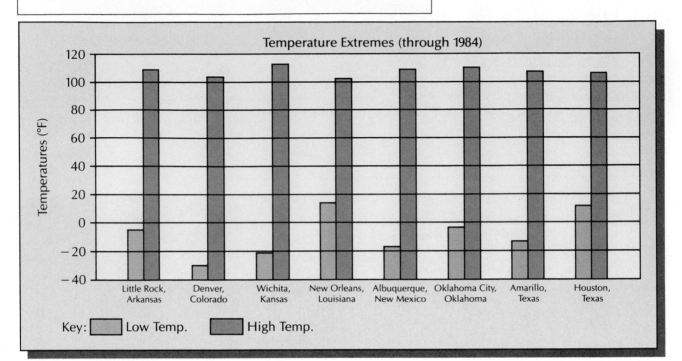

READ THE TITLE
This is a graph of the record high and low temperatures in eight cities, made from data collected through 1984.

READ THE LABELS
The key shows that blue bars represent record lows and red bars represent record highs. Temperatures are shown in degrees Fahrenheit. The cities included are Little Rock, Arkansas; Denver, Colorado; Wichita, Kansas; New Orleans, Louisiana; Albuquerque, New Mexico; Oklahoma City, Oklahoma; Amarillo, Texas; and Houston, Texas.

Think About It

1. Why is it a good practice to rewrite a problem in your own words?

2. Why is it important to know if too many facts are given?

3. In finding the facts of a problem, why is it important to know if too few facts are given?

4. What are some advantages of presenting certain types of information in the form of a graph, table, or map?

Diagnostic Exercises

1. Rewrite the following problem in your own words.

A heating engineer, when installing an oil burner, finds it necessary to use pieces of pipe measuring $4\frac{7}{8}$ inches, $7\frac{5}{8}$ inches, $3\frac{1}{2}$ inches, and 9 inches. What length of pipe does he need to be able to cut the 4 pieces, disregarding waste?

2. Find the given facts in the following.

How long was a telephone call if the charge between the two cities was $2.54 at the rate of $.38 for the first minute and $.27 for each additional minute?

3. What is the question in the following?

Find the sale price of a computer that regularly sells for $849.95, but can now be purchased at a 20% reduction.

4. Which fact is not needed to solve the following problem?

The Bells live 16.5 miles from the airport.
It takes them 25 minutes to get there.
They live 26.4 miles from the nearest railroad station.
How much closer do they live to the airport?

5. Are enough facts given to solve the problem? If not, which fact is missing?

Lisa's piano lesson lasted $1\frac{1}{2}$ hours this week. How much longer was her piano lesson last week?

6. According to its title, what is being shown in the table below? According to the labels, what specific facts are being presented?

Enrollment at Grant H.S., 1987–88					
Grade	9	10	11	12	Total
Boys	130	98	106	129	163
Girls	125	117	132	103	477
Total	255	215	238	232	940

Related Practice

1. Rewrite each of the following problems in your own words.

   a. The Lawrence family drove 424 miles to their seashore home. If it took 8 hours to drive this distance, what was the average rate of speed?

   b. At the three-game home series, the local baseball team played before 14,526 spectators for the first game, 9,277 spectators for the second game, and 12,965 spectators for the third game. Find the total attendance.

2. Find the given facts in each of the following.

   a. A ribbon 42 inches long is cut into 3 equal pieces. How long is each piece?

   b. How much will a taxi ride cost if you travel 2 miles at a cost of $.90 for the first $\frac{1}{4}$ mile and $.50 for each additional $\frac{1}{4}$ mile?

3. What is the question in each of the following?

   a. Teresa bought a bicycle that regularly sells for $125 at a 20% reduction sale. How much did she save?

   b. The Metroliner travels 68.4 miles from Wilmington to Baltimore at an average speed of 94.5 miles per hour. How long does it take the Metroliner to travel this distance?

4. Which fact is not needed to solve each of the following problems?

   a. Gary plans to see a movie at the 8:15 P.M. showing. The movie lasts $1\frac{1}{2}$ hours. It is now 6:30 P.M. How much time does he have to get to the movie theater?

   b. Charlotte's station wagon holds 18 gallons of gasoline. She drove 126 miles in one day. How many gallons are in the tank when the gauge shows $\frac{1}{2}$?

5. Are enough facts given to solve the problem? If not, which fact is missing?

   a. John and David ran for student council president at the Evan school. John received 923 votes. How many more votes did David receive?

   b. The Jordans bought 8 flowering bushes at $6.95 per bush and 12 tomato plants. What was the total amount that the Jordans spent?

6. According to its title, what is being shown in the map at the right? What specific information is presented?

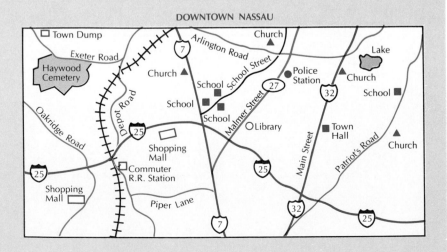

DOWNTOWN NASSAU

Using Mathematics

OPERATION INDICATORS

To help you choose the operation needed you can refer to this table of operation indicators. Look for these words and phrases as you read word problems.

Addition Indicators	Subtraction Indicators	Multiplication Indicators	Division Indicators
how much in all	how many more	total	how much is each
how many in all	how many fewer	how many in all	how many in each
sum	how much larger	how much altogether	find the average
total	how much smaller	Note: All indicators	Note: All indicators
how much altogether	amount of increase	are used when	generally refer to
Note: All indicators	amount of decrease	combining equal	equal quantities.
usually refer to	difference	quantities.	
unequal amounts.	balance		
	how many left over		
	how much left over		

Look for operation indicators to help choose the operation needed.

Examples

1. At the supermarket Selma bought hamburger in packages weighing 1.65 pounds, 2.14 pounds, and 1.85 pounds. What is the total weight of the ground meat?

OPERATION: Addition

When combining two or more quantities of unequal (or equal) size to find a sum or total, use addition.

2. On Friday the Dow Jones Average increased from 2191.45 to 2200.93. What is the amount of increase?

OPERATION: Subtraction

When finding the difference between two quantities or how many more or less or the amount of increase or decrease, use subtraction.

3. Julio bought 3 shirts, each costing $11.49. What is the total cost?

OPERATION: Multiplication

When combining two or more quantities of equal size to find a total, use multiplication.

4. A 15-foot board is cut into 6 equal pieces. What is the length of each piece?

OPERATION: Division

When separating a given quantity into a number of groups (or parts) of equal size, use division.

• Divide the given quantity by the number of groups to find the size of the group.
• Divide the given quantity by the size of the group to find the number of groups.

1-2 *Making Your Plan*

Problem Solving Step 2

2. **PLAN** how to solve the problem.

    a. Choose an appropriate strategy.

b. Develop your plan:
Think or write out your plan.
Sometimes you choose an operation.
Sometimes you write an equation or use a formula.

Problems

1. The Rose Bowl in Pasadena has a seating capacity of 106,721 and the Sugar Bowl in New Orleans has a seating capacity of 80,982. How many more seats does the Rose Bowl have than the Sugar Bowl?

    a. Choose a strategy: Solve a simpler problem.
Use simpler numbers. Think: 100,000; 80,000.

    b. Develop your plan.
In formalizing your plan, express in your own words the relationship between the given facts and the question asked. There are about 100,000 seats in the Rose Bowl and about 80,000 seats in the Sugar Bowl. How many more in the Rose Bowl?

    c. Choose an operation.
The words *how many more* indicate subtraction. After you have expressed the relationship between the question asked and the given facts, translate it into symbols.

The difference in seats between the Rose Bowl and the Sugar Bowl is how much?

$$106{,}721 - 80{,}982 \qquad\qquad = \quad n$$

---

2. The Barkers are going to drive to visit friends in a city 240 miles away. If they average 48 miles per hour, how long will the trip take?

QUESTION ASKED:
How long will the trip take?
OPERATION INDICATORS:
average speed
how long

GIVEN FACTS:
The distance is 240 miles. The car travels at an average speed of 48 miles per hour.
OPERATION NEEDED:
division

Write an equation.   The distance divided by the average speed is the time.

$$240 \qquad \div \qquad 48 \qquad = \quad t$$

1. What is meant by the term "operation indicators"?

2. How does using simpler numbers in place of given numbers help you to find the operation needed?

3. Once you have written a word relationship between given facts, why is it helpful to translate this relationship into symbols?

4. The road distance from Willimantic to Norwich is a certain number of miles. Ledyard is several miles beyond Norwich. What operation would you use to find the distance between Willimantic and Ledyard? How could you write this relationship in symbols without using numbers?

Diagnostic Exercises

State the operation indicator and determine the operation needed to solve Problems 1–4.

1. Tony repaid his loan by paying $97.50 per month for 12 months. Find the total amount paid.

2. Ben practiced the piano $2\frac{1}{2}$ hours, $1\frac{3}{4}$ hours, 2 hours, and $1\frac{1}{2}$ hours last week. How many hours did he practice in all?

3. Denise's father purchased a house costing $63,500. He paid $12,700 as a down payment. What is the balance owed?

4. A Jet helicopter flew 420 miles in 6 hours. What was its average speed?

5. Use simpler numbers to determine the operation needed to solve the problem. In one year, U.S. mints produced 666,081,544 nickels, 1,062,100,504 dimes, and 16,725,504,368 pennies. What was the total number of coins minted?

6. Express your plan in words using the facts given and the question asked. Choose the needed operation first.
   Question asked: What is the total cost?
   Given facts: skirt costs $24.95; blouse costs $19.90.

7. Write the number sentence symbolically. The product of nine and eight is equal to seventy-two.

8. Write as an equation. Fourteen increased by nine is equal to some number n.

9. Express the following sentence as an equation. The distance (4,840 miles) a Boeing 727 flies divided by its average rate of speed (605 miles per hour) is equal to the time (number of hours) of flight (n).

10. Write an equation that represents your plan of solution in the following problem. Sara's sister purchased a new car selling for $8,750. If she received $1,990 for her old car as a trade-in, what is the balance owed?

Related Practice

State the operation indicator and determine the operation needed to solve the problems in 1–4.

1. a. There are 38 rows of 24 orange trees in the grove. How many trees are there altogether?

   b. An airline has 12 flights between Atlanta and Boston each day. If each flight has the capacity to hold 226 passengers, what is the total number of passengers that the airline could fly between the two cities daily?

2. a. A baseball team had three doubles, two home runs, and seven singles in one game. What was their total number of hits?

   b. How many miles in all did Florence drive if on Friday she drove 288 miles; Saturday, 198 miles; and Sunday, 269 miles?

3. a. The Nile River is 4,145 miles long. The length of the Mississippi River is 2,348 miles. How much longer is the Nile?

   b. The barometric pressure at New York City is 29.59 inches and at Montreal is 29.94 inches. Find the difference in pressures.

4. a. Ray cut a piece of wood 20 feet long into 8 equal pieces. How long is each piece?

   b. Five pounds of lamb chops cost $16.45. What was the cost per pound?

5. Use simpler numbers to determine the operation needed to solve the following problems.

   a. Fenway Park in Boston seats 33,583 people, while the seating capacity of Wrigley Field in Chicago is 37,272 people. How many more people can be seated at Wrigley Field?

   b. The Pacific Ocean has an area of 64,186,300 square miles; Atlantic Ocean, 33,420,000 square miles; Indian Ocean, 28,350,500 square miles; and Arctic Ocean, 5,105,700 square miles. What is the total area of these four major bodies of water?

6. Express your plan in words using the given facts and the question asked. Choose the needed operation first.

   a. Question asked: How many more girls than boys are enrolled?
   Given facts: 979 girls and 896 boys are enrolled at school.

   b. Question asked: What is the total cost of the paper?
   Given facts: 6 reams of paper were purchased; each ream costs $8.95.

7. Write each of the following number sentences symbolically.

   a. The difference between twelve and seven is equal to five.

   b. The quotient of twenty-seven divided by nine is equal to three.

8. Write each of the following as an equation.

   a. Eighty-six decreased by forty-nine is equal to some number x.

   b. Fifteen percent of some amount is equal to $12.

9. Express each of the following sentences as an equation.

   a. The average rate of speed (45 miles per hour) times the travel time (5 hours) is equal to the distance traveled (d).

   b. The regular price of a dishwasher ($395) less the reduction ($72.50) is equal to the sale price (s).

10. Write an equation that represents your plan of solution in each problem.

   a. The school auditorium has 24 rows of seats. In each row there are 18 seats. Find the total number of seats.

   b. Cory Ann bought a baseball glove for $19.95 at a sale. If she saved $8.50, what was the regular price?

Problem Solving Step 3

3. **SOLVE** the problem.

   a. Choose an appropriate method of computation
   - mental arithmetic
   - estimation
   - paper and pencil (estimate and calculate)
   - calculator (estimate first)

   b. Find the solution.

Solution Method: Mental Arithmetic

Sometimes the computation can be performed mentally.

$15 + 500$	$31,680 \div 10$	$4 \times 25,000$
ANSWER: 515	**ANSWER:** 3,168	**ANSWER:** 100,000

Solution Method: Estimation

Sometimes an exact value is not required for an answer. In these cases, round the numbers to be used in order to estimate a solution.

Is 3×18 less than 40?

Estimate
$$\begin{array}{ccc} 18 & \to & 20 \\ \underline{\times\ 3} & \to & \underline{\times\ 3} \\ & & 60 \end{array}$$

ANSWER: Since 60 is a reasonable estimate, and is much greater than 40, then 3×18 is *not* less than 40.

Solution Method: Paper and Pencil

When an exact answer is required and you cannot find it using mental arithmetic, you can use paper and pencil to perform the computation. You should always estimate the answer before finding the solution.

In each of the following, *first* estimate, then find the solution.

1. 683 + 2,907

$$
\begin{array}{r}
700 \\
+\ 2,900 \\
\hline
3,600 \text{ estimate}
\end{array}
\qquad
\begin{array}{r}
683 \\
+\ 2,907 \\
\hline
3,590 \text{ solution}
\end{array}
$$

ANSWER: 3,590

2. 48,305 − 9,219

$$
\begin{array}{r}
48,000 \\
-\ 9,000 \\
\hline
39,000 \text{ estimate}
\end{array}
\qquad
\begin{array}{r}
48,305 \\
-\ 9,219 \\
\hline
39,086 \text{ solution}
\end{array}
$$

ANSWER: 39,086

3. 39 × 615

$$
\begin{array}{r}
40 \\
\times\ 600 \\
\hline
24,000 \text{ estimate}
\end{array}
\qquad
\begin{array}{r}
39 \\
\times\ 615 \\
\hline
23,985 \text{ solution}
\end{array}
$$

ANSWER: 23,985

4. $270\overline{)1890}$

$$
\begin{array}{r}
6 \text{ estimate} \\
300\overline{)1800} \\
\underline{1800}
\end{array}
\qquad
\begin{array}{r}
7 \text{ solution} \\
270\overline{)1890} \\
\underline{1890}
\end{array}
$$

ANSWER: 7

Solution Method: Calculator

If there are several numbers to be operated on, if the numbers are particularly large, or if more than one type of operation is required, you may choose to use a calculator to perform the computation. You should always estimate before you calculate.

1. 712 × 386

Estimate: 700 × 400 = 280,000

Key: 712 ⨯ 386 = 274382

ANSWER: 274,382

2. $\dfrac{88 + 93 + 75 + 68 + 84 + 90 + 87 + 79}{8}$

Estimate: 90 + 90 + 80 + 70 + 80 + 90 + 90 = 590
590 ÷ 8 = 75

Key: 88 + 93 + 75 + 68 + 84 + 90 + 87 + 79 = ÷ 8 = 83

ANSWER: 83

Study the order in which the four solution methods were presented in this lesson. Explain why it is reasonable to consider mental arithmetic, estimation, paper and pencil, and calculator in this particular order.

Diagnostic Exercises

In each of the following, choose a method of computation.
Find the answer or estimate.

1. 45 + 40

2. 5 × 30

3. Which is greater, 3 × 48 or 120?

4. 639 + 795

5. 896 − 312

6. 72 × 28

7. 810 ÷ 49

8. 926 × 85 × 4

Related Practice

In each of the following, choose a method of computation.
Find the answer or estimate.

1. a. 23 + 52 **b.** 178 − 18 **c.** 64 + 66 **d.** 97 − 43

2. a. 10 × 12 **b.** 40 ÷ 8 **c.** 77 ÷ 11 **d.** 9 × 9

3. a. Which is greater, 39 + 113 or 142? **b.** Which is less, 56 or 72 − 23?

 c. Which is less, 5 × 19 or 33 + 49? **d.** Which is greater, 81 − 27 or 64 ÷ 2?

4. a. 916 + 789 **b.** 5,824 + 683 **c.** 945 + 279 + 382 **d.** 1,321 + 582 + 907

5. a. 859 − 422 **b.** 7,811 − 968 **c.** 21,000 − 12,063 **d.** 56,302 − 9,294

6. a. 61 × 58 **b.** 605 × 93 **c.** 529 × 386 **d.** 406 × 9,117

7. a. 209 ÷ 7 **b.** 413 ÷ 50 **c.** 842 ÷ 39 **d.** 9,758 ÷ 198

8. a. 71 + 126 + 89 + 45 + 203 + 368 **b.** 12 × 6 × 31 × 8 × 25 × 2

 c. $\dfrac{95 + 82 + 87 + 94 + 92}{5}$ **d.** $\dfrac{83 + 156 + 41}{6 \times 7}$

1-4 Checking the Answer

Problem Solving Step 4

4. ☐ **CHECK** your solution.
   a. Check the accuracy. Check the units.
   b. See if your answer is reasonable, given:
      • your estimate
      • the facts of the problem

Accuracy and Reasonableness
To check the accuracy of your solution, perform the opposite operation from the one used to solve the problem. In working backward from the solution, you should arrive at the given facts. Also, check that your answer is reasonable by comparing it to your estimate.

Problem

Leaving school for a field trip are 5 buses, each carrying 38 students. How many students are going in all?

Estimate the answer:	Solution:	Check the accuracy:	Compare the answer to the estimate:
38 → 40	38	38	The answer of 190 compares reasonably with the estimate of 200.
× 5 → × 5	× 5	5)190	
200 estimate	190 solution		

ANSWER: There are 190 students going.

In solving a certain problem, Bernard made an estimate of 1500 and got 95 for his answer. Is the answer reasonable, given the estimate?

ANSWER: No. The answer 95 makes no sense if the estimate is 1500.

Diagnostic Exercise

Select the letter of the estimate that makes the answer reasonable.

1. The answer 189 is reasonable if the estimate is: **a.** 50   **b.** 200   **c.** 1000

Related Practice

Write the estimate that makes the answer reasonable.

1. a. The answer 67 is reasonable if the estimate is:	25	110	70
b. The answer 346 is reasonable if the estimate is:	400	40	4,000
c. The answer 1,529 is reasonable if the estimate is:	1,000	2,500	1,600
d. The answer 21,787 is reasonable if the estimate is:	15,000	22,000	30,000

1-5 Reading and Writing Whole Numbers

A calculator displays the number 30283146. How would you separate this number into periods? How would you read this number, or write it in words?

Separate 30283146 into periods.

$$\underbrace{3\ 0}_{}\ \underbrace{2\ 8\ 3}_{}\ \underbrace{1\ 4\ 6}_{}$$
periods

Start at the right and mark off groups of three digits.

30,283,146 or 30 283 146

ANSWER: 30,283,146 or 30 283 146

Place a comma or space between each period. This is the **standard form.**

Read 30,283,146.

READ | Start at the left, and read each period of digits; then state the name of the period.

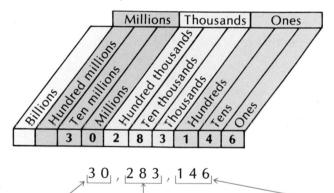

You may use a place value chart to identify the periods.

$$\underbrace{3\ 0}_{\text{thirty million}}\ ,\ \underbrace{2\ 8\ 3}_{\text{two hundred eighty-three thousand}}\ ,\ \underbrace{1\ 4\ 6}_{\text{one hundred forty-six}}$$

ANSWER: Thirty million two hundred eighty-three thousand one hundred forty-six.

Examples

1. Read 9,100 in two ways.

$$\underset{\text{nine thousand}}{9}\ ,\ \underset{\text{one hundred}}{1\ 0\ 0}$$

$$\underset{\text{ninety-one hundred}}{9\ 1\ 0\ 0}$$

Note that four-digit numerals may also be written without either a space or a comma.

ANSWER: Nine thousand one hundred, or ninety-one hundred

2. Read 2 540 000 250.

When all digits of a period are zero, do not name the period.

2 540 000 250

two billion   five hundred forty million   two hundred fifty

ANSWER: Two billion five hundred forty million two hundred fifty.

Vocabulary

The **whole numbers** are the numbers 0, 1, 2, 3, 4, 5, 6, 7, 8, 9, 10, and so on without end. The whole numbers beginning with 1 that are used in counting are called **counting,** or **natural numbers.**

A **period** is a group of three places in the number scale such as thousands, millions, etc.

A number such as 27,123 is said to be in **standard form.**

Think About It

1. Find out the names of the three periods to the left of billions in the place value chart.

2. Do you think there is a greatest whole number? Explain your answer.

Calculator Know-How

Examine your calculator and find the following keys.

Operating Keys: (Clear) $\boxed{\text{C}_{\text{E}}^{\text{C}}}$ $\boxed{\text{ON}}$ Arithmetic Keys: $\boxed{+}$ $\boxed{-}$ $\boxed{\times}$ $\boxed{\div}$ $\boxed{=}$

Special Keys: (Decimal Point) $\boxed{\cdot}$ (Percent) $\boxed{\%}$ (Add to Memory) $\boxed{\text{M}+}$

(Subtract from Memory) $\boxed{\text{M}-}$ (Recall from Memory) (Clear Memory) $\boxed{\text{M}_{\text{C}}^{\text{R}}}$

(Square Root) $\boxed{\sqrt{}}$ (Change Sign) $\boxed{+/-}$

In future lessons you will learn how to use these keys to simplify your computations.

Key the following numbers on your calculator. Read the number you see in the display. Remember to clear the display before keying a new number.

1. 369   **2.** 5,179   **3.** 62,408   **4.** 120,005

5. Work with a classmate. One person keys in a number, the other reads it. Take turns.

Diagnostic Exercises

Separate the following numerals into proper periods.

1. By using commas: 29740049265040

2. By using spaces: 817604009000

Read, or write in words, each of the following.

3. 582

4. 4,975

5. 5,003

6. 10,500

7. 823,659

8. 7,847,000

9. 3,582,942

10. 18,467,125

11. 995,028,000

12. 62 669 004 716

13. Read, or write in words, each number.

A soft drink company had a year-end sales total of $7,098,354,250.
The company had a profit of $307,818,078.

Related Practice

Separate each of the following into proper periods.

1. By using commas:
   a. 793058
   b. 96400127
   c. 592473085604
   d. 8335000000000000

2. By using spaces:
   a. 61437
   b. 483050095
   c. 216038823700000
   d. 890325000000000000

Read, or write in words, each of the following.

3. a. 426   **b.** 784   **c.** 905   **d.** 380   **4. a.** 8,278   **b.** 3,926   **c.** 4,206   **d.** 7,959

5. a. 6,004   **b.** 4,015   **c.** 9,080   **d.** 3,007   **6. a.** 27,432   **b.** 20,730   **c.** 45,063   **d.** 81,896

7. a. 150,000   **b.** 291,429   **c.** 900,281   **d.** 674,364

8. a. 9,250,000   **b.** 3,746,000   **c.** 2,006,000   **d.** 5,670,000

9. a. 7,122,843   **b.** 8,900,527   **o.** 5,416,084   **d.** 6,004,009

10. a. 32,439,784   **b.** 57,105,933   **c.** 96,863,848   **d.** 25,000,975

11. a. 700,000,000   **b.** 241,849,000   **c.** 928,376,500   **d.** 843,297,861

12. a. 5 000 000 000   **b.** 6 450 000 000   **c.** 93 884 526 767   **d.** 842 699 019 508

13. Read, or write in words, each number.
   a. The area of the Pacific Ocean is 165,246,320 square kilometers; of the Atlantic, 82,441,560 square kilometers.
   b. The total number of U.S. coins minted one year was 13,377,659,747 valued at $769,466,209.
   c. Recently Shanghai had a population of 10,820,000; Tokyo, 8,349,000; New York, 7,895,543; Moscow, 8,303,000; and Bombay, 8,202,759.
   d. The trading volume on the New York Stock Exchange one day was 149,802,130 shares making the total volume so far for the year 2,921,686,626 shares.

Using Mathematics

ROMAN NUMERALS

The ancient Romans used capital letters to write numerals. Today Roman numerals are still used on clock and watch faces, for the display of dates, and for numbering in outlines.

The Roman number symbols and their values are shown at the right.

Roman numerals are formed by writing from left to right as a sum. First the symbol for the greatest possible value is written, followed by symbols of decreasing values.

The symbols I, X, C, and M may be used as many as three times in succession.

When one of the symbols I, X, or C precedes a symbol of greater value, its value is subtracted from the greater value.

Symbol	Value
I	1
V	5
X	10
L	50
C	100
D	500
M	1,000

IV = 5 − 1 = 4	XL = 50 − 10 = 40	CD = 500 − 100 = 400
IX = 10 − 1 = 9	XC = 100 − 10 = 90	CM = 1,000 − 100 = 900

The symbols V, L, and D never precede a symbol of greater value and are never used in succession.

Examples

1. What number does MDCCXLVIII represent?

MDCCXLVIII =

M + D + C + C + XL + V + III
↓   ↓   ↓   ↓   ↓   ↓   ↓
1,000   500   100   100   (50 − 10)   5   3 = 1,748

ANSWER: 1,748

2. Write the Roman numeral representing 287.

287 = 200 + 50 + 30 + 5 + 2
↓   ↓   ↓   ↓   ↓
CC + L + XXX + V + II = CCLXXXVII

ANSWER: CCLXXXVII

Exercises

What number does each of the following Roman numerals represent?

1. XXXVI   **2.** CLXV   **3.** MDCLIX   **4.** MCCLIV   **5.** MCMXCIII   **6.** MMXLVII

Write a Roman numeral for each of the following numbers.

7. 19   **8.** 299   **9.** 847   **10.** 1,492   **11.** 1,776   **12.** 2,015

13. A cornerstone is marked MCMXIX. What date does it represent?

What number is represented by the Roman numeral in each of the following?

14. Page VII   **15.** Unit XLVI   **16.** Item CXC

What Roman numeral comes next?

17. XIII   **18.** XXIV   **19.** XCIX   **20.** XLVIII

Writing Large Whole Numbers

The land area of the United States is three million six hundred eighteen thousand seven hundred seventy square miles. How would you write this number in standard form?

Write each period. Then use a comma or space to indicate the period.

three million	six hundred eighteen thousand	seven hundred seventy
↓	↓	↓
3,	618,	770

ANSWER: The land area of the United States is 3,618,770, or 3 618 770, square miles.

Examples

1. Write in standard form: Twelve billion nine hundred twenty-six million eight hundred thousand.

twelve	billion	nine hundred twenty-six	million	eight hundred	thousand
↓	↓	↓	↓	↓	↓
12	,	926	,	800	, 000

↑
Write zeros to complete the numeral.

ANSWER: 12,926,800,000 or 12 926 800 000

2. Write in standard form: One trillion eight hundred forty million five hundred sixteen thousand seventy-one.

one trillion eight hundred forty million five hundred sixteen thousand seventy-one

1 , 000, 840 , 516 , 071

NO billions

ANSWER: 1,000,840,516,071 or 1 000 840 516 071

Think About It

1. How does the word name indicate the periods that contain all zeros in a number like 423,000,000,112?

2. True or false? When given two numbers, the word name of the greater number will be longer than the word name of the smaller. Explain your answer or give an example.

3. Describe a procedure for determining whether the number you have written for a given word name is correct.

Diagnostic Exercises

Write in standard form, using commas.

1. Six hundred fifty-two

2. Five hundred nineteen thousand eight hundred forty-seven

3. Thirty million four hundred five thousand eighteen

4. Seven billion nine hundred thirty million two hundred thousand

5. Ten trillion twenty-one billion three hundred million three hundred fifty

6. Write the number in standard form, using digits and spaces: Four hundred sixty million

Write the number in standard form, using commas.

7. In one day, the U.S. Post Office handled one hundred seventy-six million eight hundred seventy-three thousand four hundred twelve pieces of mail.

Related Practice

Write each number in standard form, using commas.

1. a. Four hundred
   c. Seven hundred nine

   b. Eight hundred thirty-three
   d. Two hundred forty-eight

2. a. Four thousand six hundred ten
   c. Seven hundred eighty-one thousand

   b. Ninety thousand fifty
   d. Five hundred sixteen thousand one hundred seventy-three

3. a. Three million nine hundred thousand

   c. Six hundred eight million one hundred fifty thousand ninety-two

   b. Eighty-five million six thousand two hundred forty
   d. Nine hundred fourteen million seven hundred thousand fifty

4. a. Forty-eight billion eighteen million seven thousand
   c. Eight hundred seventeen billion four hundred million seventy-six thousand

   b. Two hundred billion sixty-two thousand nine hundred
   d. Seven billion five hundred four million, seven hundred thousand four hundred fifty-nine

5. a. Four trillion nine hundred fifty billion

   c. Three hundred trillion forty billion five million eighty-nine thousand two hundred one

   b. Eleven trillion eight hundred billion two hundred thirty-six million
   d. One trillion six hundred eighty-two billion four hundred seven million sixteen thousand

6. Write each number in standard form, using spaces.
   a. Fourteen thousand six hundred
   b. Three million eighty thousand nine hundred twenty-three
   c. Nineteen billion seventy-five million six hundred forty thousand
   d. Eight trillion one hundred billion fifty million

7. Write each number in standard form, using commas.
   a. The area of Greenland is two million one hundred seventy-five thousand five hundred ninety-seven square kilometers.
   c. The cost of a new factory for the A & E company is estimated at five billion two hundred million dollars.

   b. The area of China is three million six hundred ninety-one thousand five hundred twenty-one square miles.
   d. The total sales of the Y & Z company last year were sixty billion forty thousand dollars.

Rounding Whole Numbers

The attendance at the baseball game was 48,628.
Round this number to the nearest thousand.
Round 48,628 to the nearest thousand.

48,628   thousands

Locate the place to be rounded.

48,628   6 is greater than 5.

If the digit to the right is 5 or more, round to the next higher digit.

49,000

Replace digits to the right with zeros.

ANSWER: 49,000

Examples

1. Round 26,385,000 to the nearest million.

     millions
26,385,000

     3 is less than 5.
26,385,000
26,000,000

If the digit to the right is less than 5, write the same digit.

ANSWER: 26,000,000

---

2. Round 6,430,859,627 to the nearest
Ten:
ANSWER: 6,430,859,630
Hundred:
ANSWER: 6,430,859,600
Thousand:
ANSWER: 6,430,860,000 ←———— Sometimes rounding to two different places can give the same result.
Ten thousand:
ANSWER: 6,430,860,000 ←
Hundred thousand:
ANSWER: 6,430,900,000 ←———— Since the digit to the right was 5, round up. Write the next higher digit.
Million:
ANSWER: 6,431,000,000
Ten million:
ANSWER: 6,430,000,000
Hundred million:
ANSWER: 6,400,000,000
Billion:
ANSWER: 6,000,000,000

Think About It

1. What is the least number that can be rounded to 45,000? What is the greatest number that can be rounded to 45,000?

2. What is the greatest possible difference between two numbers that each round to 5,000,000 to the nearest million?

3. Give three reasons for rounding numbers.

Diagnostic Exercises

Round each of the following numbers to the nearest

Ten: **1.** 57   **2.** 394

Hundred: **3.** 5,626   **4.** 890

Thousand: **5.** 4,501   **6.** 63,289

Ten thousand: **7.** 71,989   **8.** 149,000

Hundred thousand: **9.** 243,762   **10.** 1,650,059

Million: **11.** 3,728,409   **12.** 41,371,000

Billion: **13.** 24,492,568,200   **14.** 170,964,135,000

Trillion: **15.** 1,602,373,068,040   **16.** 43,089,967,000,000

17. There were 14,807 people at the basketball game. Round this number to the nearest ten thousand.

Related Practice

Round each of the following numbers to the nearest

Ten:

1. a. 28   **b.** 97   **c.** 416   **d.** 2,539

2. a. 31   **b.** 152   **c.** 683   **d.** 3,844

Hundred:

3. a. 230   **b.** 947   **c.** 3,814   **d.** 5,408

4. a. 551   **b.** 7,863   **c.** 9,092   **d.** 15,654

Thousand:

  5. a. 8,726   **b.** 24,539   **c.** 57,603   **d.** 10,954

  6. a. 9,485   **b.** 5,094   **c.** 40,159   **d.** 878,208

Ten thousand:

  7. a. 94,872   **b.** 80,056   **c.** 142,625   **d.** 6,531,009

  8. a. 56,000   **b.** 37,160   **c.** 295,928   **d.** 7,849,323

Hundred thousand:

  9. a. 345,946   **b.** 1,510,214   **c.** 3,835,837   **d.** 22,403,541

  10. a. 560,500   **b.** 2,375,908   **c.** 5,481,481   **d.** 19,792,624

Million:

  11. a. 4,500,000   **b.** 9,742,820   **c.** 8,906,437   **d.** 84,827,466

  12. a. 7,398,250   **b.** 14,487,095   **c.** 462,264,128   **d.** 8,250,175,000

Billion:

  13. a. 6,150,000,000   **b.** 39,076,524,692   **c.** 253,264,800,000   **d.** 598,337,060,489

  14. a. 8,832,000,000   **b.** 21,584,427,000   **c.** 67,940,839,148   **d.** 908,671,595,286

Trillion:

  15. a. 5,592,000,000,000   **b.** 6,460,813,289,400   **c.** 130,914,325,674,149   **d.** 815,637,453,116,588

  16. a. 7,459,540,000,000   **b.** 6,187,969,486,725   **c.** 95,019,831,645,103   **d.** 604,231,993,584,270

17. a. During the school year, the total attendance at sports competitions at State U was 874,273. Round this number to the nearest hundred thousand.

    b. The total season attendance at a major league ballpark was 1,402,684. Round this number to the nearest hundred thousand.

    c. In 1980, the population of California was 23,667,565. Round to the nearest million.

    d. In 1980, the population of New York State was 17,558,165. Round to the nearest million.

Enrichment

Decimal System of Numeration

    A system of numeration is a method of naming numbers by writing numerals.

    In the system of numeration called the decimal system, ten number symbols, 0, 1, 2, 3, 4, 5, 6, 7, 8, and 9, are used to represent all numbers. There is no single number symbol in the notation system for the number ten or for numbers greater than ten. The numerals representing numbers greater than nine are formed by writing two or more number symbols next to each other in different positions or places.

    The decimal system is built on the base, ten. The base of a system of numeration is the number it takes in any one place to make 1 in the next higher place. In the decimal system it takes ten in any one place to make 1 in the next higher place.

1-8 Adding Whole Numbers

Building Problem Solving Skills

What is the total enrollment at the County Junior High School if there are 1,075 students in the seventh grade and 881 students in the eighth grade?

Addition Indicators

- how much in all
- how many in all
- sum
- total
- how much altogether
  Note: All indicators usually refer to *unequal* amounts.

1. **READ** the problem carefully.
   a. Find the **question asked.** What is the *total* enrollment?
   b. Find the **given facts.** Two groups have already been counted: 1,075 seventh grade students and 881 eighth grade students.

2. **PLAN** how to solve the problem.
   a. Choose the **operation needed.**
   The enrollments in the seventh and eighth grades must be combined to find the *total* enrollment. Since you are combining to find a total, you must add the enrollments.
   b. **Think or write out your plan** relating the given facts to the question asked.
   The sum of the enrollments in the seventh and eighth grades is the total.
   c. Express your plan as an **equation.** $1,075 + 881 = n$

3. **SOLVE** the problem.
   a. **Estimate the answer.**

$$
\begin{array}{r}
1,075 \rightarrow 1,100 \\
+ \quad 881 \rightarrow + \quad 900 \\
\hline
2,000
\end{array}
$$

Round the numbers to the same place. Usually, this is the greatest place in the smallest number.
Add.

   b. **Solution.**

```
  1,075
+   881
      6
```

Arrange in columns. Add each column starting from the right.

```
    1
  1,075
+   881   8 + 7 = 15
  1,956
```

Regroup when necessary.
15 tens = 1 hundred 5 tens

You might use a calculator.

Key: 1075 `+` 881 `=` 1956

4. | CHECK |

a. Check the accuracy of your arithmetic. Check by adding up.

```
  1,075 ↑
+   881 |
  1,956 ✔
```

ANSWER: The total enrollment is 1,956 students.

b. Compare the answer to the estimate. The answer 1,956 students compares reasonably with the estimate, 2,000 students.

Think About It

1. Why must the total enrollment be greater than the number of students enrolled in a single grade?

2. If the enrollment at each grade had been the same, what other operation could you have used to find the total? Why?

3. When might you want to find an estimated enrollment?

Calculator Know-How

Calculators can show only a limited number of digits in the display. A maximum of 8 digits is common, so the greatest whole number that can be displayed is 99,999,999. You can add large numbers by finding the sum in parts. Here is an example.

Find the total area of the Pacific, Atlantic, and Indian Oceans using a calculator.

Estimate.

60,000,000 + 30,000,000 + 30,000,000 = 120,000,000
This is more than 8 digits. So, find the sum in parts.

Ocean	Area in Square Miles
Pacific	64,186,300
Atlantic	33,420,000
Indian	28,350,500

Separate the millions. Find the sum without the millions.
Write the figure.
Find the sum of the millions. Write the figure.
Find the combined sum.

Key: 186300 + 420000 + 350500 = 956800
Key: 64 + 33 + 28 = 125
125,956,800

Use your calculator to add.

1. 21,556,319 + 328,403,625 + 476,283,057 + 826,243,001

2. 105,327,601 + 298,353,819 + 62,395,028 + 915,690,443 + 1,381,766,891

Skills Review

1. Add: 684 + 1,292

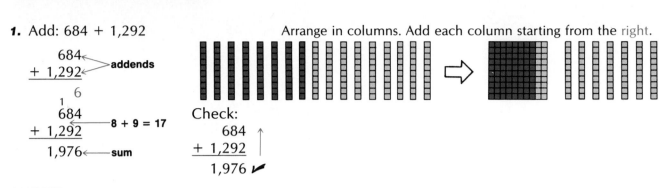

Arrange in columns. Add each column starting from the right.

$$\begin{array}{r} 684 \\ + \ 1,292 \end{array}$$ ← addends

$$\begin{array}{r} {}^{6}_{1} \\ 684 \\ + \ 1,292 \\ \hline 1,976 \end{array}$$

8 + 9 = 17

1,976 ← sum

Check:
$$\begin{array}{r} 684 \\ + \ 1,292 \\ \hline 1,976 \end{array}$$ ✔

ANSWER: 1,976

---

2. Find the sum: 2,484; 337; 4,069; 42

$$\begin{array}{r} {}^{2\,2} \\ 2,484 \\ 337 \\ 4,069 \\ + \quad 42 \\ \hline 6,932 \end{array}$$

Check:
$$\begin{array}{r} 2,484 \\ 337 \\ 4,069 \\ + \quad 42 \\ \hline 6,932 \end{array}$$ ✔

22 ones = 2 tens 2 ones
23 tens = 2 hundreds 3 tens

ANSWER: 6,932

---

3. Estimate the sum: 742,908 + 67,484 + 85,367

$$\begin{array}{rcl} 742,908 & \to & 740,000 \\ 67,484 & \to & 70,000 \\ + \quad 85,367 & \to & + \quad 90,000 \\ \hline & & 900,000 \end{array}$$

Round the numbers to the same place. Usually, this is the greatest place in the smallest number. Add.

ANSWER: 900,000

---

4. Solve the equation: 847 + 9,269 = n

$$\begin{array}{r} {}^{1\,1} \\ {}_{1}847 \\ + \ 9,269 \\ \hline 10,116 \end{array}$$

Check:
$$\begin{array}{r} 847 \\ + \ 9,269 \\ \hline 10,116 \end{array}$$ ✔

To find the value of n, add the given numbers.

ANSWER: 10,116

Vocabulary

An **addend** is a number that you add.　　　The **sum** is the answer in addition.

Diagnostic Exercises

Add and check.

1. 32 + 45	**2.** 56 + 17	**3.** 67 + 59	**4.** 9 5 + 8	**5.** 15 26 + 87	**6.** 6 2 4 + 9	**7.** 25 94 84 + 39

8. 5 7 6 8 + 9	**9.** 34 90 23 65 + 57	**10.** 389 + 459	**11.** 3,592 + 2,738	**12.** 86,056 + 44,598	**13.** 435 599 + 796	**14.** 3,598 6,487 + 5,739

15. 24,673 12,762 + 37,857	**16.** 556 479 628 + 493	**17.** 3,962 6,109 2,854 + 6,875	**18.** 79,459 68,417 75,388 + 91,754	**19.** 258 584 845 207 + 396	**20.** 4,973 9,282 3,970 2,639 + 9,789	**21.** 9,651 78 83,795 206 5,184 + 16,745

22. Add.
3,845 + 928 + 63,847 + 795 + 1,356

23. Find the sum.
2,381; 967; 29; 8,406; 750

24. Select the nearest given estimate for the sum.
29,625 + 8,370 + 41,598
a. 75,000 **b.** 80,000 **c.** 85,000

25. Solve. 726 + 93 = n

26. How many seats are there altogether at the county baseball stadium if there are 978 box seats, 19,564 general admission seats, and 3,825 bleacher seats?

Related Practice

Add and check.

1. a. 34 + 25	**b.** 67 + 32	**c.** 20 + 58	**d.** 63 + 26	**2. a.** 59 + 8	**b.** 37 + 57	**c.** 59 + 24	**d.** 28 + 62

3. a. 48 + 94	**b.** 24 + 76	**c.** 83 + 60	**d.** 95 + 99	**4. a.** 7 5 + 8	**b.** 9 7 + 7	**c.** 4 6 + 9	**d.** 9 8 + 9

5. a. 25 50 + 83	**b.** 86 92 + 58	**c.** 65 7 + 94	**d.** 34 79 + 88	**6. a.** 3 5 7 + 4	**b.** 8 4 9 + 6	**c.** 4 0 9 + 8	**d.** 7 6 2 + 9

7. a.
```
  82
  43
  92
+ 30
```
b.
```
  45
  79
  63
+ 42
```
c.
```
  94
  36
  49
+ 78
```
d.
```
  86
  59
  97
+ 95
```
8. a.
```
  5
  1
  3
  9
+ 2
```
b.
```
  8
  6
  9
  4
+ 7
```
c.
```
  8
  9
  4
  7
+ 8
```
d.
```
  9
  9
  6
  8
+ 9
```

9. a.
```
  12
  26
  21
  13
+ 22
```
b.
```
  73
   9
  24
   8
+ 87
```
c.
```
  91
  30
  54
  79
+ 68
```
d.
```
  76
  89
  97
  88
+ 53
```
10. a.
```
  231
+ 457
```
b.
```
  267
+ 385
```
c.
```
  974
+ 566
```
d.
```
  608
+  96
```

11. a.
```
  6,142
+ 3,756
```
b.
```
  2,358
+ 4,135
```
c.
```
  5,705
+   986
```
d.
```
  7,478
+ 9,757
```

12. a.
```
  37,454
+ 41,345
```
b.
```
     946
+ 30,306
```
c.
```
  35,878
+ 89,646
```
d.
```
  21,973
+ 92,027
```

13. a.
```
  582
  143
+ 214
```
b.
```
  779
  698
+ 579
```
c.
```
  926
   15
+ 658
```
d.
```
  639
  852
+ 543
```
14. a.
```
  4,214
  1,152
+ 3,403
```
b.
```
  4,043
  3,285
+ 9,772
```
c.
```
  8,588
    765
+    89
```
d.
```
  6,957
  9,799
+ 7,186
```

15. a.
```
  13,581
  61,132
+ 15,273
```
b.
```
  45,496
  27,383
+ 31,576
```
c.
```
  68,432
     257
+  9,746
```
d.
```
  73,858
  59,576
+ 38,643
```

16. a.
```
  637
  131
  528
+ 300
```
b.
```
  458
  173
  649
+ 326
```
c.
```
  682
  745
  384
+ 973
```
d.
```
   39
  896
    9
+ 372
```
17. a.
```
  2,105
  3,216
  1,143
+ 2,319
```
b.
```
  4,966
  3,859
  5,573
+ 8,426
```
c.
```
  7,581
    470
     28
+     6
```
d.
```
  6,479
  8,642
  9,736
+ 6,857
```

18. a.
```
  34,832
  11,913
  22,741
+ 10,420
```
b.
```
  56,834
  67,509
  31,915
+ 47,628
```
c.
```
  42,763
  96,833
  47,485
+ 92,739
```
d.
```
   6,445
  95,214
      87
+  1,756
```

19. a.
```
  129
  325
  356
  218
+ 497
```
b.
```
  289
  769
  427
  892
+ 976
```
c.
```
  646
   53
    8
   24
+ 593
```
d.
```
    7
  226
  589
   36
+ 328
```
20. a.
```
  1,673
  2,191
  1,814
  1,271
+ 2,353
```
b.
```
  2,765
  4,497
  2,086
  2,199
+ 7,385
```
c.
```
  7,328
    478
  9,663
     98
+ 8,976
```
d.
```
  6,494
  3,849
  8,589
  6,355
+ 2,914
```

21. a.
```
  529
   63
  784
    8
   29
+ 685
```
b.
```
  6,583
  3,476
  4,285
  3,842
  5,273
+ 2,646
```
c.
```
  95,329
     458
   7,509
      26
  84,531
+    727
```
d.
```
  82,541
  31,796
  28,865
  49,496
  22,738
+ 57,633
```

22. Add.

a. 8 + 5 + 6 + 4 + 9

b. 69 + 75 + 29 + 82 + 51

c. 47 + 419 + 72 + 5 + 27

d. 328 + 139 + 426

23. Find the sum.

    a. 86; 153; 128; 249
    b. 425; 836; 595; 26; 84
    c. 200; 175; 335; 95; 480; 45
    d. 731; 453; 846; 924; 877; 382; 502

24. Write the nearest given estimate.

a. $927 + 596 + 487$	1,000	1,500	2,000
b. $6,380 + 5,162 + 8,609$	20,000	25,000	10,000
c. $7,859 + 235 + 376$	8,000	9,000	12,000
d. $67,618 + 8953$	140,000	89,000	77,000

25. Solve.

    a. $18 + 35 = n$ **b.** $653 + 989 + 864 = n$ **c.** $1,528 + 9,396 = n$ **d.** $n = 59 + 427 + 76$

26. a. What was the total attendance at the four school league football games played by Central High School if the first game was witnessed by 2,972 spectators, the second by 2,684, the third by 3,189, and the last game by 2,708 spectators?

    b. How many farms are in the East North Central States (Ohio, Indiana, Illinois, Michigan, Wisconsin) if there are 111,322 farms in Ohio; 101,479 farms in Indiana; 123,565 in Illinois; 77,944 in Michigan; and 98,973 in Wisconsin?

    c. What is the population of New England if Maine has a population of 1,124,660; New Hampshire, 920,610; Vermont, 511,456; Massachusetts, 5,737,037; Rhode Island, 947,154; and Connecticut, 3,107,576?

    d. The school library has 2,309 fiction books, 1,894 nonfiction books, 195 reference books, and 275 magazines. How many books and magazines are there in all?

Enrichment

Reading and Writing Number Patterns

Long lists of numbers that follow a pattern can be written as shortened lists using three dots. The three dots indicate that the numbers continue in the same pattern.

To write an unending list of numbers that follow a pattern:

a. Write the first few numbers to show the pattern.

b. Place three dots after the last listed number.

1. Write the list indicating:

Zero, seven, fourteen, twenty-one, and so on, without end.

ANSWER: 0,7,14,21, . . .

To write a list of numbers that follow a pattern but with too many numbers to list:

a. Write the first few numbers to show the pattern.

b. Place three dots before the last number.

2. Write the list indicating:

One, three, five, seven, and so on, up to and including twenty-nine.

ANSWER: 1,3,5,7, . . . ,29

Read, or write in words, each of the following:

    1. 0,1,2,3, . . .   **2.** 1,3,5,7, . . .   **3.** 0,1,2,3, . . . ,25   **4.** 0,4,8,12, . . . ,64

Write the list of numbers.

    5. Zero, three, six, nine, and so on, without end.

    6. Zero, six, twelve, eighteen, and so on, up to and including seventy-two.

    7. Two, three, five, seven, and so on, up to and including eighty-nine.

    8. Three, nine, twenty-seven, eighty-one, and so on, without end.

Mixed Problems

Choose an appropriate method of computation and solve each of the following problems. For Problems 5–8 select the letter corresponding to your answer.

1. The elevation at the bottom of a ski slope is 1,017 feet. The height of the hill is 943 feet. About what is the elevation at the top of the slope?

2. Bill Fisher had weights of 10, 20, and 40 pounds on each end of a barbell. He added a 5-pound weight to each end. How much weight was on the barbell?

3. The Village Pizzeria sold 106 pizzas on Monday, 75 on Tuesday, 38 on Wednesday, 88 on Thursday, 93 on Friday, and 98 on Saturday. They are closed on Sunday. What was the total number of pizzas sold during the week?

4. The number of acres of forested land in Delaware is 392,000; Maryland, 2,653,000; New Jersey, 1,928,000; New York, 17,218,000; Pennsylvania, 16,826,000; and West Virginia, 11,669,000. What is the total number of acres of forested land in these six states?

5. What is the approximate total enrollment at Central High School if there are 492 students in the 9th grade, 415 students in the 10th grade, 397 students in the 11th grade, and 395 students in the 12th grade?

   a. 1,600 students   b. 1,400 students
   c. 1,700 students   d. 1,500 students

6. How many calories are contained in the following meal? Vegetable soup, 86; roast lamb, 175; fresh peas, 66; boiled potato, 117; two slices of white bread, 134; apple, 81; glass of milk, 170.

   a. 819 calories   b. 739 calories
   c. 929 calories   d. 829 calories

7. Hill Electric Company sold 8 air conditioners in May, 57 in June, 256 in July, and 313 in August. What were the total sales for the four months?

   a. 534 air conditioners
   b. 614 air conditioners
   c. 624 air conditioners
   d. Answer not given

8. The attendance at the county fair was 3,123 on Saturday, 4,265 on Sunday, 2,314 on Monday, 2,489 on Tuesday, and 2,982 on Wednesday. About what was the total attendance for the five-day fair?

   a. 13,000 people   b. 15,000 people
   c. 10,000 people   d. 13,500 people

Refresh Your Skills

The numbers in boxes indicate lessons where help may be found.

1. Write in words. [1–5]

   a. 6,862,925

   b. 327,952,728

2. a. Round 129,484,056 to the nearest million. [1–7]

   b. Round 7,528,653,076 to the nearest hundred thousand. [1–7]

Using Mathematics

EVEN AND ODD NUMBERS

Whole numbers may be separated into even and odd numbers.

Any whole number that can be divided exactly by two (2) is called an even number. Zero is considered an even number.

Even numbers always end in 0, 2, 4, 6, or 8.

Any whole number that cannot be divided exactly by two (2) is called an odd number.

Odd numbers always end in 1, 3, 5, 7, or 9.

Exercises

1. Which of the following are even numbers? Odd numbers?
30   72   49   97   386   565   908   2,183

2. Write all the one-digit even whole numbers.

3. Write all the odd numbers greater than 94 and less than 106.

Tell whether the sum is odd or even. Illustrate.

4. Two even numbers.

5. Two odd numbers.

6. An odd number plus an even number.

Tell whether the product is odd or even. Illustrate.

7. Two even numbers.

8. Two odd numbers.

9. An odd number times an even number.

True or false? Illustrate.

10. One more than any whole number is an even number.

11. Two more than any whole number is an even number.

12. Two times any whole number is an even number.

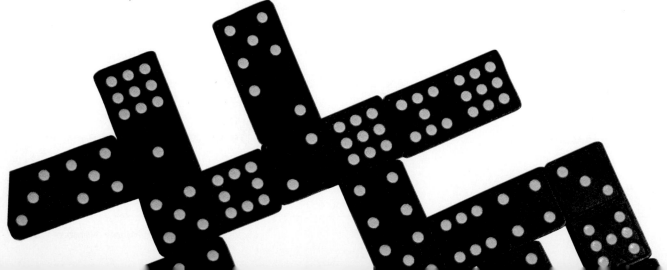

Subtracting Whole Numbers

Building Problem Solving Skills

The seating capacity of the bleachers at the new athletic field is 729 persons. The stands at the old field held only 493 persons. How many more people can now be seated?

1. ┌─ **READ** ─┐ the problem carefully.

a. Find the **question asked.** How many more people can now be seated?

b. Find the **given facts.** Two groups have already been counted: the new bleachers seat 729 persons, the old bleachers seated 493 persons.

Subtraction Indicators
• how many more
• how many fewer
• how much larger
• how much smaller
• amount of increase
• amount of decrease
• difference
• balance
• how many left over
• how much left over

2. ┌─ **PLAN** ─┐ how to solve the problem.

a. Choose the **operation needed.** To find how many more, you must find the difference in the seating capacities of the old and new bleachers. Since you are finding the difference, you must subtract.

b. **Think or write out your plan** relating the given facts to the question asked. The difference between 729 and 493 is equal to the increase in the number of seats.

c. Express your plan as an **equation.** $729 - 493 = n$

3. ┌─ **SOLVE** ─┐ the problem.

a. Estimate the answer.

$$
\begin{array}{r}
729 \rightarrow 700 \\
-\,493 \rightarrow -\,500 \\
\hline
200
\end{array}
$$

b. Solution

$$
\begin{array}{r}
729 \\
-\,493 \\
\hline
6
\end{array}
$$

Round the numbers to the same place. Usually this is the greatest place in the smallest number. Subtract.

2 is less than 9

$$
\begin{array}{r}
729 \\
-\,493 \\
\hline
236
\end{array}
$$
difference

Regroup when necessary. Regroup 1 hundred.
1 hundred = 10 tens
10 tens + 2 tens = 12 tens
$12 - 9 = 3$

You might use a calculator.

Key: 729 ☐─ 493 ☐= 236

4. `CHECK`

a. Check the accuracy of your arithmetic by adding.

$$\begin{array}{r} 493 \\ + 236 \\ \hline 729 \ \checkmark \end{array}$$

Add the number subtracted and the difference. The sum should be the number you subtracted from in your original problem.

b. Compare the answer to the estimate. The answer 236 compares reasonably with the estimate of 200.

ANSWER: 236 more people can be seated in the bleachers of the new field.

Think About It

1. What was counted in the problem above?

2. Is the difference in seating capacity less than the seating capacity of the new bleachers? of the old bleachers?

3. Why must the difference in the seating capacities of the old and new bleachers be less than the capacity of the new bleachers?

4. Will the difference in a subtraction always be greater than the number you subtract? Will it always be less than the number you subtract from? Explain your answer.

Calculator Know-How

On some calculators, `=` (the = key) can be used as an automatic constant. It will cause the calculator to repeat the last operation and its associated number.

For instance, to add 3 + 3 + 3 + 3, press 3 `+` `=` `=` `=` `=`. What did you see? You should have seen 3, 6, 9, 12.

To add 5 + 3, 10 + 3, 16 + 3, press keys:

5 `+` 3 `=` 10 `=` 16 `=`.
You should see 8, 13, 19.

To subtract 13 − 3 − 3 − 3 − 3, press keys:

13 `−` 3 `=` `=` `=` `=`. What did you see?
You should have seen 10, 7, 4, 1.

To subtract 27 − 3, 10 − 3, 32 − 3, press keys:

27 `−` 3 `=` 10 `=` 32 `=`.
You should see 24, 7, 29.

Use the constant feature to add 183 to each number. Write your answers.

1. 15   **2.** 95   **3.** 125   **4.** 1252

Use the constant feature to subtract 27 from each number. Write your answers.

5. 180   **6.** 75   **7.** 336   **8.** 261

9. Make up a problem that can be solved using the constant feature. Exchange papers with a classmate.

1. Subtract: 327 − 182

```
  327
− 182
    5
```

Arrange in columns. Subtract each column starting from the right.

```
  327    2 is less than 8
− 182
  145    difference
```

Regroup as necessary. Regroup 1 hundred as tens.
10 tens + 2 tens = 12 tens
12 − 8 = 4

Check:

```
  182
+ 145
  327 ✔
```

Check by adding.

ANSWER: 145

---

2. Find the difference: 4,926 − 1,894

```
  4,926    Check:    1,894
− 1,894            + 3,032
  3,032              4,926 ✔
```

ANSWER: 3,032

3. Take 279 from 3,856.

```
  3,856    Check:      279
−   279              + 3,577
  3,577                3,856 ✔
```

ANSWER: 3,577

---

4. From 94,500 subtract 93,264.

```
  94,500    Check:    93,264
− 93,264            +  1,236
   1,236              94,500 ✔
```

ANSWER: 1,236

---

5. Estimate the difference: 53,108 − 7,865

```
53,108 →    53,000
 7,865 → −   8,000
            45,000
```

Round the numbers to the same place. Usually, this is the greatest place in the smaller number. Subtract.

ANSWER: 45,000

---

6. Solve the equation: 925 − 467 = n

```
  925    Check:    467
− 467            + 458
  458              925 ✔
```

To find the value of n, subtract the given numbers.

ANSWER: 458

Vocabulary

The **difference** is the answer in subtraction.

Diagnostic Exercises

Subtract and check.

| 1. | 28 − 5 | 2. | 31 − 6 | 3. | 76 − 14 | 4. | 87 − 68 | 5. | 32 − 12 | 6. | 59 − 51 | 7. | 50 − 29 | 8. | 496 − 266 |

| 9. | 562 − 228 | 10. | 643 − 367 | 11. | 465 − 359 | 12. | 800 − 698 | 13. | 604 − 589 | 14. | 255 − 37 | 15. | 4,965 − 1,842 | 16. | 6,841 − 2,327 |

| 17. | 7,653 − 4,725 | 18. | 3,962 − 1,487 | 19. | 5,000 − 3,792 | 20. | 8,429 − 64 | 21. | 85,924 − 23,713 | 22. | 92,045 − 21,924 | 23. | 68,247 − 44,395 |

| 24. | 89,463 − 15,684 | 25. | 70,571 − 39,782 | 26. | 35,823 − 19,341 | 27. | 45,936 − 798 | 28. | 845,094 − 384,276 | 29. | 4,575,000 − 1,395,463 |

30. 56,070 − 984

31. 98,370 − 84,697

32. From 8,463 subtract 579.

33. Take 3,582 from 9,348.

34. Find the difference between 17,947 and 13,799.

35. Select the nearest given estimate for 50,943 − 18,499.
a. 20,000 **b.** 30,000 **c.** 40,000

36. Solve: 1,305 − 498 = n

37. The Student Council bought 3,463 banners for the pep rally. They sold 2,649. How many were left over?

Related Practice

Subtract and check.

| 1. a. | 59 − 4 | b. | 84 − 2 | c. | 96 − 3 | d. | 69 − 5 | 2. a. | 23 − 8 | b. | 94 − 6 | c. | 47 − 9 | d. | 52 − 3 |

| 3. a. | 85 − 61 | b. | 98 − 26 | c. | 49 − 15 | d. | 57 − 23 | 4. a. | 36 − 19 | b. | 64 − 25 | c. | 97 − 58 | d. | 85 − 37 |

| 5. a. | 78 − 48 | b. | 95 − 35 | c. | 29 − 9 | d. | 87 − 17 | 6. a. | 37 − 34 | b. | 86 − 82 | c. | 98 − 90 | d. | 79 − 72 |

7. a. 60
   − 36

b. 30
   − 23

c. 40
   − 8

d. 90
   − 65

8. a. 978
   − 642

b. 362
   − 150

c. 549
   − 325

d. 853
   − 721

9. a. 891
   − 379

b. 582
   − 166

c. 439
   − 284

d. 918
   − 627

10. a. 842
   − 358

b. 516
   − 179

c. 975
   − 587

d. 923
   − 476

11. a. 428
   − 218

b. 785
   − 380

c. 964
   − 357

d. 545
   − 138

12. a. 609
   − 392

b. 750
   − 273

c. 500
   − 307

d. 600
   − 485

13. a. 972
   − 965

b. 784
   − 697

c. 901
   − 838

d. 735
   − 686

14. a. 253
   − 26

b. 467
   − 58

c. 884
   − 9

d. 573
   − 83

15. a. 8,654
   − 2,431

b. 4,296
   − 3,003

c. 6,582
   − 4,172

d. 7,849
   − 3,812

16. a. 9,642
   − 6,425

b. 2,938
   − 1,682

c. 3,463
   − 2,702

d. 9,185
   − 6,345

17. a. 7,194
   − 3,457

b. 8,362
   − 4,554

c. 5,774
   − 2,968

d. 3,680
   − 2,962

18. a. 9,426
   − 3,258

b. 8,335
   − 6,452

c. 4,900
   − 3,825

d. 5,371
   − 4,691

19. a. 8,324
   − 5,975

b. 8,050
   − 4,584

c. 6,000
   − 2,505

d. 9,000
   − 8,427

20. a. 4,823
   − 8

b. 3,058
   − 65

c. 1,971
   − 986

d. 3,000
   − 479

21. a. 38,762
   − 26,521

b. 55,706
   − 52,304

c. 45,383
   − 32,162

d. 79,648
   − 36,136

22. a. 45,962
   − 12,535

b. 79,614
   − 48,532

c. 90,349
   − 70,281

d. 52,767
   − 28,343

23. a. 37,424
   − 15,258

b. 56,382
   − 42,491

c. 76,200
   − 41,350

d. 38,509
   − 32,754

24. a. 61,847
   − 25,952

b. 54,000
   − 23,475

c. 18,508
   − 10,809

d. 79,406
   − 25,837

25. a. 97,437
   − 28,769

b. 50,583
   − 48,794

c. 62,471
   − 42,587

d. 90,000
   − 82,575

26. **a.** 24,593 **b.** 57,924 **c.** 67,500 **d.** 29,070
  − 10,638   − 39,819   − 59,045   − 14,924

27. **a.** 65,925 **b.** 21,507 **c.** 86,500 **d.** 95,374
  −   386   − 4,742   −   95   − 8,298

28. **a.** 645,987 **b.** 705,961 **c.** 450,000 **d.** 532,654
  − 314,265   − 536,809   − 328,542   −   2,867

29. **a.** 8,935,299 **b.** 2,963,475 **c.** 7,694,251 **d.** 4,000,000
  − 3,411,274   − 1,724,133   −   86,206   − 2,950,966

30. **a.** 601 − 382   **b.** 4,563 − 3,275   **c.** 95,375 − 92,186   **d.** 636,059 − 278,770

31. **a.** 427 − 26   **b.** 3,788 − 909   **c.** 32,743 − 814   **d.** 545,611 − 39,768

32. **a.** From 300 subtract 152.   **b.** From 237 subtract 95.
  c. From 9,153 take 2,846.   **d.** From 84,073 take 28,536.

33. **a.** Take 497 from 520.   **b.** Take 574 from 1,582.
  c. Subtract 200 from 1,582.   **d.** Subtract 2,059 from 37,685.

34. Find the difference between the numbers.

  a. 450 and 199   **b.** 45,836 and 21,862
  c. 68,593 and 40,500   **d.** 5,000,000 and 2,750,000

35. For each of the following, write the nearest given estimate:
  a. 7,950 − 2,035   4,000   5,000   6,000
  b. 5,831 − 1,928   3,000   4,000   5,000
  c. 91,475 − 39,696   50,000   60,000   70,000
  d. 40,237 − 2,869   12,000   23,000   37,000

36. Solve.
  a. $174 − 87 = n$ **b.** $2,586 − 1,248 = n$
  c. $n = 200 − 63$ **d.** $n = 20,000 − 8,295$

37. **a.** The enrollment at the Township Junior High decreased from 3,468 students to 2,849 students. What was the decrease in enrollment?

  b. In the Student Association election, Joan Wynar received 1,206 votes and Mike Post 978 votes. How many fewer votes did Mike receive?

  c. How much larger is the state of Texas with an area of 688,617 square kilometers than the state of Rhode Island with an area of 3,232 square kilometers?

  d. The distance from the moon to the earth is 384,393 kilometers and from the sun to the earth is 149,499,812 kilometers. What is the difference in distance?

Mixed Problems

Choose an appropriate method of computation and solve each of the following problems. For Problems 1–4 select the letter corresponding to your answer.

1. A salesperson's automobile mileage for the past year was 41,266 kilometers. If 11,200 kilometers represents pleasure driving, about how many kilometers was the car driven for business purposes?

   a. 30,000 km   **b.** 50,000 km
   c. 29,000 km   **d.** 40,000 km

2. During the year Ms. Santos bought the following amounts of fuel oil to heat her house: 665 liters, 787 liters, 936 liters, 905 liters, 738 liters, and 946 liters. How many liters of fuel oil did she buy in all?

   a. 4,877 L   **b.** 4,887 L
   c. 4,977 L   **d.** 5,067 L

3. This year Washington High School presented three performances of their dramatic show. 1,058 persons saw the opening performance. On the succeeding two nights 993 and 1,196 persons attended. What was the total attendance at the school show?

   a. 3,147 persons   **b.** 3,247 persons
   c. 3,437 persons   **d.** Answer not given

4. Approximately how many fewer farms are there in the United States if the number of farms has decreased from 5,859,169 to 2,730,242?

   a. 2,000,000 farms   **b.** 3,000,000 farms
   c. 3,500,000 farms   **d.** 2,900,000 farms

5. The total attendance for the year at all athletic events at the Amerigo Vespucci High School was 48,547. The attendance for the previous year was 39,498. Find the increase in attendance.

6. Find the total area of the Pacific States if the area of California is 404,975 square kilometers; Oregon, 249,117 square kilometers; and Washington, 172,416 square kilometers.

Refresh Your Skills

The numbers in boxes indicate lessons where help may be found.

1. Write 208,070,009 in words. 1-5

2. Round 7,583,284 to the nearest thousand. 1-7

First estimate, then add. 1-8

3.
```
   45,836
   29,797
 + 75,086
```

4.
```
   21,673
    4,689
    5,401
 + 10,333
```

5.
```
     965,425
     829,517
 + 2,303,619
```

Multiplying Whole Numbers by One-Digit Numbers

Building Problem Solving Skills

At an average speed of 685 kilometers per hour, how many kilometers does an airplane fly in 6 hours?

1. **READ** the problem carefully.

a. Find the **question asked.** How many kilometers does the airplane fly?

b. Find the **given facts.** The speed of the airplane is 685 kilometers per hour. The flight time is 6 hours.

---

2. **PLAN** how to solve the problem.

a. Choose the **operation needed.** The airplane flies at the **same** average speed for each of the 6 hours, so multiplication is an appropriate operation to choose to find the length of the flight.

b. **Think or write out your plan** relating the given facts to the question asked. Multiply the time of the flight and the rate of speed to find the distance flown in kilometers.

c. Express as an **equation.** $685 \times 6 = n$

Multiplication Indicators
• total
• how many in all
• how much altogether
Note: All indicators are used when combining *equal* quantities.

3. ☐ **SOLVE** ☐ the problem.

a. Estimate the answer:

685 → 700 kilometers per hour   Round the speed to the nearest hundred.
  6 → 6 hours
700 × 6 = 4,200 kilometers   Multiply.

b. Solution:

$$\begin{array}{r} \overset{3}{685} \\ \times\ \ \ 6 \\ \hline 0 \end{array}$$ **6 × 5 = 30**

Arrange in columns. Multiply, starting from the right. Regroup when necessary.
30 ones = 3 tens 0 ones

$$\begin{array}{r} \overset{5\,3}{685} \\ \times\ \ \ 6 \\ \hline 4{,}110 \end{array}$$ **6 × 8 = 48**
48 + 3 = 51

Continue to multiply and move to the left, regrouping when necessary.

You might use a calculator. Key: 685 ☒ 6 ☐ 4110

---

4. ☐ **CHECK** ☐

a. Check the accuracy of your arithmetic.

$$\begin{array}{r} \overset{5\,3}{6} \\ \times\ 685 \\ \hline 4{,}110 ✔ \end{array}$$

Interchange the numbers and multiply.

b. Compare the answer to the estimate. The answer 4,110 kilometers compares reasonably with the estimate of 4,200 kilometers.

ANSWER: The distance is 4,110 kilometers.

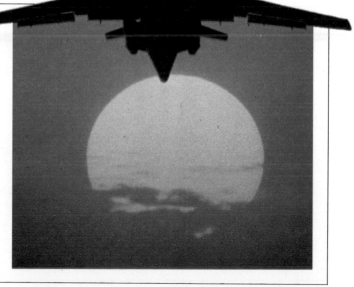

Think About It

1. If the airplane's average speed had been *different* for each hour of the flight, what operation would you have used instead of multiplication?

2. If you multiply by a one-digit number, can the product ever be more than 1,000? Explain your answer or illustrate.

3. If the answer to this problem were given in miles, would the number be greater or less than the number of kilometers? A kilometer is less than a mile.

In order to use a calculator to solve problems with a variety of mathematical operations, you must follow the rules for the order of operations. The memory keys will help you. Use of the memory keys does not affect any calculations in progress, so memory operations can be used whenever needed.

Example

Evaluate: $9 \times (3 + 4) - 15 = n$

Follow the rules for the order of operations.

1. Do any computations in parentheses first: $3 + 4 = 7$

2. Do the multiplication and division in order, from left to right: $9 \times 7 = 63$

3. Do the addition and subtraction in order, from left to right: $63 - 15 = 48$

To use a calculator to evaluate the expression above, key:

$3 \boxed{+} 4 \boxed{M+} 9 \boxed{\times} \boxed{M^R_C} \boxed{-} 15 \boxed{=} \underline{48}$

When you have completed the operations, you must clear the display.

Key: $\boxed{M^R_C} \boxed{M^R_C} \boxed{C^C_E}$ This first recalls the memory, then clears the memory, then clears the display.

Exercises

Use a calculator and the memory keys to evaluate each expression. Check Exercises 1–4 with pencil and paper. Tell which operation you did first. Clear the memory and display after you have evaluated each expression.

1. $6 + (8 - 3) - 2 \times 2$

2. $180 \div 12 + (40 - 5) \div 7$

3. $9 \times (14 - 3) + 1 - 42$

4. $32 + 8 \div (9 - 5)$

5. $12 + (9 + 26) \div 5$

6. $12 + 9 + 20 \div 5$

7. $18 \times (3 + 27) - 132 \div 6$

8. $5 + 81 \div 9 \times 11 + 23$

9. $64 \div 16 \times 9 - (8 + 3)$

10. Evaluate the expressions in Exercises 5–9 on a calculator *without* using the memory keys. Explain what you did.

1. Multiply: 3 × 34

$$
\begin{array}{r}
34 \\
\times\ \ 3 \\
\hline
102
\end{array}
$$
← factors
← product

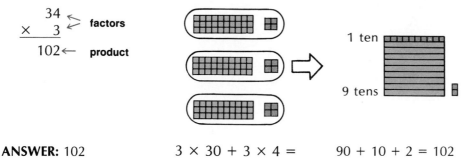

Arrange in columns. Multiply, starting from the right. Regroup when necessary. 12 = 1 ten 2 ones

1 ten

9 tens

ANSWER: 102   3 × 30 + 3 × 4 =   90 + 10 + 2 = 102

---

2. Estimate the product: 8 × 822

$$
\begin{array}{rcr}
822 & \to & 800 \\
\times\ \ 8 & \to & \times\ \ 8 \\
\hline
& & 6{,}400
\end{array}
$$

Round each number to its greatest place. Do not round one-digit numbers. Multiply.

ANSWER: 6,400

---

3. Solve the equation: 7 × 4,328 = n

$$
\begin{array}{r}
4{,}328 \\
\times\ \ \ \ 7 \\
\hline
30{,}296
\end{array}
$$

To find the value of n, multiply the given numbers.

Vocabulary

A **factor** is any of the numbers used in multiplication to form a product. The **product** is the answer in multiplication.

ANSWER: 30,296

Diagnostic Exercises

Multiply and check.

1. 23
× 3

2. 72
× 4

3. 24
× 3

4. 39
× 7

5. 231
× 2

6. 319
× 3

7. 874
× 6

8. 1,728
× 9

9. 34,267
× 6

10. 60
× 3

11. 400
× 5

12. 302
× 2

13. 5,208
× 8

14. 4,006
× 7

15. 3,000
× 5

16. 50,800
× 4

17. 6 × 985

18. Multiply 48 by 9.

19. Find the product: 314 × 8

20. Select the nearest given estimate for 8 × 687: **a.** 5,600 **b.** 4,800 **c.** 48,000

21. Solve: 5 × 1,496 = n

22. Mr. Jones, the school librarian, orders 8 dictionaries at $24 each. What is his total bill?

Related Practice

Multiply and check.

1.
a.
$$32 \times 2$$
b.
$$11 \times 8$$
c.
$$43 \times 2$$
d.
$$21 \times 3$$
2.
a.
$$63 \times 3$$
b.
$$94 \times 2$$
c.
$$52 \times 3$$
d.
$$71 \times 6$$

3.
a.
$$23 \times 4$$
b.
$$28 \times 3$$
c.
$$19 \times 5$$
d.
$$15 \times 6$$
4.
a.
$$48 \times 6$$
b.
$$65 \times 8$$
c.
$$64 \times 4$$
d.
$$59 \times 9$$

5.
a.
$$321 \times 3$$
b.
$$542 \times 2$$
c.
$$823 \times 4$$
d.
$$912 \times 4$$
6.
a.
$$217 \times 4$$
b.
$$816 \times 5$$
c.
$$191 \times 6$$
d.
$$649 \times 2$$

7.
a.
$$376 \times 2$$
b.
$$593 \times 7$$
c.
$$938 \times 9$$
d.
$$697 \times 8$$
8.
a.
$$2{,}143 \times 2$$
b.
$$1{,}728 \times 4$$
c.
$$5{,}914 \times 5$$
d.
$$4{,}789 \times 9$$

9.
a.
$$24{,}239 \times 4$$
b.
$$49{,}663 \times 7$$
c.
$$39{,}145 \times 6$$
d.
$$84{,}574 \times 8$$
10.
a.
$$20 \times 8$$
b.
$$60 \times 5$$
c.
$$70 \times 4$$
d.
$$50 \times 9$$

11.
a.
$$700 \times 3$$
b.
$$900 \times 9$$
c.
$$200 \times 5$$
d.
$$800 \times 6$$
12.
a.
$$104 \times 2$$
b.
$$304 \times 2$$
c.
$$3{,}012 \times 3$$
d.
$$34{,}032 \times 2$$

13.
a.
$$405 \times 4$$
b.
$$1{,}806 \times 7$$
c.
$$9{,}072 \times 8$$
d.
$$46{,}084 \times 5$$
14. a. $2{,}004 \times 2$ b. $4{,}008 \times 7$
c. $80{,}005 \times 6$ d. $90{,}026 \times 9$

15.
a.
$$5{,}280 \times 8$$
b.
$$9{,}500 \times 7$$
c.
$$82{,}000 \times 5$$
d.
$$30{,}000 \times 6$$
16.
a.
$$3{,}040 \times 7$$
b.
$$6{,}080 \times 5$$
c.
$$50{,}700 \times 9$$
d.
$$40{,}030 \times 8$$

17. a. 8×21 b. 6×709
c. $8 \times 5{,}030$ d. 351×9

18. Multiply.

a. 26 by 4   b. 6 by 401
c. 5,975 by 8   d. 16,000 by 5

19. Find the product.
a. 144×8 b. 5×320 c. $9 \times 2{,}446$ d. $25{,}050 \times 6$

20. For each of the following, write the nearest given estimate.

a. 4×79 360   320   300   b. 8×908 7,000   7,200   7,500
c. $9 \times 3{,}844$ 2,700   27,000   35,000   d. $5 \times 61{,}485$ 66,000   350,000   300,000

21. Solve.

a. $3 \times 93 = n$ b. $8 \times 506 = $ ▓   c. ▓ $= 5 \times 740$ d. $n = 9 \times 8{,}762$

22. a. If Jane rides her moped at an average speed of 26 miles per hour, how far will she travel in 5 hours?
b. In her first week as a cashier at Southfield Supermarket, Charlotte worked 37 hours at an hourly rate of $6. What was her gross pay for the week?

c. The scale used on a trail map is 1 inch = 2,500 feet. Roger wants to hike from his camp to Bass River. If his camp and Bass River are 7 inches apart on the map, what is the total distance he will walk?
d. Sue works in a sporting goods store. If she sells at least $250 of merchandise a day, what is the minimum amount she will sell in 5 days?

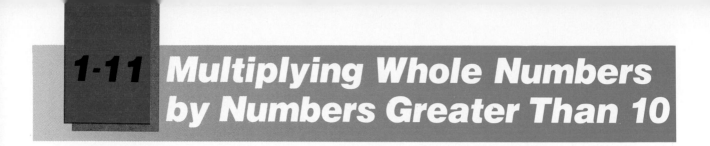

1-11 Multiplying Whole Numbers by Numbers Greater Than 10

There are 32 students in Homeroom 3A. If each student sells $63 in advertisements for the yearbook, how much would they make in all?

Multiply $63 by 32.

```
              $63
           ×   32  ← multiplier
63 × 2 →     126  ⟩ partial products
63 × 3 →     189
            $2,016      Check:    32
                               ×  63
                                  96
                                 192
                               2,016 ✔
```

Multiply by *each* digit in the multiplier to form partial products.
Place the right-hand digit of each partial product directly under the multiplier.
Add.
Check by interchanging the factors.

ANSWER: They would make $2,016 in all.

Examples

1. Find the product: 681 × 205

```
    681              Check:     205
  × 205                       × 681
   3405                         205
  13620                        1640
 139,605                       1230
                            139,605 ✔
```

When one partial product is zero, the next partial product may be written on the same line.

ANSWER: 139,605

---

2. Multiply: 38 × 4 × 284

38 × 4 = 152
        152 × 284 = 43,168

Multiply 2 factors.
Multiply the product by the remaining factor.

ANSWER: 43,168

---

3. Estimate the product: 842 × 6,559

842 → 800
6,559 → 7,000
800 × 7,000 = 5,600,000

Round each factor to its greatest place. (Do not round 1-digit factors.)
Multiply.

ANSWER: 5,600,000

4. Solve the equation: $6,297 \times 2,005 = n$

$$
\begin{array}{r}
6,297 \\
\times\ 2,005 \\
\hline
31,485 \\
1259400\ \ \\
\hline
12,625,485
\end{array}
$$

To find the value of n, multiply the given numbers.

ANSWER: 12,625,485

You might use a calculator.

Key: 6297 ☒ 2005 ☐ 12625485

Think About It

1. How could you solve the word problem without using multiplication? Which operation is more efficient?

2. If you were multiplying by a 5-digit factor in which three of the digits were zero, how many partial products would you have?

3. If 6 factors are to be multiplied, how many times must the multiplication operation be performed? What about for 11 factors? For 15 factors? Do you see a pattern?

4. If you estimated the product in the word problem, what would the answer be? By how much does this differ from the exact answer?

Diagnostic Exercises

Multiply and check.

1. $\begin{array}{r} 37 \\ \times\ 24 \\ \hline \end{array}$

2. $\begin{array}{r} 78 \\ \times\ 56 \\ \hline \end{array}$

3. $\begin{array}{r} 485 \\ \times\ \ 92 \\ \hline \end{array}$

4. $\begin{array}{r} 6,948 \\ \times\ \ \ \ 89 \\ \hline \end{array}$

5. $\begin{array}{r} 45,847 \\ \times\ \ \ \ \ \ 65 \\ \hline \end{array}$

6. $\begin{array}{r} 36 \\ \times\ 9,967 \\ \hline \end{array}$

7. $\begin{array}{r} 592 \\ \times\ 231 \\ \hline \end{array}$

8. $\begin{array}{r} 6,342 \\ \times\ \ \ 358 \\ \hline \end{array}$

9. $\begin{array}{r} 16,959 \\ \times\ \ \ \ \ 786 \\ \hline \end{array}$

10. $\begin{array}{r} 4,574 \\ \times\ 1,728 \\ \hline \end{array}$

11. $\begin{array}{r} 74,686 \\ \times\ \ \ 9,743 \\ \hline \end{array}$

12. $\begin{array}{r} 38,457 \\ \times\ 75,962 \\ \hline \end{array}$

13. $\begin{array}{r} 8,500 \\ \times\ \ \ \ 54 \\ \hline \end{array}$

14. $\begin{array}{r} 700 \\ \times\ 500 \\ \hline \end{array}$

15. $\begin{array}{r} 208 \\ \times\ 144 \\ \hline \end{array}$

16. $\begin{array}{r} 693 \\ \times\ 907 \\ \hline \end{array}$

17. $\begin{array}{r} 5,009 \\ \times\ \ \ \ 69 \\ \hline \end{array}$

18. $\begin{array}{r} 40,603 \\ \times\ \ \ \ \ \ 28 \\ \hline \end{array}$

19. $\begin{array}{r} 8,001 \\ \times\ \ \ 306 \\ \hline \end{array}$

20. $\begin{array}{r} 6,080 \\ \times\ \ \ 705 \\ \hline \end{array}$

21. 384×597

22. $36 \times 407 \times 743$

23. Multiply 75 by 49.

24. Find the product: 144 and 24

25. Select the nearest given estimate for 308 × 49:

    a. 1500 **b.** 15,000 **c.** 150,000

26. Solve: 85 × 21 = *n*

27. If Mary can type 87 words per minute, how many words can she type in 36 minutes?

Related Practice

Multiply and check.

1. a. 23   **b.** 28   **c.** 19   **d.** 53   **2. a.** 72   **b.** 93   **c.** 57   **d.** 74
    × 12   × 25   × 37   × 14   × 18   × 27   × 68   × 96

3. a. 144   **b.** 526   **c.** 967   **d.** 897
    × 23   × 42   × 36   × 88

4. a. 4,113   **b.** 6,374   **c.** 8,439   **d.** 1,728
    × 21   × 35   × 78   × 93

5. a. 93,153   **b.** 38,642   **c.** 68,459   **d.** 84,696
    × 24   × 85   × 47   × 63

6. a. 24   **b.** 53   **c.** 32   **d.** 25
    × 312   × 659   × 2,978   × 68,426

7. a. 144   **b.** 975   **c.** 347   **d.** 886   **8. a.** 2,786   **b.** 149   **c.** 8,327   **d.** 4,784
    × 324   × 638   × 231   × 697   × 231   × 9,687   × 524   × 379

9. a. 21,462   **b.** 78,356   **c.** 764   **d.** 15,647
    × 344   × 492   × 34,687   × 989

10. a. 2,467   **b.** 2,462   **c.** 8,622   **d.** 9,675
    × 1,236   × 6,374   × 7,393   × 8,326

11. a. 12,345   **b.** 56,397   **c.** 3,846   **d.** 96,749
    × 1,728   × 8,457   × 86,798   × 8,795

12. a. 23,814   **b.** 49,279   **c.** 57,625   **d.** 91,352
    × 16,523   × 82,536   × 34,719   × 73,858

13. a. 50   **b.** 300   **c.** 6,000   **d.** 3,600   **14. a.** 59   **b.** 365   **c.** 640   **d.** 5,280
    × 62   × 27   × 24   × 375   × 30   × 200   × 3,000   × 15,000

15. a.   501
      ×   26

b.   603
    ×   59

c.   803
    × 144

d.   3,205
    × 475

16. a.   529
    × 706

b.   69
    × 608

c.   2,534
    × 109

d.   3,842
    × 8,067

17. a.   4,009
    ×   87

b.   6,008
    × 365

c.   9,005
    × 2,183

d.   80,006
    ×   176

18. a.   2,050
    ×   16

b.   6,080
    ×   53

c.   60,506
    ×   125

d.   329
    × 7,070

19. a.   4,003
    × 205

b.   3,009
    × 6,082

c.   7,006
    × 7,006

d.   20,058
    × 1,009

20. a.   6,080
    × 203

b.   73,050
    × 8,007

c.   46,050
    × 2,500

d.   50,903
    × 60,704

21. a. 87×832 **b.** $56 \times 4,973$ **c.** 156×849 **d.** $523 \times 6,942$

22. a. $30 \times 605 \times 178$ **b.** $264 \times 671 \times 9$ **c.** $829 \times 72 \times 485$ **d.** $3,000 \times 20 \times 50$

23. Multiply.

    a. 32 by 16   **b.** 296 by 83   **c.** 314 by 52   **d.** 1,265 by 200

24. Find the product.

    a. 56 and 17   **b.** 253 and 85   **c.** 405 and 840   **d.** 5,280 and 487

25. For each of the following, write the nearest given estimate.

a. 33×67	1,800	2,100	2,400
b. 519×806	40,000	400,000	4,000
c. $993 \times 5,784$	600,000	450,000	6,000,000
d. $1,893 \times 725$	1,000,000	1,200,000	1,400,000

26. Solve.

    a. $24 \times 89 = n$ **b.** $72 \times 125 = n$ **c.** $n = 491 \times 163$ **d.** $n = 50 \times 845$

27. a. If a box contains 144 envelopes, how many envelopes will there be in 36 boxes?

    b. A merchant purchased 48 computer tapes at $29 each. How much was the total cost?

    c. If 1 centimeter represents 12 kilometers, how many kilometers apart are two cities 12 centimeters apart on the chart?

    d. Joan's average reading rate is 195 words per minute. How many words can she read in 30 minutes?

Mixed Problems

Select an appropriate method of computation. Then solve. For Problems 1–3, select the letter corresponding to your answer.

1. How far can a car go on a tankful of gasoline if it averages 6 kilometers on a liter and the tank holds 80 liters?

    a. 470 kilometers   **b.** 380 kilometers
    c. 480 kilometers   **d.** Answer not given

2. The present enrollment at the Township Junior High School is 849 pupils. There are 675 pupils registered at the senior high. What is the difference in their enrollments?

    a. 185 pupils   **b.** 164 pupils
    c. 174 pupils   **d.** Answer not given

3. Each of the 26 major baseball teams has 24 players on its team roster. How many major league players are there in all?

    a. 584 players   **b.** 624 players
    c. 554 players   **d.** 564 players

4. A two-story factory building with office and showroom is available for rent. The office has 87 square meters of floor space; showroom, 224 square meters; first floor, 1,986 square meters; and second floor, 1,468 square meters. What is the total floor space of the building?

5. The area of the earth is 510,100,500 square kilometers. If there are 361,149,600 square kilometers of water, how many square kilometers of land are on the earth?

6. Find the winner of a 9-hole golf match (the smaller number of strokes wins):
Jones, Carson H.S. $4|4|5|6|3|4|5|4|5 =$
Williams, Dalton H.S. $3|5|4|5|4|5|6|5|4 =$

Refresh Your Skills

The numerals in boxes indicate lessons where help may be found.

1. Write 19,207,485,000,000 in words. $\boxed{1-5}$

2. Write in digits: Four million twenty-six thousand seven hundred ninety-six. $\boxed{1-6}$

3. Write the complete number for 75 billion. $\boxed{1-6}$

4. Round 5,963,499,716 to the nearest million. $\boxed{1-7}$

First estimate, then solve each of the following.

5. $\boxed{1-8}$
    9,832
   17,496
   50,957
    8,268
 +   594

6. $\boxed{1-9}$
  105,000
−   98,634

7. $\boxed{1-10, 11}$
   807
× 609

8. $\boxed{1-10, 11}$
  1,728
× 853

Using Mathematics

EXPONENTS

An **exponent** tells how many times a number called the **base** is used as a factor. It is indicated by a small numeral written to the upper right of the factor (base).

$$5^2 \quad \leftarrow \textbf{exponent}$$
$$\leftarrow \textbf{base}$$

5^2 represents $\underline{5} \times \underline{5}$ or the product, 25. It is read "five squared" or "five to the second power."

2^3 represents $\underline{2} \times \underline{2} \times \underline{2}$ or the product, 8. It is read "two cubed" or "two to the third power."

10^4 represents $\underline{10} \times \underline{10} \times \underline{10} \times \underline{10}$ or the product, 10,000. It is read "ten to the fourth power."

3^5 represents $\underline{3} \times \underline{3} \times \underline{3} \times \underline{3} \times \underline{3}$ or the product, 243. It is read "three to the fifth power."

Numbers like 10, 10^2, 10^3, 10^4, 10^5, 10^6, ... are called **powers of 10.**

Exercises

Read, or write in words.

1. 3^4 **2.** 7^8 **3.** 2^{11} **4.** 10^9

5. 6^1 **6.** 11^5 **7.** 5^3 **8.** 20^2

Write, using exponents.

9. Six to the eighth power   **10.** Nine cubed   **11.** Fifty squared

Name the base and exponent in each.

12. 8^7 **13.** 3^9 **14.** 11^4 **15.** 6^{11}

16. 2^8 **17.** 21^{10} **18.** 5^{12} **19.** 7^6

Write, using exponents.

20. $2 \times 2 \times 2 \times 2$

21. $5 \times 5 \times 5 \times 5 \times 5 \times 5 \times 5 \times 5 \times 5 \times 5$

22. Express 36 as a power of 6.

23. Express 32 as a power of 2.

Give the square and cube of each of the following.

24. 6   **25.** 10   **26.** 9

27. 12   **28.** 30   **29.** 25

Give the value of each of the following.

30. 9^2 **31.** 6^4 **32.** 2^8

33. 3^7 **34.** $8^3 + 7^2$ **35.** $5^4 - 1^{10}$

Dividing Whole Numbers by One-digit Divisors

Building Problem Solving Skills

Pam has taken out a used car loan for $1,440. She is going to repay the loan in 8 equal monthly payments. How much is each monthly payment?

1. **READ** the problem carefully.

   a. Find the **question asked.** How much is each payment?
   b. Find the **given facts.** Pam took out a loan for $1,440. She will repay it in 8 equal payments.

Division Indicators
• how much is each
• how many in each
• find the average
• rate per unit
Note: All indicators generally refer to equal quantities.

2. **PLAN** how to solve the problem.

   a. Choose the **operation needed.** You are separating the loan into a number of equal parts, so you divide.
   b. **Think or write out your plan** relating the given facts to the question asked. The total amount of the loan, $1,440, divided by the number of payments, 8, is equal to the amount of each payment.
   c. Express your plan as an **equation.** $\$1,440 \div 8 = n$

3. **SOLVE** the problem.

   a. **Estimate** the answer.

   $\$1,440 \rightarrow \$1,600$
   $\$1,600 \div 8 = \$200$

   b. Solution:

$$
\begin{array}{r}
\$180 \leftarrow \textbf{quotient} \\
8\overline{)\$1,440} \\
\end{array}
$$

14 ÷ 8 $8\overline{)\$1,440}$
1 × 8   8
14 − 8   64   **64 ÷ 8**
            64

You might use a calculator.

Key: 1440 ÷ 8 = 180

Round the dividend so it can be divided exactly.

Divide.
Write the problem in vertical form.
Start from the left. Sometimes it is necessary to begin with more than one digit of the dividend.
8 is more than 1, so divide 8 into 14.
Multiply.
Subtract, and bring down next figure.
Repeat until there are no digits to bring down.

4. CHECK

a. Check the accuracy of your arithmetic. Check by multiplying.

$$\begin{array}{r} \$180 \\ \times \quad 8 \\ \hline \$1440 \end{array}$$ ✔

Multiply the quotient by the divisor. Your answer is the number you divided.

b. Compare the answer to the estimate. The answer $180 compares reasonably with the estimate of $200.

ANSWER: Each monthly payment is $180.

Think About It

1. In the given problem, what was the whole? What were the parts?

2. What is the largest number you can divide by a single-digit number and still get a single-digit number with no remainder for an answer?

Calculator Know-How

When you divide using a calculator, the remainder is displayed as a decimal. To find the remainder as a whole number, multiply the divisor by the whole-number part of the quotient; subtract this product from the dividend.

A warehouse shipped 3,726 cans of paint. They pack 8 cans in each box. How many boxes will be filled? How many cans will be left over? Use a calculator to solve.

Key: 3726 ÷ 8 = <u>465.75</u>

ANSWER: 465 boxes will be filled

To find how many cans are left over, express the remainder as a whole number.

a. Key: CE 465 × 8 = <u>3720</u>   **b.** Key: CE 3726 − 3720 = <u>6</u>

ANSWER: There will be 6 cans left over.

Find the whole-number remainder for each division.

1. 532 ÷ 5   **2.** 1,153 ÷ 6   **3.** 9,075 ÷ 4   **4.** 8,116 ÷ 3

Solve.

5. Bob is packing apples in bags of 14. He has 230 apples. How many bags will he fill? How many apples will be left over?

6. The bank teller is rolling quarters in rolls of 40. She has 335 quarters. How many rolls will she have? How many quarters will be left over?

1. Divide 978 by 6.

$6\overline{)978}$ or $978 \div 6$

↑ └─dividend ↑
└────divisor────┘

```
          163  ← quotient
9 ÷ 6 → 6)978
1 × 6 →   6↓|
9 − 6 →   37| ← 37 ÷ 6
          36↓
          18 ← 18 ÷ 6    Check:
          18             163 × 6 = 978 ✔
```

ANSWER: 163

Write the problem in vertical form.
Start from the left.
Divide.
Multiply.
Subtract, and bring down next figure.
Repeat until there are no digits to bring down.
Check by multiplying the quotient by the divisor.

---

2. Divide: $756 \div 7$

```
          108
        7)756
7 is greater   7↓↓
than 5. →     56        Check:
```

ANSWER: 108 $\underline{56}$ $108 \times 7 = 756$ ✔

Bring down more than one digit if necessary.
7 is greater than 5, so bring down 6.

---

3. Divide: $8\overline{)746}$

```
               93 R2
8 is greater  8)746 ↑    Check:
than 7.        72↓|      93 × 8 = 744
               26|          +    2
               24  remainder   746 ✔
                2 ←
```

ANSWER: 93 R2

Sometimes it is necessary to begin with more than one digit of the dividend.
8 is greater than 7, so divide 8 into 74.

Express the number left over as a remainder.

---

4. Estimate the quotient: $874 \div 9$

$874 \rightarrow 900$
$900 \div 9 = 100$

ANSWER: 100

Round the dividend so that it can be divided exactly by the divisor.
Divide.

---

5. Solve the equation: $14{,}511 \div 7 = n$

```
         2,073
      7)14,511
        14
         51
         49
         21
         21
```

ANSWER: 2,073

To find the value of *n*, divide the given numbers.

Check: 7 × 2,073 = 14,511 ✔

Diagnostic Exercises

Divide and check.

1. 2)68   **2.** 3)78   **3.** 2)846   **4.** 4)928

5. 6)834   **6.** 7)427   **7.** 6)552   **8.** 4)8,448

9. 3)8,526   **10.** 6)9,852   **11.** 8)3,928   **12.** 4)96,852

13. 7)36,533   **14.** 3)6,900   **15.** 4)804   **16.** 6)642

17. 2)6,008   **18.** 5)5,030

Find the quotient and remainder.

19. 7)156   **20.** 4)6,923   **21.** 6)935   **22.** 8)4,668   **23.** 7)843

24. Divide: 8,795 ÷ 5   **25.** Divide 3,870 by 9.

26. Select the nearest given estimate for 5,375 ÷ 9.   **27.** Solve: 553 ÷ 7 = n
   a. 700   **b.** 600   **c.** 500

28. Miguel bought a stereo tape deck for a total, with interest, of
   $684. He promised to pay for it in equal installments over 6
   months. How much must he pay each month?

Related Practice

Divide and check.

1. a. 3)93   **b.** 2)46   **c.** 4)84   **d.** 2)28

2. a. 2)34   **b.** 3)87   **c.** 5)85   **d.** 6)96

3. a. 3)693   **b.** 2)682   **c.** 4)488   **d.** 3)366

4. a. 6)846   **b.** 4)496   **c.** 7)798   **d.** 8)968

5. a. 4)944   **b.** 5)745   **c.** 6)918   **d.** 3)861

6. a. 6)246   **b.** 9)729   **c.** 2)128   **d.** 3)216

7. **a.** $7\overline{)504}$ **b.** $4\overline{)348}$ **c.** $5\overline{)315}$ **d.** $8\overline{)608}$

8. **a.** $2\overline{)6,448}$ **b.** $4\overline{)8,484}$ **c.** $3\overline{)9,636}$ **d.** $2\overline{)2,486}$

9. **a.** $5\overline{)8,255}$ **b.** $8\overline{)9,928}$ **c.** $3\overline{)9,738}$ **d.** $4\overline{)9,568}$

10. **a.** $3\overline{)7,491}$ **b.** $4\overline{)9,156}$ **c.** $6\overline{)8,748}$ **d.** $5\overline{)7,490}$

11. **a.** $7\overline{)5,243}$ **b.** $6\overline{)2,154}$ **c.** $8\overline{)6,792}$ **d.** $4\overline{)3,392}$

12. **a.** $3\overline{)69,876}$ **b.** $4\overline{)95,860}$ **c.** $2\overline{)76,952}$ **d.** $6\overline{)85,794}$

13. **a.** $8\overline{)72,976}$ **b.** $5\overline{)37,460}$ **c.** $7\overline{)40,992}$ **d.** $4\overline{)29,984}$

14. **a.** $4\overline{)840}$ **b.** $2\overline{)9,680}$ **c.** $3\overline{)3,600}$ **d.** $6\overline{)90,000}$

15. **a.** $2\overline{)608}$ **b.** $4\overline{)8,408}$ **c.** $2\overline{)4,082}$ **d.** $3\overline{)9,060}$

16. **a.** $5\overline{)525}$ **b.** $8\overline{)8,416}$ **c.** $6\overline{)67,218}$ **d.** $2\overline{)81,106}$

17. **a.** $3\overline{)9,003}$ **b.** $2\overline{)60,024}$ **c.** $4\overline{)40,008}$ **d.** $5\overline{)50,005}$

18. **a.** $3\overline{)6,012}$ **b.** $6\overline{)24,036}$ **c.** $4\overline{)80,032}$ **d.** $2\overline{)10,010}$

Write the quotient and remainder.

19. **a.** $5\overline{)14}$ **b.** $3\overline{)553}$ **c.** $7\overline{)27,645}$ **d.** $4\overline{)98,622}$

20. **a.** $6\overline{)65}$ **b.** $8\overline{)243}$ **c.** $3\overline{)8,582}$ **d.** $5\overline{)38,403}$

21. **a.** $9\overline{)35}$ **b.** $5\overline{)74}$ **c.** $6\overline{)4,657}$ **d.** $8\overline{)89,749}$

22. **a.** $4\overline{)78}$ **b.** $6\overline{)940}$ **c.** $8\overline{)7,534}$ **d.** $6\overline{)59,374}$

23. **a.** $6\overline{)365}$ **b.** $8\overline{)644}$ **c.** $3\overline{)812}$ **d.** $7\overline{)85,264}$

24. **a.** $438 \div 3$ **b.** $94,500 \div 4$
  c. $69,376 \div 9$ **d.** $23,583 \div 7$

25. **a.** Divide 932 by 4. **b.** Divide 3,482 by 9.
  c. Divide 5,795 by 8. **d.** Divide 83,007 by 3.

26. For each of the following, write the nearest given estimate.

a. $584 \div 6$	10	100	1,000
b. $9,104 \div 3$	3,000	5,000	6,000
c. $27,491 \div 5$	4,000	6,000	8,000
d. $40,963 \div 8$	2,000	4,000	5,000

27. Solve. **a.** $69 \div 3 = n$ **b.** $468 \div 9 = \blacksquare$
  c. $8,040 \div 5 = n$ **d.** $n = 96,544 \div 7$

28. **a.** Alfred Tomaso made 128 dozen muffins at the bakery in one day. If each muffin tin held 2 dozen, how many tins of muffins did he make?

b. If he could get 8 muffin tins in the ovens at once for a batch, how many batches did he have?

c. If it took 360 minutes to bake all the batches, how many minutes did it take to bake one batch?

d. Alfred packed 6 muffins in each box. How many boxes did he use?

Using Mathematics

TESTS FOR DIVISIBILITY

When 252 is divided by 3, the remainder is 0.

$$252 \div 3 = 84$$

Thus 3 divides 252 exactly, or 252 is **divisible** by 3.

The following tests may be used to determine whether a given number is divisible by 2, 3, 4, 5, 6, 8, 9, or 10.

Divisor	Test	Examples
2	The number must be even. Its last digit must be 0, 2, 4, 6, or 8.	30 or 154 or 518 or 96 or 372
3	The sum of the digits must be divisible by 3.	For 834: 8 + 3 + 4 = 15; 15 ÷ 3 = 5 For 4,971: 4 + 9 + 7 + 1 = 21; 21 ÷ 3 = 7
4	The number named by the last two digits must be divisible by 4.	For 92,136: 36 ÷ 4 = 9 For 7,992: 92 ÷ 4 = 23
5	The last digit of the number must be 0 or 5.	430 3,645
6	The number must be even and the sum of its digits must be divisible by 3.	For 9,558: An even number 9 + 5 + 5 + 8 = 27 27 ÷ 3 = 9
8	The number named by the last three digits must be divisible by 8. Numbers ending in 3 zeros are divisible by 8.	For 39,712: 712 ÷ 8 = 89 For 153,000: Ends in 3 zeros
9	The sum of the digits must be divisible by 9.	For 50,382: 5 + 0 + 3 + 8 + 2 = 18 18 ÷ 9 = 2
10	The last digit of the number must be 0.	80 9,670

Exercises

Determine whether the following numbers are divisible:

By 3:	**1.** 417	**2.** 2,853	**3.** 7,415	**4.** 29,538	**5.** 593,618
By 5:	**6.** 251	**7.** 8,970	**8.** 1,565	**9.** 38,006	**10.** 686,400
By 8:	**11.** 4,794	**12.** 1,352	**13.** 86,727	**14.** 75,000	**15.** 827,223
By 2:	**16.** 518	**17.** 6,359	**18.** 4,506	**19.** 83,000	**20.** 179,244
By 9:	**21.** 837	**22.** 5,706	**23.** 89,784	**24.** 40,389	**25.** 779,895
By 4:	**26.** 536	**27.** 9,252	**28.** 1,874	**29.** 92,425	**30.** 954,768
By 6:	**31.** 914	**32.** 8,349	**33.** 6,858	**34.** 27,291	**35.** 906,702
By 10:	**36.** 470	**37.** 3,000	**38.** 20,105	**39.** 90,790	**40.** 813,060

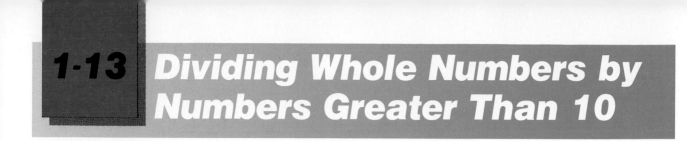

If 3 people together work 22 hours in 1 day, how many days would it take for them to do a job that takes 374 hours?

Write the problem in vertical form.

```
      17
22)374
    22
   154
   154
```

Think: 22 rounds to 20
374 rounds to 380
380 ÷ 20 = 19

Check: 17 × 22 = 374 ✔

ANSWER: It would take 17 days.

Examples

1. Find the quotient: 89)45,037

```
      506 R3
89)45,037
   44 5
     537
     534
       3
```

89 is greater than 4 and greater than 45, so divide 89 into 450.
Think: 89 rounds to 90; 450 rounds to 450;
450 ÷ 90 = 5
Try 5.
Think: 537 rounds to 540; 540 ÷ 90 = 6
Try 6.

ANSWER: 506 R3

Check: 506 × 89 = 45,034 + 3 = 45,037 ✔

---

2. Find the quotient and remainder: 15,979 ÷ 438

```
        36 R211
438)15,979
    13 14
     2 839
     2 628
       211
```

438 is greater than 1, greater than 15, greater than 159, so divide 438 into 1,597
Think: 438 rounds to 400; 1,597 rounds to 1,600
1,600 ÷ 400 = 4
Try 4. 438 × 4 = 1,752. The product is too great.
Try 3.
Think: 2,839 rounds to 2,800; 2,800 ÷ 400 = 7
Try 7. 438 × 7 = 3,066. The product is too great.
Try 6.

ANSWER: 36 R 211

You might use a calculator.

Key: 15979 ÷ 438 = 36.481735 C

438 × 36 = 15768 C 15979 − 15768 = 211

Check: 36 × 438 =   15,768
                   +   211
                      15,979 ✔

3. Estimate the quotient: 47,928 ÷ 61

$$61 \rightarrow 60$$
$$47,928 \rightarrow 48,000$$
$$48,000 \div 60 = 800$$

Round the divisor to its highest place.
(Do not round 1-digit numbers.)
Round the dividend so that it can be divided exactly
by the rounded divisor.
Divide.

ANSWER: 800

---

4. Solve the equation: 70,785 ÷ 165 = n

```
        429
165)70,785
     66 0↓
      4 78
      3 30
      1 485
      1 485
```

To find the value of n, divide the given numbers.

Check:
```
      429
×     165
   70,785 ✔
```

ANSWER: 429

Think About It

1. What is the largest number you can divide into 2,000 and still get a whole number with no remainder for an answer?

2. If you multiply an even number by an odd number, the answer is always even. If you divide an even number by an odd number and get a whole number for a quotient, is the quotient odd or even?

Diagnostic Exercises

Divide and check.

1. 24)96

2. 38)228

3. 27)567

4. 14)1,022

5. 26)8,164

6. 85)71,995

7. 48)61,584

8. 54)349,272

9. 60)28,980

10. 57)34,656

11. 36)230,400

12. 42)126,252

13. 63)27,044

14. 72)45,018

15. 144)864

16. 174)7,482

17. 298)11,026

18. 257)32,383

19. 946)810,722

20. 907)61,676

21. 506)405,306

22. 843)552,660

23. 400)338,000

24. 391)185,805

25. 144)75,984

26. 1,760)14,080

27. 1,728)101,952

28. 8,526)3,939,012

29. $5{,}280\overline{)23{,}876{,}160}$ **30.** $8{,}005\overline{)4{,}306{,}690}$ **31.** $21{,}714 \div 231$ **32.** Divide 300,960 by 5,280

33. Select the nearest given estimate for: $80{,}526 \div 92$
   a. 700   **b.** 800   **c.** 900

34. Solve: $2{,}184 \div 56 = n$

35. Rosa worked 28 days during the summer and earned \$1,232. How much did she earn per day?

Related Practice

Divide and check.

1. a. $16\overline{)80}$ **b.** $23\overline{)92}$ **c.** $27\overline{)81}$ **d.** $19\overline{)76}$

2. a. $54\overline{)378}$ **b.** $48\overline{)432}$ **c.** $83\overline{)332}$ **d.** $67\overline{)536}$

3. a. $32\overline{)896}$ **b.** $48\overline{)672}$ **c.** $12\overline{)324}$ **d.** $24\overline{)984}$

4. a. $24\overline{)1{,}824}$ **b.** $72\overline{)2{,}664}$ **c.** $95\overline{)6{,}935}$ **d.** $41\overline{)1{,}845}$

5. a. $37\overline{)4{,}551}$ **b.** $26\overline{)8{,}918}$ **c.** $17\overline{)3{,}366}$ **d.** $52\overline{)9{,}672}$

6. a. $72\overline{)39{,}456}$ **b.** $85\overline{)41{,}820}$ **c.** $46\overline{)30{,}038}$ **d.** $77\overline{)64{,}603}$

7. a. $29\overline{)70{,}035}$ **b.** $32\overline{)49{,}984}$ **c.** $17\overline{)48{,}892}$ **d.** $25\overline{)81{,}325}$

8. a. $43\overline{)224{,}417}$ **b.** $54\overline{)526{,}932}$ **c.** $64\overline{)212{,}416}$ **d.** $71\overline{)138{,}947}$

9. a. $30\overline{)5{,}670}$ **b.** $50\overline{)37{,}450}$ **c.** $70\overline{)61{,}040}$ **d.** $80\overline{)41{,}840}$

10. a. $36\overline{)25{,}380}$ **b.** $69\overline{)41{,}607}$ **c.** $37\overline{)29{,}933}$ **d.** $83\overline{)25{,}232}$

11. a. $15\overline{)3{,}300}$ **b.** $57\overline{)51{,}300}$ **c.** $30\overline{)144{,}000}$ **d.** $66\overline{)574{,}200}$

12. a. $12\overline{)24{,}048}$ **b.** $24\overline{)120{,}168}$ **c.** $69\overline{)276{,}207}$ **d.** $35\overline{)105{,}140}$

13. a. $56\overline{)48{,}297}$ **b.** $75\overline{)19{,}366}$ **c.** $48\overline{)36{,}923}$ **d.** $64\overline{)24{,}869}$

14. a. $14\overline{)2{,}412}$ **b.** $52\overline{)47{,}639}$ **c.** $24\overline{)58{,}819}$ **d.** $78\overline{)651{,}092}$

15. a. $128\overline{)896}$ **b.** $427\overline{)3{,}416}$ **c.** $640\overline{)1{,}920}$ **d.** $569\overline{)3{,}414}$

16. a. $173\overline{)8{,}996}$ **b.** $144\overline{)6{,}768}$ **c.** $234\overline{)5{,}382}$ **d.** $285\overline{)9{,}690}$

17. a. $229\overline{)10{,}992}$ **b.** $231\overline{)13{,}398}$ **c.** $929\overline{)52{,}024}$ **d.** $745\overline{)51{,}405}$

18. a. $152\overline{)95{,}608}$ **b.** $223\overline{)66{,}008}$ **c.** $475\overline{)75{,}525}$ **d.** $337\overline{)95{,}708}$

19. a. $863 \overline{)650{,}702}$ **b.** $629 \overline{)367{,}336}$ **c.** $847 \overline{)337{,}953}$ **d.** $768 \overline{)433{,}920}$

20. a. $208 \overline{)7{,}696}$ **b.** $706 \overline{)55{,}068}$ **c.** $405 \overline{)25{,}110}$ **d.** $504 \overline{)49{,}392}$

21. a. $106 \overline{)31{,}906}$ **b.** $204 \overline{)124{,}032}$ **c.** $903 \overline{)367{,}521}$ **d.** $502 \overline{)355{,}416}$

22. a. $478 \overline{)152{,}960}$ **b.** $640 \overline{)524{,}800}$ **c.** $679 \overline{)549{,}990}$ **d.** $725 \overline{)304{,}500}$

23. a. $200 \overline{)85{,}200}$ **b.** $300 \overline{)290{,}100}$ **c.** $700 \overline{)408{,}800}$ **d.** $600 \overline{)492{,}000}$

24. a. $692 \overline{)365{,}489}$ **b.** $375 \overline{)323{,}467}$ **c.** $589 \overline{)433{,}704}$ **d.** $866 \overline{)583{,}759}$

25. a. $320 \overline{)149{,}520}$ **b.** $231 \overline{)188{,}518}$ **c.** $144 \overline{)41{,}088}$ **d.** $625 \overline{)294{,}625}$

26. a. $5{,}280 \overline{)42{,}240}$ **b.** $2{,}774 \overline{)22{,}192}$ **c.** $8{,}304 \overline{)74{,}736}$ **d.** $4{,}568 \overline{)22{,}840}$

27. a. $6{,}080 \overline{)255{,}360}$ **b.** $9{,}286 \overline{)529{,}302}$ **c.** $3{,}974 \overline{)337{,}790}$ **d.** $5{,}280 \overline{)153{,}120}$

28. a. $8{,}366 \overline{)3{,}906{,}922}$ **b.** $5{,}863 \overline{)2{,}872{,}870}$ **c.** $1{,}728 \overline{)1{,}627{,}776}$ **d.** $6{,}080 \overline{)1{,}824{,}000}$

29. a. $2{,}240 \overline{)12{,}286{,}400}$ **b.** $8{,}267 \overline{)31{,}050{,}852}$ **c.** $2{,}794 \overline{)12{,}802{,}108}$ **d.** $3{,}246 \overline{)24{,}377{,}460}$

30. a. $7{,}006 \overline{)4{,}581{,}924}$ **b.** $6{,}080 \overline{)4{,}572{,}160}$ **c.** $3{,}600 \overline{)2{,}109{,}600}$ **d.** $3{,}007 \overline{)1{,}202{,}800}$

31. a. $408 \div 24$ **b.** $93{,}940 \div 35$ **c.** $173{,}712 \div 231$ **d.** $401{,}280 \div 5{,}280$

32. a. Divide 5,616 by 144.   **b.** Divide 13,398 by 231.
    c. Divide 153,600 by 320.   **d.** Divide 1,487,200 by 1,760.

33. For each of the following, write the nearest given estimate.
    a. $3{,}120 \div 48$ 40   50   60
    b. $28{,}086 \div 93$ 200   300   400
    c. $71{,}865 \div 89$ 700   800   900
    d. $219{,}480 \div 415$ 50   500   5,000

34. Solve.

    a. $72 \div 24 = n$ **b.** $7{,}344 \div 34 = n$ **c.** $n = 4{,}002 \div 87$ **d.** $n = 8{,}295 \div 105$

35. a. Stefan types an average of 55 words per minute. If a page has approximately 500 words, how long would it take him to type one page?

    b. If he kept typing at this rate, how many pages could he type in one hour?

    c. A photocopier can copy and collate 13 copies per minute. How long will it take to make 900 copies?

    d. The print shop used 28,470 sheets of paper to make 1,095 booklets for Sports Week. How many sheets were in each booklet?

Mixed Problems

Solve each of the following problems. For Problems 1–4 select the letter corresponding to your answer.

1. The deepest place thus far discovered in the world is 10,863 meters in the Marianas Trench near the island of Guam. The greatest depth in the Atlantic Ocean is 9,219 meters. Find the difference in these depths.

  a. 20,082 meters **b.** 1,054 meters
  c. 1,644 meters **d.** Answer not given

2. If a car averages 5 kilometers on a liter of gasoline, how many liters of gasoline are needed to drive from Los Angeles to San Francisco, a distance of 680 kilometers?

  a. 142 L   **b.** 136 L
  c. 132 L   **d.** 685 L

3. What is the total enrollment at the Southwest High School if there are 785 freshmen, 697 sophomores, 648 juniors, and 657 seniors?

  a. 2,697   **b.** 2,767
  c. 2,687   **d.** 2,787

4. How many sheets of writing paper are in 24 packages, each containing 500 sheets?

  a. 524 sheets   **b.** 24,500 sheets
  c. 12,000 sheets **d.** Answer not given

5. How many people attended the Baltimore-Philadelphia five-game World Series if 52,204 saw the first game, 52,132 the second, 65,792 the third, 66,947 the fourth, and 67,064 the fifth game?

6. How many more square kilometers of territory did the United States acquire by the Louisiana Purchase of 2,142,427 square kilometers than by the purchase of Alaska with 1,518,776 square kilometers?

7. An airplane flies from Atlanta to Seattle, a distance of 2,690 miles, in 5 hours. What is the average speed of the airplane?

8. On her weekly science tests Maria received grades of 87, 79, 92, 64, 88, 75, and 96. Find her average grade in science.

Using Mathematics

PRIME AND COMPOSITE NUMBERS

Whole numbers other than 0 and 1 may be classified as prime or composite numbers.

Any whole number greater than 1 that can be divided exactly by only the number itself and 1 is called a **prime number.**

23 can be divided exactly only by 23 and by 1.

23 is a prime number.

Any whole number greater than 1 that can be divided exactly by at least one whole number other than 1 and the number itself is called a **composite number.**

15 can be divided exactly not only by 15 and by 1

but also by 5 and by 3.

15 is a composite number.

Exercises

Which of the following are prime numbers?

1. 14   **2.** 23   **3.** 79   **4.** 51   **5.** 85   **6.** 97

Which of the following are composite numbers?

7. 81   **8.** 18   **9.** 2   **10.** 91   **11.** 53   **12.** 39

Write all the one-digit numbers.

13. Prime numbers   **14.** Even prime numbers   **15.** Odd prime numbers

Write the numbers.

16. All prime numbers greater than 18 and less than 32.

17. All composite numbers less than 70 and greater than 53.

18. Are all even numbers composite numbers? If not, name an even prime number.

19. Are all odd numbers prime numbers? If not, name an odd composite number.

20. Prime numbers that differ by 2 are called **twin primes.** For example, 17 and 19 are twin primes. Find four pairs of twin primes between 25 and 75.

> Goldbach, a mathematician, guessed that "Any even number greater than 4 can be expressed as the sum of two odd prime numbers."

Express each of the following even numbers as a sum of two odd prime numbers.

21. 8   **22.** 12   **23.** 36   **24.** 50   **25.** 108

26. Find an odd number that is the sum of two odd prime numbers.

Using Mathematics

PRIME FACTORIZATION

A composite number can be expressed as a product of prime numbers. This product is called the **prime factorization.**
Factor 72 as a product of prime numbers.

By using a factor tree:

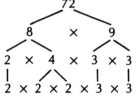

Find *any* two factors of the number. (Do not use 1 as a factor.) Continue to factor any factor that is a composite number until all factors are prime.

By division:

```
2)72
2)36
2)18
3)9
3)3
   1
```

Divide by *any* prime factor of the number. (Note that each quotient is *below* the corresponding dividend.) Continue to divide the resulting quotients by prime factors that divide exactly until the quotient is 1.

$72 = 2 \times 2 \times 2 \times 3 \times 3$

ANSWER: $2 \times 2 \times 2 \times 3 \times 3$ (or $2 \times 2 \times 3 \times 2 \times 3$) (or $3 \times 2 \times 3 \times 2 \times 2$) etc.

The divisors are the prime factors.

Though the composite number has only one set of factors, these factors may be written in any order.

After the composite number has been factored, the answer can also be expressed in exponential form.
$2 \times 2 \times 2 \times 3 \times 3 = 2^3 \times 3^2$

Exercises

Factor each of the following numbers as a product of prime numbers, using a factor tree.

1. 20 **2.** 64 **3.** 80 **4.** 48 **5.** 100 **6.** 144 **7.** 960

Factor each of the following as a product of prime numbers.

8. 14 **9.** 54 **10.** 72 **11.** 135 **12.** 168 **13.** 600 **14.** 306

Factor each of the following numbers as a product of prime numbers and write the answers in exponential form.

15. 36 **16.** 75 **17.** 135 **18.** 147 **19.** 200 **20.** 144 **21.** 800

1-14 Factors: Greatest Common Factor (GCF)

How can you determine all the possible factors of a number? Once you have factored two or more numbers, how can you find all their common factors? Is there more than one way to find the greatest common factor of two or more numbers?

Examples

1. Name all the factors of 24.

$$\frac{24}{1)24}$$ 1 divides 24 exactly, so 1 and 24 are factors.

Divide the given number by 1, 2, 3, etc. until a factor repeats.

$$\frac{12}{2)24}$$ 2 divides 24 exactly, so 2 and 12 are factors.

$$\frac{8}{3)24}$$ 3 divides 24 exactly, so 3 and 8 are factors.

$$\frac{4\ R4}{5)24}$$ 5 does not divide 24 exactly; 5 is not a factor.

$$\frac{6}{4)24}$$ 4 divides 24 exactly, so 4 and 6 are factors.

$$\frac{4}{6)24}$$ 6 divides 24 exactly but repeats, so stop.

ANSWER: 1, 2, 3, 4, 6, 8, 12, and 24 are factors of 24.

---

2. Name the common factors of 54, 72, and 108.

54: 1 2 3 6 9 18 27 54

72: 1 2 3 4 6 8 9 12 18 24 36 72

108: 1 2 3 4 6 9 12 18 27 36 54 108

List all the factors for each given number. Identify the factors that are common in each.

ANSWER: 1, 2, 3, 6, 9, and 18 are the common factors.

---

3. Name the greatest common factor (GCF) of 36 and 48.

36: 1 2 3 4 6 9 12 18 36

48: 1 2 3 4 6 8 12 16 24 48

GCF: 12

List all the factors for each given number. Then, select the greatest common factor.

Alternate method—prime factorization

$36 = 3 \times 3 \times 2 \times 2$

$48 = 3 \times 2 \times 2 \times 2 \times 2$

GCF: $3 \times 2 \times 2 = 12$

Factor each given number as a product of primes. The product of the common primes is the GCF.

ANSWER: 12 is the greatest common factor

Vocabulary

A factor is any of the numbers used in multiplication to form a product.

A common factor of two or more whole numbers is any whole number that is a factor of all the given numbers.

The greatest common factor (GCF) of two or more whole numbers is the greatest whole number that will divide all the given numbers exactly.

Think About It

1. Use a calculator to find the prime factorization of 12,155. What number would you try first? Why?

2. Joan was asked to write all the factors of 210. She factored 210 into primes: 1, 2, 3, 5, 7. Then she began to write the factors using only multiplication: 1 and 210, 2 and 105, 3 and 70, etc. Explain the technique she used to find the factors.

Diagnostic Exercises

1. Is 9 a factor of 63?

2. What is the second factor of a pair of factors when 6 is one factor of 96?

3. Name all the factors of 120.

4. Name the common factors of 56 and 84.

5. Name the common factors of 32, 80, and 128.

6. Name the greatest common factor of 27 and 72.

7. Carl claims that 9 is the greatest common factor of 90 and 108. Bob says that 18 is the greatest common factor. Who is correct?

Related Practice

1. **a.** Is 6 a factor of 24?
   b. Is 5 a factor of 40?
   c. Is 9 a factor of 56?
   d. Is 18 a factor of 90?

2. What is the second factor of a pair of factors when:

   a. 7 is one factor of 21?
   b. 8 is one factor of 72?
   c. 14 is one factor of 98?
   d. 15 is one factor of 135?

3. Name all the factors of each of the following.

   a. 48
   b. 132
   c. 225
   d. 400

4. Name the common factors of each pair of numbers.

   a. 12 and 16
   b. 63 and 84
   c. 32 and 104
   d. 105 and 150

5. Name the common factors of each group of numbers.

   a. 24, 60, and 96
   b. 39, 65, and 91
   c. 36, 81, and 108
   d. 252, 588, and 420

6. Name the greatest common factor of each pair of numbers.

   a. 10 and 16
   b. 12 and 40
   c. 30 and 75
   d. 52 and 78

7. **a.** Linda knew that she could rewrite a fraction in lowest terms if she could find the GCF of numerator and denominator and divide both numbers by this GCF. What is the GCF of the numerator and denominator of $\frac{102}{170}$?

   b. How many common factors are there for the two numbers 66 and 120?

   c. Kathy lists all the prime factors of two numbers: 2, 5, 3, 13 and 2, 7, 13, 11. What is the greatest common factor of these two numbers?

Career Application
CLERK

When merchandise—a radio, jacket, or other item—leaves the factory, it may be taken to a wholesaler's warehouse. Then it may go to a warehouse belonging to a chain of retail stores. Finally, it goes to one of the stores, where it may be sold to a consumer. The workers who keep track of merchandise in its journey from factory to consumer are called clerks. Shipping clerks pack items for shipment and prepare bills. Receiving clerks log in and distribute merchandise and deal with damaged items. Stock clerks count and mark items in a warehouse or store. Most clerks are high school graduates who can do arithmetic well. They get special training on the job. Some clerks use computers.

Before an order is shipped from a warehouse, a form, similar to the one below, must be completed by the shipping clerk to keep an accurate inventory of stock.

Help the shipping clerk determine each of the following.

1. The quantity of each item.

2. Total number of cartons.

3. Total quantity of items.

4. Number of delivery vans needed, if each van can carry 24 cartons.

Description	ID Number	Number of Cartons	Number Per Carton	Quantity
Item A	001	12	12	*1a.*
Item B	002	14	24	*1b.*
Item C	003	20	10	*1c.*
Item D	004	22	16	*1d.*
Item E	005	40	8	*1e.*
Item F	006	36	18	*1f.*
TOTALS		**2.**		**3.**

Chapter 1 Review

Vocabulary

addend p. 31	exponent p. 54	product p. 47
common factor p. 69	factor p. 47	quotient p. 58
counting number p. 20	GCF p. 69	remainder p. 58
difference p. 40	natural number p. 20	standard form p. 20
dividend p. 58	partial product p. 49	sum p. 31
divisor p. 58	period p. 20	whole number p. 20

Use the following problem to answer Exercises 1, 2, and 3.

The all-time leading scorer in the National Hockey League at the end of the 1986 season was Gordie Howe, who had scored 1,850 points in 1,767 games. The second place scorer was Marcel Dion, with 1,599 points in 1,163 games. How many more points did Howe have than Dion?

Page

7 **1.** What facts are not needed to solve the problem?

Page

12 **2.** What operation is needed to solve the problem?

15 **3.** What is the answer?

18 **4.** Is 573 a reasonable answer if the estimate is 600?

19 **5.** Write 4,652,000 in words.

23 **6.** Write six hundred seven million twenty-five thousand six hundred in standard form.

26 **7.** Round 4,728 to the nearest ten.

26 **8.** Round 682,471 to the nearest thousand.

26 **9.** Round 7,296,516 to the nearest ten thousand.

29 **10.**
$$53,467$$
$$47,811$$
$$+ \ 20,733$$

11.
$$4,682$$
$$1,735$$
$$1,028$$
$$+ \ 6,512$$

37 **12.** $368 + 145 + 226 + 1,074 + 15$

37 **13.**
$$24,094$$
$$- \ 18,244$$

14. $60,000 - 11,625 = $ ▨

37 **15.** Take 3,684 from 4,977.

44 **16.** $2,758 \times 9 = n$

47 **17.** The answer to a multiplication problem is the __?__.

49 **18.**
$$672$$
$$\times \quad 15$$

19. $186 \times 245 = n$

55 **20.** $6\overline{)8,358}$ **21.** $3\overline{)5,124}$

61 **22.** $23\overline{)3,588}$ **23.** $105\overline{)29,505}$

68 **24.** Find all the factors of 80.

68 **25.** Find the greatest common factor of 24 and 42.

Chapter 1 Test

Page
15 **1.** A flight from Dallas to Boston, a distance of about 1,700 miles, costs $116 one way. How much would it cost for two people round trip?

Page
7 **2.** Refer to Problem 1. What information is not needed?

18 **3.** Is 1,723 a reasonable answer for an estimate of 900?

19 **4.** Write 265,027,000 in words.

23 **5.** Write forty-seven million two hundred eight thousand in standard form.

26 **6.** Round 472 to the nearest hundred.

26 **7.** Round 768,462 to the nearest ten thousand.

29 **8.**
$$\begin{array}{r} 34,893 \\ 27,625 \\ + \; 12,839 \\ \hline \end{array}$$

29 **9.** 745 + 962 + 107 + 225

29 **10.** 64 + 928 + 114 + 5,313 + 228

37 **11.**
$$\begin{array}{r} 38,494 \\ - \; 27,725 \\ \hline \end{array}$$

37 **12.** 68,200 − 19,450

37 **13.** Take 1,386 from 1,972

44 **14.**
$$\begin{array}{r} 1,275 \\ \times \quad\;\; 6 \\ \hline \end{array}$$

49 **15.** 208 × 90 = ▨

49 **16.** 387 × 463 = ▨

55 **17.** $6\overline{)4,818}$

61 **18.** $28\overline{)9,128}$

61 **19.** $163\overline{)39,935}$

68 **20.** List all the factors of 33.

68 **21.** Find the greatest common factor of 28 and 70.

37 **22.** The World Trade Center in New York is 1,350 feet high and has 110 stories, while the Texas Commerce Tower is 1,002 feet high and has 75 stories. How much higher is the World Trade Center than the Texas Commerce Tower?

29 **23.** Out of a possible 538 electoral votes to be cast for a presidential election, New York has 36, Pennsylvania has 25, Texas has 29, California has 47, and Ohio has 23. How many votes do these five states cast?

49 **24.** How many ounces of soda are in a case of twenty-four 12-ounce cans?

55 **25.** Rob earned $51 for working three days last week. How much did he earn per day?

Cumulative Test

Solve each problem and select the letter corresponding to your answer.

Use the following statement to answer Questions 1 and 2.
Twenty-eight eighth-graders paid $4 each toward bus transportation
and a baseball game.

Page
12 **1.** What operation would be necessary to find the total amount paid by the students for the bus and the game?

   a. addition   **b.** subtraction
   c. multiplication **d.** division

Page
7   **2.** What fact is missing to find out how much of the $4 went toward the game?

   a. total cost   **b.** cost of bus
   c. size of bus   **d.** time left

23 **3.** Select the standard form for twenty thousand eighty-six.

   a. 20,086   **b.** 2,086
   c. 20,860   **d.** 200,086

26 **4.** Round 48,265 to the nearest hundred.

   a. 48,200   **b.** 48,300
   c. 48,270   **d.** 48,260

29 **5.** 345 + 2,678 + 2,665 + 6,384

   a. 11,962   **b.** 12,062
   c. 11,972   **d.** 12,072

37 **6.** 13,682 − 7,814

   a. 6,878   **b.** 5,868
   c. 6,272   **d.** 5,272

49 **7.** 6,273 × 47

   a. 294,831   **b.** 294,821
   c. 294,721   **d.** 295,831

61 **8.** $26\overline{)5,668}$

   a. 2,171   **b.** 218
   c. 208   **d.** 227

55 **9.** $4\overline{)1,236}$

   a. 390   **b.** 39
   c. 309   **d.** 3,900

68 **10.** Find the greatest common factor of 18 and 36.

   a. 9   **b.** 2
   c. 36   **d.** 18

29 **11.** From the years 1985 to 2000 the number of jobs in Washington, DC, is expected to increase by 509,000. If the number of jobs in 1985 was 2,113,100 how many jobs are projected for the year 2000?

   a. 2,622,100 jobs   **b.** 1,604,100 jobs
   c. 2,662,100 jobs   **d.** 1,416,100 jobs

44 **12.** The Hudson River is 306 miles long. The Pecos River is about 3 times longer than the Hudson. How long is the Pecos River?

   a. 102 miles   **b.** 309 miles
   c. 918 miles   **d.** 2,754 miles

COMPUTER ACTIVITY 1

Computers can be used to calculate. The following BASIC program can calculate average monthly sales.

The program contains statements (instructions to the computer) and names assigned by the programmer. For example, READ statements include the names of locations that hold data for the program to use. Two locations are SUM and AVG. The numbers stored in a location can change as the program is run.

Lines 330, 340, 350, and 360 contain DATA lists. Every time a statement contains READ, the computer reads a DATA statement. For the first READ statement, the computer uses the first number on the data list. At the next READ statement, the computer uses the next number in the data list, and so on.

```
1Ø   SUM = Ø                    2ØØ SUM = SUM + AUG
5Ø   READ JAN                   21Ø READ SEP
6Ø   SUM = SUM + JAN            22Ø SUM = SUM + SEP
7Ø   READ FEB                   23Ø READ OCT
8Ø   SUM = SUM + FEB            24Ø SUM = SUM + OCT
9Ø   READ MAR                   25Ø READ NOV
1ØØ SUM = SUM + MAR             26Ø SUM = SUM + NOV
11Ø READ APR                    27Ø READ DEC
12Ø SUM = SUM + APR             28Ø SUM = SUM + DEC
13Ø READ MAY                    31Ø AVG = SUM/12
14Ø SUM = SUM + MAY             33Ø DATA 3ØØ, 467, 1,232
15Ø READ JUN                    34Ø DATA 1,345, 1,1Ø1, 1,ØØ9
16Ø SUM = SUM + JUN             35Ø DATA 897, 925, 1,Ø32
17Ø READ JUL                    36Ø DATA 1,243, 1,278, 1,45Ø
18Ø SUM = SUM + JUL             37Ø PRINT AVG
19Ø READ AUG                    38Ø END
```

What happens to SUM when the program is run?

To make notes about this program, use a REM STATEMENT. REM stands for REMark. Use a line number and put down any notes you wish. These will be saved by the computer if you save the program. However, the notes do not become part of the processing if you instruct the computer to RUN the program.

For example, write a REM STATEMENT on line 320 to remind yourself that you used 12 months sales.

Using Mathematics

Introduction to Decimals

Our number scale is extended to the right of the ones place to express parts of 1. Since the decimal number scale is based on tens, the value of each place on the number scale is one-tenth of the value of the next place to its left.

The first place to the right of the ones place has the value of one-tenth of one whole unit. It is indicated as .1 and expresses "tenths." The dot is called the **decimal point.**

When a unit is divided into ten equal parts, the size of each equal part is one-tenth and is written as .1 and read "one-tenth."

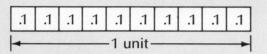

| .1 | .1 | .1 | .1 | .1 | .1 | .1 | .1 | .1 | .1 |

|← —————————— 1 unit —————————— →|

The second place to the right of the ones place expresses hundredths (.01).
The third place expresses thousandths. (.001).
The fourth place expresses ten-thousandths (.0001).
The fifth place expresses hundred-thousandths (.00001).
The sixth place expresses millionths (.000001).

$3 \times (1)$	decimal point	$9 \times (.1)$	$6 \times (.01)$	$4 \times (.001)$	$2 \times (.0001)$	$8 \times (.00001)$	$7 \times (.000001)$
3	.	9	6	4	2	8	7

A **decimal** is a fractional number such as .83 or 1.0259.

A number containing a whole number and a decimal, such as 16.5, is often called a mixed decimal. The decimal point is used to separate the whole number from the fractional part.

2-1 Reading Decimals

A piece of pipe for a machine measures 0.734 m in length.
Read this number.

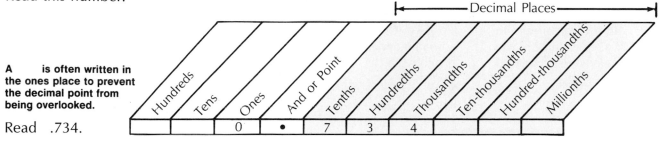

Decimal Places

Hundreds	Tens	Ones	And or Point	Tenths	Hundredths	Thousandths	Ten-thousandths	Hundred-thousandths	Millionths
		0	•	7	3	4			

A is often written in the ones place to prevent the decimal point from being overlooked.

Read .734.

0.⎣734⎦

seven hundred thirty-four thousandths

ANSWER: Seven hundred thirty-four thousandths.

Read the digits as for a whole number. Then name the place value of the last digit

Examples

1. Read 0.5291.

0.⎣5291⎦

five thousand two hundred ninety-one ten-thousandths

ANSWER: Five thousand two hundred ninety-one ten-thousandths.

2. Read 0.005 in two ways.

⎣0 005⎦

five thousandths, or zero   zero zero five

Decimals may also be read by naming each digit separately after saying the word

ANSWER: Five thousandths, or zero point zero zero five.

3. Read 84.06

84 06

eighty-four and six hundredths

For decimals greater than 1, read the word *and* for the decimal point.

ANSWER: Eighty-four and six hundredths.

---

4. Read 634.368 in two ways.

634 368

six hundred thirty-four and three hundred sixty-eight thousandths
or six hundred thirty-four point three six eight

ANSWER: Six hundred thirty-four and three
hundred sixty-eight thousandths, or six
hundred thirty-four point three six eight.

Vocabulary

In a decimal number, the places to the right of a decimal point are called **decimal places**. The word name for a decimal uses the word *and* or *point* to indicate the position of the decimal point.

Think About It

1. How does the value of a given place in a number compare with the place to its right?

2. How does the value of a given place in a number compare with the place to its left?

3. True or false? Writing zeros in a decimal right after the decimal point decreases the value of the number.

Diagnostic Exercises

Read, or write in words.

1. 0.2　　　　　　**2.** 0.06　　　　　　**3.** 0.58

4. 1.5　　　　　　**5.** 3.73　　　　　　**6.** 0.004

7. 0.076　　　　　**8.** 0.289　　　　　**9.** 14.708

10. 0.0037　　　　　　　**11.** 0.17925　　　　　　　**12.** 0.000456

Read or write in two ways.

13. 129.4261　　　　　　**14.** 0.836　　　　　　　**15.** 24.78

16. The diameter of a wire is 0.15 cm. Read this number.

Related Practice

Read, or write in words.

1. a. 0.8　　　　　　**b.** 0.1　　　　　　**c.** 0.5　　　　　　**d.** 0.4

2. a. 0.03　　　　　**b.** 0.07　　　　　**c.** 0.02　　　　　**d.** 0.08

3. a. 0.24　　　　　**b.** 0.85　　　　　**c.** 0.91　　　　　**d.** 0.60

4. a. 1.6　　　　　　**b.** 2.9　　　　　　**c.** 38.5　　　　　**d.** 126.4

5. a. 2.51　　　　　**b.** 7.37　　　　　**c.** 89.03　　　　**d.** 248.19

6. a. 0.005　　　　**b.** 0.008　　　　**c.** 0.001　　　　**d.** 0.007

7. a. 0.024　　　　**b.** 0.063　　　　**c.** 0.080　　　　**d.** 0.092

8. a. 0.832　　　　**b.** 0.946　　　　**c.** 0.253　　　　**d.** 0.798

9. a. 6.005　　　　**b.** 21.769　　　**c.** 186.528　　**d.** 200.042

10. a. 0.0007　　　**b.** 0.0089　　　**c.** 0.0574　　　**d.** 0.9350

11. a. 0.00006　　**b.** 0.00392　　**c.** 0.09413　　**d.** 0.72815

12. a. 0.000001　**b.** 0.000534　**c.** 0.080076　**d.** 0.175283

Read or write in two ways.

13. a. 8.0025　　　**b.** 23.9317　　**c.** 49.08329　**d.** 571.05875

14. a. 0.93　　　　**b.** 0.619　　　**c.** 0.2　　　　　**d.** 0.3428

15. a. 36.89　　　**b.** 17.6　　　　**c.** 5.925　　　**d.** 87.4662

16. Read each number.

　a. A piece of paper is 0.12 centimeters thick.　　**b.** A length of string measures 0.04 meters.

　c. The river is 3.016 kilometers wide at Point Belle.　　**d.** The elapsed time of the experiment was 804.217 seconds.

Writing Decimals

A chemist used fifty-two hundredths milligram of sodium for an experiment. Write this number in standard form.

Write fifty-two hundredths in standard form.

fifty-two hundredths
↓
0.52
↑
hundredths place

Write the digits as for a whole number. Then place the decimal point so that the last digit has the stated place value

ANSWER: 0.52

Examples

1. Write seven tenths in standard form.

seven tenths
↓
0.7

Write decimals with a zero in the ones place.

ANSWER: 0.7

---

2. Write zero point four nine two in standard form.

zero point	four	nine	two
↓ ↓	↓	↓	↓
0 .	4	9	2

Write a decimal point for the word *point*

ANSWER: 0.492

---

3. Write four hundred and three hundred fifty-three thousandths in standard form.

four hundred and	three hundred fifty-three	thousandths
↓	↓	
400 .	353	

Write a decimal point for the word *and*

ANSWER: 400.353

---

4. Write sixty-two point zero four five in standard form.

sixty-two	point	zero	four	five
↓	↓	↓	↓	↓
62	.	0	4	5

ANSWER: 62.045

5. Write seventy-six and eight ten-thousandths in standard form.

seventy-six and eight ten-thousandths

76 . 000 8

Write 3 zeros.

If necessary, write zeros to the right of the decimal point.

ANSWER: 76.0008

Think About It

1. In the number forty-two hundred thousandths, how many zeros to the right of the decimal point do you need?

2. True or false? The last word in a decimal name tells you the place value of the last digit of the decimal.

3. If the last word of a decimal name is ten-thousandths, how many places must the decimal have?

4. What is the smallest decimal number you can enter into your calculator?

Diagnostic Exercises

Write each of the following in standard form.

1. Three tenths

2. Four and eight tenths

3. Seven hundredths

4. Twenty-five hundredths

5. One hundred nine and eighty-four hundredths

6. Nine thousandths

7. Ninety-four thousandths

8. Five hundred twenty-seven thousandths

9. Seven hundred and ninety-three thousandths

10. Four thousand six hundred thirty-six ten-thousandths

11. Eight hundred forty-two hundred-thousandths

12. Four millionths

13. Two hundred sixty and three hundred forty-seven ten-thousandths

14. Five hundred nineteen point nine

15. The pharmacist used six hundredths of an ounce of medicine. Write this number in standard form.

Related Practice

Write each of the following in standard form.

1. **a.** Four tenths
   b. Eight tenths
   c. Two tenths
   d. One tenth

2. **a.** Six and five tenths
   b. Nine and seven tenths
   c. Five and one tenth
   d. Twenty and six tenths

3. **a.** Eight hundredths
   b. Two hundredths
   c. Four hundredths
   d. Five hundredths

4. **a.** Thirty-six hundredths
   b. Fifty-seven hundredths
   c. Seventeen hundredths
   d. Forty-nine hundredths

5. **a.** Six and four hundredths
   b. Seventy-three and eighteen hundredths
   c. Two hundred and five hundredths
   d. Four hundred seven and twenty-five hundredths

6. **a.** Three thousandths
   b. Seven thousandths
   c. Five thousandths
   d. One thousandth

7. **a.** Sixty-nine thousandths
   b. Forty-seven thousandths
   c. Sixteen thousandths
   d. Eighty-three thousandths

8. **a.** Two hundred seventy-four thousandths
   b. Four hundred thirty-nine thousandths
   c. Seven hundred twenty-one thousandths
   d. Three hundred six thousandths

9. **a.** Two and seventeen thousandths
   b. Eight hundred and thirty-five thousandths
   c. Thirty-six and two hundred fifty-three thousandths
   d. Nineteen and one hundred twenty-two thousandths

10. **a.** Eight ten-thousandths
   b. Thirty-six ten-thousandths
   c. Four hundred ninety-four ten-thousandths
   d. Three thousand five hundred sixty ten-thousandths

11. **a.** Three hundred-thousandths
   b. Forty-two hundred-thousandths
   c. Four hundred fifty-six hundred-thousandths
   d. Six thousand eight hundred twenty-two hundred-thousandths

12. **a.** Six millionths
   b. Ninety-five millionths
   c. Three hundred seven millionths
   d. Seventy-two thousand one hundred forty-nine millionths

13. **a.** Five hundred and fifty-eight ten-thousandths
   b. Seventy and two thousand five hundred twelve ten-thousandths
   c. Eighty-five and seven hundred thirty-two hundred-thousandths
   d. Six hundred forty-three and sixty-seven millionths

14. **a.** Sixty-one point two
   b. Zero point six four one
   c. Eight hundred eighty point one six one three
   d. Three thousand ten point seven nine

15. Write each number in standard form.

   a. Louis bought forty-five hundredths of an ounce of saffron.
   b. Tara bought four hundredths of an ounce of diamond dust.
   c. A machine weighs eight hundred seventy-four and five hundred thirty-five thousandths pounds.
   d. The survey noted that the road measured four hundred sixty-five and twenty-five ten-thousandths kilometers.

Rounding Decimals

A carton of milk costs $.47. Round this to the nearest dime to find its approximate cost.

a. Locate the place to be rounded.

$.47̲   tenths (a dime is one-tenth of a dollar)

b. If the digit to the right is 5 or more, round to the next higher digit

   $.47   7 is greater than 5,
   $.50   so round the 4 to a 5.

ANSWER: The cost of the carton of milk is about $.50.

Examples

1. Round 0.83249 to the nearest thousandth.

   0.83249
   0.832̲49   If the digit to the right is less than 5, write the same
     └─4 is less than 5.   digit.

   0.832   Omit all digits to the right of the place to be
                              rounded.
ANSWER: 0.832

2. Round 15.964 to the nearest tenth.

15.964
15.964

⬆
└─ **Adding 1 to the tenths place changes the ones place.**

Sometimes the digit to the left of the place to be rounded is increased by one.

16.0

ANSWER: 1 .

---

3. Round $4.9782 to the nearest cent.

$4.9782
$4.9782
$4.98

ANSWER: $4.98

4. Round 73.568 to the nearest whole number.

73.568
73.568
74

ANSWER: 74

Think About It

1. When you round to the nearest cent, what decimal place are you rounding to?

2. When you round to the nearest dollar, what place are you rounding to?

3. Which way would your estimate be closer to the actual cost of a grocery cart of food: by rounding the cost of each item to the nearest dollar or to the nearest dime? Why might this be important?

Diagnostic Exercises

Round each of the following to the nearest

Tenth:

1. 0.68   **2.** 5.425

Hundredth:

3. 8.349   **4.** 0.5146

Thousandth:

5. 1.9685   **6.** 3.24728

Ten-thousandth:

7. 0.42769   **8.** 26.58634

Hundred-thousandth:

9. 0.005924   **10.** 54.632857

Millionth:

11. 0.0000985   **12.** 2.0500302

Cent:

13. $2.386   **14.** $4.843

Whole number:

15. 57.18   **16.** 89.562

Dollar:

17. $9.84   **18.** $27.46

19. The bill for supplies for the Drama Club was $136.42. Round this to the nearest dollar.

Related Practice

Round each of the following to the nearest

Tenth:

1. a. 0.25
   b. 0.87
   c. 0.984
   d. 2.39

2. a. 0.14
   b. 0.32
   c. 1.83
   d. 5.205

Hundredth:

3. a. 0.517
   b. 0.308
   c. 5.845
   d. 15.2263

4. a. 0.323
   b. 0.934
   c. 6.544
   d. 8.3025

Thousandth:

5. a. 0.2146
   b. 0.3998
   c. 3.0815
   d. 16.76876

6. a. 0.5452
   b. 0.9293
   c. 2.0084
   d. 24.82509

Ten-thousandth:

7. a. 0.20585
   b. 0.932473
   c. 4.00619
   d. 43.020471

8. a. 0.34143
   b. 0.528129
   c. 18.781304
   d. 39.56372

Hundred-thousandth:

9. a. 0.000083
   b. 0.005142
   c. 1.079324
   d. 83.591243

10. a. 0.000079
    b. 0.030585
    c. 2.000467
    d. 23.075306

Millionth:

11. a. 0.0000029
    b. 0.0001538
    c. 0.0254985
    d. 1.0038296

12. a. 0.0000052
    b. 0.0000194
    c. 0.0039481
    d. 5.0006373

Cent:

13. a. $.267
    b. $.59
    c. $1.316
    d. $5.8682

14. a. $.643
    b. $.834
    c. $6.572
    d. $14.9615

Whole number:

15. a. 6.3
    b. 19.4
    c. 2.09
    d. 34.28

16. a. 9.6
    b. 68.5
    c. 100.81
    d. 399.704

Dollar:

17. a. $5.69
    b. $37.90
    c. $420.71
    d. $19.53

18. a. $2.46
    b. $53.17
    c. $269.39
    d. $84.25

19. a. One nautical mile is 1.1515 statute miles. Round the statute miles to the nearest hundredth.

b. One cubic inch is 16.3872 cubic centimeters. Round the cubic centimeters to the nearest tenth.

c. One statute mile is 0.8684 nautical miles. Round the nautical mile to the nearest thousandth.

d. One U.S. dollar is worth $1.3749 Canadian. Round the Canadian dollar to the nearest cent.

2-4 Comparing Decimals

Which is more, 0.3 pound of cheese, or 0.03 pound of cheese?

0.3
0.03
↓ ↓
same ↳3 > 0

Since 3 > 0, then 0.3 > 0.03.

ANSWER: 0.3 lb. of cheese is more.

To compare decimals, do this.

Compare the digits in each place, starting at the left.

Examples

1. Which is less: 1.765 or 1.76?

1. 7 6 5

1. 7 6 0

same same same 0 < 5

Since 0 < 5, then 1.760 < 1.765.

ANSWER: 1.76 is less.

Write a zero to make the same number of decimal places if needed.

2. Is the statement 32.72 > 0.983 true?
Since 32 > 0, the statement is true.

ANSWER: Yes, true.

3. Is the statement 0.16 > 0.6 true?
Since 1 < 6, 0.16 < 0.6.

ANSWER: No, false.

4. Arrange in order from least to greatest, 0.635, 0.65, 6.1, 0.069.

0.635
0.65
6.1
0.069

In the ones place:
   Since 6 > 0, 6.1 is the greatest.
In the tenths place:
   Since 0 < 6, 0.069 is the least.
In the hundredths place:
   Since 3 < 5, 0.635 < 0.65

ANSWER: 0.069, 0.635, 0.65, 6.1

Vocabulary

The symbol < means "is less than."   The symbol > means "is greater than."

Think About It

1. True or false? A whole number is always greater than a decimal. Give an example to illustrate.

2. How can you write a whole number as a decimal?

3. Compare 0.3 and 0.5 using >.

Diagnostic Exercises

1. Which is greater, 0.154 or 1.02?

2. Which is less, 5.53 or 0.553?

3. Arrange in order of size.
   a. greatest first 0.06, 1.4, 0.19, and 0.388
   b. least first 4.72, 0.493, 4.8, and 0.465

4. a. Is the statement 0.725 < 0.73 true?
   b. Is the statement 0.9 > 0.19 true?

5. Which has the same value as 0.607?   a. 0.067   b. 0.0607   c. 0.6070   d. 0.670

6. The Ching family has a pool that is 12 meters long. Their neighbors, the Wolfes, have a pool that is 12.7 meters long. Which family has the longer pool?

Related Practice

Which is greater?

1. a. 0.3 or 0.29   b. 0.04 or 0.004
   c. 0.91 or 0.893   d. 0.7 or 0.074

Which is less?

2. a. 0.89 or 0.9   b. 0.2 or 0.21
   c. 0.50 or 0.05   d. 0.0051 or 0.006

Arrange in order of size (greatest first).

3. 0.01, 0.001, 0.1, and 0.0001
   2.25, 0.253, 0.2485, and 2.249
   0.38, 1.5, 0.475, and 0.0506
   0.006, 5.02, 0.503, and 0.1483

True or false?

4. a. 0.306 < 0.36
   b. 0.3 < 0.27
   c. 0.049 < 0.5
   d. 6.7 < 0.675

Write the number that has the same value as the number at the left.

5. a. 0.06   0.6   0.60   0.060   0.0060
   b. 0.40   0.04   4.0   40.00   0.400
   c. 0.2   2.0   0.02   0.002   0.200
   d. 0.087   0.807   0.0870   0.0807   0.8700

6. a. Mrs. Goldstein's patio measured 6.7 meters across. Her neighbor's patio was 5.9 meters wide. Who had the wider patio?

   b. Juan jumped 1.63 meters and Jonathan jumped 1.6 meters during the long-jump contest. Who won?

   c. Mr. Spangler's garden is 6.3 meters long. Mr. O'Connor's garden is 6.25 meters long. Who has the longer garden?

   d. The teacher wants the girls to line up, shortest to tallest. If Carla is 1.50 meters tall, Corrine is 1.6 meters tall, and Connie is 1.55 meters tall, what should be the order?

2-5 Adding Decimals

Building Problem Solving Skills

During the five times it rained in August, 1.02 inches, 2 inches, 0.79 inches, 0.4 inches, and 1.6 inches fell. What was the total rainfall for the month?

1. **READ** the problem carefully.
   a. Find the **question asked.** What was the total rainfall in August?
   b. Find the **given facts.** It rained five times in August: 1.02 inches, 2 inches, 0.79 inches, 0.4 inches, and 1.6 inches.

---

2. **PLAN** how to solve the problem.
   a. Choose the **operation needed.** The word *total* with *unequal* quantities indicates addition.
   b. **Think or write out your plan** relating the given facts to the question asked. The sum of 1.02 inches, 2 inches, 0.79 inches, 0.4 inches, and 1.6 inches of rain is equal to the total rainfall.

Addition Indicators
• how much in all
• how many in all
• sum
• total
• how much altogether
Note: All indicators usually refer to *unequal* amounts.

3. SOLVE the problem.

a. Estimate the answer.

1.02	→	1
2	→	2
0.79	→	1
0.4	→	0
+ 1.6	→	+ 2
		6

Round each decimal to the nearest whole number. Add. The answer should be about 6 inches.

b. Solution.

```
  1 1
 1.02
 2.00
 0.79
 0.40
+ 1.60
 5.81
```

Arrange in columns, lining up the decimal points. Add, starting from the right. Regroup when necessary. Place the decimal point in the sum below the decimal points in the addends.

4. CHECK

a. Check the accuracy of your arithmetic.

```
 1.02  ↑
 2.00  |
 0.79  |   Add up.
 0.40  |
+ 1.60 |
 5.81  ✔
```

b. Compare the answer to the estimate. Check the unit. The answer 5.81 inches compares reasonably with the estimate of 6 inches.

ANSWER: The total rainfall was 5.81 inches.

Think About It

1. If each of the five rainfalls had been 1.2 inches, what operation could you have used instead of addition?

2. How could you find the average rainfall *per day* during August?

Calculator Know-How

When using your calculator to solve problems with decimals you must remember to key ⚬ .

Key the following numbers on a calculator. Clear the display before keying a new number.
1. 89.5   **2.** 75.482   **3.** 0.759   **4.** 38.14

Use a calculator to solve these addition problems.

5. What was the total snowfall in the White Mountains the year it snowed 12.7 inches in November, 38.12 inches in December, 59.08 inches in January, 63.74 inches in February, 29.09 inches in March, and 2.81 inches in April?

6. What was the total deposit made by a company if the checks deposited were for: $3,602.95, $781.92, $100.29, $1,005.78; $333.02; $2,593.46, and $2.89?

1. Add: 0.24 + 0.39

$$\overset{1}{0.24}$$
$$\underline{+\ 0.39}$$
$$0.6\underset{3}{3}$$

Check: $\overset{1}{0.24}$ ↑
$\underline{+\ 0.39}$
0.63 ✔

Arrange in columns. Line up the decimal points. Add, starting from the right. Regroup when necessary.
13 hundredths = 1 tenth 3 hundredths
Place the decimal point in the sum directly below the decimal points in the addends.
Check by adding up.

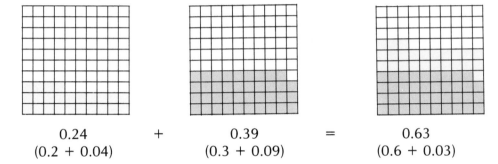

0.24	+ 0.39	= 0.63
(0.2 + 0.04)	(0.3 + 0.09)	(0.6 + 0.03)

ANSWER: 0.63

---

2. Add 0.7 to 3.46

$$\overset{1}{3.46}$$
$$\underline{+\ 0.7\underline{0}}$$
$$4.16$$

Check: $\overset{1\ 1}{3.46}$ ↑
$\underline{+\ 0.70}$
4.16 ✔

Write ending zeros so both addends have the same number of decimal places.
11 tenths = 1 one 1 tenth

ANSWER: 4.16

---

3. Add: 46 + 0.927

$$46\overset{\downarrow}{.}000$$
$$\underline{+\ \ 0.927}$$
$$46.927$$

Check: 46.000 ↑
$\underline{+\ \ 0.927}$
46.927 ✔

When adding whole numbers and decimals, place a decimal point after each whole number.

ANSWER: 46.927

4. Add: 0.26 + 0.38 + 0.06

```
    0.26           Check:    0.26 ↑
    0.38                     0.38 |
  + 0.06                   + 0.06 |
  0.70 or 0.7             0.70 or 0.7 ✔
```

When a decimal ends in one or more zeros to the right of the decimal point, the zeros may be dropped.

ANSWER: 0.70 or 0.7

---

5. Find the sum: 0.523; 4.16; 3; and 2.2

```
    0.523          Check:    0.523 ↑
    4.160                    4.160 |
    3.000                    3.000 |
  + 2.200                  + 2.200 |
    9.883                    9.883 ✔
```

ANSWER: 9.883

---

6. Estimate the sum: 0.71; 9.052; 47.53

```
    0.71   →      1.0
    9.052  →      9.0
 + 47.53   →   + 48.0
                 58.0
```

Round the numbers to the same place.
Add.

ANSWER: 58.0

---

7. Solve the equation: $\$20.56 + \$14.27 = n$

```
       1
    $20.56         Check:   $20.56 ↑
  +  14.27                 +  14.27 |
    $34.83                  $34.83 ✔
```

To find the value of n, add the given numbers.

ANSWER: $34.83

Diagnostic Exercises

Add.

1.	**2.**	**3.**	**4.**	**5.**
0.4	0.6	0.02	0.08	0.15
0.3	0.9	0.01	0.03	0.03
+ 0.2	+ 0.4	+ 0.03	0.02	0.48
			+ 0.06	+ 0.17

6. 0.52
    0.43
    0.69
+ 0.74

7. 0.85
    0.10
+ 0.05

8. 1.5
    2.1
+ 4.3

9. 6.4
    8.9
    4.4
+ 3.5

10. 3.06
    4.09
+ 2.08

11. 4.57
    2.93
+ 4.87

12. 7.49
+ 3.5

13. 2.86
    0.7
+ 0.12

14. 8
+ 0.05

15. 2.103
    4.839
+ 3.542

16. 50.48
    37.59
+ 23.84

17. 326.04
    183.75
    225.39
+ 491.26

18. 19.47
    8.46
    592.75
    74.81
    126.78
+ 91.33

19. $.50
    .28
    .79
    .84
+ .67

20. $42.85
    9.74
    223.30
    54.67
+   7.98

21. $0.08 + 1.5$

22. $0.752 + 4.53 + 6$

23. $\$1.43 + \$.89 + \$5.07 + \$.36 + \$9.58$

24. Find the sum of 6.4, 0.976, and 2.87.

25. Select the nearest given estimate for:
0.493 + 0.78 + 0.6

   a. 1.5 **b.** 2.0 **c.** 2.5

26. Solve. $0.13 + 0.7 = n$

27. Rick has the following bills for maintenance of his car during the month of January: $12.56; $35.42; $18.95; $13.50. What was the total amount he spent on his car during January?

Related Practice

Add.

1. a. 0.2
+ 0.6

b. 0.3
    0.2
+ 0.2

c. 0.2
    0.3
    0.1
+ 0.2

d. 0.1
    0.3
    0.2
    0.1
+ 0.2

2. a. 0.9
+ 0.7

b. 0.5
    0.8
+ 0.4

c. 0.4
    0.8
    0.7
+ 0.4

d. 0.5
    0.7
    0.4
    0.8
+ 0.3

3. a. 0.03
+ 0.04

b. 0.03
    0.02
+ 0.03

c. 0.01
    0.03
    0.02
+ 0.03

d. 0.03
    0.01
    0.02
    0.01
+ 0.02

4. a. 0.06
+ 0.09

b. 0.05
    0.05
+ 0.07

c. 0.02
    0.07
    0.04
+ 0.05

d. 0.09
    0.06
    0.08
    0.04
+ 0.09

5. a. 0.68 **b.** 0.57 **c.** 0.12 **d.** 0.23 **6. a.** 0.43 **b.** 0.38 **c.** 0.82 **d.** 0.27
    + 0.26   0.04   0.23   0.18   + 0.89   0.45   0.59   0.74
            + 0.18   0.45   0.29   + 0.22   0.77   0.51
                   + 0.16   0.03   + 0.31   0.93
                         + 0.25   + 0.89

7. a. 0.53 **b.** 0.48 **c.** 0.52 **d.** 0.17 **8. a.** 2.4 **b.** 6.2 **c.** 4.1 **d.** 9.2
    + 0.47   0.26   0.81   0.08   + 3.5   9.3   2.2   3.2
            + 0.16   0.49   0.49   + 8.3   1.1   4.3
                   + 0.68   0.27   + 9.2   2.1
                       + 0.99   + 7.1

9. a. 7.8 **b.** 4.1 **c.** 2.8 **d.** 8.5 **10. a.** 8.05 **b.** 9.08 **c.** 6.01 **d.** 6.08
    + 6.5   1.5   1.4   4.9   + 4.08   8.03   8.07   2.03
            + 7.4   8.2   6.3   + 1.07   4.05   9.07
                   + 7.5   7.4   + 2.09   4.09
                       + 5.6   + 7.03

11. a. 3.26 **b.** 6.27 **c.** 2.19 **d.** 3.86 **12. a.** 6.25 **b.** 6.74 **c.** 5.7 **d.** 4.92
    + 2.15   2.83   3.82   2.29   + 4.6   9.3   4.3   8.1
            + 5.76   5.27   4.57   + 1.87   9.25   9.6
                   + 1.38   5.18   + 8.75   2.43
                       + 1.33   + 1.8

13. a. 0.9 **b.** 9.5 **c.** 1.4 **d.** 5.27 **14. a.** 0.6 **b.** 0.94 **c.** 3.9 **d.** 0.36
    + 5.28   0.33   0.7   7.4   + 4.0   7.   0.49   4.68
            + 0.19   0.29   0.35   + 0.8   6.0   0.2
                   + 2.45   3.82   + 0.57   9.0
                       + 0.06   + 1.7

15. a. 5.01 **b.** 3.728 **c.** 1.516 **d.** 4.783 **16. a.** 20.56 **b.** 36.87 **c.** 28.45 **d.** 32.28
    + 2.999   3.517   6.24   1.829   + 14.27   8.26   42.83   14.85
            + 9.282   0.006   5.318   + 15.84   56.19   9.74
                   + 4.518   2.175   + 18.28   98.42
                       + 9.384   + 0.94

17. a. 138.35 **b.** 639.32 **c.** 0.00532 **d.** 335.48
    + 253.42   182.08   0.13847   618.37
            + 24.19   0.32296   407.54
                   + 0.48325   250.36
                              + 195.83

18. a.
```
   0.08
   0.57
   0.13
   0.42
   0.83
+  0.66
```
b.
```
   1.16
   2.65
   4.27
   1.04
   3.53
+  5.21
```
c.
```
   16.91
   28.38
   19.43
   37.21
    4.82
+   0.48
```
d.
```
   596.74
   289.57
   193.86
   319.29
   475.43
+  628.58
```

19. a.
```
   $.24
    .59
    .37
    .73
+   .85
```
b.
```
   $.83
    .25
    .03
    .62
+   .54
```
c.
```
   $.36
    .88
    .41
    .29
+   .37
```
d.
```
   $.84
    .27
    .49
    .56
    .99
+   .78
```

20. a.
```
   $2.33
    .96
   4.28
    .51
+  9.16
```
b.
```
   $8.25
   3.70
   9.64
   8.23
+  4.72
```
c.
```
   $45.63
    5.96
     .83
   14.71
    9.43
+  18.65
```
d.
```
   $355.95
   109.82
   481.56
   247.49
   575.37
+  281.25
```

21. a. 0.05 + 0.12   **b.** 3.6 + 5.1   **c.** 0.275 + 0.38   **d.** 0.017 + 15

22. a. 0.52 + 1.6 + 8.26
b. 0.83 + 7 + 4.45 + 0.049
c. 1.5 + 0.18 + 6.84 + 0.016 + 0.27
d. 1.4 + 26 + 0.39 + 5.98 + 9

23. a. $.17 + $.49 + $.83
b. $1.80 + $2.60 + $4.25
c. $.74 + $1.60 + $.99 + $4.88 + $.04
d. $3.42 + $6.51 + $12.54 + $9.49 + $8.68

Find the sum.

24. a. 4.23, 6.832, and 4.4
b. 0.6, 8, and 0.24
c. 2.05, 0.156, 4.69, and 0.08
d. $12.59, $9.47, $1.27, $.56, and $3.46

Write the nearest given estimate.

25. a. 6.8 + 0.42   7   10   11
b. 0.226 + 0.79   0.9   1   1.4
c. 0.023 + 0.064 + 0.019   0.1   0.2   0.09
d. 0.7 + 0.42 + 0.168   1.1   1.3   1.5

Solve.

26. a. $0.6 + 0.9 = n$
b. ▧ = 3.5 + 0.24 + 8
c. 0.09 + 0.009 = ▧
d. n = $3.50 + $.75 + $8.36

27. a. Jim has checks of $21.68; $45.90; and $73.45 to deposit. What is his total deposit?
b. Barbara can run a lap in 41.2 seconds. Susan took 6.9 seconds more to run the same lap. How long did it take Susan?
c. The school orchestra bought a violin for $169.95, a saxophone for $249.75, and a trumpet for $124.50. What is the total cost of the new instruments?
d. A mutual fund reported a dividend of 8.954 shares to a stockholder who already owns 490.573 shares. Find the new total of shares owned by the stockholder.

Subtracting Decimals

Building Problem Solving Skills

One year the winning automobile speed for the Indianapolis 500 was 162.117 miles per hour and for the Daytona 500 it was 155.979 miles per hour. How much faster was the Indianapolis winning speed?

Subtraction Indicators
• how many more
• how many fewer
• how much larger
• how much smaller
• amount of increase
• amount of decrease
• difference
• balance
• how many left over
• how much left over

1. **READ** the problem carefully.

a. Find the **question asked.**
How much faster is the Indianapolis winning speed?

b. Find the **given facts.**
Indianapolis 500: 162.117 miles per hour
Daytona 500: 155.979 miles per hour

2. **PLAN** how to solve the problem.

a. Choose the **operation needed.**
The words *how much faster* indicate subtraction.

b. **Think or write out your plan** relating the given facts to the question asked.
The difference in the two speeds is equal to the number of miles per hour faster.

c. Express your plan as an **equation.**
$162.117 - 155.979 = n$

3. **SOLVE** the problem.

a. Estimate the answer.

$$
\begin{array}{r}
162.117 \rightarrow 162 \\
- 155.979 \rightarrow - 156 \\
\hline
6
\end{array}
$$

Round the numbers to the nearest whole number. Subtract.

b. Solution.

$$
\begin{array}{r}
162.117 \\
- 155.979 \\
\hline
6.138
\end{array}
$$

Arrange in columns. Subtract each column starting from the right. Regroup as needed.

You might use a calculator to solve the problem.

Key: 162.117 $\boxed{-}$ 155.979 $\boxed{=}$ 6.138

4. ☐ **CHECK** ☐

a. Check the accuracy by adding.

```
   155.979
+    6.138
   162.117 ✔
```

ANSWER: The speed is 6.138 mi/h faster.
The unit, mi/h, is correct.

b. Compare the answer to the estimate.
The answer 6.138 compares reasonably
with the estimate 6.

To use a calculator to check,

Key: 155.979 ☐+☐ 6.138 ☐=☐ <u>162.117</u>

Think About It

1. Why must you line up the decimal points
when you add and subtract decimals?

2. What digit can you write at the end of a
decimal without changing the value of the
decimal? Explain.

3. How can you write a whole number as a
decimal?

Calculator Know-How

You can use the memory key to perform a series of computations.

You have $15 and you make these purchases: 6 oranges costing 24¢ each, 3
boxes of cereal costing $1.89 each, 2 half-gallons of milk costing $1.15
each, and 2 loaves of bread costing 87¢ each. How much change will you
receive?

Solve using a calculator.

Key: 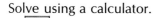 15 M+

6 ☒ • 24 M−

Enter 15 into memory.

Compute the cost of each item and subtract
from the memory.

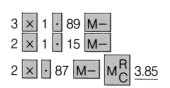
3 ☒ 1 • 89 M−
2 ☒ 1 • 15 M−

2 ☒ • 87 M− M⌐R⌐C 3.85

The M⌐R⌐C key shows the amount left in the
memory. The calculator displays <u>3.85</u>.

ANSWER: You will receive $3.85 change.

Find the change using a calculator and memory.

1. You have $10 and you buy a notebook for $2.69, 3 pencils for $.15 each,
and 2 erasers for $.29 each. How much change should you receive?

2. You have $50 and you buy 2 shirts for $9.95 each, 3 pairs of socks for $2.79
each, and a pair of slacks for $18.95. How much change will you receive?

1. Subtract: 0.513 − 0.421

```
  0.513
− 0.421
      2
```

```
   411
 0.5̸1̸3
− 0.421   > 2 is greater than 1.
 0.092  ←difference
```

Check: 0.421
      + 0.092
        0.513 ✔

Arrange in columns. Line up the decimal points. Subtract, starting from the right.

Regroup when necessary.
5 tenths 1 hundredth = 4 tenths 11 hundredths
Place the decimal point in the difference directly below the decimal points in the given numbers. Check by adding.

ANSWER: 0.092

---

2. From 10.6 subtract 5.524.

```
      5 9 10
 10.6̸0̸0̸
−  5.524
   5.076
```

Check: 5.524
      + 5.076
       10.600 = 10.6 ✔

Write ending zeros so both numbers have the same number of decimal places.
6 tenths = 5 tenths 9 hundredths 10 thousandths

For a whole number, write a decimal point and ending zeros.

ANSWER: 5.076

---

3. Find the difference: 5.61 − 0.9

```
  4 16
 5̸.6̸1
− 0.90
  4.71
```

Check: 0.90
      + 4.71
        5.61 ✔

ANSWER: 4.71

4. Subtract: $20 − $6.38

```
    19 9 10
 $2̸0̸.0̸0̸
−    6.38
 $13.62
```

Check: $ 6.38
      + 13.62
       $20.00 ✔

ANSWER: $13.62

---

5. Estimate the difference: 2.892 − 0.763

```
  2.892  →    2.9
− 0.763  → − 0.8
               2.1
```

Round the numbers to the same place. Usually, this is the greatest place in the smaller number. Subtract.

ANSWER: 2.1

---

6. Solve the equation: 18.5 − 9.647 = n

```
 18.500
−  9.647
   8.853
```

Check: 9.647
      + 8.853
       18.500 ✔

To find the value of n, subtract the given numbers.

ANSWER: 8.853

Diagnostic Exercises

Subtract.

1. 0.8
     − 0.2

2. 0.38
     − 0.24

3. 0.67
     − 0.48

4. 0.05
     − 0.02

5. 0.86
     − 0.79

6. 0.57
     − 0.37

7. 4.6
     − 3.1

8. 5.4
     − 2.8

9. 6.5
     − 2.5

10. 0.679
      − 0.398

11. 3.72
      − 1.95

12. 15.8
      − 3.9

13. 0.6593
      − 0.4978

14. 83.452
      − 49.596

15. 0.34617
      − 0.14596

16. 0.4583
      − 0.2783

17. 0.85328
      − 0.84793

18. 6.531
      − 5.975

19. 8.000
      − 1.742

20. 3.56
      − 0.8

21. 0.9
      − 0.735

22. 5.0
      − 1.43

23. 4.2
      − 0.372

24. $176.27
      − 93.48

25. 0.15 − 0.08

26. 0.375 − 0.2

27. 0.7 − 0.625

28. 4.2 − 0.83

29. 9 − 0.05

30. $16 − $1.50

31. Subtract 0.08 from 0.3

32. Select the nearest given estimate for: 8.2 − 0.82
    a. 0 **b.** 16 **c.** 7

33. Solve: $0.9 − 0.06 = n$

34. Enrique's grade point average is 3.5. Kieron has a 2.75 grade point average. How much higher is Enrique's average?

Related Practice

Subtract.

1. a. 0.9
       − 0.5
b. 0.4
     − 0.3
c. 0.6
     − 0.1
d. 0.7
     − 0.2

2. a. 0.29
       − 0.03
b. 0.73
     − 0.32
c. 0.98
     − 0.65
d. 0.64
     − 0.21

3. a. 0.36
       − 0.18
b. 0.44
     − 0.27
c. 0.25
     − 0.09
d. 0.83
     − 0.36

4. a. 0.08
       − 0.01
b. 0.09
     − 0.03
c. 0.07
     − 0.05
d. 0.09
     − 0.06

5. a. 0.85
       − 0.76
b. 0.54
     − 0.45
c. 0.25
     − 0.17
d. 0.13
     − 0.08

6. a. 0.58
       − 0.28
b. 0.46
     − 0.36
c. 0.25
     − 0.05
d. 0.92
     − 0.42

7. a. 8.4
       − 4.2
b. 4.7
     − 2.5
c. 5.6
     − 3.2
d. 9.5
     − 4.4

8. a. 7.5
       − 4.7
b. 9.3
     − 1.9
c. 4.6
     − 2.8
d. 6.1
     − 4.3

9. a. 7.6
       − 2.6
b. 9.2
     − 6.2
c. 2.8
     − 1.8
d. 8.4
     − 4.4

10. a. 0.835
        − 0.214
b. 0.967
     − 0.378
c. 0.076
     − 0.043
d. 0.749
     − 0.729

11. a. 7.64 **b.** 9.68 **c.** 4.29 **d.** 9.01 **12. a.** 18.5 **b.** 17.3 **c.** 24.5 **d.** 42.7
  − 3.53   − 5.95   − 1.76   − 7.84   − 6.2   − 5.8   − 3.5   − 5.9

13. a. 0.9355 **b.** 0.1327 **c.** 60.07 **d.** 375.3
  − 0.8492   − 0.1219   − 42.38   − 190.4

14. a. 0.25683 **b.** 34.645 **c.** 849.54 **d.** 5,986.3
  − 0.14974   − 17.859   − 258.46   − 2,894.7

15. a. 0.5844 **b.** 6.2937 **c.** 0.83572 **d.** 43.596
  − 0.2837   − 4.2843   − 0.43564   − 35.589

16. a. 0.7856 **b.** 5.8362 **c.** 0.92041 **d.** 82.475
  − 0.3256   − 3.7362   − 0.52041   − 17.275

17. a. 0.6848 **b.** 0.04965 **c.** 0.00325 **d.** 0.03741
  − 0.6827   − 0.04894   − 0.00255   − 0.03659

18. a. 4.6 **b.** 5.83 **c.** 9.786 **d.** 7.0352
  − 3.8   − 4.96   − 8.895   − 6.9446

19. a. 4.000 **b.** 9.000 **c.** 6.0000 **d.** 24.000
  − 1.753   − 4.068   − 5.9325   − 13.307

20. a. 1.4 **b.** 2.5 **c.** 8.65 **d.** 5.0046
  − 0.9   − 0.7   − 0.8   − 4.307

21. a. 0.36 **b.** 0.78 **c.** 0.6 **d.** 0.4
  − 0.034   − 0.1561   − 0.49   − 0.003

22. a. 8 **b.** 3 **c.** 7 **d.** 6
  − 7.3   − 1.2   − 2.84   − 4.005

23. a. 5 **b.** 9 **c.** 3 **d.** 4.9
  − 0.6   − 0.09   − 0.753   − 0.807

24. a. $4.85 **b.** $36.80 **c.** $476.13 **d.** $1,250.00
  − 2.60   − 17.42   − 85.75   − 975.80

25. **a.** 0.8 − 0.3
b. 0.38 − 0.09
c. 0.536 − 0.008
d. 0.1534 − 0.0976

26. **a.** 0.97 − 0.6
b. 0.384 − 0.2
c. 0.8056 − 0.74
d. 0.5842 − 0.095

27. **a.** 0.9 − 0.83
b. 0.06 − 0.043
c. 0.2 − 0.1356
d. 0.685 − 0.5903

28. **a.** 3.6 − 0.24
b. 5.82 − 0.004
c. 12.54 − 1.054
d. 3.4 − 0.0056

29. **a.** 8 − 0.5
b. 1 − 0.16
c. 7 − 3.0625
d. 10 − 0.375

30. **a.** $.78 − $0.9
b. $10 − $5.60
c. $15.42 − $9
d. $3,500 − $2,938.75

31. **a.** From 0.8 subtract 0.35
b. From $2.84 take $1.75
c. Take 0.45 from 1.5
d. Find the difference between 4.81 and 0.481

32. Write the nearest given estimate.

a. 17.8 − 3.1	13	14	15
b. 0.61 − 0.074	0.08	0.5	0.6
c. $128.50 − $51.97	$70	$80	$90
d. 0.9 − 0.09	0.01	0.8	0

33. Solve:
a. $0.8 − 0.2 = n$
b. ▨ = $75.20 − $9.60
c. ▨ = 15 − 0.625
d. n = $3.64 − $.97

34. **a.** John bought a baseball glove that regularly sold for $27.49 at a reduction of $4.98. How much did he pay?

b. Erna threw a discus 7.81 meters. Lisa threw the discus 10.2 meters. How much farther did Lisa throw the discus than Erna?

c. Selma bought a set of weights for $10.79. How much change should she receive from a $20 bill?

d. The World Trade Center in New York is 411.48 meters tall. The Sears Tower in Chicago is 443.18 meters tall. How many meters taller is the Sears Tower?

Mixed Problems

Choose an appropriate method of computation and solve each of the following problems. For Problems 1–3, select the letter corresponding to your answer.

1. The Dow Jones stock average opened at 2385.57 and during the day gained 23.46 points. What was the closing stock average?

    a. 2409.03 **b.** 2490.03
    c. 2049.93 **d.** 2309.03

2. A girl ran the 100-m dash in 15.4 seconds, then later ran the same distance in 13.7 seconds. How many seconds less did she take the second time?

    a. 2.7 s **b.** 1.7 s
    c. 2.3 s **d.** 29.1 s

3. Marilyn bought her graduation outfit. Her dress cost $49.98; shoes, $21.49; hat, $15.75; and bag, $16.95. How much did she spend?

    a. $105.17 **b.** $106.17
    c. $104.17 **d.** $96.17

4. Harry bought a new suit costing $94.95; shoes, $21.75; hat, $12.50; shirt, $10.98; and tie, $3.50. If his mother gave him $150, how much money is left over after he pays for these articles?

5. In measuring a 1-inch block of metal by a precision instrument, a student found the average of all her readings to be 0.9996 inch. Find the amount of error.

6. Robert's brother is a salesperson. Last week he received $294.50 salary and $59.68 commission. How much did he earn in all?

7. Charlotte's parents took the family on a motor trip. The expenses included gasoline, $142.90; oil, $6.36; lodging, $175.00; meals, $259.25; amusements, $64.83; miscellaneous, $39.54. How much did the trip cost?

8. The passenger ship, *United States,* on its first voyage established a record of 35.59 knots crossing the Atlantic Ocean. If the record speed of the *Queen Mary* is 30.99 knots, how much faster is the *United States*?

Refresh Your Skills

The numerals in boxes indicate lessons where help may be found.

1. Add. [1–8]

    149,526
     12,997
    500,725
     36,589
  +   8,763

2. Subtract. [1–9]

    821,427
  − 291,398

3. Multiply. [1–10, 11]

    1,760
  × 500

4. Divide. [1–12, 13]

    407)3,538,865

5. Name the greatest common factor of 15, 40, and 90. [1–14]

6. Round 30,246,701 to the nearest million. [1–7]

7. Write 800.08 in words. [2–1]

8. Write in standard form: Three and fifty-nine thousandths. [2–2]

9. Round $5.8273 to the nearest cent. [2–3]

10. Add: [2–5]
    0.693 + 4.82 + 37.6

2-7 *Multiplying Decimals*

Building Problem Solving Skills

Ceil wants to build 12 shelves. If each shelf is 1.3 meters long, how much lumber does she need in all?

> **Multiplication Indicators**
> - total
> - how many in all
> - how much altogether
>
> Note: All indicators are used when combining *equal* quantities.

1. ⬛ READ ⬛ the problem carefully.

a. Find the **question asked:** How much lumber does she need in all?

b. Find the **given facts:** Building 12 shelves, each 1.3 m long.

2. ⬛ PLAN ⬛ how to solve the problem.

a. Choose the **operation needed:** The words *in all* with *equal* quantities indicate multiplication.

b. **Think or write out your plan** relating the given facts to the question asked. The number of shelves times the lumber needed for one shelf is equal to the total lumber needed.

c. Express your plan as an **equation.**
$12 \times 1.3 = n$

3. ⬛ SOLVE ⬛ the problem.

a. Estimate the answer.

$$\begin{array}{r} 12 \rightarrow 10 \\ \underline{\times\ 1.3} \rightarrow \underline{\times\ \ 1}\ m \\ 10\ m \end{array}$$

Round to the nearest ten.
Round to the nearest whole number.
Multiply.

b. Solution:

$$\begin{array}{r} 1\ 2 \leftarrow 0\ \text{decimal places} \\ \underline{\times\ \ \ 1.3} \leftarrow 1\ \text{decimal place} \\ 3\ 6 \\ \underline{12\ } \\ 15.6 \leftarrow 1\ \text{decimal place} \end{array}$$

Arrange the factors in columns and multiply as you do with whole numbers.

The number of decimal places in the product is the sum of the number of decimal places in the factors.

To use a calculator, press these keys: 12 ⬛×⬛ 1 ⬛·⬛ 3 ⬛=⬛ <u>15.6</u>

4. **CHECK** .

a. Check the accuracy of your arithmetic.

Check: 1.3
          × 1 2
          ────
            2 6
          1 3
          ────
          1 5.6 ✔

Check by reversing the order of the factors. Double check the placement of the decimal point in the product.

b. Compare the answer to the estimate. Check the unit. The answer 15.6 meters compares reasonably with the estimate of 10. The unit, meters, is correct.

ANSWER: Ceil needs 15.6 meters of lumber.

Think About It

1. If tenths are multiplied by tenths, how many decimal places will be in the product?

2. If hundredths are multiplied by hundredths, what place value will the solution have?

3. Can the product of two decimals be a whole number? Give an example.

Calculator Know-How

You can use a calculator to round decimals. To round a decimal to the nearest whole number, add 0.5 to the decimal; nearest tenth, add 0.05; nearest hundredth, add 0.005; nearest thousandth, add 0.0005; and so on. In the sum all digits to the right of the place rounded to are dropped.

Examples

Use a calculator to round.

1. 0.856 to the nearest tenth.

Key: 0 · 856 + 0 · 05 = 0.906 → 0.900

ANSWER: 0.9

2. 0.749 to the nearest hundredth.

Key: 0 · 749 + 0 · 005 = 0.754 → 0.750

ANSWER: 0.75

1. Use a calculator to determine how far Maria traveled if she drove for 6.75 hours at an average speed of 54.26 miles per hour. Round the answer to the nearest mile.

2. At a picnic, 19.38 gallons of cider were consumed. If each gallon is equivalent to 4 quarts, how many quarts of cider were consumed? Round the answer to the nearest tenth of a quart.

1. Multiply: 43 × 2.6

Arrange in columns. Multiply, starting from the right as you do with whole numbers. Add partial products.

factors ⟨ 43 ← **0 decimal places**
× 2.6 ← **1 decimal place**

25 8 ⟩ **partial products**
86

product → 111.8 ← **1 decimal place**

Place a decimal point in the product. The number of decimal places in the product is equal to the sum of the numbers of decimal places in the factors.

Check: 2.6
× 43
7 8
104
111.8 ✔

Check by interchanging the factors.

ANSWER: 111.8

---

2. Multiply: 0.12 × 0.4

0.12 ← **2 decimal places**
× 0.4 ← **+ 1 decimal place**
Write 1 zero. → 0.048 ← **3 decimal places**

If there are not enough decimal places in the product, write a zero (or zeros) to the left of the digits in the answer.

Check: 0.4
× 0.12
0.048 ✔

ANSWER: 0.048

---

3. Multiply 2.43 by 0.56

2.43
× 0.56
1458
1 215
1.3608

Check: 0.56
× 2.43
168
224
1 12
1.3608 ✔

ANSWER: 1.3608

4. Estimate the product: 41.95 × 1.8

41.95 → 40
1.8 → 2
40 × 2 = 80

Round each factor to its greatest place. Multiply.

ANSWER: 80

5. Solve the equation: $7.62 \times 42.9 = n$

$$\begin{array}{r} 7.62 \\ \times\ 42.9 \\ \hline 6\ 858 \\ 15\ 24 \\ 304\ 8 \\ \hline 326.898 \end{array}$$

Check:
$$\begin{array}{r} 42.9 \\ \times\ 7.62 \\ \hline 858 \\ 25\ 74 \\ 300\ 3 \\ \hline 326.898\ \checkmark \end{array}$$

To find the value of n, multiply the given numbers.

ANSWER: 326.898

Diagnostic Exercises

Multiply.

1.
$$\begin{array}{r} 0.3 \\ \times\ 8 \\ \hline \end{array}$$

2.
$$\begin{array}{r} 43 \\ \times\ 0.24 \\ \hline \end{array}$$

3.
$$\begin{array}{r} 0.351 \\ \times\ 86 \\ \hline \end{array}$$

4.
$$\begin{array}{r} 0.6739 \\ \times\ 7 \\ \hline \end{array}$$

5.
$$\begin{array}{r} 75 \\ \times\ 0.48 \\ \hline \end{array}$$

6.
$$\begin{array}{r} 37 \\ \times\ 0.05 \\ \hline \end{array}$$

7.
$$\begin{array}{r} 0.03 \\ \times\ 2 \\ \hline \end{array}$$

8.
$$\begin{array}{r} 14 \\ \times\ 0.007 \\ \hline \end{array}$$

9.
$$\begin{array}{r} 0.002 \\ \times\ 4 \\ \hline \end{array}$$

10.
$$\begin{array}{r} 3.14 \\ \times\ 18 \\ \hline \end{array}$$

11.
$$\begin{array}{r} 0.6 \\ \times\ 0.2 \\ \hline \end{array}$$

12.
$$\begin{array}{r} 0.3 \\ \times\ 0.3 \\ \hline \end{array}$$

13.
$$\begin{array}{r} 0.58 \\ \times\ 0.6 \\ \hline \end{array}$$

14.
$$\begin{array}{r} 0.21 \\ \times\ 0.4 \\ \hline \end{array}$$

15.
$$\begin{array}{r} 0.56 \\ \times\ 0.37 \\ \hline \end{array}$$

16.
$$\begin{array}{r} 0.05 \\ \times\ 0.01 \\ \hline \end{array}$$

17.
$$\begin{array}{r} 16.2 \\ \times\ 0.045 \\ \hline \end{array}$$

18.
$$\begin{array}{r} 34.89 \\ \times\ 0.875 \\ \hline \end{array}$$

19.
$$\begin{array}{r} 0.147 \\ \times\ 0.03 \\ \hline \end{array}$$

20.
$$\begin{array}{r} 3.1416 \\ \times\ 0.75 \\ \hline \end{array}$$

21.
$$\begin{array}{r} 0.059 \\ \times\ 0.064 \\ \hline \end{array}$$

22. 6×0.005

23. 0.012×0.07

In examples involving money, find the product to the nearest cent.

24.
$$\begin{array}{r} \$.25 \\ \times\ 8 \\ \hline \end{array}$$

25.
$$\begin{array}{r} \$3.80 \\ \times\ 24 \\ \hline \end{array}$$

26.
$$\begin{array}{r} \$3.62 \\ \times\ 0.06 \\ \hline \end{array}$$

27.
$$\begin{array}{r} \$4.28 \\ \times\ 0.125 \\ \hline \end{array}$$

28. Find 0.46 of 150.

29. Select the nearest given estimate for: $0.29 \times \$51$ **a.** \$5   **b.** \$10   **c.** \$15

30. Solve. $0.04 \times 1.25 = n$

31. Julia bought 12 albums at a garage sale. Each one cost \$4.59.
How much did she pay altogether?

Related Practice

Multiply.

1. a. $\begin{array}{r} 4 \\ \times\ 0.1 \\ \hline \end{array}$ **b.** $\begin{array}{r} 6 \\ \times\ 0.7 \\ \hline \end{array}$ **c.** $\begin{array}{r} 12 \\ \times\ 9.8 \\ \hline \end{array}$ **d.** $\begin{array}{r} 0.9 \\ \times\ 5 \\ \hline \end{array}$ **2. a.** $\begin{array}{r} 0.23 \\ \times\ 9 \\ \hline \end{array}$ **b.** $\begin{array}{r} 7 \\ \times\ 0.56 \\ \hline \end{array}$ **c.** $\begin{array}{r} 84 \\ \times\ 0.42 \\ \hline \end{array}$ **d.** $\begin{array}{r} 39 \\ \times\ 0.61 \\ \hline \end{array}$

3. a. $\begin{array}{r} 0.247 \\ \times\ 5 \\ \hline \end{array}$ **b.** $\begin{array}{r} 0.456 \\ \times\ 34 \\ \hline \end{array}$ **c.** $\begin{array}{r} 0.572 \\ \times\ 159 \\ \hline \end{array}$ **d.** $\begin{array}{r} 28 \\ \times\ 0.707 \\ \hline \end{array}$

4. a. $\begin{array}{r} 0.9522 \\ \times\ 8 \\ \hline \end{array}$ **b.** $\begin{array}{r} 5 \\ \times\ 0.4673 \\ \hline \end{array}$ **c.** $\begin{array}{r} 46 \\ \times\ 0.5034 \\ \hline \end{array}$ **d.** $\begin{array}{r} 8725 \\ \times\ 0.2839 \\ \hline \end{array}$

5. a. $\begin{array}{r} 25 \\ \times\ 0.4 \\ \hline \end{array}$ **b.** $\begin{array}{r} 0.95 \\ \times\ 20 \\ \hline \end{array}$ **c.** $\begin{array}{r} 8 \\ \times\ 0.125 \\ \hline \end{array}$ **d.** $\begin{array}{r} 246 \\ \times\ 0.625 \\ \hline \end{array}$ **6. a.** $\begin{array}{r} 0.03 \\ \times\ 8 \\ \hline \end{array}$ **b.** $\begin{array}{r} 0.05 \\ \times\ 17 \\ \hline \end{array}$ **c.** $\begin{array}{r} 10 \\ \times\ 0.04 \\ \hline \end{array}$ **d.** $\begin{array}{r} 200 \\ \times\ 0.08 \\ \hline \end{array}$

7. a. $\begin{array}{r} 0.03 \\ \times\ 3 \\ \hline \end{array}$ **b.** $\begin{array}{r} 0.02 \\ \times\ 4 \\ \hline \end{array}$ **c.** $\begin{array}{r} 3 \\ \times\ 0.02 \\ \hline \end{array}$ **d.** $\begin{array}{r} 5 \\ \times\ 0.01 \\ \hline \end{array}$ **8. a.** $\begin{array}{r} 0.006 \\ \times\ 3 \\ \hline \end{array}$ **b.** $\begin{array}{r} 0.009 \\ \times\ 10 \\ \hline \end{array}$ **c.** $\begin{array}{r} 5 \\ \times\ 0.013 \\ \hline \end{array}$ **d.** $\begin{array}{r} 21 \\ \times\ 0.004 \\ \hline \end{array}$

9. a. $\begin{array}{r} 0.001 \\ \times\ 9 \\ \hline \end{array}$ **b.** $\begin{array}{r} 0.0015 \\ \times\ 2 \\ \hline \end{array}$ **c.** $\begin{array}{r} 13 \\ \times\ 0.0006 \\ \hline \end{array}$ **d.** $\begin{array}{r} 4 \\ \times\ 0.0014 \\ \hline \end{array}$ **10. a.** $\begin{array}{r} 8.7 \\ \times\ 6 \\ \hline \end{array}$ **b.** $\begin{array}{r} 56.17 \\ \times\ 75 \\ \hline \end{array}$ **c.** $\begin{array}{r} 460 \\ \times\ 4.8 \\ \hline \end{array}$ **d.** $\begin{array}{r} 3.1416 \\ \times\ 32 \\ \hline \end{array}$

11. a. $\begin{array}{r} 0.9 \\ \times\ 0.8 \\ \hline \end{array}$ **b.** $\begin{array}{r} 0.5 \\ \times\ 0.6 \\ \hline \end{array}$ **c.** $\begin{array}{r} 3 \\ \times\ 8.2 \\ \hline \end{array}$ **d.** $\begin{array}{r} 5.7 \\ \times\ 2.5 \\ \hline \end{array}$ **12. a.** $\begin{array}{r} 0.3 \\ \times\ 0.2 \\ \hline \end{array}$ **b.** $\begin{array}{r} 0.1 \\ \times\ 0.1 \\ \hline \end{array}$ **c.** $\begin{array}{r} 0.4 \\ \times\ 0.2 \\ \hline \end{array}$ **d.** $\begin{array}{r} 0.1 \\ \times\ 0.8 \\ \hline \end{array}$

13. a. $\begin{array}{r} 0.34 \\ \times\ 0.3 \\ \hline \end{array}$ **b.** $\begin{array}{r} 0.95 \\ \times\ 0.4 \\ \hline \end{array}$ **c.** $\begin{array}{r} 0.7 \\ \times\ 0.66 \\ \hline \end{array}$ **d.** $\begin{array}{r} 1.15 \\ \times\ 5.2 \\ \hline \end{array}$ **14. a.** $\begin{array}{r} 0.24 \\ \times\ 0.2 \\ \hline \end{array}$ **b.** $\begin{array}{r} 0.3 \\ \times\ 0.03 \\ \hline \end{array}$ **c.** $\begin{array}{r} 0.15 \\ \times\ 0.6 \\ \hline \end{array}$ **d.** $\begin{array}{r} 0.07 \\ \times\ 1.3 \\ \hline \end{array}$

15. a. $\begin{array}{r} 0.28 \\ \times\ 0.74 \\ \hline \end{array}$ **b.** $\begin{array}{r} 5.93 \\ \times\ 0.87 \\ \hline \end{array}$ **c.** $\begin{array}{r} 0.45 \\ \times\ 4.91 \\ \hline \end{array}$ **d.** $\begin{array}{r} 12.52 \\ \times\ 0.06 \\ \hline \end{array}$ **16. a.** $\begin{array}{r} 0.67 \\ \times\ 0.02 \\ \hline \end{array}$ **b.** $\begin{array}{r} 0.05 \\ \times\ 0.76 \\ \hline \end{array}$ **c.** $\begin{array}{r} 0.02 \\ \times\ 0.02 \\ \hline \end{array}$ **d.** $\begin{array}{r} 2.14 \\ \times\ 0.03 \\ \hline \end{array}$

17. a. $\begin{array}{r} 0.003 \\ \times\ 0.4 \\ \hline \end{array}$ **b.** $\begin{array}{r} 0.7 \\ \times\ 0.002 \\ \hline \end{array}$ **c.** $\begin{array}{r} 0.375 \\ \times\ 1.4 \\ \hline \end{array}$ **d.** $\begin{array}{r} 0.009 \\ \times\ 36.6 \\ \hline \end{array}$

18. a. $\begin{array}{r} 0.368 \\ \times\ 0.26 \\ \hline \end{array}$ **b.** $\begin{array}{r} 70.84 \\ \times\ 0.034 \\ \hline \end{array}$ **c.** $\begin{array}{r} 95.26 \\ \times\ 1.125 \\ \hline \end{array}$ **d.** $\begin{array}{r} 453.40 \\ \times\ 0.375 \\ \hline \end{array}$

19. a. $\begin{array}{r} 0.0002 \\ \times\ 0.2 \\ \hline \end{array}$ **b.** $\begin{array}{r} 0.057 \\ \times\ 0.38 \\ \hline \end{array}$ **c.** $\begin{array}{r} 3.14 \\ \times\ 0.002 \\ \hline \end{array}$ **d.** $\begin{array}{r} 0.04 \\ \times\ 1.225 \\ \hline \end{array}$

20. a. 0.268
× 0.924

b. 3.1416
× 6.25

c. 8.504
× 0.015

d. 2.423
× 9.146

21. a. 0.00008
× 0.6

b. 0.003
× 0.009

c. 0.01
× 0.0007

d. 0.02167
× 1.8

22. a. 5 × 0.7 **b.** 10 × .14 **23. a.** 0.2 × 0.3 **b.** 0.3 × 0.04
c. 9 × 0.0001 **d.** 18 × .05 **c.** 0.4 × 0.35 **d.** 0.04 × 0.029

In examples involving money, find the product correct to the nearest cent.

24. a. $.42
× 4

b. $.69
× 18

c. $.80
× 7

d. $.75
× 48

25. a. $4.97
× 3

b. $10.50
× 60

c. $16.31
× 96

d. $4.25
× 144

26. a. $89
× 0.04

b. $.75
× 0.06

c. $14.25
× 0.19

d. $293.28
× 0.63

27. a. $840
× 0.625

b. $15.61
× 0.045

c. $2,500
× 0.4375

d. $675.90
× 0.002

Find.

28. a. 0.25 of 60 **b.** 0.04 of 9 **c.** 0.39 of $3.40 **d.** 0.13 of $15.64

29. Write the nearest given estimate.

a. 0.21 × $19: $4 $5 $6
b. 3.98 × .007: 3 0.3 0.03
c. 0.06 × $59.85: $36 $3.60 $.36
d. 42 × $7.90: $500 $400 $300

30. Solve.

a. 6 × 0.8 = n
b. 0.004 × 0.03 = n
c. ▧ = 1.4 × 5.8
d. ▧ = 0.08 × $10.50

31. a. Joan's mother bought 25 shrubs at $7.49 each. What was the total cost?

b. Find the cost of 16.4 gallons of gasoline at $1.35 per gallon.

c. Richy earns $7.90 per hour and works 38.5 hours each week. How much are his total earnings per week?

d. A carton contains 48 cans of food each weighing 0.375 kilograms. What is the total weight of the carton?

Mixed Problems

Choose an appropriate method of computation and solve each of the problems. For Problems 1–4, select the letter corresponding to your answer.

1. A certain plane on a flight used 175.2 liters of gasoline per hour. If its flight lasted 4.5 hours, how many liters of gasoline were consumed?
  a. 7,884 liters   *b.* 78.84 liters
  c. 7.8840 liters   *d.* 788.4 liters

2. Find the distance represented by 6.7 centimeters if the scale is 1 centimeter = 50 kilometers.
  a. 325 kilometers   *b.* 415 kilometers
  c. 335 kilometers   *d.* 300.7 kilometers

3. In addition to her weekly allowance of $8.50, Wanda earned $15.65 after school. How much money should she have left at the end of the week if her expenses were: bus fare, $3.50; school lunches and supplies, $6.97; movies, $1.85; church, $1.00; and savings, $3.75?
  a. $6.08   *b.* $5.08
  c. $7.08   *d.* Answer not given

4. Find the total expenses in producing the Jacques Cartier School show if renting and making costumes cost $362.55; royalty fee, $75.00; properties, $143.66; tickets, $25.50; lighting, $52.25; and miscellaneous items, $157.19.
  a. $806.15   *b.* $796.15
  c. $826.15   *d.* $816.15

5. An airplane has a ground speed of 325 knots. What is its ground speed in statute miles per hour if 1 knot = 1.15 statute miles per hour?

6. For how much should a dealer sell a CB radio if it cost him $68.75 and he wishes to make a profit of $36.75?

7. The outside diameter of a piece of copper tubing is 2.375 millimeters and its wall thickness is 0.083 millimeters. What is the inside diameter? Draw a diagram.

8. A refrigerator costs $625 cash or $80 down and 12 payments of $52.95 each. How much do you save by paying cash?

Refresh Your Skills

The numerals in boxes indicate lessons where help may be found.

1. 1–8
```
  28,104
     625
   9,302
      29
  75,412
+    828
```

2. 1–9
```
  1,500,020
−   690,175
```

3. 1–10, 11
```
    876
×   938
```

4. 1–12, 13
```
289)285,821
```

5. 2–5

  $0.88 + 0.8 + 0.888$

6. 2–6

  $9.6 − 0.45$

2-8 Dividing Decimals by Whole Numbers

Building Problem Solving Skills

José bought 5 new tires for $298.25. What was the cost of 1 tire?

1. **READ** the problem carefully.

a. Find the **question asked.** What was the cost of 1 tire?

b. Find the **given facts.** He bought 5 tires. The total cost was $298.25.

PLAN how to solve the problem.

a. Choose the **operation needed.** Since you are looking for the cost of 1 tire when you know the cost of 5, you divide.

b. **Think or write out your plan** relating the facts to the question asked. Divide the total cost, $298.25, by the number of tires, 5, to find the cost of 1 tire.

c. Express your plan as an **equation.** $298.25 \div 5 = n$

> **Division Indicators**
> - how much is each
> - how many in each
> - find the average
> - rate per unit
>
> Note: All indicators generally refer to equal quantities.

3. [**SOLVE**] the problem.

a. Estimate the answer.

$298.25 → $300

  5 → 5

$300 ÷ 5 = $60

Round the dividend to the nearest multiple of the divisor. Divide.

b. Solution.

```
      $59.65
  5)$298.25
    25
     48
     45
      32
      30
       25
       25
```

When the divisor is a whole number, write the decimal point in the quotient directly above the decimal point in the dividend. Divide as you do with whole numbers.

You might use a calculator. Key: 298.5 [÷] 5 [=] 59.65

---

4. [**CHECK**]

a. Check the accuracy of your arithmetic.

```
    $59.65
  ×      5
   $298.25 ✔
```

Check by multiplying.

b. Compare the answer to the estimate. Check the unit. The answer $59.65 for 1 tire compares reasonably with the estimate of $60.

ANSWER: The cost for 1 tire was $59.65.

Think About It

1. In the problem above, how do you know where to place the decimal in the quotient?

2. If José had bought 4 tires for $298.25, would the price for 1 tire have been more or less than it was above? Explain your answer.

3. When you divide a decimal by a whole number, will the quotient ever be greater than the divisor? Explain.

1.
```
    0.31
  6)1.86
    1 8
      6
      6
```
Check: 0.31
              ×   6
            1.86 ✔

When the divisor is a whole number, write the decimal point in the quotient directly above the decimal point in the dividend.
Divide as you do with whole numbers.

Check by multiplying the quotient by the divisor.

ANSWER: 0.31

---

2. Find the quotient: $8.50 ÷ 4.
Round to the nearest cent.

```
    $2.125
  4)$8.500
```

To round to the nearest cent, carry out the division to the thousandths place. If necessary, write zeros in the dividend.

ANSWER: $2.13

---

3. Estimate the quotient: 184.6 ÷ 27

```
     27 → 30
  184.6 → 180
  180 ÷ 30 = 6
```

Round the divisor to its greatest place. Round the dividend so that it can be divided exactly by the rounded divisor.
Divide.

ANSWER: 6

---

4. Solve the equation:
$21 ÷ 56 = n$

```
         0.375
    56)21.000
       16 8
        4 20
        3 92
          280
          280
            0
```

Check: 0.375
               ×   56
              2 250
             18 75
             21.000 = 21 ✔

To find the value of n, divide the given numbers.

ANSWER: $n = 0.375$

Diagnostic Exercises

Divide.

1. 4)9.2   **2.** 7)8.96   **3.** 2)5.328   **4.** 3)0.8226   **5.** 8)0.736   **6.** 36)91.44   **7.** 200)4

8. Divide 3.6 by 6   **9.** Find the quotient correct to the nearest thousandth: 29)24

10. Find the answer correct to the nearest cent: 12)$2.57

11. 2 ÷ 16 (Find the quotient to 3 decimal places.)

12. Select the nearest given estimate for: 78.12 ÷ 4   **13.** Solve. 0.96 ÷ 16 = n
a. 1.3   **b.** 20   **c.** 38.12

14. Ellen has a board that measures 4.9 meters. She cuts it into 7 pieces of equal length. How long is each piece?

Related Practice

Divide.

1. a. 3)6.9　　　　**b.** 7)86.1　　　　**c.** 5)746.5　　　　**d.** 6)6,765.6

2. a. 2)0.86　　　　**b.** 4)5.88　　　　**c.** 7)92.96　　　　**d.** 8)971.44

3. a. 8)0.968　　　**b.** 5)0.865　　　**c.** 2)7.942　　　**d.** 3)73.914

4. a. 4)0.8936　　**b.** 6)0.6432　　**c.** 5)6.9185　　**d.** 7)8.9789

5. a. 6)0.228　　　**b.** 8)0.024　　　**c.** 7)0.0114　　　**d.** 5)0.00375

6. a. 12)3.36　　　**b.** 48)158.4　　　**c.** 24)11.688　　　**d.** 144)112.32

7. a. 60)3　　　　**b.** 200)6　　　　**c.** 48)54　　　　**d.** 7,000)84

8. a. 4.2 by 7　　　**b.** 0.616 by 4　　　**c.** 0.56 by 8　　　**d.** $10.50 by 10

9. Find the quotient correct to the nearest thousandth.
a. 7)285　　　**b.** 15)46　　　**a.** 12)365　　　**b.** 24)13.59

10. Find the answer correct to the nearest cent.
a. 3)$.72　　　**b.** 12)$9.60　　　**c.** 24)$8.16　　　**d.** 144)$11.52

11. a. 0.87 ÷ 3　　　　　　　　**b.** 9 ÷ 12 (2 decimal places)
c. 8 ÷ 7 (nearest thousandth)　　**d.** $2.16 ÷ 5 (nearest cent)

Write the nearest given estimate.

12. a. 13.92 ÷ 2　　0.69　　11.92　　7　　**b.** 0.5814 ÷ 3　　175　　20　　0.2
　　c. 9.486 ÷ 9　　1　　0.125　　0　　**d.** 985.6 ÷ 49　　18.2　　20　　2.5

Solve.

13. a. 0.76 ÷ 4 = ▦　　**b.** 97.2 ÷ 2 = ▦　　**c.** 125.04 ÷ 8 = n　　**d.** ▦ = 6.945 ÷ 5

14. a. The Book Fair raised $450.75. The money will be shared by 5 clubs. How much should each club receive?

b. A piece of cloth measures 2.5 meters. Jake wants to cut it into 4 equal pieces. How long will each piece be?

c. A medical lab technician needs 3 grams of hydrogen peroxide to perform a certain test. He has 37.6 grams of the peroxide. How many tests can he perform?

d. A jeweler has a strip of silver 144.72 centimeters long. She will use 12 centimeters to make one necklace. How many necklaces can she make with the silver she has?

Dividing Decimals by Decimals

A mutual fund quarterly dividend of $59.40 was reinvested in the fund at the price of $8.64 per share. How many shares of the fund (to three decimal places) should the investor receive?

Divide $59.40 by $8.64.

```
            6.875
$8.64 )$59.40 000
       51 84
        7 56 0
        6 91 2
          64 80
          60 48
           4 320
           4 320
```

To make the divisor a whole number, move the decimal point two places to the right. Move the decimal point in the dividend the same number of places.

Check by multiplying:
$8.64 × 6.875 = $59.40

ANSWER: 6.875 shares

Examples

1. Find the quotient: 35.6 ÷ 0.4

```
       8 9.
0.4 )35.6
     32
      3 6
      3 6
        0
```

Check: 89
         × 0.4
            35.6 ✔

ANSWER: 89

2. Find the quotient: 0.0015 ÷ 0.05

```
          0.03
0.05 )0.00 15
        15
         0
```

Check: 0.03
        × 0.05
          0.0015 ✔

ANSWER: 0.03

3. Divide: 0.625)15

```
           24.
0.625 )15.000  ←Write 3 zeros.
       12 50
        2 500
        2 500
            0
```

Check: 0.625
        ×   24
          15.000 = 15 ✔

If necessary, write zeros in the dividend so that the dividend has at least as many places as the divisor.

ANSWER: 24

4. Find the quotient: $8.50 ÷ 6. Round the answer to the nearest cent.

$$\begin{array}{r} \underline{\$1.41\,6} \text{ rounds to \$1.42.} \\ 6\overline{)\$8.500} \\ \end{array}$$

```
   $1.416  rounds to $1.42.
6)$8.500
  6      └─Write 1 zero.
  2 5
  2 4
    10
     6
    40
    36
     4
```

To round to the nearest cent, carry out the division to the thousandths place.

ANSWER: $1.42

---

5. Divide 28.5 by 0.87. Find the quotient to the nearest tenth.

```
          32.75  rounds to 32.8.
0.87ᴧ)28.50ᴧ00  ←Write 2 zeros.
      26 1
       2 40
       1 74
         66 0
         60 9
          5 10
          4 35
            75
```

To round to a given decimal place, carry out the division to the next decimal place, then round to the given place.
Divide to hundredths.

ANSWER: 32.8

---

6. Estimate the quotient: 119.8 ÷ 0.42

$$0.4\underline{2} \rightarrow 0.4$$
$$11\underline{9}.8 \rightarrow 120$$
$$1\overset{.}{2}0 ÷ 0.4 = 300$$

Round the divisor to its greatest place.
Round the dividend so that it can be divided exactly by the rounded divisor. Divide.

ANSWER: 300

---

7. Solve the equation: 21 ÷ 56 = *n*

```
     0.375
56)21.000   ←Write a decimal point and ending
   16 8        zeros.
    4 20
    3 92
      280    Check:    0.375
      280           ×      56
        0            21.000 = 21 ✔
```

To find the value of *n*, divide the given numbers.

ANSWER: *n* = 0.375

Think About It

1. If the price of the stock had been $12 a share instead of $8.64, would the investor receive more shares or fewer shares of the stock?

2. How do you know how many places to move the decimal point in the divisor and the dividend?

3. When dividing by a decimal, will the quotient be less than or greater than the dividend? Explain your answer.

Calculator Know-How

When you divide a decimal by a decimal, an eight-digit number may be displayed. That number is not the exact quotient. The quotient may continue without end or terminate beyond eight digits. When the displayed number is used as a factor in the check, the product will not be exactly equal to the given dividend. However, when the product is rounded to the number of places in the dividend, it would equal the dividend.

Example

Use a calculator to solve and check.

$62.5 \div 0.723 = n$

Key: 62.5 $\boxed{\div}$ 0.723 $\boxed{=}$ (86.445366) $\boxed{+}$ 0.0005 $\boxed{=}$ 86.445866 → 86.445000

ANSWER: 86.455

Check: 86.445366 $\boxed{\times}$ 0.723 $\boxed{=}$ 62.499999

Note: Rounded to the nearest tenth, 62.499999 is equal to the dividend, 62.5.

Exercises

Use a calculator to solve and check.

1. $14.53 \div 3.875 = n$ 2. $837.2 \div 0.36 = n$

3. $19.68 \div 12.35 = n$ 4. $16.8 \div 2.7 = n$

Diagnostic Exercises

Divide.

1. $0.3\overline{)247.8}$

2. $0.5\overline{)9.25}$

3. $0.8\overline{)0.896}$

4. $0.6\overline{)2.6898}$

5. $0.2\overline{)0.0034}$

6. $1.2\overline{)108.72}$

7. $0.6\overline{)12.0}$

8. $0.7\overline{)42}$

9. $0.4\overline{)2}$

10. Find the quotient correct to the nearest tenth: $2.7\overline{)18}$

11. Select the nearest given estimate for $101.92 \div 19.6$.

    a. 5 *b.* 22.3 *c.* 82

12. Solve. $3.44 \div 8 = n$

13. $0.02\overline{)521.56}$

14. $0.79\overline{)4.661}$

15. $0.56\overline{)2.4472}$

16. $0.07\overline{)0.89789}$

17. $0.03\overline{)0.0009}$

18. $1.44\overline{)135.072}$

19. $0.16\overline{)48.00}$

20. $0.39\overline{)265.2}$

21. $0.25\overline{)50}$

22. Find the quotient correct to the nearest thousandth. $0.96\overline{)8.8}$

23. $0.04\overline{)\$1.84}$

24. Select the nearest given estimate for $\$26.80 \div \$.09$ *a.* 3 *b.* 30 *c.* 300

25. Solve: $3.5 \div 0.07 = n$

26. $0.006\overline{)74.898}$

27. $0.018\overline{)0.4554}$

28. $0.007\overline{)6.53912}$

29. $0.231\overline{)0.00924}$

30. $4.375\overline{)11.8125}$

31. $0.048\overline{)60.000}$

32. $0.125\overline{)53.75}$

33. $0.052\overline{)452.4}$

34. $0.014\overline{)112}$

35. Find the quotient correct to the nearest hundredth: $0.33\overline{)249}$

36. Select the nearest given estimate for $71 \div 0.069$.

    a. 0.001 *b.* 100 *c.* 1,000

37. Solve: $0.558 \div 0.018 = n$

38. A bolt of fabric is 49.5 meters long. It is to be cut into pieces 0.45 meters long. How many such pieces can be cut from the bolt?

Related Practice

Divide.

1. a. $0.4\overline{)7.6}$ **b.** $0.5\overline{)89.5}$ **c.** $0.3\overline{)176.7}$ **d.** $0.6\overline{)1,804.2}$

2. a. $0.3\overline{)0.84}$ **b.** $0.2\overline{)4.76}$ **c.** $0.5\overline{)32.15}$ **d.** $0.4\overline{)733.08}$

3. a. $0.7\overline{).294}$ **b.** $0.4\overline{)3.024}$ **c.** $0.3\overline{)62.928}$ **d.** $0.5\overline{)18,725}$

4. a. $0.5\overline{)0.4205}$ **b.** $0.3\overline{)0.7128}$ **c.** $0.9\overline{)9.3843}$ **d.** $0.7\overline{)6.4722}$

5. a. $0.8\overline{)0.0016}$ **b.** $0.7\overline{)0.0224}$ **c.** $0.3\overline{)0.0009}$ **d.** $0.6\overline{)0.0552}$

6. a. $1.8\overline{)43.74}$ **b.** $2.6\overline{)14.976}$ **c.** $3.5\overline{)3,041.5}$ **d.** $24.3\overline{)8.8452}$

7. a. $0.4\overline{)14.0}$ **b.** $2.8\overline{)49.00}$ **c.** $6.4\overline{)104.000}$ **d.** $1.6\overline{)15.000}$

8. a. $0.5\overline{)15}$ **b.** $0.4\overline{)72}$ **c.** $1.8\overline{)36}$ **d.** $4.2\overline{)126}$

9. a. $0.6\overline{)3}$ **b.** $0.8\overline{)2}$ **c.** $5.6\overline{)14}$ **d.** $12.8\overline{)40}$

10. Find the quotient correct to the nearest thousandth.

     a. $0.8\overline{)0.45}$ **b.** $2.6\overline{)740}$ **c.** $3.9\overline{)85.3}$ **d.** $5.7\overline{)200}$

11. Write the nearest given estimate.

     a. $0.029 \div 0.6$ 5   0.5   0.05
     b. $6.8524 \div 0.7$ 1   10   30
     c. $78 \div 4.8$ 3   16   25
     d. $5.86 \div 1.9$ 0.3   30   3

12. Solve.

     a. $48.5 \div 0.5 = ?$
     b. $7.75 \div 3.1 = \blacksquare$
     c. $9.259 \div 4.7 = n$
     d. $? = 5.76 \div 9.6$

13. a. $0.04\overline{)0.68}$ **b.** $0.06\overline{)15.06}$ **c.** $0.32\overline{)16.96}$ **d.** $0.57\overline{)369.36}$

14. a. $0.07\overline{)0.812}$ **b.** $0.03\overline{)7.749}$ **c.** $0.96\overline{)6.912}$ **d.** $0.25\overline{)24.175}$

15. a. $0.02\overline{)4.6954}$ **b.** $0.08\overline{)0.7216}$ **c.** $0.43\overline{)1.9694}$ **d.** $0.75\overline{)0.4725}$

16. a. $0.09\overline{)0.08928}$ **b.** $0.05\overline{)0.15425}$ **c.** $0.22\overline{)0.21692}$ **d.** $0.36\overline{)1.32156}$

17. a. $0.06\overline{)0.0018}$ **b.** $0.01\overline{)0.0005}$ **c.** $0.03\overline{)0.00012}$ **d.** $0.68\overline{)0.04216}$

18. a. $3.65\overline{)208.05}$ **b.** $2.27\overline{)88.303}$ **c.** $4.24\overline{)7.0808}$ **d.** $8.3\overline{)404.352}$

19. a. $0.04\overline{)76.00}$ **b.** $0.36\overline{)18.00}$ **c.** $0.64\overline{)56.000}$ **d.** $1.92\overline{)120.000}$

20. a. $0.08\overline{)57.6}$ **b.** $0.65\overline{)45.5}$ **c.** $1.47\overline{)3,719.1}$ **d.** $2.83\overline{)4,952.5}$

21. a. $0.09\overline{)27}$ **b.** $0.18\overline{)90}$ **c.** $0.64\overline{)16}$ **d.** $1.36\overline{)119}$

22. Find the quotient correct to the nearest thousandth.

   a. $0.54\overline{)98}$ **b.** $0.69\overline{)8.45}$ **c.** $0.26\overline{)42.7}$ **d.** $2.43\overline{)0.162}$

23. Find the quotient correct to the nearest hundredth.

   a. $0.08\overline{)\$1.24}$ **b.** $0.25\overline{)\$3.90}$ **c.** $0.49\overline{)\$10.43}$ **d.** $2.67\overline{)\$97.90}$

24. Write the nearest given estimate.

   a. $0.98 \div 0.04$ 2.5   25   0.25
   b. $15.7 \div 2.15$ 8   0.08   0.008
   c. $\$71.86 \div \$.08$ 9   90   900
   d. $\$74.62 \div \$15.25$ 15   10   5

25. Solve.

   a. $0.9 \div 0.03 = ?$
   b. $\$17.55 \div \$.65 = $ ▨
   c. $\$54 \div \$.06 = n$
   d. ▨ $= 0.0006 \div 0.01$

26. a. $0.007\overline{)0.763}$ **b.** $0.003\overline{)17.862}$ **c.** $0.175\overline{)49.525}$ **d.** $0.526\overline{)77.322}$

27. a. $0.006\overline{)6.0282}$ **b.** $0.035\overline{)0.1715}$ **c.** $0.216\overline{)8.1864}$ **d.** $0.024\overline{)1.2144}$

28. a. $0.004\overline{)0.05964}$ **b.** $0.073\overline{)0.31828}$ **c.** $0.524\overline{)0.36156}$ **d.** $0.358\overline{)0.86278}$

29. a. $0.009\overline{)0.00081}$ **b.** $0.018\overline{)0.0009}$ **c.** $0.105\overline{)0.00735}$ **d.** $0.667\overline{)0.04002}$

30. a. $2.548\overline{)8.1536}$ **b.** $4.125\overline{)0.94875}$ **c.** $6.875\overline{)39.1875}$ **d.** $3.002\overline{)4.44296}$

31. a. $0.008\overline{)16.000}$ **b.** $0.056\overline{)140.00}$ **c.** $0.144\overline{)108.000}$ **d.** $1.375\overline{)132.000}$

32. a. $0.094\overline{)25.38}$ **b.** $0.008\overline{)90.16}$ **c.** $0.231\overline{)147.84}$ **d.** $0.382\overline{)893.88}$

33. a. $0.005\overline{)438.5}$ **b.** $0.072\overline{)691.2}$ **c.** $0.337\overline{)1819.8}$ **d.** $0.265\overline{)12{,}799.5}$

34. a. $0.059\overline{)177}$ **b.** $0.108\overline{)972}$ **c.** $0.007\overline{)210}$ **d.** $0.591\overline{)2364}$

35. Find the quotient correct to the nearest hundredth.

   a. $0.007\overline{)33}$ **b.** $0.056\overline{)0.428}$ **c.** $0.723\overline{)62.5}$ **d.** $4.007\overline{)38.25}$

36. Write the nearest given estimate.

   a. $29.7 \div 0.005$ 0.0006   600   6,000
   b. $101.25 \div 9.814$ 20   10   0.5
   c. $12 \div 0.039$ 300   0.4   0.048
   d. $1.50864 \div 0.499$ 3   11   40

37. Solve.

   a. $1.2 \div 0.006 = ?$
   b. $0.4 \div 0.016 = $ ▨
   c. $10 \div 0.125 = n$
   d. ▨ $= 0.21 \div 2.625$

38. a. Notebooks cost $.85 each. How many can Lloyd buy for $4.25?

   b. A hospital dietitian has ordered that each patient receive 0.25 pounds of meat for lunch. There are 24 pounds of one kind of meat on hand. How many patients can be served that kind of meat?

   c. A 7.2 meter length of pipe must be cut into pieces 0.003 meters long. How many pieces can be cut?

   d. A lab has 6 liters of sulphuric acid. An experiment needs 0.015 liters of the acid. How many of the experiments can be performed?

Mixed Problems

Choose an appropriate method of computation and solve each of the following problems. For Problems 1–4 select the letter corresponding to your answer.

1. The freshmen contributed $139.75 to the Red Cross; the sophomores, $98.40; the juniors, $106.15; and the seniors, $134.90. About what was the total amount of their contributions?
   a. $575   b. $375
   c. $475   d. Answer not given

2. A vessel heads N. 15 W. for 5 hours at 10.4 knots. Find the distance traveled in nautical miles. (1 knot = 1 nautical mi/h.)
   a. 52 naut. mi   b. 5.20 naut. mi
   c. 520 naut. mi   d. 50.2 naut mi

3. A butcher charged $9.52 for a certain cut of meat at $2.24 a pound. What was the weight of the meat?
   a. 5.32 lb   b. 4.76 lb
   c. 4.25 lb   d. Answer not given

4. During the month a merchant made deposits of $439.76, $180.53, $263.98, and $129.49. Checks and cash withdrawals were: $163.20, $248.00, $92.85, $310.94, and $8.52. If his previous balance was $716.91, find his new bank balance.
   a. $897.16   b. $887.16
   c. $907.16   d. Answer not given

5. A finance company can be repaid on a loan of $100 in 6 monthly payments of $18.15, 12 monthly payments of $9.75, or 18 monthly payments of $6.97. Find the amounts paid back and the interest under each plan.

6. Find the amount saved when buying in quantity: 1 dozen cans of peas for $4.80 or $.43 each.

7. If the postage for sending a certain package by first class mail was $1.85 at the rate of $.25 for the first ounce and $.20 for each additional ounce, how much did the package weigh?

Refresh Your Skills

1. $1-8$

   216,958
   133,425
   590,643
   614,299
   271,086
   + 484,137

2. $1-9$

   8,304,060
   − 8,293,851

3. $1-11$

   3,600
   × 3,600

4. $1-13$

   $5,280)\overline{2,101,400}$

5. $2-5$

   1.4 + 0.06 + 0.8

6. $2-6$

   600 − 9.08

7. $2-7$

   14.4 × 0.09

8. Round 48,263,506 to the nearest hundred thousand. $1-7$

9. Round 96.0194 to the nearest hundredth. $2-3$

10. Is the statement 0.57 < 0.6 true? $2-4$

Finding What Decimal Part One Number Is of Another

Ellen found a packet of vegetable seeds dated last year. She tested 20 of the seeds for germination. After several days, 16 of the tested seeds had sprouted. What decimal part of the seeds sprouted?

What decimal part of 20 is 16?

16 ÷ 20
↑   ↑
part whole

Write as a division problem with the given part as the dividend and the whole as the divisor.

Divide to find the **decimal part.**

$$\begin{array}{r} 0.8 \leftarrow \textbf{decimal part} \\ 20\overline{)16.0} \\ \underline{16\ 0} \end{array}$$

Check: 0.8 × 20 = 16 ✔

ANSWER: 0.8 of the seeds sprouted.

Examples

1. 0.15 is what decimal part of 2.4?

0.15 ÷ 2.4

$$\begin{array}{r} 0.0625 \\ 2.4\overline{)0.1\,5000} \\ \underline{1\ 44} \\ 60 \\ \underline{48} \\ 120 \\ \underline{120} \\ 0 \end{array}$$

Write as a division problem.

Divide.

Check: 0.0625 × 2.4 = 0.15 ✔

Remember, use the original divisor to check.

ANSWER: 0.0625

---

2. What decimal part (rounded to two places) of 3 is 2?

$$\begin{array}{r} 0.666\ \textbf{rounds to 0.67} \\ 2 \div 3 \quad 3\overline{)2.000} \end{array}$$

ANSWER: 0.67

3. What decimal part of 32 is 18? Round the answer to the nearest thousandth.

$$\begin{array}{r} 0.5625\ \textbf{rounds to 0.563} \\ 18 \div 32 \quad 32\overline{)18.0000} \end{array}$$

ANSWER: 0.563

1. What is the greatest decimal part of seeds that could have sprouted?

2. What is the least decimal part of seeds that could have sprouted?

3. Ellen also tested a package of flower seeds and 0.75 of the seeds sprouted. Did the vegetable seed package or the flower seed package have the greater part of sprouted seeds?

4. If you compute what decimal part of this year's population in Newtown was its population in a previous year, could you get a number greater than 1? What would that mean about Newtown's population over that time?

Calculator Know-How

To find what decimal part one number is of another, remember that the number after the word "of" is always the divisor.

What part is 517.92 of 62.4?

Key: 517.92 $\div$ 62.4 $=$ 8.196

Use a calculator.
1. What part is 162.7 of 256.5?

2. What decimal part of 93 is 105?

Diagnostic Exercises

1. What decimal part of 10 is 3?

2. 4 is what decimal part of 5?

3. What decimal part of 100 is 21?

4. What decimal part of 25 is 17?

5. 7 is what decimal part of 8?

6. What decimal part (2 places) of 6 is 1?

7. 27 is what decimal part of 36?

8. 24 is what decimal part (2 places) of 72?

9. What decimal part of 0.4 is 0.02?

10. 0.65 is what decimal part (2 places) of 7.8?

11. What decimal part of $2 is $.60?

12. $2.25 is what decimal part of $11.25?

13. What decimal part (to nearest thousandth) of 14 is 9?

14. What decimal part (to nearest hundredth) of 49 is 21?

15. Mrs. Baltimore will analyze 50 sheets of data. She must complete 11 of the sheets by tomorrow. What decimal part of the sheets must she complete today?

Related Practice

1. **a.** What decimal part of 10 is 7?
   b. What decimal part of 2 is 1?
   c. What decimal part of 5 is 4?
   d. What decimal part of 10 is 1?

2. **a.** 9 is what decimal part of 10?
   b. 1 is what decimal part of 5?
   c. 3 is what decimal part of 10?
   d. 1 is what decimal part of 2?

3. **a.** What decimal part of 100 is 7?
   b. 83 is what decimal part of 100?
   c. 31 is what decimal part of 100?
   d. What decimal part of 100 is 97?

4. **a.** What decimal part of 4 is 3?
   b. What decimal part of 20 is 7?
   c. 13 is what decimal part of 50?
   d. What decimal part of 25 is 22?

5. **a.** 3 is what decimal part of 8?
   b. What decimal part of 16 is 11?
   c. What decimal part of 8 is 5?
   d. 21 is what decimal part of 32?

(Round to 2 places.)

6. **a.** What decimal part of 7 is 1?
   b. 5 is what decimal part of 6?
   c. What decimal part of 12 is 7?
   d. 13 is what decimal part of 24?

7. **a.** 16 is what decimal part of 32?
   b. What decimal part of 56 is 21?
   c. 40 is what decimal part of 500?
   d. 64 is what decimal part of 200?

(Round to 2 decimal places.)

8. **a.** 45 is what decimal part of 54?
   b. What decimal part of 84 is 35?
   c. 68 is what decimal part of 96?
   d. What decimal part of 102 is 85?

9. **a.** What decimal part of 0.12 is 0.3?
   b. What decimal part of 0.96 is 0.24?
   c. 0.5 is what decimal part of 0.75?
   d. What decimal part of 0.2 is 0.18?

10. **a.** 0.4 is what decimal part of 1.6?
   b. What decimal part of 3.65 is 1.46?
   c. 0.03 is what decimal part of 300?
   d. What decimal part of 9.6 is 6?

11. **a.** What decimal part of $1 is $.05?
   b. $6 is what decimal part of $7.50?
   c. What decimal part of $3 is $1.86?
   d. $.65 is what decimal part of $13?

12. **a.** $.18 is what decimal part of $.90?
   b. What decimal part of $12.40 is $1.86?
   c. What decimal part of $27.30 is $16.38?
   d. $1.11 is what decimal part of $1.48?

(Round to nearest thousandth.)

13. **a.** What decimal part of 9 is 4?
   b. 5 is what decimal part of 7?
   c. What decimal part of 11 is 8?
   d. 2 is what decimal part of 15?

(Round to nearest hundredth.)

14. **a.** What decimal part of 84 is 24?
   b. 15 is what decimal part of 27?
   c. What decimal part of 135 is 36?
   d. 99 is what decimal part of 121?

15. **a.** A factory inspector found 12 out of 75 light bulbs tested to be defective. What decimal part of 75 is 12?

   b. The Skating Club's budget is $250. A trip to Lake Green costs $45. What decimal part of the total budget is the cost of the trip?

   c. A chemistry lab had 1.6 kilograms of an acid at the beginning of the week. 0.4 kilograms had been used. By the end of the week, what decimal part of the acid had been used?

   d. One variety of bean plants has an average height of 25 centimeters. Another variety has an average height of 11.5 centimeters. What decimal part of 25 is 11.5?

The Sports Club took a survey showing that 0.7 of the school's ninth-graders are baseball fans. If 560 ninth-graders said they were baseball fans, how many ninth-graders are there?

0.7 of what number is 560?

560 ÷ **0.7**
  ↑   ↑
part   decimal

Write as a division problem with the known part as the dividend, and the **decimal** representing this part as the divisor.
Divide to find the number representing the whole

$$
\begin{array}{r}
80\ 0 \\
0.7_\wedge\overline{)560.0_\wedge} \\
\underline{560\ 0} \\
0\ 0
\end{array}
$$

Check: 800 × 0.7 = 560 ✔

ANSWER: There are 800 ninth-graders.

Examples

1. 480 is 0.75 of what number?

480 ÷ **0.75**

$$
\begin{array}{r}
6\ 40. \\
0.75_\wedge\overline{)480.00_\wedge}
\end{array}
$$

Check: 640 × 0.75 = 480.00 = 480 ✔

ANSWER: 640

2. 0.42 of what amount is $225?
Round the answer to the nearest cent.

$225 ÷ 0.42

$$
\begin{array}{r}
\$5\ 35.714 \quad \textbf{rounds to \$535.71} \\
0.42_\wedge\overline{)\$225.00_\wedge000}
\end{array}
$$

ANSWER: $535.71

Think About It

1. Why must the number of ninth-graders found in the problem above be greater than the number of ninth-grade baseball fans?

2. How would you check the accuracy of this statement? 0.05 of 150 is 75.

3. Suppose you are told that 0.2 of a number is 11. Would you expect the number to be slightly greater than or much greater than 11? Why?

4. True or false? If the decimal part is less than 0.5, the number to be found is more than twice the number given. Explain your answer.

Diagnostic Exercises

1. 0.2 of what number is 14?

2. 20 is 0.5 of what number?

3. 0.07 of what number is 56?

4. 0.0375 of what number is 150?

5. 11 is 0.8 of what number?

6. 0.08 of what number is 100?

7. 324 is 1.08 of what number?

8. 1.15 of what amount is $9,200?

9. Of the students voting for the class trip, 0.6 voted for a boat trip. If 240 students voted for the boat trip, how many students voted?

Related Practice

1. **a.** 0.3 of what number is 12?
   b. 0.9 of what number is 540?
   c. 0.8 of what number is 400?
   d. 0.7 of what number is 63?

2. **a.** 10 is 0.2 of what number?
   b. 96 is 0.4 of what number?
   c. 300 is 0.6 of what number?
   d. 83 is 0.1 of what number?

3. **a.** 0.06 of what number is 42?
   b. 8 is 0.01 of what number?
   c. 72 is 0.05 of what number?
   d. 0.09 of what number is 630?

4. **a.** 0.3125 of what number is 1,250?
   b. 350 is 0.4375 of what number?
   c. 0.1875 of what number is 375?
   d. 0.0625 of what number is 2,500?

5. **a.** 15 is 0.4 of what number?
   b. 0.5 of what number is 28?
   c. 24 is 0.08 of what number?
   d. 0.04 of what number is 16?

6. **a.** 0.75 of what number is 150?
   b. 0.07 of what number is 63?
   c. 24 is 0.625 of what number?
   d. 0.18 of what number is 36?

7. **a.** 500 is 1.25 of what number?
   b. 2.5 of what number is 30.75?
   c. 312 is 1.04 of what number?
   d. 1.8 of what number is 9,720?

8. **a.** 1.75 of what amount is $35?
   b. 1.06 of what amount is $4.77?
   c. $7.80 is 1.2 of what amount?
   d. 2.3 of what amount is $1,978?

9. **a.** In a factory, 0.15 of the light bulbs tested were defective. There were 30 defective light bulbs. How many were tested?

   b. A technician generated a sound of 374 hertz. The recording instrument tells her that this is 0.85 of the number of hertz there are in a pure A note. How many hertz are in a pure A note?

   c. A store billed a customer a $6.25 delivery charge on the purchase of a chair. If this is 0.05 of the cost of the chair, find the cost of the chair.

   d. A public-opinion researcher determined that 0.46 of the town members surveyed are in favor of building a new swimming pool. If 230 people favored constructing the pool, how many were surveyed?

Using Mathematics

POWERS OF TEN

The power of a factor is the product determined by multiplying the repeating factor. It usually is expressed as an indicated product using exponents.

The table at right is developed using 10 as the repeated factor.

$$10 = 10 = 10^1$$
$$1\underline{00} = 10 \times 10 = 10^2$$
$$1,\underline{000} = 10 \times 10 \times 10 = 10^3$$
$$10,\underline{000} = 10 \times 10 \times 10 \times 10 = 10^4$$
$$100,\underline{000} = 10 \times 10 \times 10 \times 10 \times 10 = 10^5$$

Notice in the table that the number of zeros in the product after the digit 1 corresponds to the exponent of 10 in the indicated product.

Any numeral having zeros for all digits except the first, such as 7,000, may be expressed as a product of a digit and a power of ten.

Examples

Express 100,000 as a power of ten using exponents.
$$100,000 = 10^5$$

5 zeros   The exponent is **5**.

Express 7,000 as a product of a digit and a power of ten.
$$7,000 = 7 \times 1,000 = 7 \times 10^3$$

3 zeros   **3 zeros**   The exponent is **3**.

Any number may be written as the sum of the product of each digit in the number and its place value expressed as a power of ten.

Examples

Write the number 5,963 in expanded form as an indicated sum (polynomial).

$5,963 = 5$ thousands $+ 9$ hundreds $+ 6$ tens $+ 3$ ones
$5,963 = (5 \times 1,000) + (9 \times 100) + (6 \times 10) + (3 \times 1)$
$5,963 = (5 \times 10^3) + (9 \times 10^2) + (6 \times 10^1) + (3 \times 10^0)$

Write the number 0.9247 in expanded form.

$0.9247 = 9$ tenths $+ 2$ hundredths $+ 4$ thousandths $+ 7$ ten-thousandths
$0.9247 = (9 \times 0.1) + (2 \times 0.01) + (4 \times 0.001) + (7 \times 0.0001)$
$0.9247 = (9 \times 10^{-1}) + (2 \times 10^{-2}) + (4 \times 10^{-3}) + (7 \times 10^{-4})$

Write the number 685.379 in expanded form.

$(6 \times 10^2) + (8 \times 10^1) + (5 \times 10^0) + (3 \times 10^{-1}) + (7 \times 10^{-2}) + (9 \times 10^{-3})$

It can be shown that:
$$0.1 = 10^{-1}$$
$$0.01 = 10^{-2}$$
$$0.001 = 10^{-3}$$
$$0.0001 = 10^{-4}$$
and that:
$$1 = 10^0$$

Exercises

Express as a power of ten using exponents.

1. 1,000,000,000

2. 10,000,000,000,000,000

Express as a product of a number less than 10 and a power of ten.

3. 80,000

4. 500

5. 200,000

6. 3,000,000,000

7. 900,000,000,000

8. 60,000,000

Express each of the following numerals in expanded form.

9. 492

10. 37,026

11. 5,688,247

12. 0.29

13. 0.674

14. 0.58143

15. 215.83

16. 4,153.972

Multiplying By Powers of Ten

Booklets describing the county parks are bundled in groups of 6. The bundles are packed in boxes of 10, 100, and 1000. How many booklets are in each size box? You can answer the question using mental arithmetic.

Multiply 6 by 10; by 100; by 1,000.

Write as many zeros to the right of a whole number as there are in the power of ten.

$6 \times 10 = 60$
1 zero

$6 \times 100 = 600$
2 zeros

$6 \times 1,000 = 6,000$
3 zeros

ANSWER: There are 60, 600, or 6,000 booklets in a box.

Examples

1. Multiply 95 by 10; by 100; by 1,000.

$95 \times 10 = 950$ $95 \times 100 = 9,500$ $95 \times 1,000 = 95,000$
1 zero   **2 zeros**   **3 zeros**

ANSWER: 950; 9,500; 95,000

2. Multiply 0.7 by 10; by 100; by 1,000.

$0.7 \times 10 = 0.7 = 7$
1 zero   1 place

$0.7 \times 100 = 0.70 = 70$
2 zeros   2 places

$0.7 \times 1,000 = 0.700 = 700$
3 zeros   3 places

Move the decimal point as many places to the right in a decimal as there are zeros in the power of ten.

ANSWER: 7; 70; 700

3. Multiply 48.67 by 10; by 100; by 1,000.

48.67 × 10 = 48.67 = 486.7

    1 zero 1 place

48.67 × 100 = 48.67 = 4867

    2 zeros 2 places

48.67 × 1,000 = 48.670 = 48,670

    3 zeros   3 places

ANSWER: 486.7; 4,867; 48,670

4. Multiply 0.493 by 1,000,000,000.

0.493 × 1,000,000,000

    9 zeros

= 0.493000000

    9 places

= 493,000,000

ANSWER: 493,000,000

Think About It

1. Is it possible to multiply a number by 1,000 and have more than 3 zeros in the product? Explain or give an example.

2. Is it possible to multiply a number by 100 and have fewer than 2 zeros in the product? Explain or give an example.

3. Would a calculator be helpful in doing the exercises on this page? Why or why not?

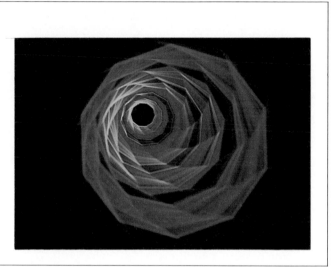

Diagnostic Exercises

Multiply each number by 10.

1. 8   **2.** 60   **3.** 0.4   **4.** 78   **5.** 0.06   **6.** 25.324

Multiply each number by 100.

7. 26   **8.** 500   **9.** 0.83   **10.** 0.5   **11.** 0.0987   **12.** 67.39

Multiply each number by 1,000.

13. 9   **14.** 420   **15.** 0.365   **16.** 0.67   **17.** 0.8574   **18.** 56.967

Multiply.

19. 574.82 × 1,000,000   **20.** 125.7 × 1,000,000,000

21. The school board distributed 1,000 flyers to each of 12 public places in the district. How many flyers were distributed?

Related Practice

Multiply each number by 10.

1. a. 5
   b. 27
   c. 85
   d. 763

2. a. 40
   b. 100
   c. 150
   d. 3,000

3. a. 0.3
   b. 0.1
   c. 0.9
   d. 0.8

4. a. 0.26
   b. 0.57
   c. 0.805
   d. 0.4326

5. a. 0.03
   b. 0.085
   c. 0.007
   d. 0.0625

6. a. 5.8
   b. 96.34
   c. 49.927
   d. 540.653

Multiply each number by 100.

7. a. 7
   b. 51
   c. 423
   d. 9,564

8. a. 20
   b. 300
   c. 590
   d. 7,400

9. a. 0.42
   b. 0.33
   c. 0.19
   d. 0.95

10. a. 0.2
   b. 0.8
   c. 0.9
   d. 0.1

11. a. 0.721
   b. 0.039
   c. 0.5257
   d. 0.0416

12. a. 8.54
   b. 36.46
   c. 5.792
   d. 25.875

Multiply each number by 1,000.

13. a. 3
   b. 62
   c. 597
   d. 2,055

14. a. 50
   b. 200
   c. 780
   d. 5,000

15. a. 0.657
   b. 0.942
   c. 0.076
   d. 0.189

16. a. 0.35
   b. 0.7
   c. 0.09
   d. 0.4

17. a. 0.2653
   b. 0.0357
   c. 0.17425
   d. 0.00072

18. a. 6.582
   b. 29.37
   c. 81.1
   d. 176.2563

Multiply.

19. By 1,000,000:
   a. 952
   b. 6.3
   c. 47.915
   d. 864.26

20. By 1,000,000,000:
   a. 83
   b. 15.7
   c. 203.06
   d. 9.658

21. Solve.

a. A local company gave 5 seniors from Central High School a college scholarship of $1,000 each. How much money did the company give in all?

b. The local mine prepares souvenir boxes of stones to be given to each visitor. Each box of stones weighs 0.35 pound. What is the weight of 100 boxes?

c. A chemical laboratory can produce 25.4 liters of a rare chemical in one day. How many liters can be produced in 10 days?

d. A recreation company is mailing a questionnaire. Printing and postage for each costs 54¢. Find the cost for 1,000.

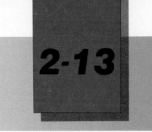

2-13 Dividing by Powers of Ten

Can a number be divided by 10, 100, or 1,000 without using long division? If a number is divided by 10, 100, or 1,000, can the answer be written in more than one way? You can do the computations using mental arithmetic.

Examples

1. a. Divide 85 by 10
$$85 \div 10 = 85. = 8.5$$
　　　↑
　　1 zero　1 place

Move the decimal point as many places to the left as there are zeros in the power of ten?

If necessary, write zeros.

b. Divide 85 by 100
$$85 \div 100 = 85. = 0.85$$
　　　　↑
　　2 zeros　2 places

c. Divide 85 by 1,000
$$85 \div 1,000 = 085. = 0.085$$
　　　　　↑
　　3 zeros　write one zero
　　　　　3 places

ANSWER: 8.5; 0.85; 0.085

---

2. a. Divide 7,600 by 10
$$7,600 \div 10 = 7,600. = 760.0 \text{ or } 760$$
　　　　　↑
　　1 zero　1 place

The answer can be written with or without the zeros to the right.

b. Divide 7,600 by 100
$$7,600 \div 100 = 7,600. = 76.00 \text{ or } 76$$
　　　　　↑
　　2 zeros　2 places

c. Divide 7,600 by 1,000
$$7,600 \div 1,000 = 7,600. = 7.600 \text{ or } 7.6$$
　　　　　↑
　　3 zeros　3 places

ANSWER: 760; 76; 7.6

Think About It

1. A teacher asks a student to choose a three-digit number, multiply it by 15, multiply the answer by 4, then divide the result by 6. The student gives the answer, and the teacher immediately tells the original number. Can you figure out how?

2. Using mental arithmetic, tell which quotient is greater, 567.83 divided by 10 or 9763.8 divided by 1,000.

3. Many corporations report their sales and profits in units of thousands of dollars. How would you express a total dollar amount as a number of thousands of dollars?

Diagnostic Exercises

Divide each of the following numbers by 10.

1. 80   **2.** 95   **3.** 6   **4.** 7   **5.** 15.683

Divide each of the following numbers by 100.

6. 400   **7.** 875   **8.** 92

9. 8   **10.** 0.34   **11.** 197.2

Divide each of the following numbers by 1,000.

12. 65,000   **13.** 2,973   **14.** 467

15. 72   **16.** 0.675   **17.** 527.3

Divide.

18. 5,900,000 by 1,000,000   **19.** 64,125,000,000 by 1,000,000,000

20. 3,720,000,000,000 by 1,000,000,000,000

21. A kilometer is equal to 1,000 meters, so to convert meters to kilometers, divide the number of meters by 1,000. Use this information to express 54,235 meters in kilometers.

Related Practice

Divide each of the following numbers by 10.

1.	**2.**	**3.**	**4.**	**5.**
a. 20	**a.** 34	**a.** 5	**a.** 0.9	**a.** 3.5
b. 600	**b.** 276	**b.** 3	**b.** 0.32	**b.** 9.82
c. 300	**c.** 408	**c.** 9	**c.** 0.08	**c.** 27.46
d. 1,000	**d.** 5,426	**d.** 4	**d.** 0.936	**d.** 39.239

Divide each of the following numbers by 100.

6.	**7.**	**8.**
a. 200	**a.** 382	**a.** 59
b. 700	**b.** 829	**b.** 32
c. 5,000	**c.** 4,520	**c.** 70
d. 2,700	**d.** 65,726	**d.** 85

9.	**10.**	**11.**
a. 4	**a.** 0.21	**a.** 29.5
b. 2	**b.** 0.60	**b.** 502.86
c. 3	**c.** 0.8	**c.** 68.24
d. 7	**d.** 0.045	**d.** 1,500.75

Divide each of the following numbers by 1,000.

12.	**13.**	**14.**
a. 8,000	**a.** 3,725	**a.** 628
b. 10,000	**b.** 2,890	**b.** 314
c. 28,000	**c.** 15,925	**c.** 200
d. 150,000	**d.** 18,464	**d.** 957

15. a. 85
   b. 93
   c. 6
   d. 8

16. a. 0.925
   b. 0.56
   c. 0.3
   d. 0.072

17. a. 284.9
   b. 500.74
   c. 1,526.1
   d. 2,963.45

Divide.

18. By 1,000,000:

   a. 617,000,000
   b. 450,000
   c. 9,200,000
   d. 13,670,000

19. By 1,000,000,000:

   a. 39,000,000,000
   b. 6,400,000,000
   c. 775,000,000
   d. 8,350,000,000

20. By 1,000,000,000,000:

   a. 8,000,000,000,000
   b. 12,500,000,000,000
   c. 470,000,000,000
   d. 1,834,000,000,000

21. a. Sean buys a new car for $12,840. If he has to pay a sales tax that is one-tenth of the price of the car, how much will he owe in sales tax?

   b. The earnings from a new product are to be divided equally among 100 investors. If the earnings in the first year were $39,524.50, how much did each investor receive?

   c. There are 100 centimeters in a meter. How many meters are there in 6,438 centimeters?

   d. Judy drove 534 miles in 10 hours. What was her average speed for the trip?

Applications

To change from watt-hours (W·h) to kilowatt-hours (kW·h), divide the number of watt-hours by 1000.

1. Change to kilowatt-hours.

   a. 20,000 W·h
   b. 135,000 W·h
   c. 27,500 W·h

2. A bag of sugar, weighing 100 pounds, costs $35.30. What is the cost per pound?

Taxes

3. If the tax rate is $9.80 per hundred dollars, how much must you pay for taxes on a house assessed for $98,900?

4. Find the amount of taxes on properties having the following assessed valuations and tax rates.

	a.	b.	c.	d.	e.
Assessed valuation	$3,400	$17,500	$51,800	$45,000	$80,200
Tax rate per $100	$5.70	$6.45	$3.60	$4.25	$5.84

Using Mathematics

SCIENTIFIC NOTATION

Scientists and mathematicians sometimes work with very large or very small numbers. To simplify their work, they often express such numbers in an abbreviated form known as scientific notation.

To express a number in scientific notation, write it as the product of a number between 1 and 10 and a power of ten. See Powers of Ten, page 127.

The required power of ten may be determined by counting the number of places the decimal point is moved to get the required whole number or mixed number between 1 and 10.

Examples

Express 9,800,000 in scientific notation.
$9\ 800\ 000. = 9.8 \times 1\ 000\ 000 = 9.8 \times 10^6$

6 places to the left.   **The exponent is ↗ positive 6.**

When you move the decimal point to the left, the exponent is positive.

Express 0.0000536 in scientific notation.
$0.0000536 = 5.36 \times 0.00001 = 5.36 \times 10^{-5}$

5 places to the right.   **The exponent is ↗ negative 5.**

When you move the decimal point to the right, the exponent is negative.

To express a number in scientific notation in standard form, multiply by the given power of ten.

$7.8 \times 10^8 = 7.8 \times 100,000,000 = 780,000,000$

$3.2 \times 10^{-3} = 3.2 \times 0.001 = 0.0032$

Exercises

Express each of the following numbers in scientific notation.

1. 60,000   **2.** 79,000   **3.** 960,000,000   **4.** 84,500,000,000,000

5. 0.0018   **6.** 0.00249   **7.** 0.000000801   **8.** 0.000000000043

Express each of the following in standard form.

9. 5.2×10^5 **10.** 4.875×10^6 **11.** 2×10^6 **12.** 3.08×10^{11}

13. 6.356×10^{-2} **14.** 9.1×10^{-6} **15.** 3.65×10^{-5} **16.** 7.9×10^{-7}

Express each number in scientific notation.

17. The sun at any second develops 500,000,000,000,000,000,000,000 horsepower.

18. The star Alpha Hercules is 3,860,000,000 kilometers in diameter.

Lisa is choosing the colors for the cover and the printing on the cover of the school yearbook. The cover of the book can be any one of 8 different colors (black, red, green, blue, yellow, purple, brown, or orange). The printing on the cover can be any one of 6 different colors (silver, gold, white, tan, grey, or burgundy). How many different color choices does Lisa have for the yearbook cover?

> **Strategy: Solve a Simpler Problem**
>
> Solve a similar but simpler problem with a less complex situation. Look for a pattern. Then apply the same pattern to the original problem.

1. **READ** the problem carefully.

a. Find the **question asked**. How many different choices does Lisa have for the yearbook cover colors?

b. Find the **given facts**. cover: 8 different colors available
printing on cover: 6 different colors available

2. **PLAN** how to solve the problem.

a. Choose a strategy: Solve a simpler problem.

b. Think or write out your plan relating the given facts to the question asked.

Consider a simpler problem: If the cover could be any one of 2 different colors (black, red) and the printing on the cover could be any one of 3 different colors (gold, silver, white), how many different color choices for the cover would there be?

List the possibilities for the simpler problem.

black, gold	red, gold
black, silver	red, silver
black, white	red, white

There are 6 possible choices.

Find a pattern to relate the number of cover colors and the number of printing colors to the number of different choices.

(number of cover colors) × (number of printing colors) =
number of different choices

133

3. **SOLVE** the problem.

Remember that there are 8 cover colors and 6 printing colors. So

$8 \times 6 = 48$

4. **CHECK**.

a. **Check the accuracy** of your arithmetic. You could make a list of all the possible choices.

$6 \times 8 = 48$

b. See if your answer is reasonable.

ANSWER: There are 48 possible choices.

Practice Problems

1. There are 16 students in the computer club. If each student plays one computer game with every other student in the club, how many games will be played?

2. Betty works at Burgers Deluxe, where hamburgers can be purchased plain or with one or more of the following "Extras": onions, mushrooms, cheese, bacon, tomatoes, lettuce, pickles. How many different burgers are available at Burgers Deluxe?

3. Josh is tacking 18 sheets of paper in a line across the top of a bulletin board. The papers overlap $\frac{1}{2}$ inch. Corners which overlap will share a tack. How many tacks will Josh need if he puts one tack in each corner of each paper?

4. Marni has 25 small square tables. Each table can seat one person on a side. If all the tables are pushed together to make one long table, how many people can sit at the long table?

5. There are 15 books containing ideas for science projects in the school library. Julie wants to check out two books. How many choices does she have?

6. How many squares are in a checkerboard? (The squares can be different sizes and they can overlap.)

The T & V Electronics company reported that it produced 98 thousand television sets. Write the standard form.

98 thousand

98 thousand = 98 × 1,000

          ↑

     1 thousand

Multiply by the power of ten equivalent to the value of the period name.

98 thousand = 98,000

ANSWER: 98,000

Examples

Write each of the following in standard form.

1. 193 million

193 million = 193 × 1,000,000

             ↑

      1 million

193 million = 193,000,000

ANSWER: 193,000,000

2. 84 billion

84 billion = 84 × 1,000,000,000

               ↑

         1 billion

84 billion = 84,000,000,000

ANSWER: 84,000,000,000

3. 7.65 million

7.65 million = 7.65 × 1,000,000 = 7.650000

           ↑

     6 zeros   **6 places**

7.65 million = 7,650,000

ANSWER: 7,650,000

4. 81.9 trillion

81.9 trillion = 81.9 × 1,000,000,000,000 = 81.900000000000

↑
12 zeros   **12 places**

ANSWER: 81,900,000,000,000

---

5. $53.71 billion

53.71 billion = 53.71 × 1,000,000,000 = 53.710000000

↑
9 zeros   **9 places**

53.71 billion = 53,710,000,000

ANSWER: $53,710,000,000

Think About It

1. Name two situations in which short names are useful.

2. Is 3.142864 million a good use of the short name technique?

3. True or false? Short names for large numbers often express numbers that have been rounded.

4. Complete: In the number 4.12 trillion, the value of the digit 2 is 2 hundredths of a ___?___.

Diagnostic Exercises

Write the standard form for the following.

1. 725 million

2. 51.9 million

3. 483.28 million

4. 6.071 million

5. $34.56 million

6. 49 billion

7. 98.4 billion

8. 2.06 billion

9. 350.742 billion

10. $691.18 billion

11. 31 trillion

12. 85.3 trillion

13. 408.57 trillion

14. 9.529 trillion

15. $70.04 trillion

16. 616 thousand

17. 18.947 thousand

18. $5.09 thousand

19. 1,826 hundred

20. 62.3 hundred

21. The total budget for a health services project is $26.73 billion. Write the standard form for this amount.

Related Practice

Write the standard form for each of the following.

1. **a.** 83 million
   b. 9 million
   c. 460 million
   d. 2,805 million

2. **a.** 6.8 million
   b. 84.7 million
   c. 358.3 million
   d. 1,098.5 million

3. **a.** 17.45 million
   b. 5.07 million
   c. 227.93 million
   d. 1,540.16 million

4. **a.** 8.562 million
   b. 13.468 million
   c. 479.704 million
   d. 3,872.835 million

5. **a.** $4.5 million
   b. $27.4 million
   c. $75.09 million
   d. $408.263 million

6. **a.** 78 billion
   b. 5 billion
   c. 609 billion
   d. 3,570 billion

7. **a.** 9.4 billion
   b. 72.8 billion
   c. 403.7 billion
   d. 2,576.3 billion

8. **a.** 12.06 billion
   b. 6.95 billion
   c. 820.73 billion
   d. 3,505.81 billion

9. **a.** 7.568 billion
   b. 31.047 billion
   c. 564.382 billion
   d. 1,067.509 billion

10. **a.** $8.3 billion
   b. $30.7 billion
   c. $54.61 billion
   d. $935.054 billion

11. **a.** 4 trillion
   b. 106 trillion
   c. 58 trillion
   d. 292 trillion

12. **a.** 3.9 trillion
   b. 207.8 trillion
   c. 83.1 trillion
   d. 6.7 trillion

13. **a.** 5.92 trillion
   b. 60.07 trillion
   c. 139.78 trillion
   d. 308.59 trillion

14. **a.** 8.227 trillion
   b. 41.058 trillion
   c. 250.829 trillion
   d. 11.256 trillion

15. **a.** $9.6 trillion
   b. $3.28 trillion
   c. $87.92 trillion
   d. $40.071 trillion

16. **a.** 9 thousand
   b. 528 thousand
   c. 77 thousand
   d. 3,863 thousand

17. **a.** 8.7 thousand
   b. 28.5 thousand
   c. 63.09 thousand
   d. 592.16 thousand

18. **a.** $19 thousand
   b. $51.6 thousand
   c. $200.52 thousand
   d. $1,174.49 thousand

19. **a.** 38 hundred
   b. 952 hundred
   c. 2,433 hundred
   d. $77 hundred

20. **a.** 8.5 hundred
   b. 63.27 hundred
   c. $83.6 hundred
   d. $4.72 hundred

21. Write the standard form for each number.

   a. The distance from the earth to the sun is 93 million miles.

   b. There are 172 million birds in this forest.

   c. The fertilizer had been used on 15.45 million acres of farmland.

   d. The distance traveled by light in one year is 5.88 trillion miles.

2-16 Writing Short Names for Numbers

Examples

Write the short number name for each of the following.

1. The scientist estimated that there were 58,000,000,000 bacteria in the culture. Write a short name for this number.

58,000,000,000 in billions

$$58{,}000{,}000{,}000 \div 1{,}000{,}000{,}000 = 58$$
↑
1 billion

58,000,000,000 = 58 billion

Divide by the power of ten equivalent to the value of the period name. Write the quotient with the period name.

ANSWER: There are 58 billion bacteria in the culture.

---

2. 8,910,000 in millions

$$8{,}910{,}000 \div 1{,}000{,}000 = 8.91$$
↑
1 million

8,910,000 = 8.91 million

ANSWER: 8.91 million

3. 556,700 in thousands

$$556{,}700 \div 1{,}000 = 556.7$$
↑
1 thousand

556,700 = 556.7 thousand

ANSWER: 556.7 thousand

---

4. 9,832,000,000,000 in trillions

$$9{,}832{,}000{,}000{,}000 \div 1{,}000{,}000{,}000{,}000 = 9.832$$
↑
1 trillion

9,832,000,000,000 = 9.832 trillion

ANSWER: 9.832 trillion

---

5. 7,250 in hundreds

$$7{,}250 \div 100 = 72.50 = 72.5$$
↑
1 hundred

7,250 = 72.5 hundred

ANSWER: 72.5 hundred

6. $68,400,000 in millions of dollars.

$$68{,}400{,}000 \div 1{,}000{,}000 = 68.4$$
↑
1 million

$68,400,000 = $68.4 million

ANSWER: $68.4 million

---

1. Can a number be expressed in only one way by a short name? Explain or give an example.

2. Are short names more often used with large numbers or small numbers? Explain your answer.

3. Can short names be used for decimals between 0 and 1? Give some examples.

Diagnostic Exercises

Write the short name for each of the following.

In hundreds: **1.** 21,400 **2.** 6,830 **3.** 8,745

In thousands: **4.** 1,632,000 **5.** 750,600 **6.** 9,050

In millions: **7.** 35,000,000 **8.** 4,700,000 **9.** 264,010,000

In billions: **10.** 167,000,000,000 **11.** 4,200,000,000

12. 379,460,000,000 **13.** 50,088,000,000

In trillions: **14.** 45,000,000,000,000 **15.** 8,100,000,000,000

16. 66,090,000,000,000 **17.** 724,525,000,000,000

18. In thousands of dollars: $27,900 **19.** In millions of dollars: $838,250,000

20. In billions of dollars: $96,080,000,000

21. In trillions of dollars: $4,202,000,000,000

22. The scientists collected 8,140,000,000 samples of fungus. Write a short name for this number.

Related Practice

Write the short name for each of the following.

In hundreds:

1. a. 900
b. 3,200
c. 7,100
d. 12,500

In hundreds:

2. a. 1,340
b. 870
c. 4,950
d. 20,820

In hundreds:

3. a. 2,568
b. 693
c. 5,756
d. 33,497

In thousands:

4. a. 6,000
b. 11,000
c. 1,437,000
d. 282,000

In thousands:

5. a. 8,500
b. 260,800
c. 53,700
d. 4,839,100

In thousands:

6. a. 7,130
b. 8,040
c. 37,290
d. 626,750

In millions:

7. a. 8,000,000
   b. 43,000,000
   c. 119,000,000
   d. 2,640,000,000

In millions:

8. a. 7,300,000
   b. 256,800,000
   c. 69,200,000
   d. 3,108,500,000

In millions:

9. a. 11,470,000
   b. 6,080,000
   c. 408,950,000
   d. 1,584,120,000

In billions:

10. a. 9,000,000,000
   b. 23,000,000,000
   c. 102,000,000,000
   d. 2,421,000,000,000

In billions:

11. a. 7,500,000,000
   b. 32,700,000,000
   c. 184,300,000,000
   d. 1,421,900,000,000

In billions:

12. a. 16,820,000,000
   b. 6,370,000,000
   c. 309,760,000,000
   d. 86,940,000,000

In billions:

13. a. 51,006,000,000
   b. 4,592,000,000
   c. 608,028,000,000
   d. 4,312,454,000,000

In trillions:

14. a. 6,000,000,000,000
   b. 263,000,000,000,000
   c. 95,000,000,000,000
   d. 107,000,000,000,000

In trillions:

15. a. 4,700,000,000,000
   b. 26,400,000,000,000
   c. 870,500,000,000,000
   d. 53,600,000,000,000

In trillions:

16. a. 32,080,000,000,000
   b. 8,450,000,000,000
   c. 97,530,000,000,000
   d. 208,410,000,000,000

In trillions:

17. a. 17,526,000,000,000
   b. 2,905,000,000,000
   c. 56,847,000,000,000
   d. 480,061,000,000,000

In thousands of dollars:

18. a. $63,000
   b. $19,500
   c. $420,930
   d. $56,078

In millions of dollars:

19. a. $84,000,000
   b. $9,360,000
   c. $450,844,000
   d. $2,655,007,000

In billions of dollars:

20. a. $58,000,000,000
   b. $15,060,000,000
   c. $8,377,000,000
   d. $142,522,000,000

In trillions of dollars:

21. a. $6,500,000,000,000
   b. $210,680,000,000,000
   c. $36,493,000,000,000
   d. $9,057,000,000,000

22. Write the short name for each number.

   a. A research project has a budget of $25,000,000,000 dollars. Write in billions of dollars.

   b. A scientist received a research prize of $52,000. Write in thousands of dollars.

   c. The company shipped 3,480 cases of ice cream. Write in hundreds.

   d. Company T sold 3,560,000 typewriters. Write in millions.

Using Mathematics

OPERATIONS WITH NUMBERS WITH SHORT NAMES

$63.45 *billion* is read:

Sixty-three point four five billion dollars and is the short name for $63,450,000,000.

We add, subtract, multiply, and divide numbers with short names just as we do with whole numbers and decimals.

Examples

Add: 82.6 million + 4.7 million
   82.6 million
+   4.7 million
   87.3 million

Subtract: 21.87 trillion − 11.78 trillion
   21.87 trillion
− 11.78 trillion
  10.09 trillion

Multiply: 8 × 4.6 billion
  4.6 billion
×   8
  36.8 billion

Divide: 148.54 billion ÷ 2
    74.27 billion
2)148.54 billion

Exercises

Read, or write in words, each short name.

1. Every day 218 million consumers in the United States spend more than 3 billion dollars.

2. An electric utility requested a $190 million increase in rates for the year.

3. Total Canadian newsprint shipments were 2.38 million metric tons.

4. In one recent year the personal income increased to 1.375 trillion dollars.

5. Earnings of a certain corporation rose to $252.3 million from $244.2 million and revenues increased to $3.8 billion from $3.45 billion.

Add.

6. 1.4 trillion + 2.36 trillion + 5.5 trillion

Subtract.

7. 56.2 thousand − 7.8 thousand

Multiply.

8. 1.9 × 5.8 billion

Divide.

9. 3.44 trillion ÷ 8

10. Earnings of a corporation this year doubled from last year. If the earnings last year were $18.2 million, what are the earnings for this year?

11. The township school budget was $9.79 million last year but is $310,000 more this year. What is the budget this year?

Many products used in industry or by the consumer require *factory workers* with a wide variety of skills to manufacture them. Pattern makers make very accurate patterns or molds for various articles that are mass-produced. Tool-and-die makers make tools, measuring devices, and other precision parts. Machinists make metal parts for machinery. Welders use heat and/or electricity to join parts together. Assemblers put together articles such as electronic components. Inspectors make sure that parts and finished products meet specifications. Industrial engineers work out ways for people and machines to work most efficiently.

Most of these factory workers are high-school graduates. They spend as long as five years as apprentices. Industrial engineers are college graduates.

1. A welder needs three pieces of I-beam with lengths of 5.85 meters, 12.6 meters, and 19.32 meters. What is the total length of I-beam needed?

2. An inspector measures a part as 2.64 centimeters long. The part should be 2.59 centimeters long. How much shorter should it be?

3. A welder uses 1.138 cubic meters of acetylene gas to make one bracket. How much gas is needed to make 15 brackets?

4. A punch press operator is paid $1.42 per molding. If the operator is paid $306.72, how many moldings were pressed?

Chapter 2 Review

Vocabulary

decimal p. 76   > symbol p. 86   < symbol p. 86   mixed decimal p. 76

Page

Read or write in words.

77   **1.** 0.007   **2.** 63.74

Write in standard form.

80   **3.** twelve thousandths   **4.** seven hundred and sixty thousand twenty-four hundred thousandths.

83   **5.** Round 0.362 to the nearest hundredth.   83   **6.** Round 1.6857 to the nearest thousandth.

86   **7.** Which is greater, 3.625 or 3.7?   86   **8.** Which is less, 0.95 or 0.096?

88   **9.**   3.7   **10.** 23.6 + 14.8 + 12.9   88   **11.** 15.2 + 17 + 12.87 + 9.6
6.3
1.8
4.3
+ 0.4

95   **12.**   6.34   **13.** 2.735 − 1.962   95   **14.** Take 3.84 from 9.62.
− 1.972

102   **15.**   5.71   **16.** 0.266 × 52   102   **17.** 0.03 × .096
×   .38

109   **18.** $3\overline{)0.1683}$ 113   **19.** $0.14\overline{)0.812}$ **20.** $5.7\overline{)0.3876}$

120   **21.** What decimal part of 50 is 7?   120   **22.** 9 is what decimal part of 45?

123   **23.** 0.7 of what number is 4.2?   123   **24.** 0.25 of what number is 74?

126   **25.** Multiply 6.825 by 100.   126   **26.** Multiply 0.02483 by 1,000.

129   **27.** Divide 2.7 by 10.   129   **28.** Divide 3.625 by 100.

95   **29.** During a recent year, two TV shows that attracted a large viewing audience were a *M.A.S.H.* special and an episode of *Dallas.* The number of households that saw *M.A.S.H.* was 50.15 million, while 41.47 million saw *Dallas.* How many more households saw *M.A.S.H.?*

102   **30.** Jupiter travels 8.12 miles per second. How far does it travel in one minute?

113   **31.** If $25 was spent on gas that costs $.969 a gallon, how many gallons were bought? (nearest tenth)

135   **32.** Write 65 billion in standard form.

138   **33.** Write 24,100,000 as millions in short form.

76   **34.** A fractional number such as 0.83 is called a ____?____ .

Chapter 2 Test

77   **1.** Read, or write 2.06 in words.

80   **2.** Write forty and seventeen thousandths in standard form.

83   **3.** Round 0.6159 to the nearest hundredth.

86   **4.** Which is greater, 0.495 or 0.5?

88   **5.**
$$\begin{array}{r} 0.36 \\ 0.87 \\ 0.74 \\ +\ 0.29 \\ \hline \end{array}$$
  6. $13.7 + 2.94 + 15 + 9.2$

95   **7.**
$$\begin{array}{r} 2.443 \\ -\ 1.97 \\ \hline \end{array}$$
  8. $9.6 - 3.84$

  9. Take 18.62 from 27.

102 **10.** 174.3×0.9 **11.** 0.36×0.07

  12. 24.6×0.72 **109** **13.** $5\overline{)11.535}$

113 **14.** $0.31\overline{)11.036}$ **113** **15.** $0.016\overline{)0.06128}$

120 **16.** What decimal part of 40 is 3?

123 **17.** 0.6 of what number is 36?

126 **18.** Multiply 3.25 by 1,000.

129 **19.** Divide 9.7 by 100.

109 **20.** How much is each can of soda if a case of 24 costs $6.40? (round to the nearest cent)

88 **21.** The 1985 circulation for the top three magazines was 17.6 million for *Readers Digest,* 16.9 million for *TV Guide,* and 10.55 million for *National Geographic.* What was the total circulation for these magazines?

102 **22.** If Andrea earns $2.20 per hour babysitting, how much did she make last week when she babysat 13.5 hours?

95 **23.** Sales tax of $3.48 brought the cost of a radio to $73.07. What was the listed price of the radio?

135 **24.** Write 7.2 billion in standard form.

138 **25.** Write 16,840,000 as millions in short form.

Cumulative Test

Solve each problem and select the letter corresponding to your answer.

Page

7

1. What missing information is needed to solve the problem?
John put down $5 and made monthly payments for a year to pay for his bike. How much did the bike cost?

   a. down payment   **b.** monthly payment
   c. amount interest   **d.** how long he had to pay

29

2. 3,592 + 6,271 + 8,994 + 5,787

   a. 23,644   **b.** 24,644
   c. 23,544   **d.** 24,544

37

3. 80,602 − 41,085

   a. 39,517   **b.** 39,527
   c. 38,517   **d.** 38,527

49

4. 82 × 49

   a. 4,118   **b.** 4,028
   c. 4,128   **d.** 4,018

61

5. 64,010 ÷ 74

   a. 785   **b.** 7,165
   c. 865   **d.** 8,641

86

6. Choose the smallest number.

   a. 0.973   **b.** 0.0944
   c. 0.0799   **d.** 0.12

88

7. 192.7 + 0.481

   a. 241.8   **b.** 673.7
   c. 193.181   **d.** 192.181

95

8. 0.095 − 0.0168

   a. 0.798   **b.** 0.073
   c. 0.0782   **d.** 0.782

102

9. 0.063 × 500

   a. 31.5   **b.** 315
   c. 3,150   **d.** 3.15

109

10. 84 ÷ 0.12

   a. 0.7   **b.** 7
   c. 70   **d.** 700

123

11. 150 is 0.75 of what number?

   a. 2   **b.** 20
   c. 200   **d.** 0.005

138

12. Write the short name for 27,310,000 in millions.

   a. 273.1 million   **b.** 27.31 million
   c. 2,731 million   **d.** 2.731 million

138

13. A Senator proposed that $1.8 billion be spent on a defense project. The sum of $2.73 billion was actually funded for the project. Find the amount of increase.

   a. $93 million   **b.** $930 million
   c. $453 million   **d.** $4.53 billion

COMPUTER ACTIVITY 2

One major advantage of a computer over a calculator is that a computer program can be modified to work in different ways. Copy and complete the chart. Use the program and its data on page 74.

LINE	SUM	JAN	FEB	MAR	APR	MAY	JUN	JUL	AUG	SEPT	OCT	NOV	DEC
10	0												
50		300											
60	300												
70			467										
80	767												
90				1,232									
100	1,999												
110					1,345								
120	?												
130						?							
140	?												
150							?						
160	?												
170								?					
180	?												
190									?				
200	?												
210										?			
220	?												
230											?		
240	?												
250												?	
260	?												
270													?
280	?												

Now calculate the average monthly sales.
Compare how long it takes you to complete the chart to how quickly the computer ran the program.

Change the program by entering a new line 260.
26Ø SALES = SALES + NOV
What do you think will happen?
RUN the program. What happens?

Modify the program to calculate average monthly sales for the first six months of the year. What is the result when you RUN the program?

Is there a way to use this program to get cumulative sales totals for portions of the year? For example, if you want to know the sales for the first quarter of the year, what do you do?
What are the results when you RUN the program?

Introduction to Fractions

When an object or unit is divided into equal parts, the number expressing the relation of one or more equal parts to the total number of equal parts is called a fraction.

If a piece of ribbon is divided into two equal parts, each part is one-half of the whole piece of ribbon and is represented by the fraction symbol $\frac{1}{2}$.

If the piece of ribbon is divided into:

Three equal parts 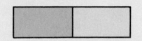 , each part is called one-third ($\frac{1}{3}$).

Four equal parts 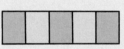 , each part is called one-fourth ($\frac{1}{4}$).

Five equal parts 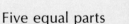 , each part is called one-fifth ($\frac{1}{5}$), etc.

The symbol for a fraction consists of a pair of numbers, one written above the other with a horizontal bar between them.

In the fraction $\frac{5}{8}$:
The numbers 5 and 8 are the terms of the fraction.
The number above the fraction bar, 5, is called the numerator.
The number below the fraction bar, 8, is called the denominator.

The denominator tells us the number of equal parts into which an object is divided. The numerator tells us how many equal parts are being used. The denominator cannot be zero.
The fraction $\frac{5}{8}$ means 5 parts of 8 equal parts.

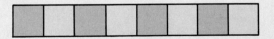

Sometimes a group of objects is divided into equal parts. The number expressing the relation of one or more of the equal parts of the group to the total number of equal parts is considered to be a fraction.

Observe above ($\frac{1}{5}$ compared to $\frac{1}{3}$) that the greater the denominator, the smaller the size of the part.

Expressing Fractions in Lowest Terms

Just as there are many different ways to name a whole number (V, 5, five), a fraction can have a variety of names. For example, $\frac{2}{4}$, $\frac{3}{6}$, $\frac{4}{8}$, and $\frac{5}{10}$ are just a few of the names for $\frac{1}{2}$. The fraction $\frac{1}{2}$ is said to be in lowest terms.

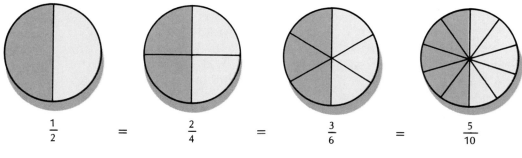

$$\frac{1}{2} \quad = \quad \frac{2}{4} \quad = \quad \frac{3}{6} \quad = \quad \frac{5}{10}$$

Of the 15 pennants on the sports field, 5 have stars. Express $\frac{5}{15}$ as a fraction in lowest terms.

numerator → $\dfrac{5}{15} = \dfrac{5 \div 5}{15 \div 5}$
denominator →
$\qquad\qquad = \dfrac{1}{3}$ ← **lowest terms**

ANSWER: $\frac{5}{15}$ in lowest terms is $\frac{1}{3}$.

Divide the numerator and the denominator of the fraction by their greatest common factor (GCF).

5: 1 5
15: 1 3 5 15

Examples

1. Express $\frac{63}{72}$ in lowest terms.
$$\frac{63}{72} = \frac{63 \div 9}{72 \div 9} = \frac{7}{8}$$
ANSWER: $\frac{63}{72}$ in lowest terms is $\frac{7}{8}$

63: 1 3 7 9 21 63
72: 1 2 3 4 6 8 9 12 18 24 36 72

2. Express $\frac{18}{29}$ in lowest terms.
ANSWER: $\frac{18}{29}$ in lowest terms is $\frac{18}{29}$

The GCF of 18 and 29 is 1. So the fraction is already in lowest terms.

Vocabulary

A fraction is in **lowest terms** when its numerator and denominator cannot be divided exactly by the same number, except by 1.

Think About It

1. In the first example above, $\frac{5}{15}$ is expressed as $\frac{1}{3}$ in lowest terms. Does this mean that $\frac{1}{3}$ is less than $\frac{5}{15}$? Explain.

2. If the numerator and the denominator of a fraction are both even numbers, is the fraction always, sometimes, or never in lowest terms?

3. Give some reasons why expressing fractions in lowest terms is useful.

4. Why do we use the GCF to find the lowest terms of a fraction?

Diagnostic Exercises

Express the following fractions in lowest terms.

1. $\frac{3}{27}$ **2.** $\frac{18}{36}$ **3.** $\frac{6}{9}$ **4.** $\frac{48}{64}$ **5.** $\frac{26}{39}$ **6.** $\frac{648}{852}$ **7.** $\frac{250}{1,000}$

8. A nail measures $\frac{12}{16}$ inches. Express this fraction in lowest terms.

Related Practice

Express the following fractions in lowest terms.

1. a. $\frac{2}{4}$ **b.** $\frac{5}{15}$ **c.** $\frac{7}{21}$ **d.** $\frac{3}{12}$ **2. a.** $\frac{8}{16}$ **b.** $\frac{4}{40}$ **c.** $\frac{8}{48}$ **d.** $\frac{6}{42}$

3. a. $\frac{8}{10}$ **b.** $\frac{4}{6}$ **c.** $\frac{6}{20}$ **d.** $\frac{10}{25}$ **4. a.** $\frac{20}{36}$ **b.** $\frac{32}{48}$ **c.** $\frac{50}{75}$ **d.** $\frac{56}{64}$

5. a. $\frac{52}{91}$ **b.** $\frac{62}{93}$ **c.** $\frac{34}{119}$ **d.** $\frac{91}{104}$ **6. a.** $\frac{180}{216}$ **b.** $\frac{420}{756}$ **c.** $\frac{405}{567}$ **d.** $\frac{680}{765}$

7. a. $\frac{300}{1,000}$ **b.** $\frac{1,200}{2,000}$ **c.** $\frac{750}{2,500}$ **d.** $\frac{2,000}{3,600}$

8. a. Frank has 18 shirts, including 8 T-shirts. Express $\frac{8}{18}$ in lowest terms.

b. A rancher bought 176 cows, of which 44 were Guernseys. Express $\frac{44}{176}$ in lowest terms.

c. A farmer put 145 tomato plants in last spring. Of these 55 were plum tomatoes. Express $\frac{55}{145}$ in lowest terms.

d. The Shoe Repair Shop has 96 pairs of shoes waiting to be fixed. Among these were 24 pairs that needed new heels. Express $\frac{24}{96}$ in lowest terms.

3-2 Improper Fractions and Mixed Numbers

When the solution to any problem with fractions is a fraction in which the numerator equals or is greater than the denominator, it is called an **improper fraction** and should be written in simplest form. If the numerator can be evenly divided by the denominator, the simplest form is a whole number. If the numerator is greater than the denominator but not evenly divisible by it, the simplest form is a whole number and a fraction in lowest terms, called a **mixed number.**

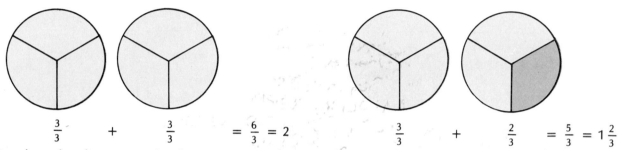

$$\frac{3}{3} \quad + \quad \frac{3}{3} \quad = \frac{6}{3} = 2 \qquad\qquad \frac{3}{3} \quad + \quad \frac{2}{3} \quad = \frac{5}{3} = 1\frac{2}{3}$$

Look at the diagrams. The first two circles illustrate an improper fraction as a whole number. The third and fourth circles illustrate an improper fraction as a mixed number.

Examples

1. Express $\frac{8}{4}$ as a whole number.

$$\frac{8}{4} = 8 \div 4 = 2$$

In a fraction, the bar signifies division. Divide the numerator by the denominator.

ANSWER: 2

---

2. Express $\frac{8}{6}$ as a mixed number.

$$\frac{8}{6} = 8 \div 6$$

$$\text{denominator} \rightarrow 6\overline{)8} \quad \begin{array}{l} 1 \leftarrow \text{whole} \\ \ \ \text{number} \end{array}$$
$$\text{of fraction} \quad \underline{6}$$
$$2 \leftarrow \begin{array}{l}\text{numerator} \\ \text{of fraction}\end{array}$$

Divide the numerator by the denominator.

The quotient is the whole number part of the mixed number, and the remainder divided by the divisor is the fraction part of the mixed number.

$$\frac{8}{6} = 1\frac{2}{6} \qquad 1\frac{2}{6} = 1\frac{1}{3}$$

Write the fraction in lowest terms. $1\frac{1}{3}$ is in simplest form.

ANSWER: $1\frac{1}{3}$

---

3. Simplify $6\frac{14}{9}$.

$$6\frac{14}{9} = 6 + \frac{14}{9}$$
$$= 6 + 1\frac{5}{9}$$
$$= 7\frac{5}{9}$$

Rewrite mixed numbers as the sum of the whole number and the fraction.
Write the fraction in lowest terms.
Add.

ANSWER: $7\frac{5}{9}$

Vocabulary

If the numerator is equal to or greater than the denominator, the fraction is called an **improper fraction.**

A number that consists of a whole number and a fraction is a **mixed number.**

Think About It

1. If 8 nubles = 1 ooble, then 26 nubles = $\frac{26}{8}$ oobles. Explain how to express the number of oobles as a mixed number.

2. Explain how 5 dollars and 17 quarters can be written completely in dollar units as a mixed number. Then explain how the mixed number can be simplified.

Diagnostic Exercises

Express each of the following as a whole number or a mixed number.

1. $\frac{18}{6}$ **2.** $\frac{5}{3}$ **3.** $\frac{12}{8}$ **4.** $\frac{19}{4}$

Express each of the following mixed numbers in simplest form.

5. $5\frac{6}{6}$ **6.** $3\frac{12}{16}$ **7.** $9\frac{13}{10}$ **8.** $7\frac{14}{8}$

9. Alice triples a recipe and finds that she needs $\frac{9}{4}$ cups of flour. Write the number of cups she needs as a mixed number.

Related Practice

Express each of the following as a whole number or a mixed number.

1. a. $\frac{8}{8}$ **b.** $\frac{12}{4}$ **c.** $\frac{15}{5}$ **d.** $\frac{36}{9}$ **2. a.** $\frac{7}{5}$ **b.** $\frac{13}{8}$ **c.** $\frac{25}{16}$ **d.** $\frac{13}{10}$

3. a. $\frac{6}{4}$ **b.** $\frac{12}{9}$ **c.** $\frac{28}{16}$ **d.** $\frac{52}{32}$ **4. a.** $\frac{9}{2}$ **b.** $\frac{12}{5}$ **c.** $\frac{25}{8}$ **d.** $\frac{67}{12}$

Express each of the following in simplest form.

5. a. $2\frac{3}{3}$ **b.** $7\frac{2}{2}$ **c.** $12\frac{32}{4}$ **d.** $15\frac{30}{5}$ **6. a.** $4\frac{6}{9}$ **b.** $6\frac{21}{24}$ **c.** $11\frac{16}{30}$ **d.** $14\frac{25}{40}$

7. a. $6\frac{8}{5}$ **b.** $11\frac{7}{6}$ **c.** $40\frac{16}{9}$ **d.** $17\frac{35}{16}$ **8. a.** $3\frac{12}{8}$ **b.** $9\frac{32}{12}$ **c.** $16\frac{15}{6}$ **d.** $45\frac{52}{24}$

9. a. Bill has computed the total length of the boards he needs for 7 stair treads to be $\frac{245}{6}$ feet. Express the number of feet as a mixed number.

b. Sarah has decided that it will take her 345 minutes to complete her carpentry project. This is equivalent to $\frac{345}{60}$ hours. Express the number of hours as a mixed number.

c. Alex finds the solution to a problem to be $\frac{39}{5}$. Bob solves another problem and gets an answer of $7\frac{1}{3}$. Which solution is greater?

d. Naomi adds 3 mixed numbers and gets an answer of $15\frac{25}{8}$. Express this answer in simplest form.

Expressing Fractions in Higher Terms

Since there are many names for a given fraction, you can express a fraction as another with a different denominator. Sometimes when you want to compare, add, or subtract fractions, you will first have to write them in higher terms. You can express a fraction as another with a higher denominator. To compare, add, or subtract fractions, express them with a common denominator.

Anna designed a necklace in which $\frac{1}{3}$ of the beads are to be blue. The necklace will require 18 beads. Express $\frac{1}{3}$ as a fraction with a denominator of 18 and tell how many beads should be blue.

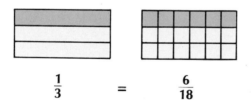

$$\frac{1}{3} \quad = \quad \frac{6}{18}$$

$\frac{1}{3} = \frac{?}{18}$

$18 \div 3 = 6$

Divide the new denominator by the denominator of the given fraction.

$\frac{1}{3} = \frac{1 \times 6}{3 \times 6} = \frac{6}{18}$

Multiply the numerator and denominator of the given fraction by the quotient.

ANSWER: $\frac{1}{3} = \frac{6}{18}$; 6 of the 18 beads in Anna's necklace should be blue.

Examples

1. Express $\frac{9}{25}$ in 100ths.

$\frac{9}{25} = \frac{?}{100}$

$100 \div 25 = 4$

$\frac{9}{25} = \frac{9 \times 4}{25 \times 4} = \frac{36}{100}$

ANSWER: $\frac{36}{100}$

2. Express 5 in 20ths.

$5 = \frac{5}{1}$

$\frac{5}{1} = \frac{?}{20}$

$20 \div 1 = 20$

$\frac{5}{1} = \frac{5 \times 20}{1 \times 20} = \frac{100}{20}$

ANSWER: $\frac{100}{20}$

Express the given whole number as a fraction with a denominator of 1.

1. Can $\frac{1}{3}$ be expressed as a fraction with a denominator of 5 and a whole number in the numerator?

2. To express $\frac{1}{3}$ in higher terms in which the numerator of the new fraction is a whole number, what must be true of the denominator?

3. True or false? One-half cannot be written as a fraction with an odd number in the denominator. Explain.

4. True or false? A whole number can be written as a fraction with any given denominator. Explain.

Diagnostic Exercises

Express the following as fractions having denominators as specified.

1. $\frac{1}{4} = \frac{?}{12}$

2. $\frac{5}{8} = \frac{?}{32}$

3. $\frac{9}{10} = \frac{?}{100}$

4. Express $\frac{3}{5}$ in 100ths.

5. Express 6 in 10ths.

6. Larry must plant 150 flowers at the nursery. The owner told him to make $\frac{2}{5}$ of the plants petunias. How many should be petunias? Express $\frac{2}{5}$ as 150ths.

Related Practice

Express the following as fractions having denominators as specified.

1. a. $\frac{1}{2} = \frac{?}{8}$ b. $\frac{1}{3} = \frac{?}{15}$ c. $\frac{1}{6} = \frac{?}{42}$ d. $\frac{1}{16} = \frac{?}{64}$

2. a. $\frac{2}{3} = \frac{?}{12}$ b. $\frac{3}{4} = \frac{?}{24}$ c. $\frac{9}{32} = \frac{?}{96}$ d. $\frac{11}{16} = \frac{?}{80}$

3. a. $\frac{7}{10} = \frac{?}{100}$ b. $\frac{3}{4} = \frac{?}{100}$ c. $\frac{3}{8} = \frac{?}{100}$ d. $\frac{17}{50} = \frac{?}{100}$

4. a. $\frac{1}{8}$ in 64ths   b. $\frac{1}{4}$ in 32nds   c. $\frac{5}{8}$ in 16ths   d. $\frac{19}{32}$ in 64ths

5. a. $\frac{2}{1}$ in 10ths   b. $\frac{4}{1}$ in 12ths   c. 7 in 100ths   d. 9 in 72nds

6. a. The library is ordering 412 new books, and $\frac{1}{4}$ of these must be children's books. How many children's books are ordered? Express $\frac{1}{4}$ as 412ths.

b. Of the tours offered by the zoo, $\frac{3}{8}$ are given by high-school students. Last weekend, 48 tours were given. How many were given by high school students? $\frac{3}{8} = \frac{?}{48}$

c. Steve baked 9 loaves of bread, and of these, 3 are whole wheat. What fraction of the loaves are *not* whole wheat?

d. In a book on boating $\frac{1}{10}$ of the pages are to have a picture. The book is to have 180 pages. How many will have pictures? Find $\frac{1}{10} = \frac{?}{180}$.

Equivalent Fractions

Fractions that look different can name the same number.
One-half of an apple and $\frac{4}{8}$ of the apple are the same amount.
$\frac{1}{2}$ and $\frac{4}{8}$ are said to be equivalent fractions.

Vocabulary

Equivalent fractions are fractions that name the same number.
The symbol $\neq$ means "is not equal to."

José and Vera each started with the same size strip of copper. José cut his into 15 equal pieces and used 6 pieces. Vera cut hers into 10 equal pieces and used 4 pieces. Did they use the same amount of copper? Are $\frac{6}{15}$ and $\frac{4}{10}$ equivalent fractions?

$$\frac{6}{15} = \frac{6 \div 3}{15 \div 3} = \frac{2}{5}$$
$$\frac{4}{10} = \frac{4 \div 2}{10 \div 2} = \frac{2}{5}$$
same

So, $\frac{6}{15} = \frac{4}{10}$

equivalent fractions

Express each fraction in lowest terms.

If the resulting fractions are the same, the given fractions are equivalent.

ANSWER: José and Vera used the same amount of copper.

Examples

1. Are $\frac{12}{20}$ and $\frac{18}{27}$ equivalent fractions?

If the resulting fractions are not the same, the given fractions are not equivalent.

$$\frac{12}{20} = \frac{12 \div 4}{20 \div 4} = \frac{3}{5}$$
$$\frac{18}{27} = \frac{18 \div 9}{27 \div 9} = \frac{2}{3}$$
not the same   So, $\frac{12}{20} \neq \frac{18}{27}$.

is not equal to

ANSWER: $\frac{12}{20}$ and $\frac{18}{27}$ are not equivalent fractions.

2. Are $\frac{9}{12}$ and $\frac{12}{16}$ equivalent fractions?

first fraction   **second fraction**

$\frac{9}{12} \times \frac{12}{16}$

$9 \times 16 = 144$
$12 \times 12 = 144$

$144 = 144$ So, $\frac{9}{12} = \frac{12}{16}$

Cross-Product Method

Multiply the numerator of the first fraction and the denominator of the second fraction.

Then multiply the numerator of the second fraction and the denominator of the first fraction.

If the products are equal, then the fractions are equivalent.

ANSWER: $\frac{9}{12}$ and $\frac{12}{16}$ are equivalent fractions.

1. In the problem above, José used 6 pieces of copper and Vera used 4 pieces. Explain how they could have used the same amount of copper.

2. You are given two fractions. The denominators of the fractions have no common factors. Are the fractions always, sometimes, or never equivalent?

Diagnostic Exercises

1. Write four fractions equivalent to $\frac{3}{8}$.

2. Write the lowest terms fraction and three other fractions equivalent to $\frac{8}{12}$.

3. Test whether $\frac{14}{35}$ and $\frac{6}{15}$ are equivalent fractions using the method of lowest terms.

4. Test whether $\frac{36}{90}$ and $\frac{10}{25}$ are equivalent fractions using the cross-product method.

5. In Leah's band, $\frac{5}{9}$ of the members sing. In Phil's band, $\frac{6}{11}$ of the members sing. Are $\frac{5}{9}$ and $\frac{6}{11}$ equivalent fractions?

Related Practice

1. Write four equivalent fractions.

  a. $\frac{1}{2}$ **b.** $\frac{3}{5}$ **c.** $\frac{17}{20}$ **d.** $\frac{13}{16}$

2. Write the lowest terms fraction and three other equivalent fractions.

  a. $\frac{7}{42}$ **b.** $\frac{9}{36}$ **c.** $\frac{35}{84}$ **d.** $\frac{48}{54}$

Test whether the following pairs of fractions are equivalent by using the specified method.

3. Lowest-terms method:

  a. $\frac{6}{16}$ and $\frac{15}{40}$ **b.** $\frac{15}{18}$ and $\frac{30}{48}$

  c. $\frac{42}{63}$ and $\frac{16}{24}$ **d.** $\frac{28}{35}$ and $\frac{35}{42}$

4. Cross-product method:

  a. $\frac{16}{28}$ and $\frac{24}{36}$ **b.** $\frac{10}{16}$ and $\frac{15}{24}$

  c. $\frac{15}{25}$ and $\frac{36}{60}$ **d.** $\frac{42}{54}$ and $\frac{28}{35}$

5. a. Of Leroy's 22 baseball cards, 6 are of pitchers. Of Kira's 24 baseball cards, 7 are of pitchers. Are $\frac{6}{22}$ and $\frac{7}{24}$ equivalent fractions?

b. Nancy and Anna each have a stamp collection. Of Nancy's stamps, $\frac{3}{5}$ are from France, while $\frac{6}{10}$ of Anna's are from France. Are $\frac{3}{5}$ and $\frac{6}{10}$ equivalent fractions?

c. In one necklace, $\frac{6}{18}$ of the beads are red. In another, $\frac{7}{21}$ of the beads are red. Are $\frac{6}{18}$ and $\frac{7}{21}$ equivalent fractions?

d. The library orders new books monthly. In May, $\frac{12}{32}$ of the books ordered were novels. In June, $\frac{21}{56}$ of the books ordered were novels. Are $\frac{12}{32}$ and $\frac{21}{56}$ equivalent fractions?

3-5 Multiples

If free popcorn is given as a prize to every 6th person who attends the movies, who gets the popcorn?

$$6 \times 0 = 0$$
$$6 \times 1 = 6$$
$$6 \times 2 = 12$$
$$6 \times 3 = 18$$
← multiples

The three dots indicate that the numbers continue in the same pattern.

To find the multiples of a number, multiply the number by the whole numbers (0, 1, 2, 3, etc.).

ANSWER: The multiples of 6 are 0, 6, 12, 18, . . . So, free popcorn was given to the holders of tickets 6, 12, 18, 24, ...

Examples

1. Find the common multiples of 6 and 8.

Zero is a multiple of every number.

Multiples of 6: 0, 6, 12, 18, 24, 30, 36, 42, 48, . . .   List all the multiples of each number.
Multiples of 8: 0, 8, 16, 24, 32, 40, 48, 56, 64, . . .   Identify the multiples that are common to both.

common multiples

ANSWER: The common multiples of 6 and 8 are: 0, 24, 48, . . .

2. Find the least common multiple (LCM) of 4 and 6.

least common multiple

Multiples of 4: 0, 4, 8, 12, 16, 20, 24, 28, . . .   The least *nonzero* common multiple is the LCM.
Multiples of 6: 0, 6, 12, 18, 24, 30, 36, 42, . . .

Alternate Method:
$$4 = 2 \times 2 = 2^2$$
$$6 = 2 \times 3$$
$$2^2 \times 3 = 4 \times 3 = 12$$

Factor the given numbers as primes. The LCM is the product of the highest power of each factor.

ANSWER: The LCM of 4 and 6 is 12.

Vocabulary

A **multiple** of a given whole number is a product of the number and a whole number. Any multiple of a number is divisible by the number.

A **common multiple** of two or more numbers is any number that is a multiple of all the numbers.

The **least common multiple (LCM)** of two or more numbers is the least nonzero number that is a multiple of all of them. It is the smallest possible whole number (excluding zero) that can be divided exactly by all the given numbers.

1. What number is a common multiple of all whole numbers?

2. What is the least number of common multiples that any two non-zero numbers have? Explain.

Diagnostic Exercises

1. Which of the following numbers are multiples of 9? **a.** 56 **b.** 27 **c.** 81 **d.** 109

2. Write all the multiples of 7, listing the first five multiples.

3. Name the first four common multiples of 20 and 25.

4. Write all the common multiples of 2, 6, and 8, listing the first six nonzero common multiples.

Write the least common multiple of each group.
5. 2 and 3   6. 8, 4, and 16   7. 12 and 18

8. A teacher insists on complete rows in her classroom and 7 desks in each row. How many desks could be in her classroom?

Related Practice

1. Which of the following are multiples?

   a. Of 3:   19   42   51   73
   b. Of 10: 25   40   135   200
   c. Of 12: 36   54   90   132
   d. Of 25: 75   150   215   340

2. Write all the multiples of each of the following, listing the first six multiples.

   a. 5   **b.** 11   **c.** 2   **d.** 24

3. **a.** Write the first three common multiples of 3 and 5.
   b. Write the first five common multiples of 8 and 4.
   c. Write the first four common multiples of 10 and 12.
   d. Write the first six common multiples of 8, 36, and 18.

4. Write all the common multiples of each of the following, listing the first four nonzero common multiples.

   a. 4 and 5   **b.** 2 and 16   **c.** 8 and 12   **d.** 6, 8, and 16

Write the least common multiple of each group.

5. **a.** 3 and 8   **b.** 7 and 2   **c.** 2, 3, and 5   **d.** 4, 5, and 7

6. **a.** 6 and 12   **b.** 25 and 100   **c.** 4, 8, and 16   **d.** 12, 3, and 24

7. **a.** 16 and 24   **b.** 15 and 20   **c.** 8, 10, and 12   **d.** 18, 45, and 54

8. **a.** Mr. Yaniz planted flowers in every 5th row of his garden. Which rows were flowers?

   b. Every 15th person in the audience was selected to be a game contestant on a TV show. Which people were contestants?

   c. At a picnic every third person was picked for the tug-of-war contest and every 10th person was chosen for the balloon contest. If there were 100 people at the picnic, which people were in both contests?

   d. Every 75th person at the concert was given a free album. Every 100th person was given a free ticket. If there were 1,250 people at the concert, how many people received both prizes?

Comparing Fractions and Rounding Mixed Numbers

Angela painted $\frac{2}{5}$ of a wall and Louise painted $\frac{1}{2}$ of the wall.
Who painted less?

1. Diagram Method

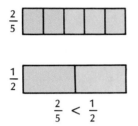

Draw a diagram.

You can see that $\frac{2}{5} < \frac{1}{2}$.

$$\frac{2}{5} < \frac{1}{2}$$

ANSWER: Angela painted less.

---

2. Equivalent Fraction Method

$\frac{2}{5} = \frac{4}{10}$

$\frac{1}{2} = \frac{5}{10}$ $\frac{4}{10} < \frac{5}{10}$

Write as equivalent fractions with the same denominator.

$4 < 5$, so $\frac{4}{10} < \frac{5}{10}$

ANSWER: Angela painted less.

---

3. Cross-Product Method

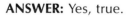 $2 \times 2 = 4$
$5 \times 1 = 5$ $4 < 5$, so $\frac{2}{5} < \frac{1}{2}$

Multiply the numerator of the first fraction by the denominator of the second. Then multiply the denominator of the first fraction by the numerator of the second. If the first product is less than the second, the first fraction is less.

ANSWER: Angela painted less.

Examples

1. Is the statement $\frac{3}{10} > \frac{1}{4}$ true?

$\frac{3}{10} = \frac{6}{20}$ and $\frac{1}{4} = \frac{5}{20}$

$\frac{6}{20} > \frac{5}{20}$, so $\frac{3}{10} > \frac{1}{4}$

ANSWER: Yes, true.

2. Is the statement $\frac{3}{5} > \frac{7}{12}$ true?

first   second
fraction   fraction

$\frac{3}{5}$ $\frac{7}{12}$ $3 \times 12 = 36$
$5 \times 7 = 35$

$36 > 35$, so the statement is true.

ANSWER: Yes, true.

---

3. Round $5\frac{3}{4}$ to the nearest whole number.
$\frac{3}{4}$ is greater than $\frac{1}{2}$
Therefore, $5\frac{3}{4}$ rounds to 6.

ANSWER: 6

If the fractional part of a mixed number is $\frac{1}{2}$ or more, round the mixed number up to the next whole number.

4. Arrange in order from least to greatest: $\frac{3}{4}, \frac{7}{8}, \frac{2}{3}$.

$\frac{3}{4} = \frac{18}{24}$

$\frac{7}{8} = \frac{21}{24}$ $\frac{16}{24} < \frac{18}{24} < \frac{21}{24}$ so $\frac{2}{3} < \frac{3}{4} < \frac{7}{8}$

$\frac{2}{3} = \frac{16}{24}$

Express the fractions as equivalent fractions having the same denominator.

ANSWER: $\frac{2}{3}, \frac{3}{4}, \frac{7}{8}$

---

5. Round $8.76\frac{1}{3}$ to the nearest cent.
$\frac{1}{3}$¢ is less than $\frac{1}{2}$¢
Therefore, $8.76\frac{1}{3}$ rounds to $8.76.

If the fractional part is less than $\frac{1}{2}$, drop the fraction.

ANSWER: $8.76

Think About It

1. If the numerators of two fractions with different denominators are equal, how do the fractions compare?

2. Rewrite the statement $\frac{1}{2} > \frac{1}{4}$ another way without using words.

Vocabulary

The symbol $\nless$ means "is not less than."

The symbol $\ngtr$ means "is not greater than."

Calculator Know-How

The cross-product method for comparing fractions is easy to do on the calculator.

Example
Which fraction is greater, $\frac{13}{17}$ or $\frac{3}{4}$?

Key: 13 $\boxed{\times}$ 4 $\boxed{=}$ $\underline{52}$

 17 $\boxed{\times}$ 3 $\boxed{=}$ $\underline{51}$

Since $52 > 51$, then $\frac{13}{17} > \frac{3}{4}$.

Multiply the numerator of the first fraction by the denominator of the second. Write the product.

Multiply the denominator of the first fraction by the numerator of the second.
Compare the products.

Exercises

Use your calculator and the cross-product method. Write a number sentence using $>$, $=$, or $<$.

1. $\frac{4}{9} \bullet \frac{3}{8}$

2. $\frac{5}{25} \bullet \frac{30}{150}$

3. $\frac{9}{27} \bullet \frac{1}{4}$

4. $\frac{15}{22} \bullet \frac{42}{60}$

5. $\frac{11}{81} \bullet \frac{24}{100}$

6. $\frac{14}{30} \bullet \frac{35}{70}$

7. $\frac{12}{81} \bullet \frac{4}{9}$

8. $\frac{18}{50} \bullet \frac{27}{63}$

9. Make up two exercises like those above. Exchange exercises with a classmate.

Diagnostic Exercises

1. Which is greater? Draw a diagram.

   a. $\frac{1}{3}$ or $\frac{1}{2}$ **b.** $\frac{5}{6}$ or $\frac{7}{8}$

2. Which is less?

   a. $\frac{1}{4}$ or $\frac{1}{10}$ **b.** $\frac{5}{8}$ or $\frac{11}{16}$

3. a. Is the statement $\frac{4}{5} < \frac{5}{6}$ true?   **b.** Is the statement $\frac{1}{3} > \frac{3}{8}$ true?

4. True or false? Use the cross-product method.

   a. $\frac{3}{4} > \frac{5}{8}$ **b.** $\frac{2}{3} < \frac{3}{5}$

5. Arrange in order of size (least first): $\frac{3}{5}$, $\frac{7}{12}$, and $\frac{1}{2}$.

6. Round to the nearest whole number or cent.

   a. $2\frac{7}{12}$ **b.** $6\frac{2}{5}$ **c.** $\$6.82\frac{4}{9}$ **d.** $\$21.09\frac{5}{6}$

7. If Alejandre has read $\frac{2}{3}$ of the assigned book and Cori has read $\frac{5}{8}$ of the book, who is closer to finishing the book?

Related Practice

1. Which is greater? Draw a diagram.

   a. $\frac{1}{4}$ or $\frac{1}{3}$ **b.** $\frac{4}{5}$ or $\frac{3}{4}$ **c.** $\frac{3}{8}$ or $\frac{1}{4}$ **d.** $\frac{13}{16}$ or $\frac{5}{6}$

2. Which is less?

   a. $\frac{1}{2}$ or $\frac{1}{6}$ **b.** $\frac{5}{16}$ or $\frac{1}{4}$ **c.** $\frac{2}{3}$ or $\frac{11}{16}$ **d.** $\frac{7}{10}$ or $\frac{3}{4}$

3. Which of the following statements are true?

   a. $\frac{1}{10} > \frac{1}{8}$ **b.** $\frac{2}{3} < \frac{7}{12}$ **c.** $\frac{3}{8} > \frac{5}{12}$ **d.** $\frac{13}{16} < \frac{19}{24}$

4. True or false? Use the cross-product method.

   a. $\frac{1}{8} > \frac{1}{6}$ **b.** $\frac{7}{12} > \frac{9}{16}$ **c.** $\frac{15}{24} > \frac{21}{30}$ **d.** $\frac{18}{48} > \frac{27}{81}$

5. Arrange in order of size (least first).

   a. $\frac{1}{4}$, $\frac{1}{2}$, and $\frac{1}{6}$ **b.** $\frac{1}{2}$, $\frac{2}{5}$, and $\frac{3}{10}$ **c.** $\frac{11}{16}$, $\frac{5}{8}$, and $\frac{3}{4}$ **d.** $\frac{5}{8}$, $\frac{3}{5}$, and $\frac{3}{4}$

6. Round to the nearest whole number or cent.

   a. $8\frac{9}{16}$ **b.** $15\frac{3}{7}$ **c.** $\$9.70\frac{1}{2}$ **d.** $\$.86\frac{5}{12}$

7. a. Jeannie ate $\frac{3}{8}$ of a pizza, and Suzanne ate $\frac{1}{4}$ of the same pizza. Who ate less pizza?

   b. Two traveling salespersons started with the same amount of merchandise. If Carrie sold $\frac{3}{7}$ of her merchandise while Ira sold $\frac{4}{9}$ of his merchandise, who sold more that day?

3-7 Least Common Denominator

One way to compare fractions with different denominators is to rename the fractions with a least common denominator (LCD).

> Barb and Jonathan earn the same amount. Barb is saving $\frac{5}{12}$ of her earnings for college, and Jonathan is saving $\frac{4}{9}$ of his earnings for college. Who is saving more?

Compare $\frac{5}{12}$ and $\frac{4}{9}$ by finding the least common denominator.

9: 0, 9, 18, 27, 36, 45

12: 0, 12, 24, 36

List the multiples of each denominator until you find a common multiple (not zero).

The LCD is 36.

$\frac{4}{9} = \frac{?}{36}$ $36 \div 9 = 4$ $\frac{5}{12} = \frac{?}{36}$ $36 \div 12 = 3$

$\frac{4}{9} = \frac{4 \times 4}{9 \times 4} = \frac{16}{36}$ $\frac{5}{12} = \frac{5 \times 3}{12 \times 3} = \frac{15}{36}$

$\frac{16}{36} > \frac{15}{36}$

ANSWER: $\frac{4}{9} > \frac{5}{12}$; Jonathan is saving more.

Examples

1. Write $\frac{3}{5}$ and $\frac{7}{8}$ with the least common denominator.

8: 0, 8, 16, 24, 32, 40, 48

5: 0, 5, 10, 15, 20, 25, 30, 35, 40

$40 \div 5$

$\frac{3}{5} = \frac{?}{40}$ $\frac{3}{5} = \frac{3 \times 8}{5 \times 8} = \frac{24}{40}$

$\frac{7}{8} = \frac{?}{40}$ $\frac{7}{8} = \frac{7 \times 5}{8 \times 5} = \frac{35}{40}$

ANSWER: LCD = 40; $\frac{3}{5} = \frac{24}{40}$, $\frac{7}{8} = \frac{35}{40}$.

2. Write $\frac{2}{3}$, $\frac{7}{8}$, and $\frac{1}{6}$ with the least common denominator.

3: 0, 3, 6, 9, 12, 15, 18, 21, 24

8: 0, 8, 16, 24

6: 0, 6, 12, 18, 24   The LCD is 24.

$\frac{2}{3} = \frac{?}{24}$ $\frac{2}{3} = \frac{2 \times 8}{3 \times 8} = \frac{16}{24}$

$\frac{7}{8} = \frac{?}{24}$ $\frac{7}{8} = \frac{7 \times 3}{8 \times 3} = \frac{21}{24}$

$\frac{1}{6} = \frac{?}{24}$ $\frac{1}{6} = \frac{1 \times 4}{6 \times 4} = \frac{4}{24}$

ANSWER: LCD = 24; $\frac{2}{3} = \frac{16}{24}$, $\frac{7}{8} = \frac{21}{24}$, and $\frac{1}{6} = \frac{4}{24}$.

Vocabulary

The **least common denominator (LCD)** is the least possible whole number (excluding zero) that can be divided exactly by the denominators of all the given fractions.

Think About It

1. If the denominator of one fraction is a factor of the denominator of the other, what is their LCD? Give an example.

2. If two denominators have no common factors other than 1, find and describe a shortcut method for finding the LCD.

Diagnostic Exercises

Find the least common denominator (LCD) of the given fractions, then express the given fractions as equivalent fractions using the (LCD) as their new denominator.

1. $\frac{1}{2}$ and $\frac{3}{8}$ **2.** $\frac{3}{4}$ and $\frac{2}{3}$ **3.** $\frac{7}{10}$ and $\frac{9}{16}$ **4.** $\frac{5}{6}$, $\frac{4}{5}$, and $\frac{1}{3}$

5. On a test, Matt answered $\frac{2}{3}$ of the questions correctly, Marita answered $\frac{7}{8}$ correctly, and Cindy answered $\frac{5}{6}$ correctly. Who had the highest score?

Related Practice

Find the least common denominator (LCD) of the given fractions, then express the given fractions as equivalent fractions using the LCD as their new denominator:

1. a. $\frac{1}{2}$ and $\frac{1}{4}$ **b.** $\frac{3}{5}$ and $\frac{9}{10}$ **c.** $\frac{1}{4}$ and $\frac{9}{16}$ **d.** $\frac{17}{20}$ and $\frac{4}{5}$

2. a. $\frac{1}{3}$ and $\frac{1}{4}$ **b.** $\frac{2}{5}$ and $\frac{5}{6}$ **c.** $\frac{8}{9}$ and $\frac{3}{4}$ **d.** $\frac{1}{6}$ and $\frac{4}{7}$

3. a. $\frac{1}{4}$ and $\frac{1}{6}$ **b.** $\frac{13}{20}$ and $\frac{7}{12}$ **c.** $\frac{19}{24}$ and $\frac{13}{18}$ **d.** $\frac{11}{12}$ and $\frac{27}{32}$

4. a. $\frac{1}{2}$, $\frac{1}{3}$, and $\frac{1}{4}$ **b.** $\frac{4}{5}$, $\frac{1}{4}$, and $\frac{2}{3}$ **c.** $\frac{9}{16}$, $\frac{3}{8}$, and $\frac{13}{24}$ **d.** $\frac{7}{8}$, $\frac{9}{10}$, and $\frac{1}{12}$

5. a. Moira plants vegetables in $\frac{5}{9}$ of her garden and fruit in $\frac{1}{5}$ of her garden. Compare the parts of her garden.

b. Elizabeth collects coins. Two-fifths of her collection is quarters and $\frac{3}{11}$ of her collection is foreign coins. Does she have more quarters or foreign coins?

Adding Fractions and Mixed Numbers

Building Problem Solving Skills

John worked on his math homework for $\frac{3}{4}$ hour and his science homework for $\frac{7}{8}$ hour. How long did he spend in all doing homework?

Addition Indicators
• how much in all
• how many in all
• sum
• total
• how much altogether
Note: All indicators usually refer to *unequal* amounts.

1. **READ** the problem carefully.

a. Find the question asked.
How long did John spend doing his homework?

b. Find the given facts.
$\frac{3}{4}$ hour and $\frac{7}{8}$ hour spent doing homework.

---

2. **PLAN** how to solve the problem.

a. Choose the operation needed.
The words "in all" indicate addition.

c. Express your plan as an equation.
$\frac{3}{4} + \frac{7}{8} = n$

b. Think or write out your plan relating the given facts to the question asked.
The sum of $\frac{3}{4}$ and $\frac{7}{8}$ hour is equal to the total number of hours spent doing homework.

---

3. **SOLVE** the problem.

a. Estimate the answer.

$$\frac{3}{4} \rightarrow \quad 1$$
$$\frac{7}{8} \rightarrow + 1$$
$$\overline{\quad 2 \text{ hours}}$$

Round each fraction to the nearest whole number. Add.

b. Solution.

$$\frac{3}{4} = \quad \frac{6}{8}$$
$$\frac{7}{8} = + \frac{7}{8}$$
$$\overline{\quad \frac{13}{8} = 1\frac{5}{8} \text{ hours}}$$

ANSWER: John spent $1\frac{5}{8}$ hours doing his homework.

Determine the LCD and write the original fractions as equivalent fractions using the LCD.

Add the new numerators. Write the sum over the LCD. Express the solution in simplest form.

You might use a calculator.

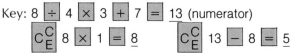

Key: 8 ÷ 4 × 3 + 7 = 13 (numerator)

C$_E^C$ 8 × 1 = 8   C$_E^C$ 13 − 8 = 5

4. **CHECK** your solution.

a. Check the accuracy of your arithmetic.
Check the units.

$$\frac{3}{4} = \frac{6}{8}$$
$$\frac{7}{8} = +\frac{7}{8}$$
$$\frac{13}{8} = 1\frac{5}{8} ✔$$

Check by adding up.

Check that the answer is in simplest form.

b. See if your answer is reasonable.
The answer $1\frac{5}{8}$ hours compares reasonably
with the estimate of 2 hours.

Think About It

1. When is the simplest form of an improper fraction a whole number?

2. How do you decide if a fraction is in simplest form?

3. How do you add fractions with unlike denominators?

Calculator Know-How

Since it is easy to compute with large numbers on a calculator, you do *not* have to find the LCD before adding unlike fractions or mixed numbers.

$$\frac{1}{4} + \frac{3}{8} + \frac{5}{9} = n$$

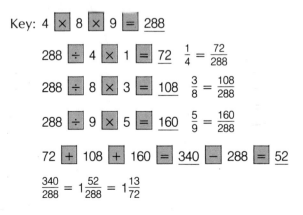

Key: 4 × 8 × 9 = 288

288 ÷ 4 × 1 = 72 $\frac{1}{4} = \frac{72}{288}$

288 ÷ 8 × 3 = 108 $\frac{3}{8} = \frac{108}{288}$

288 ÷ 9 × 5 = 160 $\frac{5}{9} = \frac{160}{288}$

72 + 108 + 160 = 340 − 288 = 52

$$\frac{340}{288} = 1\frac{52}{288} = 1\frac{13}{72}$$

Multiply *all* the denominators to find a common denominator.
Use the common denominator to rename the fractions.
Write the new fractions.

Add the numerators. Express your answer in simplest form.

1. Make up an exercise adding 3 fractions. Find the sum using your calculator.

2. Do the example above on a calculator using the memory. Write the keying sequence.

1. Add: $\frac{5}{8} + \frac{1}{8}$

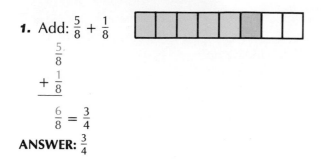

$$\begin{array}{r} \frac{5}{8} \\ + \frac{1}{8} \\ \hline \frac{6}{8} = \frac{3}{4} \end{array}$$

To add fractions that have a common denominator, add the numerators. Then write the sum over the common denominator.

Express the answer in lowest terms.

ANSWER: $\frac{3}{4}$

2. Add: $\frac{11}{12} + \frac{5}{12}$

$$\begin{array}{r} \frac{11}{12} \\ + \frac{5}{12} \\ \hline \frac{16}{12} = 1\frac{4}{12} = 1\frac{1}{3} \end{array}$$

Simplify.

ANSWER: $1\frac{1}{3}$

3. Add: $\frac{4}{5} + \frac{7}{10}$

$$\begin{array}{l} \frac{4}{5} = \frac{8}{10} \\ \frac{7}{10} = \frac{7}{10} \\ \quad \frac{8}{10} \\ + \frac{7}{10} \\ \hline \frac{15}{10} = 1\frac{5}{10} = 1\frac{1}{2} \end{array}$$

To add fractions that do *not* have a common denominator, find the LCD. Express the fractions as equivalent fractions using the LCD.
Add.

Simplify.

ANSWER: $1\frac{1}{2}$

4. Add: $2\frac{1}{2} + 5\frac{2}{3}$

$$\begin{array}{r} 2\frac{1}{2} = 2\frac{3}{6} \\ + 5\frac{2}{3} = 5\frac{4}{6} \\ \hline \frac{7}{6} \end{array}$$

To add mixed numbers, first add the fractions.

$$\begin{array}{r} 2\frac{3}{6} \\ + 5\frac{4}{6} \\ \hline 7\frac{7}{6} = 8\frac{1}{6} \end{array}$$

Then add the whole numbers. Simplify.

ANSWER: $8\frac{1}{6}$

5. Add: $9\frac{7}{8} + \frac{13}{16}$

$$9\frac{7}{8} = 9\frac{14}{16}$$
$$+ \ \frac{13}{16} = \ \frac{13}{16}$$
$$9\frac{27}{16} = 9 + 1\frac{11}{16} = 10\frac{11}{16}$$

ANSWER: $10\frac{11}{16}$

6. Add: $13\frac{1}{6} + 5\frac{1}{3} + 9\frac{1}{2}$

$$13\frac{1}{6} = 13\frac{1}{6}$$
$$5\frac{1}{3} = \ 5\frac{2}{6}$$
$$+ \ 9\frac{1}{2} = \ 9\frac{3}{6}$$
$$27\frac{6}{6} = 27 + 1 = 28$$

ANSWER: 28

---

7. Solve the equation: $7\frac{3}{5} + 8\frac{2}{3} = n$

$$7\frac{3}{5} = \ 7\frac{9}{15}$$
$$+ \ 8\frac{2}{3} = \ 8\frac{10}{15}$$
$$15\frac{19}{15} = 16\frac{4}{15}$$

To find the value of n, add the given numbers.

ANSWER: $16\frac{4}{15}$

Diagnostic Exercises

Add.

1. $\dfrac{1}{3}$
$+ \dfrac{1}{3}$

2. $\dfrac{5}{16}$
$+ \dfrac{7}{16}$

3. $\dfrac{5}{6}$
$+ \dfrac{1}{6}$

4. $\dfrac{3}{5}$
$+ \dfrac{4}{5}$

5. $\dfrac{3}{4}$
$+ \dfrac{3}{4}$

6. $\dfrac{1}{2}$
$+ \dfrac{3}{8}$

7. $\dfrac{2}{5}$
$+ \dfrac{3}{4}$

8. $\dfrac{7}{8}$
$+ \dfrac{5}{6}$

9. $\dfrac{3}{10}$
$\dfrac{1}{2}$
$+ \dfrac{4}{5}$

10. 3
$+ \dfrac{1}{8}$

11. $2\dfrac{3}{4}$
$+ 5$

12. $3\dfrac{1}{5}$
$+ 4\dfrac{3}{5}$

13. $2\dfrac{3}{10}$
$+ 3\dfrac{1}{10}$

14. $16\dfrac{5}{8}$
$+ 23\dfrac{3}{8}$

15. $7\dfrac{2}{5}$
$+ 9\dfrac{4}{5}$

16. $5\dfrac{7}{12}$
$+ 6\dfrac{11}{12}$

17. $8\dfrac{1}{6}$
$+ \dfrac{5}{6}$

18. $12\dfrac{2}{3}$
$+ 5\dfrac{1}{4}$

19. $6\dfrac{1}{12}$
$+ 8\dfrac{1}{6}$

20. $24\dfrac{9}{10}$
$+ 17\dfrac{5}{8}$

21. $15\dfrac{2}{3}$
$+ 9\dfrac{5}{6}$

22. $6\dfrac{7}{8}$
$+ \dfrac{11}{12}$

23. $\dfrac{1}{2}$
$3\dfrac{3}{4}$
$+ \dfrac{7}{8}$

24. $4\dfrac{2}{3}$
$6\dfrac{1}{2}$
$+ 5\dfrac{5}{8}$

25. $\dfrac{7}{8} + \dfrac{7}{12} + \dfrac{1}{6}$

26. $2\dfrac{5}{6} + 3\dfrac{1}{10} + 4\dfrac{1}{2}$

27. Find the sum of $6\frac{1}{4}$, $3\frac{13}{16}$, and $7\frac{3}{8}$.

28. Solve: $2\dfrac{9}{10} + 4\dfrac{1}{2} = n$

29. Marsha bought $1\frac{3}{4}$ pounds of cashews, $\frac{5}{8}$ pound of almonds, and $1\frac{1}{2}$ of peanuts. How many pounds of nuts did she buy?

Related Practice

Add.

1. a. $\frac{2}{5}$ $+\frac{2}{5}$ **b.** $\frac{3}{7}$ $+\frac{2}{7}$ **c.** $\frac{6}{25}$ $+\frac{8}{25}$ **d.** $\frac{3}{6}$ $+\frac{2}{6}$ **2. a.** $\frac{3}{8}$ $+\frac{1}{8}$ **b.** $\frac{7}{12}$ $+\frac{1}{12}$ **c.** $\frac{13}{32}$ $+\frac{15}{32}$ **d.** $\frac{27}{64}$ $+\frac{21}{64}$

3. a. $\frac{1}{2}$ $+\frac{1}{2}$ **b.** $\frac{1}{3}$ $+\frac{2}{3}$ **c.** $\frac{27}{32}$ $+\frac{5}{32}$ **d.** $\frac{11}{20}$ $+\frac{9}{20}$ **4. a.** $\frac{2}{3}$ $+\frac{2}{3}$ **b.** $\frac{4}{9}$ $+\frac{7}{9}$ **c.** $\frac{9}{16}$ $+\frac{10}{16}$ **d.** $\frac{4}{10}$ $+\frac{9}{10}$

5. a. $\frac{5}{6}$ $+\frac{5}{6}$ **b.** $\frac{3}{8}$ $+\frac{7}{8}$ **c.** $\frac{17}{24}$ $+\frac{19}{24}$ **d.** $\frac{15}{32}$ $+\frac{29}{32}$ **6. a.** $\frac{1}{2}$ $+\frac{1}{4}$ **b.** $\frac{5}{6}$ $+\frac{5}{12}$ **c.** $\frac{1}{10}$ $+\frac{2}{5}$ **d.** $\frac{13}{32}$ $+\frac{3}{4}$

7. a. $\frac{2}{3}$ $+\frac{3}{4}$ **b.** $\frac{1}{2}$ $+\frac{1}{3}$ **c.** $\frac{2}{5}$ $+\frac{1}{6}$ **d.** $\frac{7}{8}$ $+\frac{3}{5}$ **8. a.** $\frac{5}{12}$ $+\frac{3}{12}$ **b.** $\frac{3}{4}$ $+\frac{5}{6}$ **c.** $\frac{1}{8}$ $+\frac{7}{12}$ **d.** $\frac{13}{24}$ $+\frac{9}{16}$

9. a. $\frac{3}{8}$ $\frac{4}{8}$ $+\frac{1}{8}$ **b.** $\frac{2}{3}$ $\frac{3}{4}$ $+\frac{5}{6}$ **c.** $\frac{3}{16}$ $\frac{1}{8}$ $+\frac{1}{4}$ **d.** $\frac{7}{24}$ $\frac{3}{16}$ $+\frac{5}{12}$ **10. a.** $8\frac{7}{8}$ $+$ **b.** $12\frac{1}{6}$ $+$ **c.** $7\frac{3}{16}$ $+$ **d.** $18\frac{4}{7}$

11. a. $6\frac{3}{5}$ $+7$ **b.** $9\frac{5}{8}$ $+4$ **c.** 8 $+7\frac{1}{2}$ **d.** 23 $+8\frac{9}{32}$ **12. a.** $4\frac{1}{4}$ $+3\frac{2}{4}$ **b.** $7\frac{2}{9}$ $+2\frac{5}{9}$ **c.** $10\frac{4}{8}$ $+9\frac{3}{8}$ **d.** $32\frac{3}{10}$ $+14\frac{6}{10}$

13. a. $5\frac{1}{8}$ $+8\frac{5}{8}$ **b.** $6\frac{7}{32}$ $+3\frac{9}{32}$ **c.** $10\frac{7}{16}$ $+16\frac{7}{16}$ **d.** $17\frac{1}{2}$ $+11\frac{7}{12}$ **14. a.** $6\frac{3}{4}$ $+5\frac{1}{4}$ **b.** $7\frac{5}{8}$ $+7\frac{3}{8}$ **c.** $13\frac{1}{6}$ $+12\frac{5}{6}$ **d.** $17\frac{3}{16}$ $+22\frac{13}{16}$

15. a. $1\frac{2}{3}$ $+2\frac{2}{3}$ **b.** $4\frac{3}{5}$ $+7\frac{4}{5}$ **c.** $32\frac{5}{9}$ $+13\frac{8}{9}$ **d.** $4\frac{7}{16}$ $+5\frac{12}{16}$ **16. a.** $8\frac{3}{4}$ $+4\frac{3}{4}$ **b.** $9\frac{8}{10}$ $+8\frac{7}{10}$ **c.** $5\frac{5}{6}$ $+24\frac{5}{6}$ **d.** $6\frac{3}{8}$ $+16\frac{7}{8}$

17. a. $6\frac{1}{5}$ **b.** $8\frac{5}{8}$ **c.** $\frac{1}{4}$ **d.** $\frac{9}{16}$ **18. a.** $8\frac{1}{8}$ **b.** $6\frac{1}{6}$ **c.** $4\frac{3}{10}$ **d.** $15\frac{3}{4}$
$+\ \frac{1}{5}$ $+\ \frac{7}{8}$ $+\ 7\frac{3}{4}$ $+\ 18\frac{5}{16}$ $+\ 2\frac{1}{3}$ $+\ 9\frac{5}{8}$ $+\ 8\frac{5}{12}$ $+\ 29\frac{3}{16}$

19. a. $3\frac{1}{3}$ **b.** $4\frac{1}{4}$ **c.** $8\frac{1}{5}$ **d.** $22\frac{7}{12}$ **20. a.** $9\frac{1}{2}$ **b.** $2\frac{2}{3}$ **c.** $8\frac{3}{4}$ **d.** $16\frac{11}{12}$
$+\ 3\frac{1}{6}$ $+\ 8\frac{5}{12}$ $+\ 6\frac{3}{10}$ $+\ 17\frac{1}{6}$ $+\ 5\frac{3}{4}$ $+\ 6\frac{4}{5}$ $+\ 3\frac{7}{8}$ $+\ 14\frac{5}{8}$

21. a. $4\frac{4}{5}$ **b.** $3\frac{3}{4}$ **c.** $4\frac{5}{6}$ **d.** $21\frac{17}{20}$ **22. a.** $7\frac{1}{4}$ **b.** $9\frac{2}{3}$ **c.** $\frac{3}{8}$ **d.** $\frac{5}{6}$
$+\ 6\frac{7}{10}$ $+\ 7\frac{5}{12}$ $+\ 3\frac{1}{2}$ $+\ 37\frac{2}{5}$ $+\ \frac{7}{16}$ $+\ \frac{1}{2}$ $+\ 14\frac{9}{10}$ $+\ 19\frac{2}{3}$

23. a. $\frac{5}{6}$ **b.** $7\frac{3}{4}$ **c.** $\frac{1}{3}$ **d.** $6\frac{7}{12}$ **24. a.** $5\frac{1}{8}$ **b.** $3\frac{2}{3}$ **c.** $4\frac{1}{3}$ **d.** $18\frac{5}{8}$
$\frac{2}{3}$ $\frac{5}{8}$ $8\frac{2}{5}$ $\frac{3}{8}$ $1\frac{1}{4}$ $2\frac{1}{6}$ $8\frac{3}{8}$ $13\frac{11}{12}$
$+\ 2\frac{1}{2}$ $+\ 3\frac{13}{16}$ $+\ \frac{1}{4}$ $+\ 5\frac{1}{6}$ $+\ 4\frac{1}{2}$ $+\ 2\frac{7}{12}$ $+\ 2\frac{3}{4}$ $+\ 42\frac{3}{16}$

25. a. $\frac{1}{4} + \frac{1}{8}$ **b.** $\frac{2}{3} + \frac{1}{6}$ **c.** $\frac{1}{2} + \frac{9}{10} + \frac{3}{5}$ **d.** $\frac{2}{3} + \frac{5}{6} + \frac{5}{8}$

26. a. $2\frac{3}{8} + 1\frac{1}{4}$ **b.** $6\frac{1}{4} + 3\frac{3}{8} + 1\frac{7}{16}$ **c.** $7\frac{2}{3} + 5\frac{1}{6} + 3\frac{1}{12}$ **d.** $23\frac{9}{10} + 12\frac{3}{8} + 9\frac{3}{4}$

27. Find the sum.

  a. $\frac{2}{3}$ and $\frac{7}{12}$ **b.** $9\frac{1}{5}$, $6\frac{3}{10}$, and $4\frac{1}{2}$ **c.** $3\frac{7}{8}$, $4\frac{3}{4}$, and $2\frac{9}{16}$ **d.** $4\frac{3}{4}$, $5\frac{9}{10}$, and $2\frac{1}{6}$

28. Solve.

  a. $\frac{2}{3} + \frac{5}{6} = ?$ **b.** $n = 10\frac{1}{2} + 6\frac{2}{3}$ **c.** $n = 3\frac{1}{4} + 2\frac{5}{12} + 4\frac{2}{3}$ **d.** ▓ $= 4\frac{3}{8} + 5\frac{13}{32}$

29. a. A tailor made a two-piece dress, requiring $2\frac{7}{8}$ yards of material for one part and $1\frac{3}{4}$ yards for the other. How much material did he use?

  b. Carmen works after school. Last week, she worked $3\frac{3}{4}$ hours on Monday, $2\frac{1}{2}$ hours on Wednesday, and 4 hours on Friday. How many hours did she work altogether?

  c. What is the overall length of a certain machine part consisting of 3 joined pieces measuring $2\frac{9}{16}$ inches, $1\frac{27}{32}$ inches, and $\frac{7}{8}$ inch respectively?

  d. In installing an oil burner, Sara finds it necessary to use pieces of pipe measuring $4\frac{11}{16}$ inches, $7\frac{5}{8}$ inches, $3\frac{1}{2}$ inches, and 9 inches. What length of pipe does she need to cut the four pieces, disregarding waste?

Subtracting Fractions and Mixed Numbers

Building Problem Solving Skills

When Helen left her driveway, the gas tank in her car was $\frac{3}{4}$ full. When she returned home, the tank was $\frac{1}{3}$ full. What fraction of a tank did she use on her trip?

1. **READ** the problem carefully.

a. Find the question asked.
What fraction of a tank of gas did Helen use?

b. Find the given facts.
The tank was $\frac{3}{4}$ full when Helen started; it was $\frac{1}{3}$ full when she returned home.

Subtraction Indicators
· how many more
· how many fewer
· how much larger
· how much smaller
· amount of increase
· amount of decrease
· difference
· balance
· how many left over
· how much left over

2. $\boxed{\textbf{PLAN}}$ how to solve the problem.

a. Choose the operation needed.
The words "what fraction used" indicate subtraction.

b. Think or write out your plan relating the given facts to the
question asked.
The difference between $\frac{3}{4}$ and $\frac{1}{3}$ is the amount of gas used.

c. Express your plan as an equation.
$$\frac{3}{4} - \frac{1}{3} = n$$

---

3. $\boxed{\textbf{SOLVE}}$ the problem.

a. Estimate the answer. Since you are subtracting two quantities
less than 1, the answer will be less than 1 and less than the
greater quantity.

b. Solution.

$$\frac{3}{4} = \frac{9}{12}$$
$$-\frac{1}{3} = -\frac{4}{12}$$
$$\frac{5}{12}$$

Determine the LCD and write the original fractions
as equivalent fractions using the LCD.

You might use a calculator.

Key: 12 $\boxed{\div}$ 3 $\boxed{\times}$ 1 $\boxed{\text{M+}}$ 12 $\boxed{\div}$ 4 $\boxed{\times}$ 3 $\boxed{\text{M−}}$

$\boxed{\text{M}^{\text{R}}_{\text{C}}}$ 5 (numerator)

ANSWER: Helen used $\frac{5}{12}$ of a tank of gas.

---

4. $\boxed{\textbf{CHECK}}$ the solution.

Check the accuracy of your arithmetic. Check the units.

$$\frac{5}{12} = \frac{5}{12}$$
$$+\frac{1}{3} = +\frac{4}{12}$$
$$\frac{9}{12} = \frac{3}{4}$$

Check by adding. The amount used plus the amount
remaining equals the amount Helen started with.

Think About It

1. A quick way to tell whether a fraction is more
or less than $\frac{1}{2}$ is to double the numerator. If
the doubled numerator is greater than the
denominator, the fraction is greater than $\frac{1}{2}$.
Explain.

2. If the doubled numerator equals the
denominator, what do you know about the
fraction?

3. Think of a rule like the one above for telling
whether a fraction is less than or greater than
$\frac{1}{3}$.

1. Subtract: $\frac{4}{5} - \frac{1}{5}$

$$\frac{4}{5} - \frac{1}{5} = \frac{4-1}{5} = \frac{3}{5}$$

ANSWER: $\frac{3}{5}$

To subtract fractions that have a common denominator, subtract the numerators. Then write the difference over the common denominator.

---

2. Subtract: $\frac{11}{16} - \frac{5}{16}$

$$\begin{array}{r} \frac{11}{16} \\ -\ \frac{5}{16} \\ \hline \frac{6}{16} = \frac{3}{8} \end{array}$$

ANSWER: $\frac{3}{8}$

Simplify.

---

3. Subtract: $\frac{7}{8} - \frac{1}{4}$

$$\begin{array}{r} \frac{7}{8} = \frac{7}{8} \\ -\ \frac{1}{4} = \frac{2}{8} \\ \hline \frac{5}{8} \end{array}$$

ANSWER: $\frac{5}{8}$

To subtract fractions that do not have a common denominator, find the LCD. Express the fractions as equivalent fractions using the LCD. Subtract.

---

4. Subtract: $7\frac{5}{6} - 3\frac{1}{6}$

$$\begin{array}{r} 7\frac{5}{6} \\ -\ 3\frac{1}{6} \\ \hline \frac{4}{6} \end{array}$$

To subtract mixed numbers, first subtract the fractions

$$\begin{array}{r} 7\frac{5}{6} \\ -\ 3\frac{1}{6} \\ \hline 4\frac{4}{6} = 4\frac{2}{3} \end{array}$$

Then subtract the whole numbers

Simplify.

ANSWER: $4\frac{2}{3}$

---

5. Subtract: $6\frac{4}{5} - 3\frac{1}{2}$

$$\begin{array}{r} 6\frac{4}{5} = 6\frac{8}{10} \\ -\ 3\frac{1}{2} = 3\frac{5}{10} \\ \hline 3\frac{3}{10} \end{array}$$

Express each fraction as an equivalent fraction using the LCD. Subtract.

ANSWER: $3\frac{3}{10}$

6. Subtract: $8 - 1\frac{7}{16}$

$$8 \quad = 7\frac{16}{16}$$
$$- 1\frac{7}{16} = 1\frac{7}{16}$$
$$\overline{\qquad 6\frac{9}{16}}$$

Regroup when necessary.
$$8 = 7 + 1 = 7 + \frac{16}{16} = 7\frac{16}{16}$$
Subtract.

ANSWER: $6\frac{9}{16}$

7. Subtract: $7\frac{2}{3} - 4$

$$7\frac{2}{3}$$
$$- 4$$
$$\overline{3\frac{2}{3}}$$

ANSWER: $3\frac{2}{3}$

8. Subtract: $6\frac{1}{8} - \frac{7}{8}$

$$6\frac{1}{8} = 5\frac{9}{8}$$
$$- \frac{7}{8} = \frac{7}{8}$$
$$\overline{\qquad 5\frac{2}{8} = 5\frac{1}{4}}$$

ANSWER: $5\frac{1}{4}$

9. Subtract: $9\frac{1}{3} - 4\frac{5}{6}$

$$9\frac{1}{3} = 9\frac{2}{6} = 8\frac{8}{6}$$
$$- 4\frac{5}{6} = 4\frac{5}{6} = 4\frac{5}{6}$$
$$\overline{\qquad\qquad 4\frac{3}{6} = 4\frac{1}{2}}$$

Regroup.
$$9\frac{2}{6} = 8 + 1 + \frac{2}{6} = 8 + \frac{6}{6} + \frac{2}{6} = 8\frac{8}{6}$$

ANSWER: $4\frac{1}{2}$

10. Solve the equation: $14\frac{1}{4} - 12\frac{4}{5} = n$

$$14\frac{1}{4} = 14\frac{5}{20} = 13\frac{25}{20}$$
$$- 12\frac{4}{5} = 12\frac{16}{20} = 12\frac{16}{20}$$
$$\overline{\qquad\qquad\qquad 1\frac{9}{20}}$$

To find the value of n, subtract the given numbers.

ANSWER: $n = 1\frac{9}{20}$

Diagnostic Exercises

Subtract.

1. $\frac{2}{3}$ $-\frac{1}{3}$	**2.** $\frac{5}{8}$ $-\frac{1}{8}$	**3.** $\frac{5}{6}$ $-\frac{1}{2}$	**4.** $\frac{7}{8}$ $-\frac{1}{5}$	**5.** $\frac{3}{4}$ $-\frac{7}{10}$	**6.** $4\frac{4}{5}$ $-3\frac{2}{5}$	**7.** $8\frac{13}{16}$ $-5\frac{3}{16}$	**8.** $45\frac{1}{3}$ $-32\frac{2}{3}$	**9.** $12\frac{3}{8}$ $-7\frac{7}{8}$
10. $38\frac{9}{16}$ $-36\frac{9}{16}$	**11.** $5\frac{15}{32}$ -3	**12.** 6 $-2\frac{3}{4}$	**13.** 9 $-\frac{4}{5}$	**14.** $8\frac{11}{32}$ $-5\frac{1}{16}$	**15.** $9\frac{3}{4}$ $-3\frac{1}{3}$	**16.** $13\frac{7}{16}$ $-7\frac{5}{12}$		

17. $14\frac{5}{8}$ **18.** $18\frac{1}{5}$ **19.** $6\frac{3}{10}$ **20.** $14\frac{1}{8}$ **21.** $1\frac{1}{4}$ **22.** $\frac{15}{16} - \frac{3}{4}$
$-\ 5\frac{3}{4}$ $-\ 2\frac{1}{3}$ $-\ 3\frac{9}{16}$ $-13\frac{1}{2}$ $-\ \frac{5}{6}$

23. $14\frac{1}{4} - 5\frac{2}{3}$ **24.** Subtract $1\frac{5}{8}$ from 6.   **25.** Solve: $7\frac{5}{12} - 2\frac{1}{6} = n$

26. Scott's baseball game took $2\frac{1}{3}$ hours, while his basketball game took only $1\frac{1}{2}$ hours. How much longer did Scott play baseball than basketball?

Related Practice

Subtract.

1. a. $\frac{4}{5}$ **b.** $\frac{5}{7}$ **c.** $\frac{3}{4}$ **d.** $\frac{27}{32}$ **2. a.** $\frac{3}{4}$ **b.** $\frac{9}{16}$ **c.** $\frac{5}{6}$ **d.** $\frac{7}{10}$
$-\frac{3}{5}$ $-\frac{3}{7}$ $-\frac{2}{4}$ $-\frac{18}{32}$ $-\frac{1}{4}$ $-\frac{3}{16}$ $-\frac{1}{6}$ $-\frac{3}{10}$

3. a. $\frac{7}{8}$ **b.** $\frac{2}{3}$ **c.** $\frac{3}{4}$ **d.** $\frac{17}{20}$ **4. a.** $\frac{1}{2}$ **b.** $\frac{3}{4}$ **c.** $\frac{7}{8}$ **d.** $\frac{3}{5}$
$-\frac{3}{16}$ $-\frac{1}{6}$ $-\frac{5}{12}$ $-\frac{3}{5}$ $-\frac{1}{3}$ $-\frac{2}{5}$ $-\frac{2}{3}$ $-\frac{7}{12}$

5. a. $\frac{1}{4}$ **b.** $\frac{5}{6}$ **c.** $\frac{7}{12}$ **d.** $\frac{5}{8}$ **6. a.** $6\frac{3}{5}$ **b.** $8\frac{8}{9}$ **c.** $15\frac{2}{4}$ **d.** $53\frac{7}{8}$
$-\frac{1}{6}$ $-\frac{3}{8}$ $-\frac{3}{16}$ $-\frac{1}{12}$ $-\ 2\frac{1}{5}$ $-\ 7\frac{3}{9}$ $-\ 4\frac{1}{4}$ $-25\frac{2}{8}$

7. a. $3\frac{5}{8}$ **b.** $6\frac{11}{16}$ **c.** $9\frac{7}{12}$ **d.** $11\frac{9}{10}$ **8. a.** $9\frac{2}{5}$ **b.** $5\frac{3}{7}$ **c.** $16\frac{9}{16}$ **d.** $18\frac{1}{4}$
$-\ 1\frac{3}{8}$ $-\ 3\frac{5}{16}$ $-\ 7\frac{5}{12}$ $-\ 6\frac{3}{10}$ $-\ 4\frac{4}{5}$ $-\ 1\frac{6}{7}$ $-\ 7\frac{12}{16}$ $-14\frac{2}{4}$

9. a. $7\frac{1}{6}$ **b.** $15\frac{5}{8}$ **c.** $13\frac{3}{10}$ **d.** $17\frac{5}{32}$ **10. a.** $6\frac{1}{2}$ **b.** $11\frac{3}{4}$ **c.** $43\frac{3}{10}$ **d.** $32\frac{11}{12}$
$-\ 1\frac{5}{6}$ $-\ 6\frac{7}{8}$ $-10\frac{9}{10}$ $-13\frac{29}{32}$ $-\ 5\frac{1}{2}$ $-\ 4\frac{3}{4}$ $-28\frac{3}{10}$ $-23\frac{11}{12}$

11. a. $4\frac{1}{2}$ **b.** $15\frac{3}{8}$ **c.** $43\frac{9}{10}$ **d.** $39\frac{25}{32}$ **12. a.** 9 **b.** 3 **c.** 13 **d.** 42
$-\ 2$ $-\ 9$ -27 -14 $-\ 2\frac{1}{4}$ $-\ 1\frac{3}{5}$ $-\ 6\frac{5}{8}$ $-17\frac{7}{20}$

13. a. 5 **b.** 9 **c.** 12 **d.** 10 **14. a.** $8\frac{5}{6}$ **b.** $11\frac{11}{12}$ **c.** $18\frac{9}{10}$ **d.** $42\frac{7}{8}$
$-\ \frac{3}{8}$ $-\ \frac{7}{12}$ $-\ \frac{9}{10}$ $-\ \frac{2}{3}$ $-\ 6\frac{1}{3}$ $-\ 8\frac{1}{4}$ $-15\frac{2}{5}$ $-\ 9\frac{3}{16}$

15. a. $7\frac{1}{3}$ **b.** $20\frac{3}{5}$ **c.** $11\frac{2}{3}$ **d.** $40\frac{4}{5}$ **16. a.** $8\frac{3}{4}$ **b.** $3\frac{7}{12}$ **c.** $7\frac{15}{16}$ **d.** $13\frac{9}{10}$
$\quad\;\; -4\frac{1}{5}$ $\quad\; -12\frac{7}{16}$ $\quad\; -5\frac{1}{2}$ $\quad\; -8\frac{3}{4}$ $\qquad -5\frac{1}{6}$ $\quad\; -1\frac{3}{8}$ $\quad\; -2\frac{5}{6}$ $\quad\; -9\frac{3}{4}$

17. a. $6\frac{1}{2}$ **b.** $9\frac{3}{8}$ **c.** $16\frac{3}{10}$ **d.** $17\frac{5}{32}$ **18. a.** $10\frac{1}{3}$ **b.** $13\frac{1}{4}$ **c.** $9\frac{1}{6}$ **d.** $25\frac{3}{4}$
$\quad\;\; -3\frac{3}{4}$ $\quad\; -5\frac{9}{16}$ $\quad\; -8\frac{1}{2}$ $\quad\; -12\frac{9}{16}$ $\qquad -4\frac{1}{2}$ $\quad\; -8\frac{2}{3}$ $\quad\; -2\frac{3}{5}$ $\quad\; -21\frac{4}{5}$

19. a. $9\frac{3}{4}$ **b.** $7\frac{1}{6}$ **c.** $18\frac{5}{6}$ **d.** $24\frac{7}{12}$ **20. a.** $8\frac{11}{16}$ **b.** $4\frac{7}{8}$ **c.** $7\frac{1}{3}$ **d.** $13\frac{3}{4}$
$\quad\;\; -1\frac{9}{10}$ $\quad\; -5\frac{3}{8}$ $\quad\; -12\frac{15}{16}$ $\quad\; -14\frac{7}{10}$ $\qquad -8\frac{7}{16}$ $\quad\; -4\frac{1}{2}$ $\quad\; -6\frac{2}{3}$ $\quad\; -12\frac{5}{6}$

21. a. $5\frac{1}{2}$ **b.** $3\frac{9}{16}$ **c.** $1\frac{7}{10}$ **d.** $1\frac{1}{3}$
$\quad\;\; -\frac{3}{4}$ $\quad\;\; -\frac{7}{8}$ $\quad\;\; -\frac{5}{6}$ $\quad\;\; -\frac{1}{2}$

22. a. $\frac{3}{5}-\frac{1}{5}$ **b.** $\frac{1}{2}-\frac{1}{6}$ **c.** $\frac{4}{5}-\frac{3}{4}$ **d.** $2\frac{3}{8}-1$

23. a. $6\frac{7}{8}-2\frac{3}{8}$ **b.** $8\frac{1}{4}-4\frac{2}{5}$ **c.** $11\frac{1}{2}-2\frac{1}{3}$ **d.** $14\frac{3}{8}-9\frac{1}{10}$

24. a. From $\frac{7}{8}$ subtract $\frac{5}{16}$. **b.** From $1\frac{4}{5}$ take $\frac{9}{10}$.

   c. From 7 subtract $2\frac{11}{16}$. **d.** Take 3 from $4\frac{5}{8}$.

25. Solve.

   a. $\frac{2}{3}-\frac{5}{8}=$ ▓ **b.** $10-9\frac{3}{4}=$ ▓ **c.** $2\frac{1}{2}-\frac{7}{8}=n$ **d.** ▓ $=5\frac{1}{2}-1\frac{5}{6}$

26. a. Wes used to weigh $123\frac{1}{4}$ pounds. Now he weighs $116\frac{3}{4}$ pounds. How many pounds did he lose?

   b. If normal body temperature is $98\frac{3}{5}$ degrees Fahrenheit, how many degrees above normal is a temperature of 101 degrees?

   c. One mechanic can assemble a motor in $6\frac{1}{2}$ hours while another mechanic can do the same job in $7\frac{5}{6}$ hours. In how much less time can the first mechanic do the job?

   d. A $2\frac{1}{2}$-inch nail is driven through a $1\frac{1}{8}$-inch thick piece of wood supporting a joist. How far into the joist did the nail extend?

Mixed Problems

Choose an appropriate method of computation and solve each of the following problems. For Problems 1–4, select the letter corresponding to your answer.

1. A merchant sold $7\frac{5}{8}$ yards of cloth to a customer. If it was cut from a bolt that contained $18\frac{2}{3}$ yards, what length remained on the bolt?

   a. $10\frac{11}{24}$ yards   **b.** $11\frac{1}{24}$ yards

   c. $11\frac{1}{6}$ yards   **d.** $10\frac{5}{6}$ yards

2. Find the net change in a stock if it opened at $63\frac{7}{8}$ and closed at $65\frac{1}{4}$.

   a. $2\frac{3}{8}$ points gain   **b.** $1\frac{7}{8}$ points gain

   c. $1\frac{3}{8}$ points gain   **d.** Answer not given

3. Find the total thickness of two pieces of wood that Donald glued together if one is $\frac{5}{16}$-inch thick and the other $\frac{7}{8}$-inch thick.

   a. $1\frac{5}{16}$ inches   **b.** $1\frac{1}{2}$ inches

   c. $1\frac{3}{16}$ inches   **d.** $1\frac{1}{4}$ inches

4. The running time of a train from Chicago to San Francisco was changed to $49\frac{1}{3}$ hours. If this schedule saves $13\frac{3}{4}$ hours, how long did the trip take before the change was made?

   a. $64\frac{1}{4}$ hours   **b.** $63\frac{2}{3}$ hours

   c. $63\frac{1}{12}$ hours   **d.** $63\frac{5}{6}$ hours

5. A pilot finds that winds slow her progress in reaching her destination. Her trip out takes $2\frac{3}{4}$ hours. How long should her return trip take if she must return at the end of $4\frac{1}{2}$ hours flying time?

6. Team B is $4\frac{1}{2}$ games behind team A in the standings. Team C is $2\frac{1}{2}$ games behind team B. How many games behind team A is team C?

7. What is the outside diameter of tubing when the inside diameter is $2\frac{5}{8}$ inches and the wall thickness is $\frac{3}{16}$ inch?

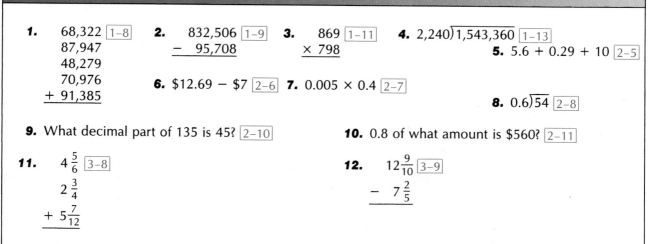

Refresh Your Skills

1. $\begin{array}{r} 68,322 \\ 87,947 \\ 48,279 \\ 70,976 \\ + 91,385 \end{array}$ $\boxed{1-8}$

2. $\begin{array}{r} 832,506 \\ - 95,708 \end{array}$ $\boxed{1-9}$

3. $\begin{array}{r} 869 \\ \times 798 \end{array}$ $\boxed{1-11}$

4. $2{,}240\overline{)1{,}543{,}360}$ $\boxed{1-13}$

5. $5.6 + 0.29 + 10$ $\boxed{2-5}$

6. $\$12.69 - \$7$ $\boxed{2-6}$ **7.** 0.005×0.4 $\boxed{2-7}$

8. $0.6\overline{)54}$ $\boxed{2-8}$

9. What decimal part of 135 is 45? $\boxed{2-10}$

10. 0.8 of what amount is $560? $\boxed{2-11}$

11. $\begin{array}{r} 4\frac{5}{6} \\ 2\frac{3}{4} \\ + 5\frac{7}{12} \end{array}$ $\boxed{3-8}$

12. $\begin{array}{r} 12\frac{9}{10} \\ - 7\frac{2}{5} \end{array}$ $\boxed{3-9}$

Writing Mixed Numbers as Improper Fractions

It is often convenient to write a mixed number as an improper fraction.

Examples

1. Write $4\frac{2}{3}$ as an improper fraction.

$$4\frac{2}{3} = 4 + \frac{2}{3}$$
$$= \frac{12}{3} + \frac{2}{3}$$
$$= \frac{14}{3}$$

Write the whole number and the fraction as a sum.
Express the whole number as a fraction with the same denominator as the fraction.
Add the two fractions.

ANSWER: $\frac{14}{3}$

Look at this diagram. $4\frac{2}{3}$ circles are shaded. $4\frac{2}{3} = \frac{14}{3}$

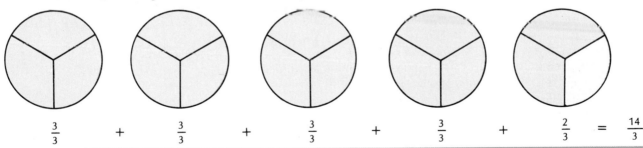

$$\frac{3}{3} \quad + \quad \frac{3}{3} \quad + \quad \frac{3}{3} \quad + \quad \frac{3}{3} \quad + \quad \frac{2}{3} \quad = \quad \frac{14}{3}$$

2. Write $5\frac{2}{7}$ as an improper fraction.

$$5\frac{2}{7} = \frac{35 + 2}{7} = \frac{37}{7}$$
$$5 \times 7 = 35$$

Another way is to multiply the whole number by the denominator of the fraction. Add the product to the numerator of the fraction. Write the sum over the denominator of the fraction.

ANSWER: $\frac{37}{7}$

Writing Mixed Numbers as Improper Fractions **177**

1. A student wants to write $6\frac{3}{5}$ as an improper fraction and then rewrite the fraction with a denominator of 35. Explain the steps to do this.

2. A person has 8 one-dollar bills and 2 quarters. Explain, using the techniques just discussed, how to convert the entire amount to quarters, and then how to convert the amount in quarters to an amount in half-dollars.

Diagnostic Exercises

Write the following mixed numbers as improper fractions.

1. $1\frac{1}{3}$ **2.** $1\frac{3}{4}$ **3.** $7\frac{1}{2}$ **4.** $5\frac{7}{8}$

5. If it takes $\frac{1}{8}$ yard of material to make a certain necktie, how many similar neckties can be made from a bolt of material containing $15\frac{3}{8}$ yards?

Related Practice

Write the following mixed numbers as improper fractions.

1. **a.** $1\frac{1}{5}$ **b.** $1\frac{1}{4}$ **c.** $1\frac{1}{2}$ **d.** $1\frac{1}{16}$

2. **a.** $1\frac{5}{8}$ **b.** $1\frac{2}{3}$ **c.** $1\frac{7}{12}$ **d.** $1\frac{13}{16}$

3. **a.** $8\frac{1}{5}$ **b.** $3\frac{1}{7}$ **c.** $9\frac{1}{6}$ **d.** $16\frac{1}{3}$

4. **a.** $5\frac{4}{7}$ **b.** $4\frac{13}{32}$ **c.** $10\frac{9}{16}$ **d.** $13\frac{3}{5}$

5. **a.** Mr. Damon has only a $\frac{1}{2}$-cup measure. How many times would he have to fill it to measure out $9\frac{1}{2}$ cups of flour?

b. How many minutes are there in $5\frac{7}{60}$ hours?

c. The scale of a map is $\frac{1}{8}$ inch = 1 mile. Jen measures a distance of $6\frac{7}{8}$ inches between two towns on the map. What is the actual distance between the two towns?

d. How many pieces of wood $\frac{1}{3}$-foot long can be cut from a board $6\frac{2}{3}$ feet long?

Multiplying Fractions and Mixed Numbers

Building Problem Solving Skills

Luke rode his bike to the playground in $\frac{1}{3}$ hour. The return trip is downhill, and it took him only $\frac{3}{4}$ as long. What fraction of an hour was the return trip?

Multiplication Indicators
· total · how many in all · how much altogether Note: All indicators are used when combining *equal* quantities.

1. **READ** the problem carefully.

a. Find the question asked.

What fraction of an hour did it take Luke to return?

b. Find the given facts.

It took Luke $\frac{1}{3}$ hour to get to the playground.
It took him $\frac{3}{4}$ as long to get home.

2. **PLAN** how to solve the problem.

a. Choose the operation needed.

You can use multiplication to find a fractional part of a fraction.

b. Think or write out your plan relating the given facts to the question asked.

$\frac{3}{4}$ of $\frac{1}{3}$ is the fractional part of an hour it took Luke to return.

c. Express your plan as an equation.

$$\frac{3}{4} \times \frac{1}{3} = n$$

3. **SOLVE** the problem.

a. Estimate the answer.

Since you are multiplying two quantities less than 1, the answer will be less than either fraction. So your answer will be less than $\frac{1}{3}$.

b. Solution.

$$\frac{3}{4} \times \frac{1}{3} = n$$

Write an equation.

$$\frac{\overset{1}{\cancel{3}}}{4} \times \frac{1}{\underset{1}{\cancel{3}}} = n$$

Divide numerator and denominator by their greatest common factor.

$$\frac{1}{4} \times \frac{1}{1} = \frac{1}{4}$$

Multiply the numerators and the denominators.

4. [CHECK].

a. Check the accuracy of your arithmetic. Check the units.

$$\frac{3}{4} \times \frac{1}{3} = \frac{3}{12} = \frac{1}{4}$$

This time multiply the denominators and the numerators and *then* write the answer in lowest terms.

b. Compare the answer to the estimate.

$\frac{1}{4} < \frac{1}{3}$, so your answer is reasonable. The unit is hours.

ANSWER: It took Luke $\frac{1}{4}$ hour to return.

Think About It

1. What kind of number would you have to use instead of $\frac{3}{4}$ so that the return trip would be *longer* than it took going?

2. If you multiply a whole number by a fraction less than 1, will the answer be more or less than the whole number?

Calculator Know-How

You can use a calculator to multiply fractions and mixed numbers.

$5\frac{3}{8} \times 9\frac{5}{7} = n$ Key: 5 ⊠ 8 ⊞ 3 ⊟ <u>43</u>   Express as improper fractions.

                              9 ⊠ 7 ⊞ 5 ⊟ <u>68</u>

$\frac{43}{8} \times \frac{68}{7} = n$ Key: 43 ⊠ 68 ⊟ <u>2924</u>   Multiply the numerators and denominators.

                              8 ⊠ 7 ⊟ <u>56</u>

$\frac{43}{8} \times \frac{68}{7} = \frac{2924}{56}$

$\frac{2924}{56} = 52\frac{12}{56} = 52\frac{3}{14}$ Key: 2924 ⊡ 56 ⊟ <u>52.214285</u>   Express the answer as a mixed number in lowest terms.

                           56 ⊠ 52 ⊟ <u>2912</u>

                         2924 ⊟ 2912 ⊟ <u>12</u>

Multiply using a calculator or mental arithmetic. Explain your choice.

1. $2 \times \frac{1}{2}$ **2.** $\frac{3}{4} \times 8$ **3.** $6\frac{5}{6} \times 12$ **4.** $\frac{9}{16} \times \frac{4}{9}$

5. $\frac{7}{8} \times \frac{3}{8}$ **6.** $7\frac{4}{9} \times 3\frac{9}{16}$ **7.** $1\frac{1}{2} \times \frac{3}{8}$ **8.** $\frac{13}{20} \times 80$

1. Multiply: $\frac{1}{2} \times \frac{3}{8}$

$\frac{1}{2} \times \frac{3}{8} = \frac{3}{16}$

Multiply the numerators and multiply the denominators.

$$\frac{3}{8}$$

$$\frac{1}{2}$$

ANSWER: $\frac{3}{16}$

---

2. Multiply: $\frac{8}{20} \times \frac{15}{7}$

$\overset{2}{\underset{\underset{1}{\cancel{20}}}{\cancel{8}}} \times \overset{3}{\cancel{\frac{15}{7}}} = \frac{2 \times 3}{1 \times 7}$

If possible, divide *any* numerator and denominator by their greatest common factor before multiplying. This is the same as dividing by 1.

Divide 8 and 20 by 4 (GCF).

Divide 5 and 15 by 5 (GCF).

ANSWER: $\frac{6}{7}$

---

3. Multiply: $\frac{2}{3} \times 5$

$\frac{2}{3} \times 5 = \frac{2}{3} \times \frac{5}{1}$

$\phantom{\frac{2}{3} \times 5} = \frac{10}{3}$

$\phantom{\frac{2}{3} \times 5} = 3\frac{1}{3}$

Rewrite whole numbers as fractions with a denominator of 1.

Multiply.

Simplify.

ANSWER: $3\frac{1}{3}$

---

4. Solve the equation: $2\frac{1}{2} \times 1\frac{1}{5} = n$

$2\frac{1}{2} \times 1\frac{1}{5} = \frac{5}{2} \times \frac{6}{5} = n$

$\phantom{2\frac{1}{2} \times 1\frac{1}{5}} = \frac{3}{1} = 3 = n$

Rewrite mixed numbers as improper fractions.

ANSWER: $3 = n$

---

5. Find the product of 21 and $3\frac{1}{2}$.

$$
\begin{array}{r}
21 \\
\times \;\; 3\frac{1}{2} \\
\hline
63 \\
+ \; 10\frac{1}{2} \\
\hline
73\frac{1}{2}
\end{array}
$$

$\leftarrow$ **21 × 3**

$\leftarrow$ **21 × $\frac{1}{2}$**

When multiplying a mixed number and a whole number, the vertical form may also be used.

ANSWER: $73\frac{1}{2}$

6. Multiply: $1\frac{4}{5} \times \frac{2}{3} \times 3\frac{1}{8}$

$$1\frac{4}{5} \times \frac{2}{3} \times 3\frac{1}{8} = \frac{\overset{3}{\cancel{9}}}{\cancel{5}} \times \frac{\overset{1}{\cancel{2}}}{\cancel{3}} \times \frac{\overset{5}{\cancel{25}}}{\cancel{8}}$$
$$= \frac{15}{4} = 3\frac{3}{4}$$

ANSWER: $3\frac{3}{4}$

---

7. Simplify: $\dfrac{5{,}280 \times 60}{3{,}600}$

$$\frac{5{,}280 \times 60}{3{,}600} = \frac{\overset{88}{\cancel{5{,}280}} \times \overset{1}{\cancel{60}}}{\underset{1}{\underset{60}{\cancel{3{,}600}}}} = \frac{88}{1} \text{ or } 88$$

ANSWER: 88

Division by the GCF may be used to simplify an expression involving both multiplication and division.

---

8. Find $\frac{5}{8}$ of 48.

$$\frac{5}{8} \text{ of } 48 = \frac{5}{8} \times 48 \quad = \frac{5}{\cancel{8}} \times \frac{\overset{6}{\cancel{48}}}{1} \quad = \frac{30}{1} = 30$$

ANSWER: 30

To find a fractional part of a number, multiply the given number and the fraction.

---

9. Find $\frac{5}{6}$ of $4.19 correct to the nearest cent.

$$\frac{5}{6} \text{ of } \$4.19 = \frac{5}{6} \times \$4.19$$
$$= \frac{\$20.95}{6}$$
$$= \$3.49\tfrac{1}{6} \text{ **rounds to \$3.49**}$$

ANSWER: $3.49

10. Find $0.37\frac{1}{2}$ of $28.96.

$$\begin{array}{r} \$28.96 \\ \times \quad 0.37\tfrac{1}{2} \\ \hline 202\ 72 \\ 868\ 8\ \\ \hline 1071\ 52 \\ \underline{14\ 48} \quad \leftarrow \tfrac{1}{2} \times \textbf{2{,}896} \\ \$10.86\ 00 \end{array}$$

ANSWER: $10.86

Diagnostic Exercises

Multiply.

1. $\frac{1}{5} \times \frac{1}{3}$ **2.** $\frac{3}{4} \times \frac{5}{8}$ **3.** $\frac{1}{2} \times \frac{2}{3}$ **4.** $\frac{3}{8} \times \frac{4}{5}$

5. $\frac{3}{4} \times \frac{8}{15}$ **6.** $\frac{9}{16} \times \frac{5}{6}$ **7.** $\frac{5}{2} \times \frac{10}{3}$ **8.** $\frac{7}{8} \times 8$

9. $\frac{2}{3} \times 6$ **10.** $\frac{3}{4} \times 2$ **11.** $\frac{5}{6} \times 10$ **12.** $\frac{3}{5} \times 7$

13. $10 \times \frac{9}{10}$ **14.** $48 \times \frac{7}{12}$ **15.** $4 \times \frac{7}{8}$ **16.** $12 \times \frac{5}{8}$

17. $5 \times \frac{3}{16}$ **18.** $4\frac{1}{2} \times 4$ **19.** $1\frac{7}{12} \times 8$ **20.** $3\frac{1}{3} \times 5$

21. $12 \times 1\frac{5}{6}$ **22.** $10 \times 2\frac{9}{16}$ **23.** $7 \times 3\frac{1}{4}$ **24.** $2\frac{1}{2} \times \frac{4}{5}$

25. $6\frac{1}{4} \times \frac{3}{8}$ **26.** $\frac{5}{16} \times 9\frac{3}{5}$ **27.** $\frac{5}{6} \times 1\frac{9}{16}$ **28.** $5\frac{1}{3} \times 1\frac{1}{8}$

29. $2\frac{5}{8} \times 2\frac{2}{5}$ **30.** $4\frac{1}{2} \times 2\frac{1}{4}$ **31.** $\frac{1}{2} \times \frac{8}{15} \times \frac{5}{6}$ **32.** $1\frac{3}{4} \times 3\frac{1}{7} \times 1\frac{3}{5}$

33. $\begin{array}{r} 18 \\ \times\ 7\frac{1}{3} \\ \hline \end{array}$ **34.** $\begin{array}{r} 12\frac{5}{6} \\ \times\ 8 \\ \hline \end{array}$ **35.** Find the product of $2\frac{7}{8}$ and $1\frac{3}{4}$.

36. $\frac{5}{100} \times 900$ **37.** Simplify. $\frac{132 \times 3600}{5280}$

38. Correct to the nearest cent.   **39.** Correct to the nearest cent.   **40.** Solve. $3\frac{3}{4} \times 1\frac{1}{5} = n$

$\begin{array}{r} \$15.38 \\ \times\ \ 0.62\frac{1}{2} \\ \hline \end{array}$ $\begin{array}{r} \$23.89 \\ \times\ \ 0.33\frac{1}{3} \\ \hline \end{array}$

In each of the following, find the indicated fractional part of the given number.

41. Find $\frac{3}{4}$ of 18.   **42.** Find $\frac{1}{2}$ of \$.84.

43. Find $\frac{7}{8}$ of \$3.45 correct to nearest cent.

44. Mr. Garcia hires Juan and Felicia to take care of his yard for the summer. Juan and Felicia agree to divide the pay according to what part of the job each one does. If Felicia does $\frac{3}{5}$ of the work, what part of the total amount of \$138.50 should she receive?

Related Practice

Multiply.

1. a. $\frac{1}{4} \times \frac{1}{2}$ **b.** $\frac{1}{8} \times \frac{1}{3}$ **c.** $\frac{1}{5} \times \frac{1}{4}$ **d.** $\frac{1}{2} \times \frac{1}{10}$

2. a. $\frac{1}{2} \times \frac{3}{5}$ **b.** $\frac{5}{6} \times \frac{7}{8}$ **c.** $\frac{7}{16} \times \frac{3}{4}$ **d.** $\frac{5}{12} \times \frac{1}{3}$

3. a. $\frac{4}{5} \times \frac{3}{4}$ **b.** $\frac{3}{10} \times \frac{1}{3}$ **c.** $\frac{2}{3} \times \frac{1}{2}$ **d.** $\frac{5}{6} \times \frac{6}{7}$

4. a. $\frac{5}{6} \times \frac{3}{8}$ **b.** $\frac{6}{7} \times \frac{11}{12}$ **c.** $\frac{4}{5} \times \frac{7}{8}$ **d.** $\frac{1}{3} \times \frac{9}{10}$

5. a. $\frac{2}{5} \times \frac{5}{12}$ **b.** $\frac{9}{10} \times \frac{2}{3}$ **c.** $\frac{5}{8} \times \frac{16}{25}$ **d.** $\frac{8}{9} \times \frac{3}{4}$

6. a. $\frac{5}{6} \times \frac{4}{5}$ **b.** $\frac{3}{16} \times \frac{6}{7}$ **c.** $\frac{7}{8} \times \frac{12}{21}$ **d.** $\frac{10}{12} \times \frac{14}{15}$

7. a. $\frac{5}{4} \times \frac{2}{3}$ **b.** $\frac{7}{2} \times \frac{4}{5}$ **c.** $\frac{4}{3} \times \frac{3}{4}$ **d.** $\frac{10}{9} \times \frac{15}{8}$

8. a. $\frac{5}{6} \times 6$ **b.** $\frac{11}{8} \times 8$ **c.** $\frac{7}{16} \times 16$ **d.** $\frac{3}{5} \times 5$

9. a. $\frac{3}{4} \times 8$ **b.** $\frac{7}{6} \times 72$ **c.** $\frac{5}{12} \times 36$ **d.** $\frac{9}{16} \times 64$

10. a. $\frac{3}{8} \times 4$ **b.** $\frac{7}{16} \times 2$ **c.** $\frac{5}{12} \times 3$ **d.** $\frac{11}{24} \times 12$

11. a. $\frac{3}{8} \times 6$ **b.** $\frac{9}{16} \times 20$ **c.** $\frac{19}{12} \times 8$ **d.** $\frac{5}{6} \times 4$

12. a. $\frac{1}{3} \times 7$ **b.** $\frac{3}{5} \times 9$ **c.** $\frac{9}{8} \times 5$ **d.** $\frac{3}{10} \times 21$

13. a. $4 \times \frac{3}{4}$ **b.** $8 \times \frac{5}{8}$ **c.** $12 \times \frac{13}{12}$ **d.** $16 \times \frac{15}{16}$

14. a. $12 \times \frac{5}{6}$ **b.** $24 \times \frac{3}{8}$ **c.** $32 \times \frac{17}{16}$ **d.** $30 \times \frac{2}{5}$

15. a. $2 \times \frac{3}{8}$ **b.** $2 \times \frac{7}{4}$ **c.** $5 \times \frac{3}{10}$ **d.** $6 \times \frac{13}{24}$

16. a. $12 \times \frac{7}{16}$ **b.** $18 \times \frac{15}{32}$ **c.** $16 \times \frac{19}{12}$ **d.** $15 \times \frac{3}{10}$

17. a. $4 \times \frac{2}{3}$ **b.** $9 \times \frac{4}{5}$ **c.** $3 \times \frac{9}{8}$ **d.** $7 \times \frac{5}{12}$

18. a. $2\frac{1}{8} \times 16$ **b.** $1\frac{2}{3} \times 9$ **c.** $7\frac{1}{2} \times 6$ **d.** $4\frac{3}{5} \times 10$

19. a. $2\frac{5}{6} \times 3$ **b.** $5\frac{7}{8} \times 6$ **c.** $3\frac{11}{12} \times 8$ **d.** $1\frac{9}{10} \times 12$

20. a. $2\frac{1}{5} \times 4$ **b.** $5\frac{2}{3} \times 2$ **c.** $4\frac{5}{8} \times 3$ **d.** $3\frac{9}{16} \times 5$

21. a. $8 \times 5\frac{1}{4}$ **b.** $16 \times 2\frac{5}{8}$ **c.** $48 \times 1\frac{13}{16}$ **d.** $24 \times 5\frac{11}{12}$

22. a. $15 \times 1\frac{7}{10}$ **b.** $8 \times 2\frac{9}{16}$ **c.** $18 \times 5\frac{5}{8}$ **d.** $4 \times 3\frac{1}{6}$

23. a. $2 \times 6\frac{1}{3}$ **b.** $3 \times 9\frac{1}{2}$ **c.** $9 \times 5\frac{7}{16}$ **d.** $7 \times 1\frac{3}{4}$

24. a. $3\frac{1}{3} \times \frac{3}{5}$ **b.** $6\frac{3}{4} \times \frac{2}{3}$ **c.** $3\frac{1}{8} \times \frac{8}{15}$ **d.** $4\frac{1}{2} \times \frac{2}{9}$

25. a. $2\frac{1}{2} \times \frac{3}{4}$ **b.** $5\frac{2}{3} \times \frac{5}{8}$ **c.** $1\frac{3}{8} \times \frac{1}{2}$ **d.** $3\frac{7}{16} \times \frac{3}{4}$

26. a. $\frac{1}{2} \times 3\frac{1}{5}$ **b.** $\frac{5}{6} \times 2\frac{3}{10}$ **c.** $\frac{7}{8} \times 3\frac{1}{7}$ **d.** $\frac{9}{16} \times 1\frac{1}{3}$

27. a. $\frac{7}{8} \times 1\frac{1}{4}$ **b.** $\frac{1}{2} \times 3\frac{3}{8}$ **c.** $\frac{1}{5} \times 4\frac{1}{3}$ **d.** $\frac{9}{10} \times 1\frac{1}{2}$

28. a. $1\frac{1}{4} \times 1\frac{3}{5}$ **b.** $5\frac{1}{3} \times 4\frac{1}{2}$ **c.** $2\frac{5}{8} \times 1\frac{5}{7}$ **d.** $2\frac{2}{3} \times 3\frac{3}{8}$

29. a. $2\frac{1}{3} \times 1\frac{1}{5}$ **b.** $2\frac{1}{6} \times 2\frac{2}{3}$ **c.** $2\frac{2}{5} \times 1\frac{3}{16}$ **d.** $1\frac{3}{4} \times 3\frac{1}{3}$

30. a. $2\frac{1}{8} \times 1\frac{1}{2}$ **b.** $3\frac{3}{4} \times 2\frac{7}{8}$ **c.** $1\frac{9}{16} \times 4\frac{1}{3}$ **d.** $2\frac{5}{6} \times 1\frac{3}{8}$

31. a. $\frac{1}{4} \times \frac{5}{6} \times \frac{2}{5}$ **b.** $\frac{2}{3} \times \frac{5}{8} \times \frac{3}{10}$ **c.** $\frac{5}{12} \times \frac{3}{16} \times \frac{4}{5}$ **d.** $\frac{3}{4} \times \frac{3}{5} \times \frac{1}{2}$

32. a. $2\frac{3}{4} \times 1\frac{1}{8} \times 3\frac{5}{6}$ **b.** $1\frac{1}{2} \times \frac{4}{5} \times 2\frac{1}{6}$ **c.** $3\frac{1}{5} \times 1\frac{1}{4} \times 1\frac{1}{3}$ **d.** $1\frac{5}{16} \times 2\frac{2}{3} \times 3\frac{1}{7}$

33. a.
$$\begin{array}{r} 28 \\ \times\ 4\frac{1}{2} \\ \hline \end{array}$$
b.
$$\begin{array}{r} 16 \\ \times\ 5\frac{3}{4} \\ \hline \end{array}$$
c.
$$\begin{array}{r} 35 \\ \times\ 7\frac{5}{8} \\ \hline \end{array}$$
d.
$$\begin{array}{r} 29 \\ \times\ 2\frac{2}{3} \\ \hline \end{array}$$

34. a.
$$\begin{array}{r} 32\frac{1}{4} \\ \times\ 8 \\ \hline \end{array}$$
b.
$$\begin{array}{r} 24\frac{3}{8} \\ \times\ 6 \\ \hline \end{array}$$
c.
$$\begin{array}{r} 17\frac{3}{5} \\ \times\ 10 \\ \hline \end{array}$$
d.
$$\begin{array}{r} 5\frac{5}{6} \\ \times\ 9 \\ \hline \end{array}$$

35. a. Multiply $4\frac{7}{8}$ by 6. **b.** Multiply $2\frac{3}{4}$ by $4\frac{1}{2}$.

  c. Find the product of $1\frac{1}{8}$ and $5\frac{1}{3}$. **d.** Find the product of $2\frac{3}{16}$ and $2\frac{2}{5}$.

36. a. $\frac{3}{100} \times 400$ **b.** $\frac{4}{100} \times 500$ **c.** $\frac{2}{100} \times 1,000$ **d.** $\frac{6}{100} \times 2,500$

37. Simplify.

  a. $\frac{1760 \times 60}{60}$ **b.** $\frac{75 \times 32}{25 \times 24}$ **c.** $\frac{18 \times 12 \times 16}{144}$ **d.** $\frac{5280 \times 120}{3600}$

38. Find the product correct to the nearest cent.

  a.
$$\begin{array}{r} \$80 \\ \times\ 0.12\frac{1}{2} \\ \hline \end{array}$$
b.
$$\begin{array}{r} \$3,000 \\ \times\ 0.04\frac{1}{2} \\ \hline \end{array}$$
c.
$$\begin{array}{r} \$0.54 \\ \times\ 0.37\frac{1}{2} \\ \hline \end{array}$$
d.
$$\begin{array}{r} \$18.48 \\ \times\ 0.05\frac{3}{4} \\ \hline \end{array}$$

39. Find the product correct to the nearest cent.

  a.
$$\begin{array}{r} \$246 \\ \times\ 0.33\frac{1}{3} \\ \hline \end{array}$$
b.
$$\begin{array}{r} \$6,000 \\ \times\ 0.08\frac{1}{3} \\ \hline \end{array}$$
c.
$$\begin{array}{r} \$1.15 \\ \times\ 0.66\frac{2}{3} \\ \hline \end{array}$$
d.
$$\begin{array}{r} \$45.71 \\ \times\ 0.83\frac{1}{3} \\ \hline \end{array}$$

40. Solve.

  a. $\frac{2}{3} \times \frac{9}{4} = n$ **b.** $\blacksquare = 2\frac{1}{2} \times 7\frac{1}{5}$ **c.** $\blacksquare = 1\frac{5}{8} \times 16$ **d.** $n = 8\frac{3}{4} \times 3\frac{1}{7}$

41. Solve.

  a. $\frac{3}{5}$ of $1\frac{2}{3}$ **b.** $\frac{2}{5}$ of 12 **c.** $\frac{7}{10}$ of 180 **d.** $\frac{3}{10}$ of $1\frac{1}{2}$

42. a. $\frac{1}{2}$ of $4.64 **b.** $\frac{3}{4}$ of $5.20 **c.** $2\frac{2}{3}$ of $38.16 **d.** $4\frac{3}{5}$ of $51.40

43. Round to hundredths or the nearest cent.

  a. $\frac{1}{2}$ of $.39 **b.** $\frac{1}{6}$ of $1.22 **c.** $\frac{7}{12}$ of $10 **d.** $4\frac{2}{5}$ of 3.64

44. a. How much wood is needed to make 15 shelves each $6\frac{2}{3}$ feet long?

  b. Find the cost of $2\frac{1}{4}$ pounds of bananas at 32¢ per pound.

  c. The cooking class is divided into six teams. If each team's recipe requires $2\frac{1}{2}$ cups cake flour, $2\frac{1}{4}$ teaspoons baking powder, $\frac{1}{2}$ cup shortening, and $1\frac{1}{4}$ cups sugar, how much of each ingredient does the entire class need?

  d. Mr. Rogers earns $9 per hour. If his overtime rate is $1\frac{1}{2}$ times the regular rate, what is his hourly rate for overtime work?

Mixed Problems

Choose an appropriate method of computation and solve each of the following problems. For Problems 1–4 select the letter corresponding to your answer.

1. A sewing class is making costumes for the school play. If each costume requires $3\frac{5}{8}$ yards of goods, about how many yards are needed to make 26 costumes?

   a. 80 yards   **b.** 100 yards

   c. 70 yards   **d.** Answer not given

2. The XYZ stock opened Monday morning at the price of $27\frac{5}{8}$. During the 5 business days of the week, the stock gained $\frac{7}{8}$ point, $1\frac{1}{4}$ points, $\frac{5}{8}$ point, $1\frac{1}{2}$ points, and $\frac{3}{4}$ point respectively. What was its closing price on Friday?

   a. $32\frac{7}{8}$ **b.** $32\frac{3}{4}$

   c. $32\frac{5}{8}$ **d.** Answer not given

3. Find the net change in a stock if it opened at $16\frac{1}{8}$ and closed at $15\frac{1}{4}$.

   a. $\frac{3}{8}$ point loss   **b.** $\frac{7}{8}$ point loss

   c. $1\frac{7}{8}$ points loss   **d.** Answer not given

4. A house worth \$52,800 is assessed at $\frac{2}{3}$ of its value. What is the assessed value?

   a. \$33,600   **b.** \$37,500

   c. \$35,200   **d.** \$30,500

5. What is the perimeter (distance around) of a triangle if its three sides measure $6\frac{3}{8}$ inches, $4\frac{11}{16}$ inches, and $5\frac{3}{4}$ inches respectively?

6. If the scale on a chart is 1 inch = 30 miles, how many miles do $6\frac{1}{2}$ inches represent?

7. A plumber, in installing water pipes, used pieces measuring $5\frac{1}{2}$ feet, $3\frac{3}{4}$ feet, and $1\frac{2}{3}$ feet. If they were cut from a 15-foot length of pipe, how many feet of pipe remained? Disregard waste.

8. A family budgets $\frac{1}{4}$ of its annual income of \$18,600 for food, $\frac{3}{10}$ for shelter including operating expenses and furnishings, $\frac{1}{10}$ for transportation, $\frac{1}{20}$ for clothing, $\frac{1}{8}$ for savings, and the remainder for miscellaneous. How much is allowed for each item annually?

Refresh Your Skills

1. `1-8`
$$627$$
$$4,963$$
$$81,896$$
$$97,988$$
$$+\ \ 8,569$$

2. `1-9`
$$800,000$$
$$-\ 706,094$$

3. `1-11`
$$5,280$$
$$\times\ 5,280$$

4. `1-13`
$$365\overline{)149,285}$$

5. `2-5`
$$0.66 + 0.6 + 0.666$$

6. $0.62 - 0.2$ `2-6`

7. $0.08 \times \$19.75$ `2-7`

8. $0.02\overline{)0.007}$ `2-8`

9. Multiply 81.24 by 100. `2-7`

10. Divide 0.6 by 1,000. `2-8`

11. Express $\frac{36}{54}$ in lowest terms. `3-1`

12. `3-8`
$$3\frac{7}{10}$$
$$+\ 6\frac{5}{12}$$

13. `3-9`
$$4\frac{1}{3}$$
$$-\ 3\frac{5}{6}$$

14. $4\frac{3}{8} + 3\frac{1}{5}$ `3-8`

Using Mathematics

MULTIPLICATIVE INVERSE

If the product of two numbers is 1, then each factor is called the **multiplicative inverse**, or **reciprocal**, of the other.

8 and $\frac{1}{8}$ are multiplicative inverses of each other because $8 \times \frac{1}{8} = 1$.

$\frac{4}{3}$ and $\frac{3}{4}$ are multiplicative inverses of each other because $\frac{4}{3} \times \frac{3}{4} = 1$.

When w divide 12 by 4, the quotient is 3.
$$12 \div 4 = 3$$

When we multiply 12 by $\frac{1}{4}$ the product is 3.
$$12 \times \frac{1}{4} = 3$$

Thus, to divide a number by another number, you may multiply the first number by the multiplicative inverse of the divisor.

EXAMPLE: $\frac{2}{3} \div \frac{3}{4} = \frac{2}{3} \times \frac{4}{3} = \frac{8}{9}$

Notice that the numerator of one fraction is the denominator of its reciprocal, and its denominator is the numerator of its reciprocal.

Zero has no multiplicative inverse.

Notice that dividing a number by another number gives the same result as multiplying the first number by the multiplicative inverse of the divisor.

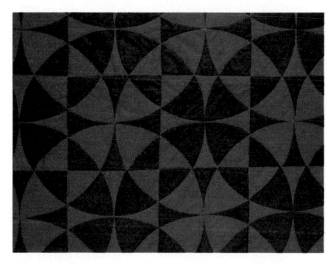

Exercises

Write the multiplicative inverse, or reciprocal, of each of the following.

1. 6   **2.** 3   **3.** 1   **4.** 9   **5.** 25   **6.** 0   **7.** 12

8. $\frac{1}{4}$ **9.** $\frac{1}{10}$ **10.** $\frac{1}{7}$ **11.** $\frac{1}{2}$ **12.** $\frac{1}{3}$ **13.** $\frac{1}{15}$ **14.** $\frac{1}{20}$

15. $\frac{5}{6}$ **16.** $\frac{7}{12}$ **17.** $\frac{3}{8}$ **18.** $\frac{2}{5}$ **19.** $\frac{9}{10}$ **20.** $\frac{11}{16}$ **21.** $\frac{23}{24}$

22. $\frac{8}{5}$ **23.** $\frac{5}{4}$ **24.** $\frac{22}{7}$ **25.** $\frac{9}{2}$ **26.** $\frac{13}{6}$ **27.** $\frac{19}{8}$ **28.** $\frac{21}{12}$

29. $2\frac{1}{2}$ **30.** $6\frac{2}{3}$ **31.** $1\frac{3}{4}$ **32.** $4\frac{3}{8}$ **33.** $1\frac{5}{12}$ **34.** $6\frac{2}{5}$ **35.** $3\frac{9}{16}$

Find the missing numbers.

36. $\frac{3}{8} \times \frac{8}{3} = $ ▩

37. $12 \times $ ▩ $= 1$

38. ▩ $\times \frac{1}{9} = 1$

39. $\frac{5}{6} \times $ ▩ $= 1$

40. ▩ $\times \frac{11}{16} = 1$

41. $\frac{7}{10} \times \frac{10}{7} = $ ▩

42. $\frac{9}{5} \times $ ▩ $= 1$

43. ▩ $\times \frac{15}{2} = 1$

Dividing Fractions and Mixed Numbers

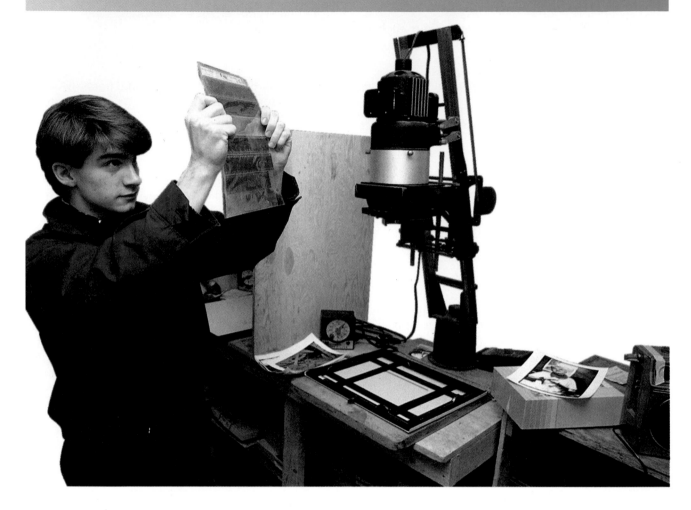

Building Problem Solving Skills

Jamie is an amateur photographer with his own darkroom for developing his film. Each roll of film takes $\frac{1}{8}$ bottle of developer solution. If the bottle is $\frac{3}{4}$ full, how many rolls of film can Jamie develop?

---

1. **READ** the problem carefully.

a. Find the question asked. How many rolls of film can Jamie develop?

b. Find the given facts. Bottle of developer solution is $\frac{3}{4}$ full. Each roll of film takes $\frac{1}{8}$ bottle.

Division Indicators
• how much is each
• how many in each
• find the average
• rate per unit
Note: All indicators generally refer to equal quantities.

2. [**PLAN**] how to solve the problem.

a. Choose the operation needed.
In order to find how many parts in the whole, you must divide.

b. Think or write out your plan relating the given facts to the question asked.

The amount of solution available divided by the amount needed for each roll of film equals the number of rolls that can be developed.

c. Express your plan as an equation.
$\frac{3}{4} \div \frac{1}{8} = n$

---

3. [**SOLVE**] the problem.

a. Estimate the answer.

$\frac{3}{4} \rightarrow 1$ — Round the dividend to the nearest whole number.

$1 \div \frac{1}{8} = 1 \times 8 = 8$ — Invert the divisor. Multiply.

b. Solution.

$\frac{3}{4} \div \frac{1}{8} = n$ — Dividing by a number gives the same result as multiplying by its reciprocal.

$\frac{3}{\overset{}{\underset{1}{4}}} \times \frac{\overset{2}{8}}{1} = n$ — Invert divisor to obtain its reciprocal.
Divide numerator and denominator by their greatest common factor.

$3 \times 2 = 6$ — Multiply.

You might use a calculator.
Key: 4 [×] 1 [M+] 3 [×] 8 [÷] [M$\overset{R}{C}$] [=] 6

---

4. [**CHECK**] .

a. Check the accuracy of your arithmetic.

$6 \times \frac{1}{8} = \frac{6}{8} = \frac{3}{4}$ — Multiply the number of rolls you computed by the amount of solution needed for each roll.

b. Compare the answer to the estimate.
If a whole bottle of solution will develop 8 rolls of film, it is reasonable that $\frac{3}{4}$ bottle will develop 6 rolls of film.

ANSWER: Jamie can develop 6 rolls of film. The unit, rolls of film, is correct.

Think About It

1. When you divide two fractions less than 1, will the quotient be less than or greater than the given fractions? Give an example.

2. When you divide a fraction less than 1 by a whole number greater than 1, will the answer be less than or greater than the fraction? Give an example.

3. When you divide a whole number by a fraction, will the answer be less than or greater than the whole number? Explain.

Calculator Know-How

You can divide fractions and mixed numbers on a calculator. Change a mixed number to an improper fraction before dividing.

Examples

1. $\frac{1}{2} \div \frac{2}{5} = n$

$\frac{1}{2} \times \frac{5}{2} = n$ Key: 2 $\boxed{\times}$ 2 $\boxed{M+}$ 1 $\boxed{\times}$ 5 $\boxed{\div}$ $\boxed{M\frac{R}{C}}$ $\boxed{=}$ <u>1.25</u>

4 $\boxed{\times}$ 1 $\boxed{=}$ 4

5 $\boxed{-}$ 4 $\boxed{=}$ 1

$\frac{1}{2} \div \frac{2}{5} = \frac{1}{2} \times \frac{5}{2} = 1.25 = 1\frac{1}{4}$

2. $3\frac{2}{3} \div 1\frac{6}{7} = n$

$\frac{11}{3} \div \frac{13}{7} = n$

$\frac{11}{3} \times \frac{7}{13} = n$ Key: 3 $\boxed{\times}$ 13 $\boxed{M+}$ 11 $\boxed{\times}$ 7 $\boxed{\div}$ $\boxed{M\frac{R}{C}}$ $\boxed{=}$ <u>1.9743589</u>

39 $\boxed{\times}$ 1 $\boxed{=}$ 39

77 $\boxed{-}$ 39 $\boxed{=}$ <u>38</u>

$\frac{11}{3} \times \frac{7}{13} = \frac{77}{39} = 1\frac{38}{39}$

Exercises

Divide on a calculator. Write the answers in simplest form.

1. $\frac{1}{4} \div \frac{3}{10}$ **2.** $\frac{2}{5} \div \frac{5}{8}$ **3.** $4\frac{7}{8} \div 2\frac{3}{5}$ **4.** $5\frac{3}{8} \div 3\frac{5}{7}$

5. Which of the exercises above is easier with paper and pencil? Why?

1. Divide: $\frac{2}{5} \div \frac{5}{8}$

reciprocals

$$\frac{2}{5} \div \frac{5}{8} = \frac{2}{5} \times \frac{8}{5} = \frac{16}{25}$$

ANSWER: $\frac{16}{25}$

To divide by a fraction, multiply by the reciprocal of the divisor.

---

2. Divide: $\frac{7}{8} \div 3$

$$\frac{7}{8} \div 3 = \frac{7}{8} \div \frac{3}{1} = \frac{7}{8} \times \frac{1}{3} = \frac{7}{24}$$

Express as a fraction.

ANSWER: $\frac{7}{24}$

---

3. Divide: $2\frac{1}{2} \div \frac{3}{4}$

$$2\frac{1}{2} \div \frac{3}{4} = \frac{5}{2} \div \frac{3}{4}$$

$$= \frac{5}{\overset{1}{2}} \times \frac{\overset{2}{4}}{3} = \frac{10}{3} = 3\frac{1}{3}$$

ANSWER: $3\frac{1}{3}$

Rewrite mixed numbers as improper fractions.

---

4. Divide: $8 \div 2\frac{4}{5}$

$$8 \div 2\frac{4}{5} = \frac{8}{1} \div \frac{14}{5}$$

$$= \frac{\overset{4}{8}}{1} \times \frac{5}{\underset{7}{14}} = \frac{20}{7} = 2\frac{6}{7}$$

ANSWER: $2\frac{6}{7}$

5. Divide $2\frac{3}{16}$ by $1\frac{1}{4}$.

$$2\frac{3}{16} \div 1\frac{1}{4} = \frac{35}{16} \div \frac{5}{4}$$

$$= \frac{\overset{7}{35}}{\underset{4}{16}} \times \frac{\overset{1}{4}}{\underset{1}{5}} = \frac{7}{4} = 1\frac{3}{4}$$

ANSWER: $1\frac{3}{4}$

---

6. Simplify: $\dfrac{\frac{9}{16}}{\frac{3}{8}}$

complex fraction

$$\frac{\frac{9}{16}}{\frac{3}{8}} = \frac{9}{16} \div \frac{3}{8}$$

$$= \frac{\overset{3}{9}}{\underset{2}{16}} \times \frac{\overset{1}{8}}{\underset{1}{3}} = \frac{3}{2} = 1\frac{1}{2}$$

ANSWER: $1\frac{1}{2}$

The fraction bar means "divided by."
Divide the numerator by the denominator

Vocabulary

If the product of two numbers is 1, then each factor is the **reciprocal** of the other.

A **complex fraction** is a fraction in which the numerator, the denominator, or both are fractions or decimals.

Diagnostic Exercises

Divide.

1. $\frac{1}{3} \div \frac{3}{4}$

2. $\frac{2}{3} \div \frac{5}{16}$

3. $\frac{3}{5} \div \frac{9}{10}$

4. $\frac{3}{4} \div \frac{3}{8}$

5. $\frac{5}{6} \div \frac{7}{12}$

6. $\frac{3}{4} \div 6$

7. $\frac{7}{8} \div 2$

8. $8 \div \frac{1}{2}$

9. $5 \div \frac{15}{16}$

10. $2 \div \frac{3}{5}$

11. $4\frac{1}{2} \div 18$

12. $4\frac{2}{3} \div 2$

13. $1\frac{2}{3} \div 4$

14. $15 \div 1\frac{7}{8}$

15. $6 \div 4\frac{1}{2}$

16. $4 \div 5\frac{1}{3}$

17. $7 \div 2\frac{3}{4}$

18. $2\frac{1}{2} \div \frac{5}{6}$

19. $2\frac{5}{8} \div \frac{3}{5}$

20. $1\frac{7}{16} \div \frac{2}{3}$

21. $\frac{7}{8} \div 1\frac{3}{4}$

22. $\frac{3}{4} \div 1\frac{3}{5}$

23. $11\frac{1}{3} \div 2\frac{5}{6}$

24. $3\frac{3}{16} \div 2\frac{1}{8}$

25. $1\frac{1}{6} \div 9\frac{1}{3}$

26. $5\frac{3}{8} \div 1\frac{2}{5}$

27. $6\frac{3}{5} \div 6\frac{3}{5}$

28. Divide $9\frac{1}{3}$ by $3\frac{1}{7}$.

29. Simplify: $\dfrac{2\frac{3}{5}}{8}$

30. Solve: $1\frac{3}{4} \div 8\frac{1}{6} = n$

31. A large serving bowl can hold 30 cups of punch. How many $1\frac{1}{5}$-cup helpings can be served from the bowl?

Related Practice

Divide.

1. a. $\frac{1}{4} \div \frac{1}{3}$　　b. $\frac{3}{4} \div \frac{4}{5}$　　c. $\frac{1}{2} \div \frac{2}{3}$　　d. $\frac{3}{5} \div \frac{11}{12}$

2. a. $\frac{1}{2} \div \frac{1}{5}$　　b. $\frac{7}{8} \div \frac{2}{3}$　　c. $\frac{9}{10} \div \frac{1}{3}$　　d. $\frac{4}{9} \div \frac{7}{10}$

3. a. $\frac{1}{4} \div \frac{3}{4}$　　b. $\frac{5}{8} \div \frac{5}{6}$　　c. $\frac{3}{16} \div \frac{5}{12}$　　d. $\frac{5}{6} \div \frac{7}{8}$

4. a. $\frac{5}{6} \div \frac{5}{12}$　　b. $\frac{2}{3} \div \frac{2}{9}$　　c. $\frac{5}{8} \div \frac{5}{8}$　　d. $\frac{4}{5} \div \frac{4}{15}$

5. a. $\frac{1}{2} \div \frac{7}{16}$　　b. $\frac{7}{8} \div \frac{5}{12}$　　c. $\frac{4}{5} \div \frac{7}{10}$　　d. $\frac{5}{12} \div \frac{3}{16}$

6. a. $\frac{2}{3} \div 4$　　b. $\frac{7}{8} \div 7$　　c. $\frac{9}{10} \div 6$　　d. $\frac{3}{5} \div 9$

7. a. $\frac{3}{5} \div 8$　　b. $\frac{1}{2} \div 10$　　c. $\frac{11}{12} \div 5$　　d. $\frac{5}{8} \div 4$

8. a. $6 \div \frac{1}{3}$　　b. $7 \div \frac{1}{8}$　　c. $8 \div \frac{4}{5}$　　d. $10 \div \frac{5}{16}$

9. a. $18 \div \frac{9}{10}$　　b. $6 \div \frac{4}{5}$　　c. $10 \div \frac{2}{3}$　　d. $12 \div \frac{15}{16}$

10. a. $5 \div \frac{3}{4}$　　b. $13 \div \frac{2}{3}$　　c. $9 \div \frac{7}{8}$　　d. $7 \div \frac{9}{10}$

11. a. $1\frac{1}{2} \div 3$ **b.** $4\frac{2}{3} \div 14$ **c.** $5\frac{3}{5} \div 7$ **d.** $2\frac{7}{16} \div 13$

12. a. $7\frac{1}{2} \div 5$ **b.** $8\frac{2}{5} \div 6$ **c.** $18\frac{3}{4} \div 10$ **d.** $11\frac{7}{8} \div 5$

13. a. $1\frac{5}{8} \div 2$ **b.** $2\frac{1}{16} \div 4$ **c.** $5\frac{3}{4} \div 5$ **d.** $3\frac{4}{5} \div 8$

14. a. $6 \div 1\frac{1}{2}$ **b.** $8 \div 1\frac{1}{3}$ **c.** $68 \div 3\frac{2}{5}$ **d.** $57 \div 2\frac{3}{8}$

15. a. $8 \div 2\frac{2}{5}$ **b.** $10 \div 1\frac{7}{8}$ **c.** $14 \div 1\frac{5}{16}$ **d.** $40 \div 5\frac{1}{3}$

16. a. $5 \div 6\frac{2}{3}$ **b.** $8 \div 9\frac{3}{5}$ **c.** $6 \div 7\frac{7}{8}$ **d.** $2 \div 4\frac{4}{5}$

17. a. $6 \div 1\frac{2}{3}$ **b.** $4 \div 6\frac{3}{5}$ **c.** $1 \div 2\frac{1}{3}$ **d.** $3 \div 1\frac{5}{8}$

18. a. $1\frac{1}{2} \div \frac{9}{16}$ **b.** $1\frac{1}{8} \div \frac{3}{32}$ **c.** $8\frac{3}{4} \div \frac{7}{8}$ **d.** $9\frac{1}{3} \div \frac{7}{24}$

19. a. $1\frac{3}{5} \div \frac{2}{3}$ **b.** $1\frac{5}{6} \div \frac{5}{12}$ **c.** $3\frac{1}{4} \div \frac{5}{6}$ **d.** $2\frac{1}{8} \div \frac{9}{10}$

20. a. $4\frac{1}{4} \div \frac{3}{5}$ **b.** $2\frac{3}{5} \div \frac{5}{8}$ **c.** $3\frac{1}{7} \div \frac{3}{4}$ **d.** $2\frac{2}{3} \div \frac{9}{16}$

21. a. $\frac{5}{12} \div 8\frac{1}{3}$ **b.** $\frac{5}{6} \div 1\frac{1}{9}$ **c.** $\frac{3}{5} \div 2\frac{2}{5}$ **d.** $\frac{13}{16} \div 1\frac{7}{32}$

22. a. $\frac{2}{3} \div 1\frac{1}{4}$ **b.** $\frac{7}{8} \div 3\frac{1}{3}$ **c.** $\frac{3}{16} \div 1\frac{3}{5}$ **d.** $\frac{1}{3} \div 2\frac{9}{16}$

23. a. $14\frac{3}{8} \div 2\frac{7}{8}$ **b.** $7\frac{1}{2} \div 1\frac{1}{4}$ **c.** $18\frac{1}{3} \div 1\frac{5}{6}$ **d.** $50\frac{1}{4} \div 4\frac{3}{16}$

24. a. $3\frac{1}{5} \div 1\frac{1}{3}$ **b.** $1\frac{5}{8} \div 1\frac{7}{32}$ **c.** $4\frac{2}{3} \div 1\frac{3}{5}$ **d.** $11\frac{1}{4} \div 2\frac{1}{2}$

25. a. $2\frac{1}{4} \div 3\frac{3}{8}$ **b.** $1\frac{3}{5} \div 3\frac{1}{5}$ **c.** $1\frac{13}{16} \div 2\frac{1}{4}$ **d.** $2\frac{1}{2} \div 3\frac{3}{4}$

26. a. $1\frac{2}{5} \div 2\frac{2}{3}$ **b.** $3\frac{3}{8} \div 3\frac{1}{5}$ **c.** $1\frac{1}{2} \div 1\frac{7}{9}$ **d.** $1\frac{2}{3} \div 1\frac{7}{16}$

27. a. $\frac{3}{8} \div \frac{3}{8}$ **b.** $1\frac{5}{12} \div 1\frac{5}{12}$ **o.** $2\frac{2}{3} \div 2\frac{2}{3}$ **d.** $6\frac{13}{16} \div 6\frac{13}{16}$

28. Divide.

   a. $4\frac{1}{2}$ by $\frac{3}{5}$ **b.** $6\frac{1}{4}$ by 8 **c.** $2\frac{2}{3}$ by $7\frac{1}{2}$ **d.** $1\frac{7}{8}$ by $1\frac{1}{3}$

29. Simplify.

   a. $\dfrac{3\frac{3}{8}}{4\frac{1}{2}}$ **b.** $\dfrac{9\frac{1}{2}}{4}$ **c.** $\dfrac{\frac{1}{2}+\frac{1}{4}}{\frac{3}{8}}$ **d.** $\dfrac{\frac{3}{4}+\frac{7}{8}}{\frac{11}{16}-\frac{1}{2}}$

30. Solve.

   a. $\frac{2}{5} \div \frac{9}{16} = n$ **b.** $\frac{7}{8} \div \frac{7}{16} = \blacksquare$ **c.** $3\frac{3}{8} \div 9 = n$ **d.** $\blacksquare = \frac{7}{12} \div 3\frac{3}{4}$

31. a. If an airplane flies 1,440 kilometers in $2\frac{1}{4}$ hours, what is its average ground speed in kilometers per hour?

   b. If a board $10\frac{1}{2}$ feet long was cut into 6 pieces of equal length, what would be the length of each piece? Disregard waste.

   c. What is the cost of 1 pound of apples if $2\frac{5}{8}$ pounds cost 84¢?

   d. How many athletic association membership cards $1\frac{3}{8}$ inches wide can be cut from stock $24\frac{3}{4}$ inches wide?

Mixed Problems

Choose an appropriate method of computation. For Problems 1–4 select the letter corresponding to your answer.

1. If each costume for the school show requires $3\frac{1}{3}$ yards of material, how many costumes can be made from a 30-yard bolt of material?

   a. 12 costumes   **b.** 9 costumes
   c. 4 costumes   **d.** 7 costumes

2. What are the actual dimensions of a porch that, drawn to the scale of $\frac{1}{4}$ inch = 1 foot, measures $1\frac{3}{4}$ inches by $2\frac{1}{2}$ inches?

   a. 8 feet by 10 feet   **b.** 7 feet by 10 feet
   c. 7 feet by 11 feet   **d.** Answer not given

3. In arranging an $8\frac{1}{2}$ inch by 11 inch piece of paper in the drawing class, pupils were directed to draw a line $\frac{1}{4}$ inch from each edge. What are the inside dimensions between the lines?

   a. $8\frac{1}{4} \times 10\frac{3}{4}$ inches   **b.** $8 \times 10\frac{3}{4}$ inches
   c. $8 \times 10\frac{1}{2}$ inches   **d.** Answer not given

4. Paul wishes to buy a cassette recorder priced at $60. He pays $\frac{1}{5}$ of the price in cash and the rest in 6 equal monthly installments. How much must he pay each month?

   a. $8   **b.** $9
   c. $10   **d.** Answer not given

5. How many lengths of pipe $3\frac{1}{2}$ feet long can be cut from a pipe 28 feet long? Disregard waste in cutting.

6. If 1 cubic foot of water weighs $62\frac{1}{2}$ pounds, find the weight of a column of water containing 32 cubic feet.

7. Carmela bought two tapes for $7.90 each and one tape for $6.57. How much change did she get from $25?

8. A recipe calls for $2\frac{1}{4}$ cups of flour. If you want to double the recipe, how many cups of flour will you need?

Refresh Your Skills

1. $\begin{array}{r} 92,838 \\ 79,665 \\ 83,756 \\ 44,973 \\ + 58,786 \\ \hline \end{array}$ 1–8

2. $\begin{array}{r} 156,072 \\ - 75,483 \\ \hline \end{array}$ 1–9

3. $\begin{array}{r} 3,600 \\ \times 905 \\ \hline \end{array}$ 1–11

4. $1,728\overline{)1,394,496}$ 1–13

5. $0.429 + 3.85 + 73.9$ 2–5

6. $8.7 - 0.23$ 2–6   **7.** 500×0.06 2–7   **8.** $\$1.10\overline{)\$13.20}$ 2–9

9. Write 407.9 million as a complete numeral. 2–15

10. Write the short name for 53,700,000 in millions. 2–16

11. $\begin{array}{r} 2\frac{3}{8} \\ \frac{11}{12} \\ + 9\frac{1}{6} \\ \hline \end{array}$ 3–8

12. $\begin{array}{r} 11\frac{1}{2} \\ - 5\frac{2}{3} \\ \hline \end{array}$ 3–9

13. $2\frac{7}{10} \times 3\frac{3}{4}$ 3–11

14. $\begin{array}{r} 2\frac{2}{3} \\ \times 24 \\ \hline \end{array}$ 3–11

3-13 Fractional Parts of Numbers

Sometimes you will have to find what fractional part one number is of another, and to find a number when a fractional part is known.

> Ming Ling correctly answered 9 out of 12 questions. What fraction of the questions did she answer correctly?

9 is what part of 12?

$\dfrac{9}{12}$ ← **part**
    ← **whole**

$= \dfrac{3}{4}$

Write a fraction.
Use the **given part** as the numerator and the number of parts in the **whole** as the denominator.

Write the fraction in lowest terms.

ANSWER: Ming Ling answered $\frac{3}{4}$ of the questions correctly.

Examples

1. Compare 10 with 16.

Write a fraction.
Use the number being compared as the numerator, and the number it is compared with as the denominator.

$\dfrac{10}{16}$ ← **number being compared**
     ← **number it is compared with**

$\dfrac{10}{16} = \dfrac{5}{8}$

ANSWER: 10 is $\frac{5}{8}$ of 16.

2. Ramon had $30 deducted from his pay. This was $\frac{1}{9}$ of the total he had earned. How much had he earned?

$\frac{1}{9}$ of what number is $30?

$\$30 \div \frac{1}{9}$

$\$30 \div \frac{1}{9} = \frac{30}{1} \div \frac{1}{9}$

$= \frac{30}{1} \times \frac{9}{1}$

$= \$270$

Divide the known part by the given fraction representing this part.

ANSWER: Ramon earned $270.

3. $\frac{3}{4}$ of what number is 21?

$21 \div \frac{3}{4} = \frac{21}{1} \div \frac{3}{4}$

$= \frac{\overset{7}{\cancel{21}}}{1} \times \frac{4}{\underset{1}{\cancel{3}}}$

$= 28$

ANSWER: 21 is $\frac{3}{4}$ of 28.

Think About It

1. When you ask " What part of 6 is 5?" what is the whole? What is the part? Make a drawing to show this.

2. How is asking "What part of 6 is 5?" different from comparing 6 with 5?

3. If 0.25 of a number is 48, you could find the number by dividing 48 by 0.25. Give another method for finding the number and explain why it works.

Diagnostic Exercises

1. What part of 6 is 1?

2. 40 is what part of 100?

3. What part of 56 is 35?

4. Compare 3 with 5.

5. Compare 12 with 16.

6. Compare 15 with 9.

7. Compare 8 with 1.

8. Compare 24 with 24.

9. $\frac{1}{4}$ of what number is 12?

10. $\frac{5}{8}$ of what number is 45?

11. 42 is $\frac{3}{5}$ of what number?

12. 560 is 0.875 of what number?

13. If 175 out of 200 freshmen passed their physical examination, what part of the freshmen passed?

14. Shay weighs 42 kilograms, or $\frac{3}{4}$ as much as her brother Stuart. How much does Stuart weigh?

Related Practice

1. a. What part of 4 is 1?
   c. 7 is what part of 8?

b. 1 is what part of 5?
d. What part of 24 is 13?

2. a. What part of 100 is 28?
   c. $37\frac{1}{2}$ is what part of 100?

b. 75 is what part of 100?
d. What part of 100 is $66\frac{2}{3}$?

3. a. What part of 12 is 6?
   c. 45 is what part of 54?

b. What part of 75 is 25?
d. 66 is what part of 108?

Compare.

4. a. 1 with 8   **b.** 3 with 7   **c.** 5 with 6   **d.** 12 with 25

5. a. 6 with 10   **b.** 21 with 28   **c.** 84 with 96   **d.** 25 with 30

6. a. 7 with 6   **b.** 13 with 8   **c.** 24 with 10   **d.** 40 with 16

7. a. 2 with 1   **b.** 5 with 1   **c.** 10 with 1   **d.** 28 with 1

8. a. 8 with 8   **b.** 10 with 5   **c.** 24 with 6   **d.** 96 with 4

9. a. $\frac{1}{2}$ of what number is 500?
   c. $\frac{1}{4}$ of what number is 250?

b. $\frac{1}{3}$ of what number is 18?
d. $\frac{1}{16}$ of what number is 21?

10. a. $\frac{3}{4}$ of what number is 9?
   c. $\frac{2}{3}$ of what number is 110?

b. $\frac{4}{5}$ of what number is 55?
d. $\frac{5}{8}$ of what number is 20?

11. a. 6 is $\frac{3}{4}$ of what number?
   c. 27 is $\frac{2}{3}$ of what number?

b. 30 is $\frac{5}{8}$ of what number?
d. 56 is $\frac{7}{16}$ of what number?

12. a. 27 is $0.37\frac{1}{2}$ of what number?
   c. 24 is $0.66\frac{2}{3}$ of what number?

b. $0.09\frac{3}{4}$ of what number is 780?
d. $0.08\frac{1}{3}$ of what number is 960?

13. a. Meyer made two errors in 25 fielding chances. In what part of his chances did he field the ball cleanly?

b. Janet, a pitcher, won 8 games and lost 4. What part of the games did she win?

c. At bat 36 times, Sue made 7 singles, 2 doubles, 1 triple, and 2 home runs. What part of the time did she hit safely?

d. Ramirez faced 35 batters. He struck out 9, 10 grounded out, 11 flied out, and the rest reached base. What fraction of the batters reached base?

14. a. Charlotte received $\frac{5}{6}$ of all the votes cast in the election for school treasurer. If she received 885 votes, how many students voted?

b. If 138 students, or $\frac{3}{8}$ of the graduating class, selected the college preparatory course, how many pupils were in the graduating class?

c. Paul purchased a radio that was reduced by $\frac{1}{4}$ from its regular price. If he paid $25.50 for the radio, what was the regular price?

d. If the school baseball team won 16 games, or $\frac{2}{3}$ of the games played, how many games were lost?

Problem Solving Strategy: Guess and Test

Jerome is buying supplies for his jewelry making. He bought some large beads that cost $.55 each, and some small beads that cost $.35 each. He bought 24 beads and spent $10.20 in all. How many beads of each kind did he buy?

Strategy: Guess and Test
Make a reasonable guess at the answer, then test it with the given facts. Keep guessing and testing until the correct answer is found.

1. **READ** the problem carefully.

a. Find the question to be answered.
How many beads of each kind did he buy?
The answer will be a number of small beads and a number of large beads.

b. Find the given facts.
Large beads cost $.55.
Small beads cost $.35.

2. **PLAN** how to solve the problem.

a. Choose a strategy: guess and test

b. Think or write out your plan relating the given facts to the question asked.
The cost of the large beads plus the cost of the small beads must total $10.20.
(large beads × $.55) + (small beads × $.35) must equal $10.20.
The number of large beads plus the number of small beads must equal 24.

3. **SOLVE** the problem.
Choose pairs of numbers whose sum is 24.
Make a table to keep track of your tries.
Test 10 large and 14 small beads:
(10 × $.55) + (14 × $.35) = $10.40   Too much.

Test 8 large and 16 small beads:
(8 × $.55) + (16 × $.35) = $10.00   Too little.
Test 9 large and 15 small beads:
(9 × $.55) + (15 × $.35) = $10.20   Right!

4. CHECK .

a. Check the accuracy of your arithmetic.

$$
\begin{array}{r}
9 \\
\times\ \$.55 \\
\hline
45 \\
4\ 5 \\
\hline
\$4.95
\end{array}
\qquad
\begin{array}{r}
15 \\
\times\ \$.35 \\
\hline
75 \\
4\ 5 \\
\hline
\$5.25
\end{array}
\qquad
\begin{array}{r}
\$5.25 \\
+\ \ 4.95 \\
\hline
\$10.20
\end{array}
$$

b. See whether the answer is reasonable.

$9 + 15 = 24$ $\$5.25 + \$4.95 = \$10.20$

The answer fits the facts given.

ANSWER: He bought 9 large beads and 15 small beads.
The units of the answer, large beads and small beads, are correct.

Practice Exercises

Use guess and test to solve each problem.

1. There are 7 more girls than boys in the science class of 31. How many boys are there?

2. What two consecutive numbers add up to 35?

3. First-class mail costs $.25 for the first ounce and $.20 for each additional ounce. If you were charged $1.55, what is the weight of the package?

4. How long is a telephone call that cost $1.70 at the rate of $.30 for the first minute and $.20 for each additional minute?

5. How many prints were made if you paid $3.25 at the rate of $1.00 for processing the roll of film and $.15 per print?

6. What distance did you travel by taxi if the rate is $.90 for the first $\frac{1}{4}$ mile and $.50 for each additional $\frac{1}{4}$ mile and you paid $3.40?

7. Find three consecutive even numbers whose sum is 144.

8. Find two consecutive numbers whose product is 1,260.

9. Twice as many full-size cars as compact cars are in the parking lot. The total number of cars parked is 162. How many full-size cars are there?

10. Jana has 12 coins: pennies, nickels, dimes, and quarters. The total value is $1.43. How many of each coin does she have?

Writing Fractions as Decimals

Cindy got 2 hits out of her 5 times at bat in the baseball game against the Wildkits. She was, therefore, on base $\frac{2}{5}$ of the time. What is her batting average? Write $\frac{2}{5}$ as a decimal.

$$5\overline{)2.0} \quad \begin{array}{r} .4 \\ \hline 2\,0 \end{array}$$

Divide the numerator by the denominator.
The zero before the decimal point is not used with batting averages.

ANSWER: Her batting average is .400

Examples

1. Write $\frac{2}{7}$ as a two-place decimal.

$$7\overline{)2.00} \quad \begin{array}{r} 0.28\ \frac{4}{7} \\ \hline 1\,4 \\ \hline 60 \\ 56 \\ \hline 4 \end{array}$$

ANSWER: $0.28\frac{4}{7}$

2. Write $\frac{7}{6}$ as a two-place decimal.

$$6\overline{)7.00} \quad \begin{array}{r} 1.16\ \frac{2}{3} \\ \hline 6 \\ \hline 1\,0 \\ 6 \\ \hline 40 \\ 36 \\ \hline 4 \end{array}$$

ANSWER: $1.16\frac{2}{3}$

3. Write $\frac{19}{100}$ as a decimal.

$$\frac{19}{100} = 0.\underset{\curvearrowleft}{19}$$

 2 zeros 2 places

ANSWER: 0.19

If a fraction has a denominator that is a power of ten, 10 or 100 or 1,000 and so on, write the numerator and move the decimal point as many places to the left as there are zeros in the denominator.

4. Write $\frac{1,650}{1,000}$ as a decimal.

$$\frac{1,650}{1,000} = 1.\underset{\curvearrowleft}{650}$$

3 zeros   3 places

ANSWER: 1.65

---

5. Write $\frac{1}{4}$ as a decimal.

$$\frac{1}{4} = \frac{1 \times 25}{4 \times 25} = \frac{25}{100}$$

$$\frac{25}{100} = 0.25$$

ANSWER: 0.25

If a fraction has a denominator that can be written as an equivalent fraction with denominator a power of ten, then write as an equivalent fraction.

---

6. Write $1\frac{7}{8}$ as a decimal.

$$1\frac{7}{8} = 1 + \frac{7}{8} = 1 + .087\frac{1}{2} = 1.87\frac{1}{2} \text{ or } 1.875$$

ANSWER: $1.87\frac{1}{2}$ or 1.875

Write the mixed number as a sum. Then rewrite the fraction as a decimal. Add the whole number and the decimal part.

Think About It

1. Do you think there is any fraction that you could not write as a decimal?

2. How many hits out of 16 times at bat would Arthur need to average better than .350?

3. During a game, Cindy struck out each of her first 3 times at the plate. If she was at bat a total of 7 times during the game, what are all of her possible batting averages?

4. Cindy needs a .450 batting average to make the All-Star Team. Her season's record to date is 15 hits out of 42 times at bat. If there is only one more game, can she possibly make the All-Star Team? Assume that she will bat 4 times in the last game.

Diagnostic Exercises

Write the following fractions or mixed numbers as decimals. (Carry out Exercises 5, 6, and 9–16 to 2 places.)

1. $\frac{9}{10}$ **2.** $\frac{1}{2}$ **3.** $\frac{27}{100}$ **4.** $\frac{1}{4}$ **5.** $\frac{7}{8}$ **6.** $\frac{5}{6}$

7. $\frac{125}{100}$ **8.** $\frac{37\frac{1}{2}}{100}$ **9.** $\frac{8}{9}$ **10.** $\frac{20}{25}$ **11.** $\frac{49}{56}$ **12.** $\frac{24}{28}$

13. $1\frac{3}{4}$ **14.** $\frac{12}{8}$ **15.** $\frac{18}{14}$ **16.** $2\frac{7}{16}$ **17.** $\frac{893}{1000}$ **18.** $\frac{7429}{10000}$

19. The length of one bolt is $\frac{5}{8}$ inch. The length of another bolt is 0.619 inch. Which bolt is longer?

Related Practice

Write the following fractions or mixed numbers as decimals. (Carry out sets 5, 6, and 9–16 to 2 places.)

1. a. $\frac{1}{10}$ **b.** $\frac{7}{10}$ **c.** $\frac{2}{10}$ **d.** $\frac{8}{10}$ **2. a.** $\frac{4}{5}$ **b.** $\frac{1}{5}$ **c.** $\frac{2}{5}$ **d.** $\frac{3}{5}$

3. a. $\frac{39}{100}$ **b.** $\frac{3}{100}$ **c.** $\frac{6}{100}$ **d.** $\frac{91}{100}$ **4. a.** $\frac{3}{4}$ **b.** $\frac{17}{20}$ **c.** $\frac{14}{25}$ **d.** $\frac{41}{50}$

5. a. $\frac{3}{8}$ **b.** $\frac{5}{8}$ **c.** $\frac{1}{8}$ **d.** $\frac{9}{16}$ **6. a.** $\frac{1}{3}$ **b.** $\frac{1}{6}$ **c.** $\frac{2}{3}$ **d.** $\frac{1}{12}$

7. a. $\frac{115}{100}$ **b.** $\frac{175}{100}$ **c.** $\frac{150}{100}$ **d.** $\frac{234}{100}$ **8. a.** $\frac{33\frac{1}{3}}{100}$ **b.** $\frac{62\frac{1}{2}}{100}$ **c.** $\frac{16\frac{2}{3}}{100}$ **d.** $\frac{5\frac{3}{4}}{100}$

9. a. $\frac{4}{7}$ **b.** $\frac{3}{11}$ **c.** $\frac{7}{9}$ **d.** $\frac{13}{15}$ **10. a.** $\frac{18}{36}$ **b.** $\frac{21}{28}$ **c.** $\frac{30}{75}$ **d.** $\frac{56}{80}$

11. a. $\frac{15}{40}$ **b.** $\frac{45}{54}$ **c.** $\frac{34}{51}$ **d.** $\frac{84}{96}$ **12. a.** $\frac{42}{54}$ **b.** $\frac{27}{63}$ **c.** $\frac{28}{105}$ **d.** $\frac{24}{108}$

13. a. $1\frac{1}{2}$ **b.** $1\frac{5}{8}$ **c.** $1\frac{2}{5}$ **d.** $1\frac{7}{16}$ **14. a.** $\frac{8}{5}$ **b.** $\frac{14}{8}$ **c.** $\frac{12}{9}$ **d.** $\frac{57}{48}$

15. a. $\frac{14}{9}$ **b.** $\frac{10}{7}$ **c.** $\frac{76}{60}$ **d.** $\frac{65}{35}$ **16. a.** $2\frac{3}{8}$ **b.** $\frac{96}{36}$ **c.** $2\frac{5}{6}$ **d.** $\frac{66}{21}$

17. a. $\frac{571}{1000}$ **b.** $\frac{49}{1000}$ **c.** $\frac{8}{1000}$ **d.** $\frac{647}{1000}$ **18. a.** $\frac{9514}{10000}$ **b.** $\frac{457}{10000}$ **c.** $\frac{26}{10000}$ **d.** $\frac{5933}{10000}$

19. a. Matt can run a mile in $5\frac{3}{5}$ minutes. Brian can run a mile in 5.74 minutes. Who is faster, Matt or Brian?

b. Rich lives $\frac{7}{8}$ of a mile from school. Jennifer lives 0.83 mile from school. Who lives closer to the school?

c. Mary needs $3\frac{5}{8}$ yards of material to make a dress. If she bought 3.73 yards of material will she have enough?

d. To find the fielding average, first find what fractional part of the total chances (the sum of the put-outs, assists, and errors) are the chances properly handled (the sum of the put-outs and assists). Then change the fraction to a decimal correct to three places.

	Put-Outs	Assists	Errors	Average
Dave	32	37	6	?
Karen	40	15	5	?
Felipe	14	49	7	?
Frank	17	31	8	?
Rosa	29	27	10	?

Using Mathematics

REPEATING DECIMALS

Decimals that have a digit or a group of digits repeating endlessly are called **repeating decimals**.

Three dots at the end of a decimal indicate that the sequence of digits repeats endlessly. A bar over a digit or a group of digits indicates the sequence of digits that repeats.

> 0.333. . . or $0.\overline{3}$ 0.7272. . . or $0.\overline{72}$

When writing the fraction $\frac{1}{3}$ as a decimal, divide the numerator by the denominator. You will notice that the division is not exact. The remainder at each step is the same (**1**) and the digit **3** repeats in the quotient.

$$\frac{1}{3} = 3)\overline{1.000} \quad \begin{array}{r} 0.333 \\ \underline{9} \\ 10 \\ \underline{9} \\ 10 \\ \underline{9} \\ 1 \end{array}$$

$\frac{1}{3} = 0.333. . .$ or $0.\overline{3}$

When writing the fraction $\frac{8}{11}$ as a decimal, divide the numerator by the denominator. Again, the division is not exact, but the remainder is the same (**8**) in every other step and the digits **72** repeat in the quotient.

$$\frac{8}{11} = 11)\overline{8.000} \quad \begin{array}{r} 0.7272 \\ \underline{7\,7} \\ 30 \\ \underline{22} \\ 80 \\ \underline{77} \\ 30 \\ \underline{22} \\ 8 \end{array}$$

$\frac{8}{11} = 0.7272. . .$ or $0.\overline{72}$

When $\frac{1}{4}$ is written as a decimal the division process is exact. The quotient is exactly **0.25** and the remainder is **0.** This type of decimal is called a **terminating decimal.**

$$\frac{1}{4} = 4)\overline{1.00} \quad \begin{array}{r} 0.25 \\ \underline{8} \\ 20 \\ \underline{20} \\ 0 \end{array}$$

$\frac{1}{4} = 0.25$

Since 0.25 can be written in repeating form, $0.25\overline{0}$, terminating decimals may also be called repeating decimals.

Exercises

Write each of the following fractions as a repeating decimal.

1. $\frac{2}{3}$ **2.** $\frac{7}{8}$ **3.** $\frac{1}{6}$ **4.** $\frac{3}{7}$ **5.** $\frac{5}{9}$

6. $\frac{11}{16}$ **7.** $\frac{5}{12}$ **8.** $\frac{13}{33}$ **9.** $\frac{7}{11}$ **10.** $\frac{9}{13}$

11. $\frac{4}{5}$ **12.** $\frac{11}{18}$ **13.** $\frac{7}{15}$ **14.** $\frac{19}{24}$ **15.** $\frac{8}{27}$

16. $\frac{20}{21}$ **17.** $\frac{9}{11}$ **18.** $\frac{3}{4}$ **19.** $\frac{14}{15}$ **20.** $\frac{2}{9}$

21. $\frac{11}{12}$ **22.** $\frac{23}{32}$ **23.** $\frac{17}{30}$ **24.** $\frac{28}{33}$ **25.** $\frac{5}{14}$

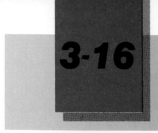

3-16 Writing Decimals as Fractions

Last week it rained 0.15 inch. What fractional part of an inch is that?

Write 0.15 as a fraction.

$$0.15 = \frac{15}{100} = \frac{3}{20}$$

hundredths

ANSWER: $\frac{3}{20}$

Use the digits of the given decimal as the numerator. As the denominator, choose from powers of ten (10, 100, 1,000, and so on) the number that corresponds to the place value of the last digit of the decimal. Simplify.

Examples

1. Write 0.625 as a fraction.

$$0.625 = \frac{625}{1,000} = \frac{5}{8}$$

thousandths

ANSWER: $\frac{5}{8}$

2. Write 0.0005 as a fraction.

$$0.0005 = \frac{5}{10,000} = \frac{1}{2,000}$$

ten thousandths

ANSWER: $\frac{1}{2,000}$

3. Write 5.875 as a mixed number.

$$5.875 = 5 + 0.875 = 5 + \frac{875}{1,000}$$
$$= 5 + \frac{7}{8} = 5\frac{7}{8}$$

ANSWER: $5\frac{7}{8}$

Write the mixed decimal as a sum. Then write the decimal as a fraction and simplify. Add the whole number and the fraction.

4. Write $0.12\frac{1}{2}$ as a fraction.

$$0.12\frac{1}{2} = \frac{12\frac{1}{2}}{100} = 12\frac{1}{2} \div 100$$
$$= \frac{25}{2} \div 100$$
$$= \frac{\overset{1}{\cancel{25}}}{2} \times \frac{1}{\underset{4}{\cancel{100}}} = \frac{1}{8}$$

ANSWER: $\frac{1}{8}$

Think About It

1. Do you think monthly rainfall data should be measured in decimal or fractional parts? Why?

2. Name three things that are best measured using fractions rather than decimals.

3. Is $\frac{1}{15}$ inch more or less than 0.15 inch?

4. Name three fractions between $\frac{1}{15}$ and 0.15.

Diagnostic Exercises

Write the following decimals as fractions or mixed numbers.

1. 0.3   **2.** 0.25   **3.** 0.04   **4.** 0.60   **5.** $0.66\frac{2}{3}$

6. 1.9   **7.** 2.85   **8.** $1.37\frac{1}{2}$ **9.** 0.672   **10.** 0.028

11. 3.125   **12.** 0.4375   **13.** 0.0075   **14.** 7.8125

15. Gary needs $\frac{2}{3}$ pound of concrete to patch his driveway. Will a 0.65 pound bag of concrete be enough?

Related Practice

Write the following decimals as fractions or mixed numbers.

1. a. 0.6   **2. a.** 0.75   **3. a.** 0.02   **4. a.** 0.40   **5. a.** $0.87\frac{1}{2}$
   b. 0.2   **b.** 0.45   **b.** 0.07   **b.** 0.70   **b.** $0.06\frac{1}{4}$
   c. 0.5   **c.** 0.52   **c.** 0.01   **c.** 0.10   **c.** $0.83\frac{1}{3}$
   d. 0.9   **d.** 0.87   **d.** 0.08   **d.** 0.80   **d.** $0.62\frac{1}{2}$

6. a. 1.2   **7. a.** 1.25   **8. a.** $1.33\frac{1}{3}$ **9. a.** 0.125   **10. a.** 0.036
   b. 1.5   **b.** 2.42   **b.** $1.12\frac{1}{2}$ **b.** 0.875   **b.** 0.085
   c. 2.8   **c.** 2.67   **c.** $2.66\frac{2}{3}$ **c.** 0.946   **c.** 0.004
   d. 1.7   **d.** 1.32   **d.** $4.08\frac{1}{3}$ **d.** 0.384   **d.** 0.048

11. a. 1.375   **12. a.** 0.3125   **13. a.** 0.0025   **14. a.** 3.5625
    b. 1.248   **b.** 0.5625   **b.** 0.0054   **b.** 2.0625
    c. 2.964   **c.** 0.9375   **c.** 0.0068   **c.** 5.0084
    d. 3.755   **d.** 0.15625   **d.** 0.0075   **d.** 6.6875

15. a. Find the total weight if a metal can weighs $2\frac{7}{8}$ ounces and its contents weigh 0.798 ounces. Express your answer as a mixed number.

b. There are $5\frac{1}{2}$ cups of sugar in a canister. Clare uses 2.75 cups for one recipe. How much sugar is left in the canister? Express your answer as a mixed number.

c. Tom ran 5.6 miles on Monday and $5\frac{7}{8}$ miles on Tuesday. Which day did he run a longer distance? How much longer? Express your answer as a fraction.

d. Mrs. Jones plans to plant some trees in her backyard. She wants 0.35 of the trees to be pines. What fraction of the trees will be pines? Express the fraction in simplest form.

Problem Solving Strategy: Hidden Question

For lunch, Harry bought a sandwich for $2.95, a beverage for $.65, and a dessert for $1.25. How much change should he get back from $10?

1. **READ** the problem carefully.

a. Find the question asked. How much change should he get from $10?

b. Find the given facts. Sandwich costs $2.95; beverage, $.65; dessert, $1.25.

Think: To find amount of change, you must first find total cost of lunch.

Hidden Question: What is the total cost of lunch?

2. **PLAN** how to solve the problem.

a. Choose the appropriate problem solving strategy.
Choose the operations needed.

Step 1: The word *total* with *unequal* quantities indicates *addition*. Use the given facts to find the missing fact (answer the hidden question needed to solve the problem).

Step 2: The words *how much . . . from* indicate *subtraction*.

b. Think or write out your plan in words. The amount of change is equal to the amount of money offered in payment (Step 2) minus the total cost of the lunch (Step 1).

c. Express the plan as an equation. The two steps may be translated into a single equation.

$$n = \$10 - (\$2.95 + \$.65 + \$1.25)$$

3. **SOLVE** the problem.

a. Estimate the answer.

Step 1: $2.95 + $.65 + $1.25
  ↓   ↓   ↓
  $3   + $1   + $1 = $5

Step 2: $10 − $5 = $5, estimate

b. Find the solution.

Step 1: $2.95 + $.65 + $1.25 = $4.85

Step 2: $10.00 offered
          − $ 4.85 cost of lunch
            $ 5.15   change

ANSWER: $5.15 change

You might use a calculator.

Key:

2.95 **+** .65 **+** 1.25 **M+** 10 **−** **M R C** 5.15

4. **CHECK** .

a. Check the accuracy of your arithmetic.

$1.25 + $.65 + $2.95 = $4.85

   $4.85 cost of lunch
+ $5.15 change
  $10.00 offered ✔

b. Compare the answer to the estimate.

The answer $5.15 compares reasonably with the estimate of $5.

Think About It

1. How can you distinguish a two-step problem from a one-step problem?

2. In your own words, state the meaning of the phrase "hidden question."

3. In Lesson 1–1 you learned to identify the facts of a problem. What is the difference between a problem with a missing fact and a problem with a hidden question?

4. Create a two-step problem of your own. Provide enough information to solve the problem, but be sure it includes a hidden question. Have a classmate identify the hidden question and solve the problem.

Diagnostic Exercises

1. Write the question that, when answered, will provide the information needed to solve the following problem.

If a table sells for $575 and each of 6 chairs sells for $125, what is the total cost of the furniture?

2. Use the given facts to find the information needed to solve the following problem.

Debbie works 20 hours per week at $5.50 per hour. How much did she earn in all during a week in which she received a bonus of $48?

3. Solve.

The Silhouette Stationery Store had 1,764 loose stickers. They sold 524. The remaining stickers were placed in boxes each holding 40. How many boxes of stickers are there?

Related Practice

1. For each problem, write the question that, when answered, will provide the information needed to solve the problem.

a. Jill spent $17.95 for a blouse and $29.55 for a skirt. How much change should she receive from $50?

b. Bill needs to rent a tuxedo and shirt for his school prom. Tuxedo Unlimited's rental price for these items is $45.50 and $6.00, respectively. Frank's Formal Wear offers everything at $54.95. Which rental is lower? How much lower?

c. A sofa and armchair set sells for $660. The sofa alone costs $465. How much more does the sofa cost than the armchair?

d. Ronald loses an average of 4 balls per round of golf. If he plays twice a week and each golf ball costs $1.95, what is the cost of the golf balls lost each week?

2. For each problem, use the given facts to answer the hidden question.

a. Mimi wears basketball sneakers all year round. If she wears out a pair of sneakers in 13 weeks and they cost $39 a pair, how much does it cost to keep Mimi in sneakers for the year?

b. Steve bought two shirts for $29.90, saving $5 on the cost of each shirt. What is the regular price of the two shirts?

c. Tanya scored 87 on her first test, scored 12 points lower on her second test, and scored 93 on her third test. What is the difference between her best and worst scores?

d. If car rental costs $20 per day and $.05 for each mile traveled, how much would it cost to keep the car one day and drive 175 miles?

3. Solve.

a. Mrs. Perkins has $1,248.73 in her checking account. During the week she deposited $897.67 to her account and wrote a check in the amount of $346.58. What was the new balance in her checking account?

b. Frank is developing five rolls of photographic film, each with 24 exposures. How many prints will Frank make if he prints every third exposure?

c. Natasha bought three cassette tapes for $2.49 each. How much change did she receive from a $10 bill?

d. Ralph's Pizza Shop had a total revenue of $1,398 this week. If his expenses were salaries $485, rent $125, supplies $292, and utilities $140, how much profit did Ralph make?

Using Mathematics

FINDING AVERAGES

The **arithmetic mean,** commonly called **average,** of two or more quantities is equal to the sum of the quantities divided by the number of quantities. Finding an average is a two-step problem, where the hidden question is "What is the *sum* of the quantities?"

Example

Maria scored 89, 74, 93, 79, and 85 in five math tests. Find her average score.

To find the average score, find the sum of the scores (hidden question) and divide that sum by the number of scores.

$$(89 + 74 + 93 + 79 + 85) \div 5 = n$$

The sum is:

$$89 + 74 + 93 + 79 + 85 = 420$$

The average is:

$$\begin{array}{r} 84 \\ 5\overline{)420} \\ \underline{40} \\ 20 \\ \underline{20} \\ 0 \end{array}$$

ANSWER: Her average score is 84.

Exercises

1. What is the average temperature for the full day if the daytime temperature is 23°C and the night-time temperature is 15°C?

2. The front four defensive linemen of the school football team weigh 91 kilograms, 95 kilograms, 102 kilograms, and 88 kilograms respectively. What is the average weight of the linemen?

3. Andrew can type an average of 48 words per minute. His report contains 2,136 words. Will he be able to type the report in 45 minutes?

4. Find the average number of points scored per game when the local basketball team scored 97 points in game 1, 89 points in game 2, and 108 points in game 3.

For every twelve students going on the field trip, four adults must come along. What is the ratio of students to adults on the field trip?

Express the ratio of 12 to 4.

$$12 \text{ to } 4 = \frac{12}{4}$$
$$= \frac{12 \div 4}{4 \div 4}$$
$$= \frac{3}{1}, \text{ or } 3:1$$

Read "three to one."

ANSWER: $\frac{3}{1}$, or 3:1

Write a fraction. Use the number being compared as the numerator and the number to which it is compared as the denominator
Simplify.
Ratios may be expressed with a fraction bar or with a colon (:).

Examples

1. Express the ratio of 4 to 12.

$$4 \text{ to } 12 = \frac{4}{12} = \frac{4 \div 4}{12 \div 4} = \frac{1}{3}$$

ANSWER: $\frac{1}{3}$, or 1:3

For each circle, there are three triangles.

2. Express the ratio of 3 meters to 8 meters.

$$3 \text{ m to } 8 \text{ m} = \frac{3}{8}$$

ANSWER: $\frac{3}{8}$, or 3:8

If the quantities compared are denominate numbers, they must first be expressed in the same units. The ratio contains no unit of measurement.

3. Express 400 kilometers in 5 hours as a ratio.

$$\frac{400}{5} = \frac{400 \div 5}{5 \div 5}$$
$$= \frac{80}{1}$$

ANSWER: $\frac{80}{1}$, or 80 kilometers per hour

Sometimes a ratio is used to express a rate.

4. Are $\frac{10}{20}$ and $\frac{18}{36}$ equivalent ratios?

$$\frac{10}{20} = \frac{1}{2} \searrow$$
$$\frac{18}{36} = \frac{1}{2} \nearrow \text{ same}$$

ANSWER: Yes, $\frac{10}{20}$ and $\frac{18}{36}$ are equivalent ratios.

Write each ratio in lowest terms. If the simplest forms of the ratios are the same, the ratios are equivalent.

Vocabulary

A **ratio** is a comparison of two quantities by division.

A **rate** is a ratio comparing two different kinds of quantities, such as kilometers and hours.

Equivalent ratios are ratios that are the same in simplest form.

Think About It

Refer to the word problem about the field trip to answer these questions.

1. What would be the ratio of students to adults on the field trip if twice as many adults were needed?

2. If students and adults must attend in a 3:1 ratio, is it possible to have 50 people on the field trip? Why or why not?

3. If only 2 adults can go on the field trip, how many students will be able to go?

4. If a total of 80 people are on the field trip, how many are students and how many are adults?

Diagnostic Exercises

1. Use a fraction bar to express the ratio of 6 to 11.

2. Use a colon to express the ratio of 9 to 5.

Express each ratio in simplest form.

3. 2 to 5   **4.** 6 to 24   **5.** 8 to 3   **6.** 30 to 6

7. 27 to 12   **8.** n to 10   **9.** 25 cm to 75 cm

10. 20 kg to 12 kg   **11.** A nickel to a quarter.   **12.** A dozen to 8 things.

13. Express the rate 240 miles on 15 gallons of gasoline as a ratio in simplest form.

14. Are $\frac{18}{42}$ and $\frac{24}{56}$ equivalent ratios?

15. A gear ratio is the ratio of the number of teeth on one gear to the number of teeth on another. If one gear has 12 teeth and the other has 36 teeth, what is the gear ratio?

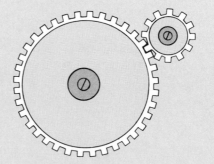

Related Practice

1. Use the fraction bar to express each ratio.
   a. Eight to fifteen **b.** One to seven   **c.** Twelve to three   **d.** Eight to four

2. Use a colon to express each ratio.
   a. Two to nine   **b.** Fourteen to three   **c.** Six to one   **d.** Five to eleven

Express each ratio in simplest form.

3. a. 5 to 8 **b.** 6 to 17 **c.** 1 to 4 **d.** 2 to 15

4. a. 4 to 12 **b.** 10 to 25 **c.** 8 to 14 **d.** 18 to 54

5. a. 11 to 2 **b.** 8 to 5 **c.** 20 to 9 **d.** 14 to 3

6. a. 24 to 3 **b.** 36 to 12 **c.** 30 to 6 **d.** 42 to 14

7. a. 6 to 4 **b.** 15 to 9 **c.** 28 to 10 **d.** 18 to 16

8. a. n to 7 **b.** d to 11 **c.** 18 to b **d.** x to y

9. a. 8 h to 10 h **b.** 48 in. to 72 in. **c.** 2 ft to 5 yd **d.** 18 min to 1 h

10. a. 105 m to 30 m **b.** 40 min to 8 min **c.** 6 yr to 18 mo **d.** 3 L to 50 L

11. a. A dollar to a nickel **b.** A dime to a half-dollar
   c. A quarter to a dime **d.** 3 quarters to 4 nickels

12. a. 3 things to 1 dozen **b.** 2 dozen to 18 things
   c. 4 dozen to 6 things **d.** 9 things to 3 dozen

13. Express each of the following rates as a ratio in simplest forms.
   a. 264 kilometers in 3 hours **b.** 3,600 liters in 20 minutes
   c. 9,000 revolutions in 5 minutes **d.** 800 feet in 16 seconds

14. Which of the following are pairs of equivalent ratios?
   a. $\frac{16}{24}$ and $\frac{10}{15}$ **b.** $\frac{4}{10}$ and $\frac{18}{45}$
   c. $\frac{27}{45}$ and $\frac{45}{72}$ **d.** $\frac{16}{28}$ and $\frac{36}{63}$

15. a. There are 18 girls and 12 boys in our class. Give each ratio.
   boys to girls girls to boys
   girls to entire class boys to entire class
   b. The aspect ratio of an airplane wing is the ratio of its length to its width. What is the aspect ratio of a wing 32 meters long and 4 meters wide?
   c. What is the ratio of rising stocks to declining stocks if 750 stocks rose and 625 stocks declined during the day?
   d. The compression ratio is the ratio of the total volume of a cylinder of an engine to its clearance volume. Find the compression ratio of a cylinder if its total volume is 1,200 cubic centimeters and its clearance is 25 cubic centimeters.

3-19 Proportion

A proportion is an equation stating that two ratios are equal.

Examples

1. Write as a proportion: 6 compared to 9 is the same as 8 compared to 12.

$\frac{6}{9} = \frac{8}{12}$ or 6:9 = 8:12

ANSWER: $\frac{6}{9} = \frac{8}{12}$ or 6:9 = 8:12

Write a fraction bar for "is compared to" and the equal sign for "is the same as." Sometimes colons are used instead of fraction bars.

---

2. Write as a proportion: Some number compares to 45 as 8 compares to 10.

$\frac{n}{45} = \frac{8}{10}$

ANSWER: $\frac{n}{45} = \frac{8}{10}$

Use *n*, or some other placeholder, for the unknown number.

---

3. Is $\frac{2}{3} = \frac{10}{15}$ a proportion?

first 2 ⤫ 10 third
second 3 ⤫ 15 fourth

first   fourth
$2 \quad \times \quad 15 = 30$
extremes

second   third
$3 \quad \times \quad 10 = 30 \qquad 30 = 30$
means

ANSWER: Yes, $\frac{2}{3} = \frac{10}{15}$ is a proportion.

Find the cross products by multiplying the extremes and means.

If the product of the extremes equals the product of the means, the statement is a proportion.

---

4. Is $\frac{1}{2} = \frac{9}{12}$ a proportion?

$\frac{1}{2}$ ⤫ $\frac{9}{12}$ $1 \times 12 = 12$
          $2 \times 9 = 18 \qquad 12 \neq 18$

ANSWER: No, $\frac{1}{2} = \frac{9}{12}$ is not a proportion.

---

5. Solve the proportion $\frac{n}{32} = \frac{3}{4}$.

$\frac{n}{32}$ ⤫ $\frac{3}{4}$

$4 \times n = 32 \times 3$
$4n = 96$

$\frac{4n}{4} = \frac{96}{4}$ Check: $\frac{24}{32} = \frac{3}{4}$?
$n = 24$ $\frac{3}{4} = \frac{3}{4}$ ✔

ANSWER: $n = 24$

To find the value of *n*, divide both sides of the equation by 4.

Check by substituting the value for *n*. Simplify and compare values.

6. Solve the proportion $\frac{8}{n} = \frac{3}{225}$.

$$\frac{8}{n} \qquad \frac{3}{225}$$

$$n \times 3 = 8 \times 225 \qquad \text{Check: } \frac{8}{600} = \frac{3}{225}?$$
$$3n = 1,800 \qquad\qquad \frac{1}{75} = \frac{1}{75} \; \text{✔}$$
$$\frac{3n}{3} = \frac{1,800}{3}$$
$$n = 600$$

ANSWER: $n = 600$

Vocabulary

A **proportion** is a mathematical sentence stating that two ratios are equivalent.
The **terms** of a proportion are:

$$\begin{array}{l} \text{first} \rightarrow a \\ \text{second} \rightarrow b \end{array} = \begin{array}{l} c \leftarrow \text{third} \\ d \leftarrow \text{fourth} \end{array}$$

The **extremes** are the first and fourth terms of a proportion.
The **means** are the second and third terms of a proportion.
The **cross products** are the product of the extremes and the product of the means.

Think About It

1. Write the keying sequence to solve the proportion $\frac{16}{n} = \frac{24}{36}$ on a calculator.

2. Are the reciprocals of a true proportion also a true proportion?

Diagnostic Exercises

Write each of the following as a proportion.

1. 9 compared to 6 is the same as 24 compared to 16.

2. Some number n is to 8 as 9 is to 12.

3. Is $\frac{9}{15} = \frac{4}{10}$ a true proportion?

Solve and check.

4. $\frac{n}{15} = \frac{4}{5}$
5. $\frac{12}{21} = \frac{y}{14}$
6. $\frac{16}{x} = \frac{2}{7}$
7. $\frac{5}{8} = \frac{15}{a}$
8. $\frac{2\frac{1}{2}}{x} = \frac{\frac{1}{2}}{10}$

9. Using a proportion, solve: What number compared to 72 is the same as 7 compared to 18?

10. A recipe calls for 4 cups of flour to 6 tablespoons of shortening. How many tablespoons of shortening are needed for 6 cups of flour?

Related Practice

Write each of the following as a proportion.

1. a. 26 compared to 13 is the same as 6 compared to 3.
   b. 4 is to 12 as 5 is to 15.
   c. 45 compared to 80 is the same as 18 compared to 32.
   d. 56 is to 14 as 76 is to 19.

2. a. Some number x compares to 30 as 9 compares to 54.
   b. 24 is to y as 6 is to 11.
   c. 35 compared to 500 is the same as n compared to 60.
   d. 42 is to 63 as 12 is to n.

3. Which of the following are true proportions?

   a. $\dfrac{2}{3} = \dfrac{3}{2}$ **b.** $\dfrac{15}{90} = \dfrac{2}{12}$ **c.** $18:14 = 27:21$ **d.** $8:20 = 30:100$

Solve and check.

4. a. $\dfrac{t}{32} = \dfrac{3}{4}$ **b.** $\dfrac{x}{9} = \dfrac{49}{63}$ **c.** $\dfrac{n}{0.01} = \dfrac{0.2}{0.16}$ **d.** $\dfrac{b}{5} = \dfrac{2}{3}$

5. a. $\dfrac{90}{54} = \dfrac{a}{3}$ **b.** $\dfrac{5}{6} = \dfrac{c}{78}$ **c.** $\dfrac{1}{6} = \dfrac{s}{5}$ **d.** $\dfrac{1.5}{0.3} = \dfrac{r}{8}$

6. a. $\dfrac{9}{n} = \dfrac{3}{38}$ **b.** $\dfrac{24}{y} = \dfrac{108}{18}$ **c.** $\dfrac{0.005}{c} = \dfrac{1.4}{0.28}$ **d.** $\dfrac{3}{b} = \dfrac{13}{6}$

7. a. $\dfrac{6}{7} = \dfrac{54}{x}$ **b.** $\dfrac{105}{35} = \dfrac{27}{n}$ **c.** $\dfrac{17}{24} = \dfrac{17}{a}$ **d.** $\dfrac{0.04}{0.12} = \dfrac{0.6}{x}$

8. a. $\dfrac{y}{3\frac{1}{2}} = \dfrac{8}{7}$ **b.** $\dfrac{24}{b} = \dfrac{2\frac{2}{3}}{5\frac{5}{6}}$ **c.** $\dfrac{n}{2\frac{1}{4}} = \dfrac{15}{\frac{3}{4}}$ **d.** $\dfrac{21}{x} = \dfrac{\frac{7}{10}}{\frac{3}{5}}$

9. Using proportions, solve each of the following.

   a. What number compared to 10 is the same as 28 compared to 5?
   b. 12 compared to 3 is the same as what number compared to 18?
   c. 7 compared to what number is the same as 42 compared to 54?
   d. 6 compared to 11 is the same as 84 compared to what number?

10. Solve by using proportions.

   a. At the rate of 3 items for 16¢, how many items can you buy for 80¢?
   b. A motorist traveled 240 kilometers in 3 hours. How long will it take at that rate to travel 400 kilometers?
   c. If the scale distance of 3.5 centimeters on a map represents an actual distance of 175 kilometers, what actual distance does a scale distance of 5.7 centimeters represent?
   d. How many square feet of lawn will 10 pounds of grass seed cover if 3 pounds of seed cover 450 square feet?

Making Decisions

CHOOSING A BASEBALL LINEUP

You and three of your classmates are co-managers of a baseball team. Working together, it is your job to pick the starting lineup for each game. To help you decide whom to pick for each position, you make a table to show some information about each player.

	At Bats	Runs	Hits	RBI	Batting Avg.	Doubles	Triples	HRs	Walks	Stolen Bases
Rivera	285	26	74	18	0.257	11	0	6	7	2
Talbott	274	46	78	33	0.285	13	2	15	12	10
Mao	203	14	55	20	0.271	7	1	4	19	5
Jackson	267	56	89	30	0.322	15	3	7	16	37

You have pretty well decided on the last five places in the lineup, but you now need to decide on the first four places. Use the table above and the following questions to help you decide on a lineup. Explain each answer.

1. You feel it is very important that the lead-off batter be able to get on base and move around into scoring position. Answer the following questions with this in mind.
   a. Do you think a good batting average is important?

   b. Do you think number of bases stolen is important?

   c. Do you think the number of home runs is important?

   d. Do you think it is important for the lead-off batter to have a lot of RBIs?

   e. Which one or two statistics about the lead-off batter are most important?

2. What do Jackson's 37 stolen bases indicate to you? Does that statistic make Jackson a good choice for lead-off?

3. Choose a lead-off batter. Make a table like this.

Position	Name	Reasons
1		
2		
3		
4		

As you make your choices, fill in the table.

4. What do you think are some important statistics to use in choosing the second batter?

5. You also know that Rivera has hit into 55 double plays and Mao has hit into 24. If you were choosing between them for the second spot, how would that affect your decision?

6. Compare the number of walks for Mao and Talbott. How do you think a player's number of walks might affect your choice for second batter?

7. Choose a batter for the second position. Write the name in your table.

8. What characteristics are important for the batters in third and fourth places?

9. You decided to pick the fourth place, or clean-up, batter next. What do you think the clean-up batter should do?

10. Pick a clean-up batter and write the name on your chart. Then pick a batter for third place.

11. Compare your batting lineups with those written by other groups. Be prepared to defend your choices!

In Your Community

Look up some statistics on your favorite baseball team, either local or major league, and make out a batting order based on the statistics you find.

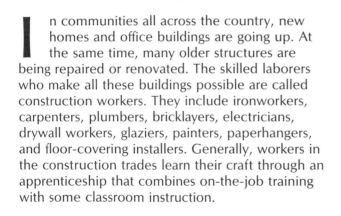

Career Application
CONSTRUCTION WORKER

In communities all across the country, new homes and office buildings are going up. At the same time, many older structures are being repaired or renovated. The skilled laborers who make all these buildings possible are called construction workers. They include ironworkers, carpenters, plumbers, bricklayers, electricians, drywall workers, glaziers, painters, paperhangers, and floor-covering installers. Generally, workers in the construction trades learn their craft through an apprenticeship that combines on-the-job training with some classroom instruction.

1. A carpenter is given a blueprint for a building. The lengths of all the sections of a wall are marked separately: $1\frac{5}{8}$ inches, $\frac{3}{4}$ inches, $\frac{7}{8}$ inches, and $2\frac{3}{16}$ inches. What is the total length of the wall on the blueprint?

2. An electrician has a wiring job that calls for 32 pieces of hollow pipe, each $7\frac{1}{2}$ feet long. What is the total number of feet of pipe needed for the job?

3. A carpenter needs to cut 22 pieces of wood, each $14\frac{3}{8}$ inches long, for trim. What is the total length of wood needed?

4. One brick, plus mortar, is $8\frac{1}{2}$ inches long. A wall will be 34 feet long. How many bricks are needed for each row?

Chapter 3 Review

Vocabulary

common multiple p. 157
complex fraction p. 191
cross product p. 214
denominator p. 148
equivalent fractions p. 155
equivalent ratios p. 211
extremes p. 214
fractions p. 148
improper fraction p. 152

lowest terms p. 150
least common denominator p. 163
least common multiple p. 157
means p. 214
mixed number p. 152
multiple p. 157
≠ symbol p. 155
≯ symbol p. 170

≮ symbol p. 160
numerator p. 148
proportion p. 214
rate p. 211
ratio p. 211
reciprocal p. 191
simplest form p. 151
terms (proportion) p. 214

Page
149 Express in lowest terms.

1. $\frac{72}{81}$

2. $\frac{144}{168}$

153 Express as an equivalent fraction having the denominator specified.

6. $\frac{1}{5} = \frac{n}{15}$

7. $\frac{7}{8} = \frac{n}{64}$

157 **9.** Find the least common multiple of 5, 7, and 28.

159 **11.** Round $2.76\frac{3}{5}$ to the nearest cent.

162 **13.** Find the LCD of $\frac{2}{9}$, $\frac{17}{36}$, and $\frac{11}{18}$.

170 **16.** $9\frac{2}{5} - 5\frac{1}{3}$ **17.** Subtract $8\frac{13}{16}$ from 20.

179 **19.** $5\frac{1}{3} \times 3\frac{3}{8}$ **20.** $\frac{7}{8} \times \frac{12}{17}$

195 **23.** What part of 12 is 7?

200 **25.** Write $\frac{9}{20}$ as a decimal.

204 **26.** Write 0.056 as a fraction in lowest terms.

206 **28.** When Marie bought her moped last year she paid $75 for a down payment and made 6 payments of $65.50 each. How much did she pay for the moped?

210 **30.** A recipe calls for 3 tablespoons of butter to be added to 5 tablespoons of flour. What is the ratio of butter to flour?

213 **32.** Is $\frac{7}{15} = \frac{28}{45}$ a true proportion?

213 **34.** An aquarium with a volume of 1 cubic foot will hold about $7\frac{1}{2}$ gallons of water. If 75 gallons of water are needed to fill another aquarium, how many cubic feet is it?

Page
151 Rewrite as a whole number or mixed number.

3. $\frac{63}{7}$ **4.** $\frac{47}{6}$ **5.** $10\frac{4}{4}$

155 **8.** Are $\frac{12}{20}$ and $\frac{9}{15}$ equivalent fractions?

159 **10.** Arrange in order from least to greatest:
$\frac{3}{8}$, $\frac{2}{5}$, and $\frac{1}{3}$

162 **12.** Find the LCD of $\frac{5}{8}$ and $\frac{1}{6}$.

164 **14.** $6\frac{7}{8} + 2\frac{3}{5}$ **15.** $24\frac{3}{4} + 19\frac{5}{8} + 10\frac{1}{6}$

177 **18.** Write $2\frac{7}{20}$ as an improper fraction.

188 **21.** $2\frac{2}{9} \div \frac{5}{6}$ **22.** $2\frac{1}{4} \div 6$

195 **24.** $\frac{4}{5}$ of what number is 32?

191 **27.** If the product of two numbers is 1, then each factor is the ____?____ of the other.

209 **29.** The price of a stock dropped from $28\frac{1}{4}$ to $21\frac{3}{4}$ during the last year. What was the average drop per month?

210 **31.** Express as a ratio in lowest terms:
3 quarters to 2 nickels

213 **33.** Solve. $\frac{0.7}{n} = \frac{21}{12}$

Chapter 3 Test

Page
149 **1.** Express $\frac{49}{63}$ in lowest terms.

Page
151 **2.** Write $\frac{75}{7}$ as a mixed number.

151 **3.** Write $4\frac{5}{5}$ in simpler form.

153 **4.** Solve. $\frac{3}{5} = \frac{n}{20}$

155 **5.** Are $\frac{6}{28}$ and $\frac{9}{42}$ equivalent fractions?

157 **6.** Find the least common multiple of 16 and 20.

159 **7.** Which is greater $\frac{3}{16}$ or $\frac{1}{5}$?

164 **8.** $\frac{3}{5} + \frac{7}{8}$ 164 **9.** $1\frac{3}{4} + 9\frac{2}{3} + 6\frac{5}{8}$

170 **10.** $7 - 2\frac{2}{3}$ 170 **11.** $15\frac{3}{10} - 4\frac{5}{6}$

177 **12.** Write $7\frac{2}{9}$ as an improper fraction.

179 **13.** $4\frac{3}{5} \times 5\frac{1}{3}$

188 **14.** $6 \div 1\frac{1}{8}$ 188 **15.** $3\frac{1}{2} \div 1\frac{1}{4}$

195 **16.** What part of 60 is 45?

195 **17.** $\frac{3}{7}$ of what number is 21?

200 **18.** Write $\frac{21}{25}$ as a decimal.

204 **19.** Write 0.64 as a fraction in lowest terms.

206 **20.** Which is a better buy, 4 tires at a sale price of $69.20 each or 3 tires at the regular price of $83.90 each and getting the fourth tire free? What is the difference in the cost?

206 **21.** A coat that sells for $78.60 is reduced by $\frac{1}{3}$. How much will the coat cost?

206 **22.** On their last test $\frac{1}{5}$ of the class received an A, and $\frac{1}{3}$ of the class received a B. If there are 30 students in the class how many received either an A or B?

210 **23.** In order to make 275 kilograms of plate glass you need 198 kilograms of sand. Express the ratio of sand to plate glass.

213 **24.** Solve: $\frac{3}{5} = \frac{n}{17}$

213 **25.** Solve by proportion.

Toshiro works for 14 hours and earns $77. How long must he work to earn $121?

Cumulative Test

Solve each problem and select the letter corresponding to your answer.

Page
29 **1.** Find the sum. 739 + 5,936 + 827

    a. 7,402   **b.** 7,492
    c. 7,502   **d.** 7,592

Page
37 **2.** Subtract. 3,061 − 1,957

    a. 1,904   **b.** 904
    c. 2,904   **d.** 1,104

49 **3.** Multiply. 804 × 360

    a. 289,440   **b.** 290,440
    c. 289,340   **d.** 290,340

61 **4.** Divide. 77,830 ÷ 86

    a. 95   **b.** 905
    c. 950   **d.** 815

88 **5.** Add. 2.7 + 0.081 + 10

    a. 13.781   **b.** 12.781
    c. 2.791   **d.** 2.781

95 **6.** Subtract. 0.48 − 0.219

    a. 0.279   **b.** 0.699
    c. 0.161   **d.** 0.261

102 **7.** Multiply. 0.07 × 0.025

    a. 0.175   **b.** 0.17500
    c. 0.00175   **d.** 0.0175

109 **8.** Divide. 4 ÷ 0.8

    a. 0.5   **b.** 0.05
    c. 50   **d.** 5

164 **9.** Add. $\frac{3}{8} + \frac{5}{12}$

    a. $\frac{19}{24}$ **b.** $\frac{5}{24}$
    c. $\frac{8}{20}$ **d.** $\frac{2}{5}$

170 **10.** Subtract. $1\frac{1}{4} - \frac{2}{3}$

    a. $1\frac{5}{12}$ **b.** $\frac{7}{12}$
    c. $1\frac{11}{12}$ **d.** $1\frac{7}{12}$

179 **11.** Multiply. $2\frac{5}{6} \times 3$

    a. $8\frac{1}{2}$ **b.** $\frac{17}{18}$
    c. $1\frac{1}{17}$ **d.** $5\frac{2}{3}$

188 **12.** Divide. $3 \div 8\frac{1}{2}$

    a. $\frac{6}{17}$ **b.** $2\frac{5}{6}$
    c. $25\frac{1}{2}$ **d.** $3\frac{2}{5}$

170 **13.** How much milk should be added to $\frac{2}{3}$ quart to make 6 quarts?

    a. $6\frac{2}{3}$ quarts   **b.** $5\frac{2}{3}$ quarts
    c. $6\frac{1}{3}$ quarts   **d.** $5\frac{1}{3}$ quarts

102 **14.** In 1986 the average per mile cost of owning a car was $.32. If there are 131.6 million passenger cars and the average mileage was 9,304 for that year, how much did it cost the average driver to operate a car in 1986? Round your answer to the nearest dollar.

    a. $42   **b.** $300
    c. $2,977   **d.** $391,810

COMPUTER ACTIVITY 3

Programmers often use flow charts to outline what a computer should do. A flow chart shows the sequence of steps the computer will follow. Flow charts usually make it easy to see the choices at different steps and what happens when each choice is made.

Diamond-shaped boxes indicate that the computer must make a choice. There are usually two paths leading out of a decision box—a *yes* path and a *no* path. In the flow chart below, the *no* choice leads to a loop. The loop causes the computer to repeat a portion of the program.

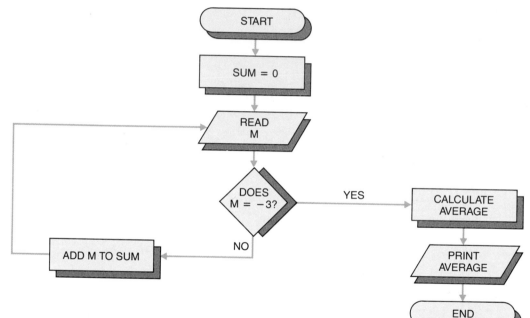

The program based on this flow chart is at the right. Compare this program with the program on page 74. This program uses one READ M statement instead of a READ statement for each month. It is much shorter than the program on page 74.

In this program, lines 20 to 50 form a loop. The computer will repeat them over and over. The IF . . . THEN statement ends the loop when the data equals −3, the last number on the DATA list. The number −3 signals that all the valid data have been read. When the computer reads −3, it goes to line 60 and averages the numbers. Then line 70 tells it to print the statement and the average.

Run the program. What is the screen display?

```
10 SUM = 0
20 READ M
30 IF M = -3 THEN GOTO 60
40 SUM = SUM + M
50 GOTO 20
60 AVG = SUM/12
70 PRINT "AVERAGE MONTHLY TOTAL IS $", AVG
80 DATA 300, 467, 1,232, 1,345, 1,101, 1,009, 897,
   925, 1,032, 1,243, 1,278, 1,450, -3
90 END
```

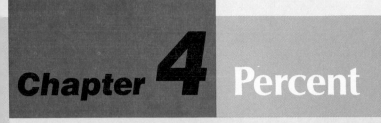

Chapter 4 Percent

Introduction to Percent

Earners, consumers, and business people use **percent** (symbol %) extensively in their daily affairs.

Earners may find that 14% withholding tax, 7.15% social security tax, and perhaps a 3% state or city income tax are deducted from their paychecks. Salespeople may earn 5% commission on what they sell.

Homeowners may receive a 2% discount for paying their real estate taxes in advance, a 3% discount on their gas and electric bills or a 5% discount on their water bills if they pay before the discount period ends. Consumers may buy merchandise at a department store advertising a 40% reduction on certain sales items. They may pay a 12% carrying charge on a new automobile. They may buy jewelry and cosmetics on which they pay a state sales tax of perhaps 1% to 7%. Labels on clothing may indicate the content of cloth like 30% mohair or 70% wool.

In business, a storekeeper may make a 35% profit on sales. A bank may pay $5\frac{1}{2}$% interest on deposits and charge $14\frac{1}{2}$% interest on loans.

Students and teachers also use percent. A student may receive a mark of 83%. A teacher may find daily class attendance averages 94%.

Percent is very useful in giving a quick comparative picture on a scale from 1 to 100. For example, when a basketball player has a 69% success rate in making foul shots, we immediately understand that the player is successful at the rate of 69 out of every 100 attempts.

4-1 Meaning of Percent

Percent (%) means *per hundred,* or *hundredths.* When you read in the newspaper that 75% of the voters voted, it means that 75 out of 100 eligible citizens voted. A percent can be considered as a ratio of a number to 100. 100% of something is all of it.

Examples

1. Express 35 hundredths as a percent.

35 hundredths = 35%

        ↑
        percent sign

ANSWER: 35%

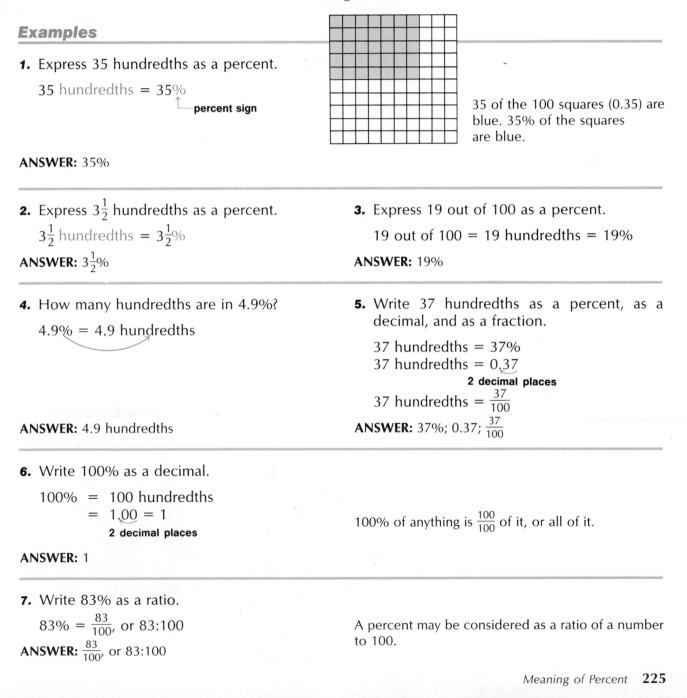

35 of the 100 squares (0.35) are blue. 35% of the squares are blue.

---

2. Express $3\frac{1}{2}$ hundredths as a percent.

$3\frac{1}{2}$ hundredths = $3\frac{1}{2}$%

ANSWER: $3\frac{1}{2}$%

3. Express 19 out of 100 as a percent.

19 out of 100 = 19 hundredths = 19%

ANSWER: 19%

---

4. How many hundredths are in 4.9%?

4.9% = 4.9 hundredths

ANSWER: 4.9 hundredths

5. Write 37 hundredths as a percent, as a decimal, and as a fraction.

37 hundredths = 37%
37 hundredths = 0.37
                    2 decimal places
37 hundredths = $\frac{37}{100}$

ANSWER: 37%; 0.37; $\frac{37}{100}$

---

6. Write 100% as a decimal.

100% = 100 hundredths
     = 1.00 = 1
        2 decimal places

ANSWER: 1

100% of anything is $\frac{100}{100}$ of it, or all of it.

---

7. Write 83% as a ratio.

83% = $\frac{83}{100}$, or 83:100

ANSWER: $\frac{83}{100}$, or 83:100

A percent may be considered as a ratio of a number to 100.

1. A suit is on sale at a 25% reduction. Use the definition of a percent to explain how much a customer would save on a $100 suit.

2. If there are 100 questions on a test and you know the number of questions you have answered correctly, how would you determine your percent grade?

3. Explain how you could decide which of the two numbers, 65% or $\frac{6.5}{100}$, represents the smaller ratio.

Diagnostic Exercises

1. Write eighteen hundredths as a percent.

2. Write 81 out of 100 as a percent.

3. How many hundredths are in 35%?

4. Write four hundredths as a percent, a decimal, and a fraction.

5. Write 36% as a ratio.

6. A teacher states that the ratio of multiple-choice questions to total questions on a test will be 35 to 100. What percent of the questions on the test will be multiple choice?

Related Practice

1. Write each of the following as a percent.

   a. 5 hundredths
   b. 37 hundredths
   c. 62.5 hundredths
   d. 0.25 hundredths

2. Write each of the following as a percent.

   a. 47 out of 100
   b. 95 out of 100
   c. 66 out of 100
   d. 12.5 out of 100

3. Give the number of hundredths in each of the following.

   a. 140%
   b. 87%
   c. 3.5%
   d. 0.5%

4. Write each of the following as a percent, a decimal, and a fraction.

   a. Twenty-three hundredths
   b. Two and one-fourth hundredths
   c. Three hundred hundredths
   d. Three-fourths hundredth

5. Write each of the following as a ratio.

   a. 18%
   b. 45%
   c. 6%
   d. 73%

6. a. The ratio of salt to water in a solution is 43:100. What percent of the solution is salt?
   b. There are 2.3% more boys than girls in the class. Write this percent as a ratio.
   c. The weather forecast predicts a 55% chance of rain. Write this percent as a ratio.
   d. A professional basketball team won 64 hundredths of the games it has played. What percent of its games has it won?

Writing Percents as Decimals

The Pacheco family has budgeted 15% of its income for food. What decimal part of their income do they spend for food?

Write 15% as a decimal.

$15\% = 0.15$

2 decimal places

ANSWER: 0.15

To write a percent as a decimal write the number without the percent sign (%) and move the decimal point two places to the left.

Examples

1. Write 125% as a decimal.

$125\% = 1.25$

ANSWER: 1.25

2. Write 5% as a decimal.

Write 1 zero.

$5\% = 0.05$

ANSWER: 0.05

Add zeros if necessary when you move the decimal point.

3. Write 0.2% as a decimal.

Write 2 zeros.

$0.2\% = 0.002$

ANSWER: 0.002

4. Write 8.34% as a decimal.

Write 1 zero.

$8.34\% = 0.0834$

ANSWER: 0.0834

5. Write $1\frac{7}{8}\%$ as a decimal.

$1\frac{7}{8}\% = 0.01\frac{7}{8}$

$0.01\frac{7}{8} = 0.01875$

ANSWER: 0.01875

Move the decimal point two places to the left. Then write the fraction as a decimal.

Remember $1\frac{7}{8}\% = 1\% + \frac{7}{8}\%$

Think About It

1. Is 38% more or less than $\frac{3}{8}$? ($\frac{3}{8} = 0.375$)

2. Is 5.7% equivalent to 0.57? Explain why or why not.

3. Is 0.251 equivalent to $25\frac{1}{2}$%? Why or why not?

4. To write a percent as a decimal, write the number without the percent sign and move the decimal point two places to the left. What number are you using as a divisor when you move the decimal point two places to the left? Explain why you would use this divisor.

Diagnostic Exercises

Write each percent as a decimal.

1. 8% **2.** 53% **3.** 90% **4.** 119%

5. 500% **6.** $12\frac{1}{2}$% **7.** $5\frac{1}{4}$% **8.** $\frac{3}{4}$%

9. 8.6% **10.** 4.82% **11.** 352.875% **12.** 0.6%

13. Janice read in the newspaper that all the sportswear at Bibi's Boutique was going to be on sale at a 15% reduction. Express this percent as a decimal.

Related Practice

Write each percent as a decimal.

1. a. 6% **2. a.** 16% **3. a.** 40% **4. a.** 134%
   b. 9%   **b.** 83%   **b.** 70%   **b.** 157%
   c. 1%   **c.** 45%   **c.** 20%   **c.** 106%
   d. 4%   **d.** 67%   **d.** 10%   **d.** 175%

5. a. 100% **6 a.** $37\frac{1}{2}$% **7. a.** $4\frac{1}{2}$% **8. a.** $\frac{1}{2}$%
   b. 200%   **b.** $62\frac{1}{2}$%   **b.** $3\frac{3}{4}$%   **b.** $\frac{1}{4}$%
   c. 600%   **c.** $83\frac{1}{3}$%   **c.** $2\frac{7}{8}$%   **c.** $\frac{5}{8}$%
   d. 400%   **d.** $18\frac{3}{4}$%   **d.** $5\frac{4}{5}$%   **d.** $\frac{2}{3}$%

9. a. 3.5% **10. a.** 1.25% **11. a.** 26.375% **12. a.** 0.7%
   b. 2.8%   **b.** 5.33%   **b.** 31.625%   **b.** 0.02%
   c. 9.3%   **c.** 2.08%   **c.** 128.333%   **c.** 0.9%
   d. 6.4%   **d.** 7.19%   **d.** 895.667%   **d.** 0.85%

13. a. Jim has to pay a sales tax of 7%. How would he enter this percent in his calculator as a decimal?

b. The average price of a medium priced 2-door sedan increased $4\frac{3}{5}$%. Write this percent as a decimal.

c. Sue pays interest of 12.5% on a loan. What decimal is equivalent to this percent?

d. Alex is not sure how to write $\frac{7}{25}$% as a decimal. Do this conversion for him.

Writing Decimals as Percents

If you have found the answer to a problem to be 0.07, how can you write this decimal as a percent? How can you write a mixed decimal such as 1.34 as a percent?

Examples

1. 0.07 = 0.07 hundredths
     2 decimal places

   7 hundredths = 7%

ANSWER: 7%

Move the decimal point 2 places to the right to express as hundredths.
Write % in place of hundredths.

---

2. 1.34 = 1.34 hundredths

   134 hundredths = 134%

ANSWER: 134%

3. Write 0.5 as a percent.

       Write 1 zero.

   0.5 = 0.50 hundredths

   50 hundredths = 50%

ANSWER: 50%

---

4. Write 8 as a percent.

     Write 2 zeros.

   8 = 8.00 hundredths

   800 hundredths = 800%

ANSWER: 800%

5. Write 3.687 as a percent.

   3.687 = 3.687 hundredths

   368.7 hundredths = 368.7%

ANSWER: 368.7%

---

6. Write $0.06\frac{1}{2}$ as a percent.

   $0.06\frac{1}{2} = 6\frac{1}{2}\%$

ANSWER: $6\frac{1}{2}\%$

7. Write $0.4\frac{1}{2}$ as a percent.

   $0.4\frac{1}{2} = 0.45 = 45\%$

ANSWER: 45%

Think About It

1. Does 2.3 = .23%? Explain.

2. The answer to a problem is 9 thousandths. If this answer is written as a percent, will the result be more or less than one percent?

Diagnostic Exercises

Write each decimal as a percent.

1. 0.06   **2.** 0.32   **3.** 0.7   **4.** 1.12

5. 1.4   **6.** $0.37\frac{1}{2}$ **7.** $0.04\frac{1}{2}\%$ **8.** $0.60\frac{3}{4}$

9. $1.66\frac{2}{3}$ **10.** 0.625   **11.** 0.0467   **12.** $0.1\frac{1}{4}$

13. 2.875   **14.** 2   **15.** $0.00\frac{3}{8}$ **16.** 0.0025

17. For a baseball team to have a standing of 0.625, what percent of the games it plays must it win?

Related Practice

Write each decimal as a percent.

1. a. 0.01   **b.** 0.08   **c.** 0.03   **d.** 0.05   **2. a.** 0.28   **b.** 0.75   **c.** 0.16   **d.** 0.93

3. a. 0.6   **b.** 0.3   **c.** 0.8   **d.** 0.1   **4. a.** 1.39   **b.** 1.92   **c.** 1.18   **d.** 1.50

5. a. 1.2   **b.** 1.8   **c.** 1.7   **d.** 1.3   **6. a.** $0.12\frac{1}{2}$ **b.** $0.83\frac{5}{6}$ **c.** $0.42\frac{2}{7}$ **d.** $0.18\frac{3}{4}$

7. a. $0.01\frac{1}{2}$ **b.** $0.03\frac{3}{4}$ **c.** $0.04\frac{2}{3}$ **d.** $0.06\frac{5}{6}$ **8. a.** $0.10\frac{1}{2}$ **b.** $0.30\frac{2}{3}$ **c.** $0.70\frac{3}{4}$ **d.** $0.50\frac{1}{2}$

9. a. $1.37\frac{1}{2}$ **b.** $1.00\frac{1}{2}$ **c.** $1.16\frac{2}{3}$ **d.** $1.05\frac{3}{4}$ **10. a.** 0.875   **b.** 0.125   **c.** 0.078   **d.** 0.989

11. a. 0.2625   **b.** 0.0875   **c.** 0.0233   **d.** 0.65125   **12. a.** $0.2\frac{1}{2}$ **b.** $0.6\frac{1}{2}$ **c.** $0.1\frac{3}{4}$ **d.** $0.8\frac{7}{8}$

13. a. 1.245   **b.** 1.375   **c.** 2.667   **d.** 3.7275   **14. a.** 1   **b.** 3   **c.** 5   **d.** 6

15. a. $0.00\frac{1}{4}$ **b.** $0.00\frac{1}{2}$ **c.** $0.00\frac{3}{5}$ **d.** $0.00\frac{2}{3}$ **16. a.** 0.005   **b.** 0.0075   **c.** 0.008   **d.** 0.00875

17. a. Over a period of 10 years, the value of a house increased 2.5 times. Express the increase in value as a percent.

   b. A baseball player has a batting average of .425. What percent of the times he has been at bat has he had hits?

   c. John has computed that 0.38 of his total pay is deducted for taxes, medical insurance, pension, etc. What percent of his pay is deducted?

   d. In a certain town 0.357 of the total land area is wetlands. What percent of the land area is wetlands?

4-4 Writing Percents as Fractions

You have learned to write a percent as a decimal representing hundredths. Remember that a percent is a ratio per hundred. How can you write a percent as a fraction?

Examples

1. Write 8% as a fraction.

$8\% = \dfrac{8}{100}$

$\quad\;\; = \dfrac{2}{25}$

ANSWER: $\dfrac{2}{25}$

Since percent means hundredths, the given number is the numerator and 100 is the denominator. Simplify.

---

2. Write 25% as a fraction.

$25\% = \dfrac{1}{4}$

or

$25\% = \dfrac{25}{100}$

$\quad\;\;\; = \dfrac{1}{4}$

ANSWER: $\dfrac{1}{4}$

If the fractional equivalent is known, write it directly.

or

If the fractional equivalent is not known, rewrite as hundredths. Simplify.

---

3. Write 130% as a mixed number.

$130\% = \dfrac{130}{100}$

$\quad\;\;\;\; = 1\dfrac{30}{100} = 1\dfrac{3}{10}$

ANSWER: $1\dfrac{3}{10}$

If the percent is greater than 100%, first write it as an improper fraction, then simplify.

4. Write $14\frac{1}{2}\%$ as a fraction.

$14\frac{1}{2}\% = \frac{29}{2} \div 100$

$\frac{29}{2} \times \frac{1}{100} = \frac{29}{200}$

ANSWER: $\frac{29}{200}$

Write a mixed number as a fraction. Divide by 100.

Think About It

1. Is 40% equivalent to the ratio 2:5? Explain.

2. Which is larger, 210% or $\frac{21}{100}$? Explain.

3. Explain the technique you would use in order to determine if 24% is more or less than $\frac{6}{27}$.

4. Before a newspaper went to press a proofreader found a typographical error in the following weather forecast: "There is a 350% chance of rain on Wednesday." What was the error?

Diagnostic Exercises

Write each percent as a fraction or mixed number.

1. 75% **2.** $16\frac{2}{3}\%$ **3.** 2% **4.** 125%

5. What part of a graduating class is planning to go to college if 40% filed applications for entrance?

Related Practice

Write each percent as a fraction or mixed number.

1. a. 50% **b.** 25% **c.** 10% **d.** 80% **2. a.** $33\frac{1}{3}\%$ **b.** $87\frac{1}{2}\%$ **c.** $83\frac{1}{3}\%$ **d.** $12\frac{1}{2}\%$

3. a. 6% **b.** 5% **c.** 9% **d.** 46% **4. a.** 110% **b.** 150% **c.** $133\frac{1}{3}\%$ **d.** $162\frac{1}{2}\%$

5. a. In a certain area $16\frac{2}{3}\%$ of all the crops were destroyed by floods. What part of the harvest was saved?

   b. If Genna Romano pays 25% down on a house, what part of the purchase price is her down payment?

   c. If $87\frac{1}{2}\%$ of a class received a passing mark in a science test, what part of the class failed the test?

   d. If a shortstop made 70% of his plays without an error, in what part of the total plays did he make errors?

Writing Fractions and Mixed Numbers as Percents

Sometimes you need to write a fraction or a mixed number as a percent. What technique would you use to write a fraction as a percent? How can an improper fraction be written as a percent? Are there other ways of writing a fraction as a percent?

Examples

1. Write $\frac{18}{25}$ as a percent.

$$\frac{18}{25} = 25\overline{)\begin{array}{r} 0.72 = 72\% \\ 18.00 \\ \underline{17\ 5} \\ 50 \\ \underline{50} \end{array}}$$

Divide the numerator by the denominator in order to write as hundredths.
Rewrite hundredths as percent.

ANSWER: 72%

2. Write $\frac{2}{9}$ as a percent.

$$\frac{2}{9} = 9\overline{)2.00}^{\,0.22\frac{2}{9}} = 22\frac{2}{9}\%$$

$$\underline{1\ 8}$$
$$20$$
$$\underline{18}$$
$$2$$

ANSWER: $22\frac{2}{9}\%$, or 22.2%

A fractional percent can be written as a decimal.

---

3. Write $\frac{55}{40}$ as a percent.

$$\frac{55}{40} = \frac{11}{8}$$

$$40\overline{)50.00}^{\,1.37\frac{4}{8}} = 137\frac{1}{2}\%, \text{ or } 137.5\%$$

Write as percent and write fraction in lowest terms or write its decimal equivalent.

ANSWER: $137\frac{1}{2}\%$, or 137.5%

---

4. Write $\frac{3}{4}$ as a percent.

$$\frac{3}{4} = 75\%$$

If the percent equivalent is known, write it directly.

ANSWER: 75%

---

5. Write $\frac{18}{100}$ as a percent.

$$\frac{18}{100} = 18\%$$

If the denominator is 100, write the **numerator** followed by the percent sign.

ANSWER: 18%

---

6. Write $\frac{7}{50}$ as a percent.

$$\frac{7}{50} = \frac{7 \times 2}{50 \times 2} = \frac{14}{100}$$
$$= 14\%$$

Write an equivalent fraction with a denominator of 100 by either raising to higher terms or reducing to lower terms.

ANSWER: 14%

---

7. Write $\frac{24}{40}$ as a percent.

$$\frac{24}{40} = \frac{3}{5} = 60\%$$

Simplify the given fraction.

ANSWER: 60%

---

8. Write $1\frac{1}{4}$ as a percent.

$$1\frac{1}{4} = 1.25 = 125\% \text{ or } 1\frac{1}{4} = 1 + \frac{1}{4} = 100\% + 25\% = 125\%$$

ANSWER: 125%

1. A student knows how many questions he has answered correctly on a test with 25 questions. Describe two methods that he can use to compute his percent grade on the test.

2. The numerator of a fraction is 28. What can you conclude about the denominator of the fraction if the percent equivalent of the fraction is greater than 100?

3. Is it correct to say that $\frac{3}{5} > 60\%$? If not, how can this statement be changed so that it is correct?

4. If $\frac{3}{500}$ is converted to a percent, will it be more or less than 1%? Explain.

Diagnostic Exercises

Write each fraction and mixed number as a percent.

1. $\frac{3}{4}$　　　2. $\frac{2}{3}$　　　3. $\frac{3}{100}$　　　4. $\frac{19}{50}$　　　5. $\frac{8}{400}$

6. $\frac{9}{36}$　　　7. $\frac{7}{9}$　　　8. $\frac{24}{56}$　　　9. $\frac{18}{18}$　　　10. $1\frac{1}{2}$

11. $\frac{7}{4}$　　　12. $\frac{72}{64}$　　　13. $2\frac{2}{3}$　　　14. $\frac{12}{5}$

15. If Maureen received $\frac{3}{5}$ of all votes cast in a homeroom election, what percent of the votes did she get?

Related Practice

Write each fraction and mixed number as a percent.

1. a. $\frac{1}{4}$　　b. $\frac{2}{5}$　　c. $\frac{1}{2}$　　d. $\frac{7}{10}$　　　2. a. $\frac{5}{6}$　　b. $\frac{3}{8}$　　c. $\frac{1}{3}$　　d. $\frac{1}{6}$

3. a. $\frac{7}{100}$　　b. $\frac{39}{100}$　　c. $\frac{145}{100}$　　d. $\frac{87\frac{1}{2}}{100}$　　　4. a. $\frac{27}{50}$　　b. $\frac{4}{25}$　　c. $\frac{13}{20}$　　d. $\frac{16}{25}$

5. a. $\frac{18}{200}$　　b. $\frac{35}{700}$　　c. $\frac{12}{300}$　　d. $\frac{60}{500}$　　　6. a. $\frac{16}{24}$　　b. $\frac{39}{65}$　　c. $\frac{40}{48}$　　d. $\frac{63}{72}$

7. a. $\frac{5}{7}$　　b. $\frac{8}{11}$　　c. $\frac{4}{9}$　　d. $\frac{7}{15}$　　　8. a. $\frac{20}{36}$　　b. $\frac{45}{99}$　　c. $\frac{42}{49}$　　d. $\frac{65}{75}$

9. a. $\frac{45}{45}$　　b. $\frac{7}{7}$　　c. $\frac{84}{84}$　　d. $\frac{156}{156}$　　　10. a. $1\frac{3}{4}$　　b. $1\frac{2}{5}$　　c. $1\frac{1}{3}$　　d. $1\frac{5}{8}$

11. a. $\frac{5}{3}$　　b. $\frac{13}{8}$　　c. $\frac{8}{7}$　　d. $\frac{9}{5}$　　　12. a. $\frac{57}{38}$　　b. $\frac{65}{52}$　　c. $\frac{70}{49}$　　d. $\frac{90}{48}$

13. a. $2\frac{5}{8}$　　b. $3\frac{4}{5}$　　c. $4\frac{2}{3}$　　d. $2\frac{3}{4}$　　　14. a. $\frac{8}{3}$　　b. $\frac{14}{4}$　　c. $\frac{17}{6}$　　d. $\frac{23}{10}$

15. a. A department store advertises a $\frac{1}{4}$ reduction on all merchandise. What is the percent markdown in this sale?

b. Lee's cat had 8 kittens, of which 3 are black. What percent of the kittens are black?

c. If $\frac{7}{10}$ of the student body participated in the interclass athletic games, what percent of the students did not participate?

d. Salespeople receive a commission of $\frac{3}{20}$ of the amount of their sales. What percent commission do they receive?

Using Mathematics

The table below shows some frequently used equivalents of percents, decimals, and fractions. You may find it useful in studying other lessons in this book.

Table of Equivalents Percents, Decimals, and Fractions		
Percent	Decimal	Fraction
5%	0.05	$\frac{1}{20}$
$6\frac{1}{4}$%	$0.06\frac{1}{4}$	$\frac{1}{16}$
$8\frac{1}{3}$%	$0.08\frac{1}{3}$	$\frac{1}{12}$
10%	0.10 or 0.1	$\frac{1}{10}$
$12\frac{1}{2}$%	$0.12\frac{1}{2}$ or 0.125	$\frac{1}{8}$
$16\frac{2}{3}$%	$0.16\frac{2}{3}$ or $0.16\overline{6}$	$\frac{1}{6}$
20%	0.20 or 0.2	$\frac{1}{5}$
25%	0.25	$\frac{1}{4}$
30%	0.30 or 0.3	$\frac{3}{10}$
$33\frac{1}{3}$%	$0.33\frac{1}{3}$ or $0.3\overline{3}$	$\frac{1}{3}$
$37\frac{1}{2}$%	$0.37\frac{1}{2}$ or 0.375	$\frac{3}{8}$
40%	0.40 or 0.4	$\frac{2}{5}$
50%	0.50 or 0.5	$\frac{1}{2}$
60%	0.60 or 0.6	$\frac{3}{5}$
$62\frac{1}{2}$%	$0.62\frac{1}{2}$ or 0.625	$\frac{5}{8}$
$66\frac{2}{3}$%	$0.66\frac{2}{3}$ or $0.6\overline{6}$	$\frac{2}{3}$
70%	0.70 or 0.7	$\frac{7}{10}$
75%	0.75	$\frac{3}{4}$
80%	0.80 or 0.8	$\frac{4}{5}$
$83\frac{1}{3}$%	$0.83\frac{1}{3}$ or $0.83\overline{3}$	$\frac{5}{6}$
$87\frac{1}{2}$%	$0.87\frac{1}{2}$ or 0.875	$\frac{7}{8}$
90%	0.90 or 0.9	$\frac{9}{10}$
100%	1.00 or 1	$\frac{1}{1}$ or $\frac{100}{100}$

Finding a Percent of a Number

Building Problem Solving Skills

A camera that regularly sells for $58 is advertised at 20% off. How much will the savings be? In order to find a percent of a quantity, write the percent as a decimal and multiply.

1. **READ** the problem carefully.

a. Find the **question asked.** How much do you save?

b. Find the **given facts.** The regular price of a camera is $58. It has been purchased at a 20% reduction.

---

2. **PLAN** how to solve the problem.

a. Choose the **operation needed.** You are finding a percent of a number so you must change the percent to a decimal and multiply.

b. **Think or write out your plan** relating the given facts to the question asked. The rate of reduction times the regular price is equal to the amount saved.

c. Express as an **equation.** $20\% \times \$58 = n$

---

3. **SOLVE** the problem.

a. Estimate the answer.

$20\% = 0.20 = 0.2$	Write the percent as a decimal.
$\$58 \rightarrow \$60$	Round the number to the greatest place.
$0.2 \times \$60 = \$12$	Multiply using mental arithmetic.

b. Solution.

$$\begin{array}{r} \$58 \\ \times\ 0.20 \leftarrow \textbf{2 places} \\ \hline \$11.60 \end{array}$$

Arrange in columns and multiply.
Count the number of decimal places.

Start from the right and count off the same number of decimal places.

You might use a calculator.

Key: 58 $\boxed{\times}$ .20 $\boxed{=}$ $\underline{11.60}$

4. [CHECK]

a. Check the accuracy of your arithmetic.
Check by reversing the order of the factors and multiplying.

$$
\begin{array}{r}
0.20 \\
\times\ \$\ 58 \\
\hline
1\ 60 \\
10\ 0 \\
\hline
\$11.60\ \text{✔}
\end{array}
$$

This is the same as your original answer.

b. Compare the answer to the estimate.

The answer $11.60 compares reasonably with the estimate of $12.

ANSWER: You save $11.60 by buying the camera on sale.

Think About It

1. If you got an answer to the problem above of $116, how would you know it was wrong? What mistake would you have made?

2. Will a discount ever be more than the original price?

3. Find a quick method to calculate 10% of any number in your head. (Remember what you learned about dividing by 10.) How can you find 20% of a number quickly?

4. If you double a number, what percent of the number is this? If you triple a number?

Calculator Know-How

Most calculators have a percent key, %. This reduces the number of keys you have to press when computing percentages. Here are some examples. Note the order of entries.

Examples

1. Find 40% of 30.
Key: 30 ⨯ 40 % 12

2. Find 30 out of 40 as a percent.
Key: 30 ÷ 40 % 75

Estimate, then use % to calculate the following. Round answers to the nearest tenth.

1. 75% of 60

2. 12% of 45

3. 16% of 85

4. 95% of 250

5. 15 out of 90

6. 28 out of 140

7. 97 out of 485

8. 133 out of 1,064

1. Find 23% of 64.

23% = 0.23

$$
\begin{array}{r}
64 \\
\times\ 0.23 \\
\hline
1\ 92 \\
12\ 8\ \\
\hline
14.72
\end{array}
$$

Write the percent as a decimal. Multiply the given number by the decimal.

ANSWER: 14.72

2. Find 3% of 18.

3% = 0.03

$$
\begin{array}{r}
18 \\
\times\ 0.03 \\
\hline
0.54
\end{array}
$$

ANSWER: 0.54

---

3. Find 114% of 240.

114% = 1.14

$$
\begin{array}{r}
2\ 40 \\
\times\ 1.14 \\
\hline
273.60
\end{array}
$$

ANSWER: 273.6

4. Find 4% of $200.

$4\% = \dfrac{4}{100}$

$\dfrac{4}{100} \times 200 = \dfrac{4}{\cancel{100}} \times \dfrac{\cancel{200}^{\,2}}{1} = 8$

Sometimes it is easier to write the percent as a fraction.

ANSWER: $8

---

5. Find $3\frac{1}{2}$% of 40.

$3\frac{1}{2}\% = 0.03\frac{1}{2} = 0.035$

$$
\begin{array}{r}
40 \\
\times\ 0.035 \\
\hline
200 \\
1\ 20\ \\
\hline
1.400 = 1.4
\end{array}
$$

or

$$
\begin{array}{r}
40 \\
\times\ 0.03\frac{1}{2} \\
\hline
1\ 20 \\
20\ \ \leftarrow 40 \times \frac{1}{2} \\
\hline
1.40
\end{array}
$$

ANSWER: 1.4

6. Find $\frac{3}{4}$% of 650.

$\frac{3}{4}\% = 0.00\frac{3}{4} = 0.0075$

$$
\begin{array}{r}
650 \\
\times\ 0.0075 \\
\hline
3250 \\
4\ 550\ \\
\hline
4.8750 = 4.875
\end{array}
$$

ANSWER: 4.875

---

7. Find 0.6% of 52.

0.6% = 0.006

$$
\begin{array}{r}
52 \\
\times\ 0.006 \\
\hline
0.312
\end{array}
$$

ANSWER: 0.312

8. Find 7% of $12.96 to the nearest cent.

7% = 0.07

$$
\begin{array}{r}
\$12.96 \\
\times\ \ \ 0.07 \\
\hline
\$.9072 = \$.91
\end{array}
$$

ANSWER: $.91

---

9. Solve the equation: 15% of 35 = n

$0.15 \times 35 = 5.25$

To find the value of n, multiply the given numbers.

ANSWER: $n = 5.25$

Diagnostic Exercises

Give the following.

1. 18% of 46

2. 6% of 24

3. 39% of 6.75

In Exercises 4, 5, and 6, write the percents as fractions.

4. 3% of 5,000

5. 25% of 36

6. $83\frac{1}{3}$% of 582

7. 140% of 295

8. 200% of 75

9. $4\frac{1}{2}$% of 624

10. $\frac{1}{4}$% of 300

11. 7.8% of 45

12. 0.3% of 160

13. 2% of $4.68 (to nearest cent)

14. Solve. 30% of $75 = n

15. The Stow Tigers lost 20% of the 40 games they played this season. How many games did they win?

Related Practice

1. a. 24% of 52

b. 63% of 75

c. 16% of 240

d. 92% of 48

2. a. 2% of 18

b. 3% of 200

c. 8% of 1,540

d. 4% of 8,462

3. a. 5% of 8.24

b. 3% of 9.62

c. 18% of 4.7

d. 38% of 251.69

4. a. 4% of 4,000

b. 6% of 650

c. 29% of 10,000

d. 83% of 7,250

5. a. 50% of 62

b. 75% of 92

c. 25% of 328

d. 80% of 1,225

6. a. $12\frac{1}{2}$% of 800

b. $33\frac{1}{3}$% of 768

c. $16\frac{2}{3}$% of 804

d. $6\frac{1}{4}$% of 9,328

7. a. 120% of 45

b. 125% of 848

c. $133\frac{1}{3}$% of 246

d. $137\frac{1}{2}$% of 5,640

8. a. 100% of 72

b. 500% of 400

c. 200% of 325

d. 300% of 3,854

9. a. $2\frac{1}{2}$% of 30

b. $1\frac{1}{2}$% of 2,500

c. $130\frac{1}{2}$% of 245

d. $98\frac{7}{8}$% of 5,000

10. a. $\frac{1}{2}$% of 48

b. $\frac{1}{3}$% of 930

c. $\frac{3}{4}$% of 219

d. $\frac{5}{8}$% of 600

11. a. 2.5% of 36

b. 4.7% of 840

c. 6.75% of 725

d. 5.625% of 192

12. a. 0.2% of 8

b. 0.1% of 95

c. 0.25% of 500

d. 0.375% of 848

13. Find to the nearest cent.

a. 5% of $4.60

b. 25% of $285.96

c. $87\frac{1}{2}$% of $.56

d. 6.05% of $58

14. a. 10% of 98 = n

b. 8% of 650 = ▨

c. 75% of 300 = n

d. n = 12.63% of $10,000

15. a. Richard received a grade of 60% on a spelling test of 25 words. How many words did he spell correctly?

b. If the sales tax is 6%, what would be the tax on a TV set costing $495?

c. Mrs. Espinosa stored her fur coat for the summer and was charged 2% of its value. If the coat is valued at $1,275, how much did she pay?

d. Betty's mother bought a house for $69,500 and made a down payment of 20%. What was the amount of the down payment?

Mixed Problems

Choose an appropriate method of computation for each problem. Solve.

1. The Manuelito's power bill is $34.50. They get a 4% discount if they pay by the end of the month. How much can they save by paying on time?

2. The seating capacity of a baseball stadium is about 45,000. To make a profit, 60% of the seats must be occupied. Approximately how many seats must they sell?

3. A watermelon is about 97% water. What is the weight of the water in a watermelon that weighs 8.96 lb.?

4. A record sells for $11.98. Pete bought one on sale and saved about 25%. How much did he save on the record?

5. Of the 160 students entering school this year, 55% are girls. How many are girls?

6. In a class of 45 students, 20% wear glasses. How many wear glasses?

7. At a recycling depot, about 3% of the returned bottles are broken or unusable. If 1,200 bottles are returned, how many are likely to be unusable?

8. Rita saves 10% by picking up a pizza. She orders a large one for $12.50. How much will she save?

9. There are 60 problems on the test. Sam got 75% correct. How many problems did he answer correctly?

10. In a can of mixed nuts, about 85% are peanuts. In a handful of 20 nuts, how many are probably peanuts?

Refresh Your Skills

1. 29 [1-8]
    6,842
    98,386
    2,594
    + 73,967

2. 1,000,000 [1-9]
    − 908,147

3. 846 [1-11]
    × 259

4. $1,728\overline{)1,655,424}$ [1-13]

5. Add: 6.03 + 2.39 + 3.26 + 7.32 [2-5]

6. Subtract: $9 − $1.32 [2-6]

7. (nearest cent) $70.29 [2-7]
    × $0.01\frac{1}{2}$

8. $144\overline{)\$12.96}$ [2 8]

9. $4\frac{3}{8} + 5\frac{9}{16}$ [3-8]

10. $8\frac{1}{4} − 2\frac{1}{3}$ [3-9]

11. $3\frac{1}{7} × 4\frac{2}{3}$ [3-11]

12. $7\frac{1}{2} ÷ 6\frac{3}{4}$ [3-12]

13. Rewrite $\frac{33}{7}$ as a mixed number. [3-2]

14. Express $\frac{7}{25}$ in hundredths [3-3]

15. Are $\frac{6}{14}$ and $\frac{18}{42}$ equivalent fractions? [3-4]

16. Which is greater, $\frac{3}{8}$ or $\frac{5}{16}$? [3-5]

17. Write 0.18 as a percent. [4-3]

18. Write 60% as a fraction. [4-4]

19. Write 56% as a decimal. [4-2]

20. Express 26.7 hundredths as a percent. [4-1]

21. Write $\frac{13}{20}$ as a decimal. [3-15]

4-7 Finding What Percent One Number Is of Another

Building Problem Solving Skills

Jean answered 31 questions correctly on a test of 40 questions. What percent of the questions did she answer correctly? To find what part one number is of another you compare the part to the whole. Write the whole as the denominator and the part as the numerator.

1. **READ** the problem carefully.

a. Find the **question asked.** What percent of the questions did she answer correctly?

b. Find the **given facts.** The test included 40 questions, and Jean answered 31 correctly.

2. **PLAN** how to solve the problem.

a. Choose the **operation needed.** Percent means *per hundred.* The fraction $\frac{31}{40}$ shows what part 31 is of 40. Write $\frac{31}{40}$ as a decimal by *dividing* 31 by 40. This shows you what decimal part 31 is of 40. Then write the decimal quotient as a percent.

b. **Think or write out your plan** relating the given facts to the question asked. The number of questions answered correctly divided by the total number of questions is equal to the decimal or fractional part of the test answered correctly. Express this quotient as a percent.

c. Express as an **equation.** $31 \div 40 = n$, or $n = \frac{31}{40}$

3. SOLVE the problem.

a. Estimate the answer.

$31 \rightarrow 30$ $\frac{30}{40} = \frac{3}{4} = 75\%$, estimate
$40 \rightarrow 40$

Round the numbers to the same place. Notice that 40 does not need to be rounded.

b. Solution.

$$\frac{31}{40} = 40\overline{)31.000}^{\;0.775\; = 77.5\%}$$
$$\underline{28\,0}$$
$$3\,00$$
$$\underline{2\,80}$$
$$200$$
$$\underline{200}$$
$$0$$

Divide the numerator by the denominator. Write the decimal quotient as a percent.

You might use a calculator.

Key: 31 ÷ 40 = 0.775

Change the decimal to a percent.

ANSWER: 77.5%

---

4. CHECK

a. Check the accuracy of your arithmetic. Check by multiplying.

$40 \times 0.775 = 31$

b. Compare the answer to the estimate. The answer 77.5% compares reasonably with the estimate 75%.

Think About It

1. State what 77.5% in the problem above represents.

2. What is the greatest percent correct a student could have gotten on the test?

3. If Jean had gotten 31 correct answers on a 50-question test, would her percent correct have been higher or lower?

Skills Review

1. What percent of 36 is 27?

$$\frac{27}{36} = \frac{27 \div 9}{36 \div 9}$$

$$= \frac{3}{4}$$

$$\frac{3}{4} = 75\%$$

ANSWER: 75%

The grid has 100 squares, and 75 of them are blue. You can see that $\frac{3}{4}$ of the large square is blue.

2. What percent of 48 is 30?

$$\frac{30}{48} = \frac{30 \div 6}{48 \div 6} = \frac{5}{8}$$

$$\begin{array}{r} 0.625 \\ 8\overline{)5.000} \\ \underline{4\,8} \\ 20 \\ \underline{16} \\ 40 \\ \underline{40} \end{array} = 62.5\% \text{ or } 62\frac{1}{2}\%$$

Divide the numerator by the denominator to 2 decimal places.

Write the decimal quotient as a percent.

ANSWER: 62.5%

3. 18 is what percent of 16?

$$\frac{18}{16} = \frac{18 \div 2}{16 \div 2} = \frac{9}{8}$$

$$\begin{array}{r} 1.125 \\ 8\overline{)9.00} \end{array} = 112.5\%$$

Since 18 is greater than 16, it is equal to more than 100% of 16.

ANSWER: 112.5%

4. 9 is what percent of 20?

$$\frac{9}{20} = \frac{45}{100} = 45\%$$

ANSWER: 45%

5. What percent of 26 is 26?

$$\frac{26}{26} = 1 = 100\%$$

ANSWER: 100%

6. $1.50 is what percent of $7.50?

$$\frac{\$1.50}{\$7.50} = \frac{1.50}{7.50} = \frac{1}{5} = 20\%$$

ANSWER: 20%

7. What percent of 5 is $\frac{3}{8}$?

$$\frac{\frac{3}{8}}{5} = \frac{\frac{3}{\cancel{8}} \times \cancel{8}}{5 \times 8} = \frac{3}{40} = 7.5\%$$

ANSWER: 7.5%

Diagnostic Exercises

1. 4 is what percent of 5?

2. What percent of 12 is 6?

3. What percent of 8 is 7?

4. 45 is what percent of 54?

5. 2 is what percent of 7?

6. What percent of 18 is 10?

7. 37 is what percent of 37?

8. 8 is what percent of 4?

9. 25 is what percent of 20?

10. What percent of 36 is 48?

11. 561 is what percent of 935?

12. What percent of $17 is $3.40?

13. $3\frac{1}{2}$ is what percent of $10\frac{1}{2}$?

14. 9 is what percent of 100?

15. 13 is what percent of 25?

16. What percent of 400 is 16?

17. Kristen picked 11 bushels of McIntosh apples and 9 bushels of Delicious. What percent were McIntosh? Hint: What is the total number of bushels Kristen picked?

Related Practice

1. a. 3 is what percent of 5?
   c. What percent of 5 is 1?
   b. 3 is what percent of 4?
   d. What percent of 2 is 1?

2. a. What percent of 36 is 9?
   c. 5 is what percent of 50?
   b. What percent of 80 is 56?
   d. 39 is what percent of 52?

3. a. What percent of 6 is 5?
   c. 3 is what percent of 8?
   b. What percent of 3 is 2?
   d. 5 is what percent of 8?

4. a. 9 is what percent of 27?
   c. 35 is what percent of 42?
   b. 16 is what percent of 24?
   d. What percent of 72 is 63?

5. a. 3 is what percent of 7?
   c. What percent of 13 is 8?
   b. What percent of 9 is 7?
   d. 17 is what percent of 18?

6. a. What percent of 14 is 4?
   c. 15 is what percent of 84?
   b. What percent of 54 is 30?
   d. 42 is what percent of 49?

7. a. 8 is what percent of 8?
   c. What percent of 82 is 82?
   b. What percent of 53 is 53?
   d. 250 is what percent of 250?

8. a. 8 is what percent of 2?
   c. What percent of 10 is 30?
   b. 60 is what percent of 12?
   d. What percent of 12 is 72?

9. a. 6 is what percent of 5?
   c. What percent of 50 is 65?

   b. What percent of 24 is 54?
   d. 63 is what percent of 35?

10. a. What percent of 3 is 4?
   c. 90 is what percent of 48?

   b. What percent of 12 is 32?
   d. 85 is what percent of 75?

11. a. 492 is what percent of 656?
   c. What percent of 1,225 is 98?

   b. 645 is what percent of 1,032
   d. 1,014 is what percent of 2,535?

12. a. What percent of 10 is 2.5?
   c. 0.54 is what percent of 9?

   b. What percent of 7.5 is 3?
   d. 4.3 is what percent of 6.45?

13. a. $2\frac{1}{4}$ is what percent of 9?
   c. What percent of $1\frac{7}{8}$ is $1\frac{1}{4}$?

   b. What percent of $4\frac{2}{3}$ is $3\frac{1}{2}$?
   d. $4\frac{1}{8}$ is what percent of $6\frac{7}{8}$?

14. a. 15 is what percent of 100?
   c. What percent of 100 is 125?

   b. 4.5 is what percent of 100?
   d. What percent of 100 is $33\frac{1}{3}$?

15. a. 7 is what percent of 20?
   c. What percent of 25 is 21?

   b. What percent of 50 is 34?
   d. 2 is what percent of 25?

16. a. What percent of 200 is 6?
   c. 120 is what percent of 1,000?

   b. What percent of 900 is 60?
   d. 32 is what percent of 800?

17. Solve.

   a. One afternoon the Juice Bar sold 150 cups of papaya juice, and 90 cups of pineapple juice. What percent of the total cups sold were pineapple juice?

   b. In the Student Government election, 455 votes were cast. Jenny received 182 votes. What percent of the votes did she receive?

   c. Last year, John worked in the office for 207 days. For 63 days, he was on the road visiting clients. What percent of his work days were travel days? (Round to the nearest percent.)

   d. There are 1,064 students in the local high school. Of these, 585 participate in at least one after-school activity. What percent of the students participate in after-school activities? (Round to the nearest percent.)

Mixed Problems

Choose an appropriate method of computation and solve each of the following problems. For Problems 1–4, select the letter corresponding to your answer.

1. There are 40 pupils in a class. On a certain day 38 pupils were present. What percent of the class attended school?
 a. 38% **b.** 2% **c.** 95% **d.** Answer not given

2. There are 18 women and 27 men in a class. What percent of the students are men?
 a. $66\frac{2}{3}$% **b.** 60% **c.** 50% **d.** 40%

3. How many problems did Joan have right if she received a grade of 85% in a mathematics test of 20 problems?
 a. 15 problems **b.** 17 problems
 c. 3 problems **d.** Answer not given

4. The enrollment in the Weston H.S. is 850. If the attendance for June was 92%, how many absences were there?
 a. 88 absences **b.** 782 absences
 c. 68 absences **d.** Answer not given

5. Charles answered 19 questions correctly and missed 6 questions. What percent of the questions did he answer correctly?

6. A certain manganese bronze contains 59% copper. How many pounds of copper are in 300 lb of manganese bronze?

7. The price of a radio is reduced by 20%. If the original price was $30, what is the sale price?

8. Fred sold 140 TVs. Of these, 65% were color. How many were color televisions?

9. Sales at Johnson's shoe store last year included 4,850 pairs of sneakers. This year the number of sneakers sold increased 2%. How many pairs of sneakers were sold this year?

10. Phil sold a shirt for $5.50. His profit was 17% of this. Find his profit.

Refresh Your Skills

1. 1–8
```
   7,124
  83,259
  61,435
     816
   9,527
+ 84,349
```

2. 1–9
```
  815,008
- 747,598
```

3. 1–11
```
    403
  × 807
```

4. 1–13
$$794\overline{)235,818}$$

5. 2–5
$$3.8 + 0.47 + 25$$

6. 2–6
$$6.825 - 0.97$$

7. 2–7
$$3.14 \times 80$$

8. 2–9
$$1.15\overline{)46}$$

9. 3–8
$$7\frac{2}{3}$$
$$+ 4\frac{5}{6}$$

10. 3–9
$$8\frac{3}{4}$$
$$- 2\frac{4}{5}$$

11. 3–11
$$4\frac{3}{8} \times 2\frac{3}{10}$$

12. 3–12
$$\frac{13}{16} \div 1\frac{1}{12}$$

13. Name the least common multiple (LCM) of 12, 24, and 36. 3–5

14. Find the least common denominator (LCD) and write equivalent fractions for: $\frac{7}{16}$, $\frac{13}{24}$, and $\frac{15}{32}$. 3–6

15. Write $6\frac{3}{7}$ as an improper fraction. 3–10

16. What part of 52 is 39? 3–13

17. Write 0.35 as a fraction and as a percent.
 3–16, 4–3

18. Luis spent 40% of his allowance on records. What fraction of his allowance is that? 4–4

Percent of Increase or Decrease

Martin received a raise from $275 to $297 a week. What percent of his original pay was his increase?

Amount of increase: $297 − $275 = $22

Divide the amount of increase by the original amount.

Percent of increase: $22 ÷ $275 = 0.08

$$0.08 = 8\%$$

Change the decimal to a percent.

Examples

1. Find the percent of decrease.

Original price: $39
Sale Price: $34.32
$39 − $34.32 = $4.68
$4.68 ÷ $39 = 0.12 = 12%

ANSWER: 12%

2. Find the percent of increase.

From 60 to 100

$$\frac{100 - 60}{60} = \frac{40}{60} = \frac{2}{3} = 66\frac{2}{3}\%$$

Remember that the denominator is always the original amount.

ANSWER: $66\frac{2}{3}\%$

3. Find the percent of decrease.

From 100 to 60

$$\frac{100 - 60}{100} = \frac{40}{100} = 40\%$$

ANSWER: 40%

4. Find the percent of increase.

From $62 to $124.

$$\frac{\$124 - \$62}{\$62} = \frac{\$62}{\$62} = 100\%$$

ANSWER: 100%

1. If a number is increased by 100%, what is the new number, compared to the old number?

2. If a number is decreased by 100%, what is the new number?

3. Can you have an increase of more than 100%? Explain.

4. Can you have a decrease of more than 100%? Explain.

Diagnostic Exercises

Find the percent of increase.

1. From 10 to 11

2. From $9 to $12

3. From 15 to 16.5

Find the percent of decrease.

4. From 80 to 60

5. From 6 to 5

6. From $30 to $22.50

7. Karl weighed 72 kilograms when school started. By April, he weighed 76 kilograms. What was the percent of increase in his weight?

Related Practice

Find the percent of increase.

1. a. 8 to 10   **b.** 12 to 15   **c.** 150 to 180   **d.** 300 to 600

2. a. $12 to $18   **b.** $36 to $42   **c.** $200 to $300   **d.** $120 to $138

3. a. 16 to 22   **b.** 90 to 210   **c.** 5 to 7.5   **d.** 125 to 137.5

Find the percent of decrease.

4. a. 100 to 90   **b.** 60 to 30   **c.** 120 to 105   **d.** 300 to 240

5. a. 9 to 6   **b.** 180 to 150   **c.** 250 to 10   **d.** 45 to 30

6. a. $60 to $44.50   **b.** $10.50 to $5.25   **c.** $81 to $74.52   **d.** $125.50 to $94.12

7. a. A basketball that regularly sells for $25 was purchased for $20. What was the rate (percent) of reduction?

b. School enrollment increased from 250 to 300 students. What is the percent of increase?

c. Mrs. Miyako's salary was raised from $28,800 per year to $30,210 per year. What is the rate of increase?

d. A stereo originally costing $115 is on sale for $97.75. What is the percent of reduction?

4-9 Finding a Number When a Percent Is Known

Building Problem Solving Skills

The school baseball team won 39 games, or 52% of the games it played. How many games did it play? In order to find the number of games played, divide the number of games won by the percent of the games won.

1. **READ** the problem carefully.

   a. Find the **question asked.** How many games did the team play?

   b. Find the **given facts.** The baseball team won 39 games, which was 52% of the games it played.

2. **PLAN** how to solve the problem.

   a. Choose the **operation needed.**
   Division is indicated when you know a percent of a number and want to find the number.

   b. **Think or write out your plan** relating the given facts to the question asked.
   52% of how many games played is 39 games? *or* The number of games won divided by the decimal equivalent of the percent is equal to the games played.

   c. Express as an **equation.** $0.52n = 39$ *or* $n = 39 \div 0.52$.

3. **SOLVE** the problem.

   a. **Estimate the answer.**

   $0.52 \rightarrow 0.5$

   $39 \rightarrow 40$

   $$0.5\!\!\underset{\wedge}{\overline{)40.0_\wedge}} \quad \underset{}{\overset{80}{}}\text{ games played (estimate)}$$

   Round the decimal equivalent of the percent to the nearest tenth.
   Round the number to a multiple of 5.
   Divide.

   b. **Solution.**

   $$0.52_\wedge\!\overline{)39.00_\wedge} \quad \overset{75}{}\text{ games played}$$
   $$\underline{364}$$
   $$260$$
   $$\underline{260}$$

   Divide the number of games won by the decimal equivalent of the percent.

You might use a calculator.

Key: 39 ÷ 0.52 = 75

4. CHECK.

a. Check the accuracy of your arithmetic.

$$
\begin{array}{r}
75 \\
\times\ 0.52 \\
\hline
1\ 50 \\
37\ 5 \\
\hline
39.00 \text{ games won} \checkmark
\end{array}
$$

Check the unit.
Multiply the number of games you found by the decimal equivalent of the percent

b. Compare the answer to the estimate. The answer 75 games compares reasonably with the estimate of 80 games.

ANSWER: 75 games played. The unit, games, is correct.

Think About It

1. If the baseball team had won the same number of games (39) but this was 60% of the games played instead of 52%, would the total number of games have been greater or less than 75?

2. If the baseball team had won 39, or 75%, of its games, what fraction could you have divided by to find the total number of games played?

3. Number A is a percent of number B. In the problem about the baseball team, A must be less than or equal to B. Explain what would have to be true about the percent for A to be greater than B.

4. List the key-press sequence on your calculator, including %, to find the number of games the baseball team played *last* year. The 44 games it won this year was 10% more than it won last year, and last year the team won $41\frac{2}{3}$% of its games.

1. 48 is 16% of what number?

$48 \div \mathbf{16\%}$

↑   ↖

part   **percent representing part**

16% = 0.16

$$0.16_{\wedge}\overline{)48.00_{\wedge}} \atop \underline{48}}^{\textstyle 300}$$

Divide the given **part** by the **percent** representing this part.

Write the percent as a decimal before dividing.

ANSWER: 300

---

2. 4.2% of what number is 32.55?

$32.55 \div \mathbf{4.2\%} = 32.55 \div 0.042$

Write the percent as a decimal. Write zeros if necessary.

$$\begin{array}{r} 775 \\ 0.042_{\wedge}\overline{)32.550_{\wedge}} \\ \underline{29\ 4} \\ 3\ 15 \\ \underline{2\ 94} \\ 210 \\ \underline{210} \end{array}$$

ANSWER: 775

---

3. 50% of what number is 40?

$40 \div \mathbf{50\%}$

$40 \div \frac{1}{2} = 40 \times \frac{2}{1}$

$\phantom{40 \div \frac{1}{2} } = 80$

Sometimes it is easier to write the percent as a fraction.

ANSWER: 80

---

4. 15 is 125% of what number?

$15 \div \mathbf{125\%} = 15 \div \frac{125}{100}$

$\phantom{15 \div \mathbf{125\%} } = 15 \div \frac{5}{4}$

$\phantom{15 \div \mathbf{125\%} } = \frac{\overset{3}{\cancel{15}}}{1} \times \frac{4}{\underset{1}{\cancel{5}}}$

$\phantom{15 \div \mathbf{125\%} } = 12$

Write the percent as a fraction and simplify it before dividing.

ANSWER: 12

5. $\frac{3}{8}$% of what number is 15?

$$15 \div \frac{3}{8}\% = 15 \div 0.00\frac{3}{8}$$
$$= 15 \div 0.00375$$

$$\begin{array}{r} 4000 \\ 0.00375_\wedge\overline{)15.00000_\wedge} \\ \underline{15\ 00} \end{array}$$

Move the decimal point 2 places to the left to write the percent as a decimal.

Write a fractional percent as a decimal equivalent.

ANSWER: 4000

Diagnostic Exercises

Find the missing numbers.

1. 12% of what number is 24?

2. 18 is 36% of what number?

3. 25% of what number is 6?

4. $66\frac{2}{3}$% of what number is 14?

5. 6% of what number is 12?

6 20 is 20% of what number?

7. 100% of what number is 70?

8. 120% of what number is 108?

9. 40% of what number is 12.6?

10. 2.5% of what number is 2?

11. $4\frac{1}{2}$% of what number is 90?

12. $187\frac{1}{2}$% of what number is 105?

13. 0.5% of what number is 4?

14. $\frac{3}{4}$% of what number is 27?

15. 8% of what amount is $180?

16. On a science test, Fred gets 34 problems correct and receives a grade of 85%. How many questions were on the test?

Related Practice

Find the missing numbers.

1. a. 45% of what number is 90?
   c. 74% of what number is 370?

   b. 65% of what number is 260?
   d. 15% of what number is 18?

2. a. 12 is 24% of what number?
   c. 9 is 15% of what number?

   b. 44 is 55% of what number?
   d. 17 is 85% of what number?

3. a. 20% of what number is 3?
   c. 18 is 60% of what number?

   b. 40% of $n = 48$
   d. $91 = 70$% of n

4. a. $33\frac{1}{3}$% of what number is 78?
   c. 4 is $16\frac{2}{3}$% of what number?

   b. $62\frac{1}{2}$% of $n = 200$
   d. $462 = 66\frac{2}{3}$% of n

5. a. 2% of what number is 10?
   c. $15 = 1$% of n

   b. 4% of $n = 26$
   d. 5% of what number is 45?

6. a. 50 is 50% of what number? **b.** 30 = 30% of n
   c. 79% of what number is 79? **d.** 43% of n = 43

7. a. 100% of what number is 59? **b.** 300% of n = 240
   c. 5 is 100% of what number? **d.** 36 = 200% of n

8. a. 160% of what number is 72? **b.** 175% of n = 42
   c. 513 is 114% of what number? **d.** 78 = 156% of n

9. a. 60% of n = 46.8 **b.** 102.7 is 65% of what number?
   c. 371.2 = 58% of n **d.** 70% of what number is 667.8?

10. a. 4.75% of what number is 38? **b.** 8.1 is 8.1% of what number?
   c. 16.5 = 4.125% of n **d.** 26.3% of n = 18.41

11. a. $4\frac{3}{4}$% of what number is 19? **b.** $6\frac{1}{8}$% of n = 12.25
   c. 18 is $2\frac{1}{2}$% of what number? **d.** 72 = $28\frac{4}{5}$% of n

12. a. $116\frac{2}{3}$% of what number is 21? **b.** $287\frac{1}{2}$% of n = 230
   c. 39 is $162\frac{1}{2}$% of what number? **d.** 644 = $233\frac{1}{3}$% of n

13. a. 0.4% of what number is 2? **b.** 0.875% of n = 70
   c. 2 is 0.1% of what number? **d.** 12 = $0.33\frac{1}{3}$% of n

14. a. $\frac{1}{2}$% of what number is 10? **b.** $\frac{5}{12}$% of n = 100
   c. 9 is $\frac{3}{20}$% of what number? **d.** 60 is $\frac{2}{3}$% of n.

15. a. 75% of what amount is $150?
   b. 6% of what amount is $78?
   c. $.12 is 10% of what amount?
   d. $5,075 is 101.5% of what amount?

16. a. How much money must be invested at 9% to earn $4,500 per year?
   b. Find the regular price of a computer that sold for $1,025 at an 18%-reduction sale.
   c. If an ore contains 16% copper, how many metric tons of ore are needed to get 20 metric tons of copper?
   d. A real estate agent received a $2,700 commission, which is 6% of the selling price of a house. At what price did she sell the house?

Mixed Problems

Choose an appropriate method of computation and solve each problem. For Problems 1–4, select the letter corresponding to your answer.

1. A house worth $47,900 is insured for 80% of its value. How much would the owner receive if the house were destroyed by fire?

    a. $47,900   **b.** $9,580
    c. $30,000   **d.** $38,320

2. Betty bought a camera at a 20%-off sale. If she paid $36 for it, what was the regular price?

    a. $18.20   **b.** $42.50
    c. $45.00   **d.** $60.00

3. School lateness dropped from 42 cases per month to 35 cases. The decrease is:

    a. 20%   **b.** 10%
    c. $16\frac{2}{3}$%   **d.** 7%

4. If 45% of the students are boys, and the girls number 858, how many boys are enrolled?

    a. 386 boys   **b.** 702 boys
    c. 1,560 boys   **d.** 1000 boys

5. Leon saved $7.50 by buying a shirt at a reduction of 25%. What was the regular price?

6. At what price should a dealer sell a TV set that costs $420 to make a profit of 30% on the selling price?

7. A large corporation increased its annual dividend from $28 per share to $29.40 per share. What was the percent increase in dividend per share?

8. Mr. Caruso sold 4%, or $200, more in computer equipment this week than last week. If his rate of commission is 12%, how much commission did he earn last week?

Refresh Your Skills

1. Add: $8.29 + $.58 + $46.75 [2–4]

2. Subtract: $49 − $.89 [2–5]

3. Multiply: 2.8 × 0.003 [2–7]

4. Divide: 20)0.1 [2–8]

5. Add: [3–8]

$$4\frac{5}{8}$$
$$\frac{3}{4}$$
$$+\ 8\frac{11}{16}$$

6. Subtract: [3–9]

$$1\frac{1}{6}$$
$$-\ \frac{1}{2}$$

7. Multiply: [3–11]

$$6\frac{2}{3} \times 7\frac{7}{8}$$

8. Divide: [3–12]

$$4\frac{1}{6} \div 2\frac{13}{16}$$

9. Write $3.83\frac{1}{3}$ as a fraction. [3–16]

10. Express in simplest form the ratio of 30 min to 2 h. [3–17]

11. Solve the proportion: $\frac{1\frac{1}{3}}{12} = \frac{\frac{1}{3}}{x}$ [3–18]

12. Write $\frac{19}{20}$ as a percent. [4–5]

13. Find $2\frac{1}{2}$% of $6000. [4–6]

14. What percent of 120 is 75? [4–7]

15. School enrollment this year is 450. Last year it was 420. What is the rate of increase? [4–8]

Using Mathematics

PERCENT: SOLVING BY PROPORTION

The three basic types of percent problems may be treated as one through the use of the *proportion*.

To find 8% of 25 means to determine *the number (n) that, compared to 25, is the same as 8 compared to 100.*

Solve the proportion.

$$\frac{n}{25} = \frac{8}{100}$$
$$100n = 200$$
$$n = 2$$

ANSWER: 2

To find the percent of 25 that 2 is means to find *the number (n) per 100, or the ratio of a number to 100, that has the same ratio as 2 to 25.*

Solve the proportion.

$$\frac{n}{100} = \frac{2}{25}$$
$$25n = 200$$
$$n = 8$$

ANSWER: 8%

To find the number of which 8% is 2 means to determine *the number (n) such that 2 compared to this number is the same as 8 compared to 100.*

Solve the proportion.

$$\frac{2}{n} = \frac{8}{100}$$
$$8n = 200$$
$$n = 25$$

ANSWER: 25

Exercises

Write each of the following as a ratio to 100.

1. 6%   **2.** 18%   **3.** 70%   **4.** $62\frac{1}{2}\%$ **5.** 2.4%

Find each of the following using proportions.

6. 37% of 16   **7.** 60% of $1,200   **8.** 3% of $940

9. 180% of 685   **10.** $16\frac{2}{3}\%$ of 732   **11.** 5.9% of 28

12. 2 is what percent of 5?   **13.** What percent of 84 is 63?

14. 49 is what percent of 56?   **15.** What percent of 8 is 10?

16. What percent of $.45 is $.27?   **17.** $1.50 is what percent of $9?

18. 9% of what number is 72?   **19.** 24 is 75% of what number?

20. 7.8 is 4% of what number?

Problem

Mrs. Romero can buy a stereo system for the cash price of $849.95, or pay the installment price of $85 down and $88.48 a month for 10 mo. How much can she save by paying cash?

1. **READ** the problem carefully.

a. Find the question asked. How much can be saved by paying cash?

b. Find the **given facts.** Cash price: $849.95; installment price: $85 down and $88.48 per month for 10 mo.

Think:

To find the amount saved by paying cash, you must find the installment price.

To find the installment price, you first must find the total amount of the monthly payments and add it to the down payment.

Hidden Questions:

What is the total amount of the monthly payments? What is the total installment price?

2. **PLAN** how to solve the problem.

a. Choose the appropriate problem solving strategy. Choose the operations needed.

Step 1: The words *total amount* with *equal* quantities indicate *multiplication.*

Step 2: The words *total installment price* indicate *addition.*

Step 3: The words *how much . . .save* indicate *subtraction.*

b. Think or write out your plan in words. The number of payments times the monthly payment (Step 1) plus the down payment (Step 2) is equal to the installment price. The installment praice (Step 2) minus the cash price (Step 3) is equal to the savings.

c. Express the plan as an equation. The three steps may be translated into a single equation.

installment price − cash price = savings

$(10 \times \$88.48 + \$85) - \$849.95 = n$

3. **SOLVE** the problem.

a. Estimate the answer.

$(10 \times \$88.48 + \$85) - \$848.85$ Remember to multiply first, then add.

$(10 \times \$90 + \$90) - \$850 =$

$\$900 + \$90 - \$850 = \$140$, estimate

b. Solution.

Step 1 $10 \times \$88.48 = \$884.80$

Step 2 $\$884.80 + \$85 = \$969.80$, installment price

Step 3 $\$969.80$ installment price   You might use a calculator.
$$- \underline{849.95} \text{ cash price}$$
        $\$119.85$ savings   Key: 10 ⊠ 88.48 ⊞ 85 ⊟ 849.95 ⊟ <u>119.85</u>

---

4. **CHECK** .

a. Check the accuracy of your arithmetic.

$\$88.48 \times 10 = \$884.80$ ✔ $\$849.95$ cash price

$\$85 + \$884.80 = \$969.80$ ✔ $+ \underline{119.85}$ savings
                                  $\$969.80$ installment price ✔

b. Compare the answer to the estimate.

The answer $\$119.85$ compares reasonably with the estimate, $\$140$.

ANSWER: $\$119.85$ savings.

Think About It

1. Although Mrs. Romero can save $119.85 by paying cash, what advantages might she see in using the installment plan?

2. Create a multistep problem of your own. Provide enough information to solve the problem, but be sure to include two hidden questions. Have a classmate identify the hidden questions and solve the problem.

Diagnostic Exercises

1. What are the hidden questions that, when answered, will provide the necessary facts to solve the following problem?

A $35 vest is on sale at a 40% reduction. If the sales tax is 5%, what is the total cost of the vest?

2. A dealer bought 24 softballs for $48. At what price must she sell each one to realize a profit of 35% on the cost?

Related Practice

1. What are the hidden questions that, when answered, will provide the necessary facts to solve each of the following problems?

a. Helen scored 82, 91, and 76 in three tests, while Miguel scored 75, 88, and 89 in the same tests. Who has the higher average and how much higher?

b. Bart found a $48.95 briefcase on sale at 20% off. He can buy the same briefcase at another store for $39.95. Which is the better buy?

c. What is the cost of a 7-minute telephone call if the rate is $.35 for the first minute and $.24 for each additional minute?

d. Ann can purchase an air conditioner for the cash price of $1550 or $350 down and 12 equal monthly payments of $115 each. How much can be saved by paying cash?

2. Solve each of the following problems.

a. David's team scored 348 points in the relay races and 589 points in the field events. Gloria's team scored 476 points and 469 points in the same events. Which team had the higher score and how many points higher?

b. A parking lot charges $1.25 for the first hour and $.75 for each additional hour. How much will it cost to park for 4 hours?

c. A VCR was bought by a merchant for $275 and marked to sell for $495. It was sold at a discount of 20% on the marked price. What was the amount of profit?

d. Mrs. Jordan bought a suitcase priced at $79.95 and a tote bag for $15.85. She had to pay a 4% sales tax. How much change did she get back from $100?

Refresh Your Skills

1.
```
   638  1–8
 5,976
+  259
```

2.
```
 3,506  1–9
－  952
```

3.
```
    207  1–11
 ×  490
```

4. $24\overline{)840}$ 1–13

5. 0.62 + 0.58 2–5

6. 9.6 − 2.8 2–6

7. 0.52 × 0.7 2–7

8. $0.7\overline{)4.9}$ 2–8

9. $2\frac{1}{2} + 1\frac{3}{4}$ 3–8

10. $8\frac{1}{3} - 1\frac{1}{4}$ 3–9

11. $6 \times 2\frac{2}{3}$ 3–11

12. $\frac{5}{8} \div \frac{3}{4}$ 3–12

13. What percent of 20 is 17? 4–7

14. Find 8% of $40. 4–6

15. 9 is 30% of what number? 4–9

Problem Solving Review

READ PLAN SOLVE CHECK

These four problem-solving steps are a guide to help you solve any word problem. Be sure you always use all four steps when you solve a problem. Each step is important in finding the correct solution. If you follow all four steps and still get an incorrect answer, begin the process again by rereading the problem.

Solve a Simpler Problem and Guess and Test are two strategies that can be used to solve a problem. To review how these strategies are used, follow the four steps in the two examples below.

SOLVE A SIMPLER PROBLEM
The owner of Video Outlet paid $2,064.15 for 9 VCRs and $1536.45 for 3 cameras. She sold the VCRs for $489.99 each. What was the owner's profit on the sale of the 9 VCRs?

1. **READ**

a. *Question:* What was the profit on the sale of the 9 VCR's?

b. *Facts:*
Paid $2,064.15 for 9 VCRs
Sold VCRs for $489.99 each

2. **PLAN**

a. Solve a simpler problem

b. Simpler problem: If owner paid $2,000 for 10 VCRs and sold the VCRs for $500 each, what would her profit be?

amount owner charged for 10 VCRs −
amount owner paid for 10 VCRs = profit

$(500 \times 10) - (2,000) = 3,000$

GUESS AND TEST
Joel has dimes and quarters in his pocket. He has 9 coins with a money value of $1.65. How many of each coin does he have?

1. **READ**

a. *Question:* How many dimes and how many quarters does Joel have?

b. *Facts:*
9 coins (some quarters, some dimes) money values of $1.65

2. **PLAN**

a. Guess and test

b. Relate facts to question asked:

number of quarters + number of dimes = 9
(number of quarters × .25) +
   (number of dimes × .10) = 1.65

Choose a pair of numbers with a sum of 9 for the number of quarters and the number of dimes.

Make a table to record the results.

3. | SOLVE |

a. Estimate.
  ($500 × 9) − $2,000 =
    $4,500 − $2,000 = $2500

b. Solve.
  ($439.99 × 9) − $2,064.15 =
    $4,409.91 − $2,064.15 = $2,345.76

4. | CHECK |

a. Your arithmetic
  $2,345.76 + $2,064.15 = $4,409.91

b. Is the answer reasonable?
  The answer $2,345.76 is close to the estimate, $2,500.

ANSWER: The owner's profit was $2,345.76.

3. | SOLVE |

Test 4 quarters and 5 dimes
  4(.25) + 5(.10) = 1.50
Test 6 quarters and 3 dimes
  6(.25) + 3(.10) = 1.80
Test 5 quarters and 4 dimes
  5(.25) + 4(.10) = 1.65

Number of Quarters	Number of Dimes	Money Value
4	5	$1.50
6	3	$1.80
5	4	$1.65

4. | CHECK |

a. Your arithmetic
  4($.10) + 5($.25) = $1.65
  4 + 5 = 9

b. Is the answer reasonable? The answer fits the facts.

ANSWER: Joel has 5 quarters and 4 dimes.

Practice Problems

Use Solve a Simpler Problem or Guess and Test to solve each problem.

1. Becky bought some 14-cent stamps and some 22-cent stamps. She paid $2.78 for 17 stamps. How many of each kind of stamp did she buy?

2. Phil has $32.95. Rock tapes are on sale for $4.79 each and classical tapes are on sale for $5.59 each. How many rock tapes can Phil buy?

3. The sum of two numbers is 20. Their product is 96. What are the two numbers?

4. The sum of two numbers is 26. The difference between the two numbers is 8. What are the two numbers?

5. Marni has been offered two part-time jobs. She can work at the bike shop $21\frac{1}{2}$ hours per week at an hourly rate of $5.32. Or, she can work at Burger Deluxe $27\frac{1}{4}$ hours per week at an hourly rate of $4.83. Which job will pay more per week?

6. The Wilson's bought 7 movie tickets. Adults' tickets cost $3.25 each. Children's tickets cost $1.75 each. If the Wilsons spent $19.75 in all, how many adults' tickets and how many children's tickets did they buy?

7. Juan saw the 3-speed bicycle he wants at the Outlet Mart for $159.99. The same bicycle is at Bert's Bike Shop for $209.99 less a 25% end-of-the-season discount. Which store has the better buy on this bicycle?

8. Twelve students signed up to play in a ping-pong tournament. If each student plays every other student once, how many ping-pong games will be played?

9. Fred, Ned, and Ted have 23 pencils in all. Ted has 3 more than Fred. Ned has twice as many as Fred. How many pencils does each one have?

Career Application

RETAILING

When you shop at a large retail store, you benefit from the work of people in many different jobs. *Buyers* decide what merchandise the store will sell. *Salespeople* assist customers with their purchases. Salespeople may be paid a commission, which is a percentage of the purchase price. *Cashiers* wrap customers' purchases and collect the money. *Credit-department clerks* use computers to prepare bills for customers who have charge accounts.

Many buyers have college degrees in marketing. Salespeople, cashiers, and credit-department clerks are usually high-school graduates who have had some special training.

1. A cashier rings up a $48.80 purchase. The sales tax rate is 5%. What is the total cost?

2. A credit-department clerk charges a customer a finance charge of $1.20. The finance-charge rate is 1.5% of the previous balance. What is the previous balance?

3. A buyer estimates that the store will sell 10,000 books next year. This year, 9,150 books were sold. What will be the percent increase to the nearest tenth of a percent?

4. A salesperson receives a commission of $52 for selling a refrigerator. The commission rate is 8%. What is the selling price of the refrigerator?

Chapter 4 Review

Vocabulary   percent p. 224

Page

225 Write each of the following as a percent, a decimal, and a fraction.

1. nine hundredths   **2.** forty-three hundredths   **3.** 87 out of 100

227 Express each percent as a decimal.

4. 2%   **5.** 23%   **6.** 213%   **7.** 3.4%

229 Express each decimal as a percent.

8. 0.04   **9.** 0.7   **10.** 2.26   **11.** 0.814

231 Express each percent as a fraction or mixed number.

12. 27%   **13.** $8\frac{1}{3}$%   **14.** 375%

233 Express each fraction or mixed number as a percent.

15. $\frac{9}{10}$ **16.** $\frac{11}{8}$ **17.** $\frac{94}{94}$ **18.** $7\frac{1}{6}$

237 Find each amount.

19. 60% of 80   **20.** 92% of 4000   **21.** $8\frac{1}{2}$% of $45.82   **22.** 140% of 670
                                                  (to the nearest cent)

242 **23.** 225 is what percent of 675?   **24.** What percent of 400 is 36?

  25. $39.20 is what percent of $560?   **26.** What percent of 12 is 11? (nearest tenth)

248 Find the percent of increase or decrease.

27. 16 to 20   **28.** 100 to 90   **29.** 75 to 150   **30.** 150 to 75

250 **31.** 30% of what is 27?   **32.** $34 is 17% of what amount?   **33.** 612 is 8.5% of what number?

---

257 **34.** To send a package air parcel post to Brazil costs $8.10 for the first 4 ounces and $1.60 for each additional 4 ounces up to 5 pounds. Over 5 pounds, each additional 8 ounce costs $3.00. How much would it cost to send a 6-pound package? (16 ounces = 1 pound)

35. Whole milk has 4% butterfat, while skim milk has 0.5% butterfat. How much more butterfat would be in a 250-milliliter glass of whole milk than in a 250-milliliter glass of skim milk?

36. A $500 TV is discounted 10%. If there is a 5% sales tax, how much will the TV cost?

37. Lisa, Eric, Karen, and Joyce had lunch together at a restaurant. The bill came to $18.60. Lisa paid 15% of the bill. How much did Eric, Karen and Joyce pay if each contributed equally?

224 **38.** Percent is a part of a whole expressed in ____?____.

Chapter 4 Test

Page
225 **1.** Express seven hundredths as a percent, a decimal, and a fraction.

2. Express nineteen out of one hundred as a percent, a decimal, and a fraction.

227 **3.** Express 4.5% as a decimal.

4. Express 125% as a decimal.

229 **5.** Write 0.27 as a percent.

6. Write 0.9 as a percent.

7. Write 0.048 as a percent.

231 **8.** Write 6% as a fraction in lowest terms.

9. Write 75% as a fraction in lowest terms.

10. Write 155% as a mixed number in lowest terms.

233 Write each of the following as a percent.

11. $\frac{3}{8}$ **12.** $1\frac{1}{5}$ **13.** $\frac{1}{25}$

237 **14.** Find 30% of 1765.   **15.** What is 94% of 76?

16. Find 15% of $40.27 to the nearest cent.

242 **17.** What percent of 16 is 4?

18. 70 is what percent of 80?

19. 13 is what percent of 9?

248 **20.** Before his diet, Max weighed 64.5 kg. Afterward, he weighed 62 kg. By what percent did his weight decrease?

250 **21.** 35% of what is 560?

22. 5% of what amount is 74?

257 **23.** A meal tax of 5% was added to the cost of two pizzas. One pizza cost $8, the other $6.75. What was the total bill?

24. A telephone call cost $.52 for the first three minutes and $.11 for each additional minute. How much will a 15-min. call cost?

242 **25.** Teenagers watch about 23.5 h of TV a week. If 10 h are after 8:00 P.M., what percent of time viewed is after 8:00 P.M.?

26. The usual tip is about 15% of the cost of the meal. If a meal for two people cost $28.30, what would be a suitable tip?

Cumulative Test

Solve each problem and select the letter corresponding to your answer.

Page 29
1. Add. 5,936 + 81,935 + 48,066 + 6,999

    a. 132,836   **b.** 142,936
    c. 142,846   **d.** 132,935

Page 37
2. Subtract. 5,770,004 − 651,920

    a. 5,118,084   **b.** 5,117,984
    c. 5,117,920   **d.** 5,119,920

49
3. Multiply. 587 × 703

    a. 413,551   **b.** 313,651
    c. 513,561   **d.** 412,661

61
4. Divide. $409\overline{)1,472,400}$

    a. 36   **b.** 306
    c. 3,060   **d.** 3,600

88
5. Find the sum. 0.7 + 7.77 + 0.777

    a. 2.254   **b.** 9.247
    c. 2.247   **d.** 9.777

95
6. Subtract. $94 − $7.71

    a. $87.71   **b.** $86.29
    c. $86.39   **d.** $87.39

102
7. Multiply. 0.37 × 0.006

    a. 0.222   **b.** 0.0222
    c. 0.00222   **d.** 0.22200

8. Divide. $\$.04\overline{)\$56}$

    a. 14   **b.** 1,400
    c. 104   **d.** 0.14

164
9. Add. $5\frac{5}{7} + 3\frac{13}{28} + 8\frac{9}{14}$

    a. $17\frac{1}{2}$ **b.** $17\frac{3}{4}$
    c. $19\frac{11}{14}$ **d.** $17\frac{23}{28}$

170
10. Subtract. $19 − 7\frac{7}{8}$

    a. $18\frac{1}{8}$ **b.** $12\frac{7}{8}$
    c. $11\frac{1}{8}$ **d.** $12\frac{1}{8}$

179
11. Multiply. $8 \times 2\frac{1}{4}$

    a. 18   **b.** $\frac{1}{18}$
    c. 36   **d.** $\frac{1}{36}$

188
12. Divide. $1\frac{1}{3} \div 7$

    a. $5\frac{1}{4}$ **b.** $\frac{4}{21}$
    c. $\frac{4}{10}$ **d.** $9\frac{1}{3}$

237
13. Find 5% of $335.

    a. $17   **b.** $.17
    c. $167.50   **d.** $16.75

242
14. What percent of 1,440 is 72?

    a. 5%   **b.** .05%
    c. 50%   **d.** 500%

250
15. 12% of what number is 84?

    a. 10.32   **b.** 0.7
    c. 700   **d.** 70

237
16. Roberta received 65% of the 1,200 votes cast for student council president. How many votes did she receive?

    a. 780   **b.** 680
    c. 78   **d.** Answer not given

COMPUTER ACTIVITY 4

A computer can perform repetitive processes with great speed. The program on page 222 uses an IF . . . THEN statement to create a loop. The program below is similar to that program, but it uses a FOR . . . NEXT loop.

Monthly sales figures could be called MONTH(1), MONTH(2), MONTH(3), . . . , MONTH(12). The general name for a month's data could be MONTH(N). N is called the subscript. N would take on the values 1, 2, 3, . . . , 12 depending on the month.

```
1Ø SUM = Ø
2Ø FOR N = 1 TO 12
3Ø READ MONTH(N)
4Ø SUM = SUM + MONTH(N)
5Ø NEXT(N)
6Ø AVG = SUM/12
7Ø PRINT "AVERAGE MONTHLY SALES = $", AVG
8Ø DATA 3ØØ, 467, 1,232, 1,345, 1,1Ø1, 1,ØØ9, 879, 925, 1,Ø32,
     1,243, 1,278, 1,45Ø
```

In this program, lines 20 to 50 form a FOR . . . NEXT loop. Line 20 tells you that the first value of N is 1 and the last value of N is 12. Line 50 increases the value of N by 1. The loop is repeated 12 times. When N equals 12, the computer does not execute line 50. Instead, it moves to line 60 and calculates the average. If there were more than 12 data items on line 80, the computer would read only the first 12 items.

Type this program into your computer. Give the computer the command RUN. What is the output?

You might want the computer to print the monthly sales figures. To do that, add a line to the program. Type:

45 PRINT "Total sales for month ", N, " are $", MONTH(N)

What do you expect to see? Run the program again.

If you insert this PRINT statement as line 55 instead of as line 45, would you get the same results?

Try it and run the program.

Achievement Test

7 **1.** What is the question?

Eleanor saved $2,386 one year. She spent $1,159 on a vacation. How much money did she have left?

19 **3.** Read or write in words: 85,267,428.

26 **5.** Round 581,560 to the nearest thousand.

37 **7.** Take 638 from 4,000.

61 **9.** Divide: $48\overline{)33,744}$

77 **11.** Read 0.0626 or write it in words.

83 **13.** Round to nearest hundredth: 26.8539

88 **15.** Add: $6.57 + $.36 + $12.40 + $.82

102 **17.** Multiply: 3.14 × 15

120 **19.** What decimal part of 20 is 13?

126 **21.** Multiply: 10 × 8.69

135 **23.** Write 14.6 million in standard form.

149 **25.** Express $\frac{28}{48}$ in lowest terms.

153 **27.** Express $\frac{9}{25}$ as a fraction with the denominator 100.

157 **29.** Which is less: $\frac{5}{8}$ or $\frac{5}{6}$?

162 **31.** Find the least common denominator of the fractions: $\frac{3}{0}$ and $\frac{11}{16}$

170 **33.** Subtract: $4\frac{1}{4} - 2\frac{1}{3}$

179 **35.** Multiply: $3\frac{7}{8} \times 2$

195 **37.** What part of 32 is 20?

200 **39.** Write $\frac{11}{16}$ as a decimal (2 places).

210 **41.** Find the ratio of 15 to 35.

225 **43.** Write sixty hundredths as: a percent, a decimal, and a fraction.

229 **45.** Express 0.49 as a percent.

233 **47.** Express $\frac{3}{25}$ as a percent.

242 **49.** 52 is what percent of 64?

2. Which fact is not needed?

Maria spent $16.06 on 22¢ stamps. She also mailed a package by express mail that cost $10.75. How many stamps did she buy?

23 **4.** Write in standard form: Ninety-two million four hundred six thousand fifty.

29 **6.** Add: 472 + 5,973 + 15,682 + 33 + 280

49 **8.** Multiply: 14 × 1,728

68 **10.** Name the greatest common factor of 42 and 70.

80 **12.** Write in standard form: Fifteen and seven tenths.

86 **14.** Which is greater: 0.46 or 0.462?

95 **16.** Subtract: 54.3 − 8.28

113 **18.** Divide: $.06\overline{)$7.44}$

123 **20.** 0.06 of what number is 18?

129 **22.** Divide: 1.95 ÷ 100

138 **24.** Write the short name for 63,700,000,000 in billions.

151 **26.** Rewrite $\frac{9}{2}$ as a mixed number.

155 **28.** Are $\frac{24}{36}$ and $\frac{18}{24}$ equivalent fractions?

159 **30.** Find the least common multiple of 8 and 18.

164 **32.** Add: $9\frac{1}{2} + 3\frac{2}{5}$

177 **34.** Write $6\frac{2}{3}$ as an improper fraction.

188 **36.** Divide: $\frac{3}{4} \div 6$

195 **38.** $\frac{9}{20}$ of what number is 36?

204 **40.** Write 0.65 as a fraction.

213 **42.** Solve and check: $\frac{n}{56} = \frac{9}{14}$

227 **44.** Express 7% as a decimal.

231 **46.** Express 20% as a fraction.

237 **48.** Find 35% of 560.

250 **50.** 75% of what number is 60?

Unit 2

Consumer Applications

PLE CIDER

Inventory Test

Page
272 **1.** How much does Ricard earn for a 42-hour week at $6.60 per hour and time and a half for overtime (over 40 hours)?

Page
278 **2.** Use the table of withholding tax (p. 279) to compute the amount of income tax to be withheld when a married person's weekly wages are $402.50 and 4 allowances are claimed.

281 **3.** Ann Williams earns $345 a week. How much social security tax does she pay if the social security tax rate is 7.15% of earnings?

282 **4.** If the tax rate is $2\frac{1}{2}$%, how much city tax is paid by a person earning $269 weekly?

283 **5.** Use the table of withholding tax (p. 279) and the social security tax rate of 7.15% to find the take-home pay when a married person's weekly wages are $375 and 2 allowances are claimed.

285 **6.** Use the Federal tax table (p. 288) to find how much Federal income tax you still owe if you are single, your taxable income is $14,225 and your withholding tax payments are $1,980.

290 **7.** Show two ways to solve the following problem. The local baseball team won 75% of the 48 games it played. How many games did it lose?

298 **8.** Find the unit price (cost per quart) of a pint of vinegar costing 49¢. What is the cost of 4 quarts of vinegar?

301 **9.** Find the sale price of a bicycle that regularly sells for $129.95 and is now reduced 20%.

305 **10.** The meal check for Tim's party at the restaurant amounted to $146.50. If a 6% sales tax is added on, how much is the total bill?

306 **11.** The electric meter reading at the end of the previous month was 3569 and this month 3724. At 10.1¢ per kilowatt hour, what is the cost of the electricity consumed?

309 **12.** Julio's father traveled 1,158 miles by automobile using 60 gallons of gasoline. How many miles per gallon did the automobile average?

310 **13.** Laura's mother could purchase a refrigerator for $499 cash or pay $49 down and 12 equal payments of $42.75. How much can she save by paying cash? What is the interest rate?

314 **14.** Fred's house was purchased for $73,500 and is assessed for $56,500. If the tax rate is $6.25 per $100, what are the property taxes?

324 **15.** Write a check payable to the Town Center Store in the amount of $418.45 for the purchase of a television set. Use today's date.

329 **16.** What is the amount due on $5,000 borrowed at 12% annual interest and repaid at the end of 3 years?

337 **17.** Excluding broker's fees, how much profit do you make if you buy 200 shares of stock at $13\frac{3}{4}$ and sell them at $16\frac{1}{2}$?

Income, Take-Home Pay, Income Tax

5-1 Computing Income

Most people seek employment in order to provide themselves with a source of income. Occupations vary, as do the methods of payment.

HOURLY WAGES
In some occupations salaries or wages are based on hourly rates of pay, which vary depending on the job.

Example

Find the weekly wages for a person who works $45\frac{1}{2}$ hours at $4.80 per hour.

To find the weekly wages when the hourly rate and the number of working hours are known, follow these steps.

a. Multiply the number of hours of work by the hourly rate. (If the number of hours is more than 40, multiply only the first 40 hours by this rate.)

$40 \times \$4.80 = \$192.00 \leftarrow$ first 40 h

b. Multiply the number of hours more than 40 by the overtime rate.
$5\frac{1}{2} \times \$7.20 = \$39.60 \leftarrow$ overtime
$(1\frac{1}{2} \times \$4.80)$

c. Add steps (a) and (b) above.

$\$192.00 \leftarrow$ wage/40 h
$+\quad 39.60 \leftarrow$ overtime wage
$\$231.60 \leftarrow$ total wage

You might use a calculator.
Key: 40 $\boxed{\times}$ 4.80 $\boxed{\text{M+}}$ 4.80 $\boxed{\times}$ 1.5 $\boxed{\times}$ 5.5 $\boxed{\text{M+}}$ $\boxed{\text{M}^\text{R}_\text{C}}$ 231.60

ANSWER: $231.60 weekly wage

Think About It

1. Define income.

2. Name some occupations that have hourly wages.

3. In what ways other than hourly are people paid?

4. Given an annual salary, explain how to use a calculator to find the weekly, semimonthly, monthly, and semiannual earnings.

Practice Exercises

1. What are the weekly wages of a person who works 40 hours per week at each of the following rates?
   a. $4.10   **b.** $7.95   **c.** $12.80   **d.** $5.65

Find the total earnings for each of the following.

2. a. 28 hours at $6.90 per hour   **b.** $37\frac{1}{2}$ hours at $10.00 per hour
   c. 31 hours at $11.65 per hour   **d.** $18\frac{1}{2}$ hours at $8.25 per hour

3. Find the weekly wages for each.

		Hours Worked					Total Hours	Hourly Rate	Wages
		M	T	W	T	F			
a.	A	8	7	8	8	8		$6.35	
b.	B	7	7	8	7	7		$8.15	
c.	C	8	$6\frac{1}{2}$	8	4	7		$9.10	
d.	D	5	5	7	$6\frac{1}{2}$	8		$5.75	

4. For each timecard below, compute the number of hours worked each day and the total hours for the week. Then find the amount of wages due each employee.

a.

			Timecard		
No. _25_		NAME: ___LISA ROSSI___			
WEEK ENDING _____					
DAY	IN	OUT	IN	OUT	HOURS
M.	8:00	12:00	1:00	5:00	
T.	8:30	12:00	1:00	5:00	
W.	8:00	11:00	12:30	5:00	
T.	9:00	12:30	1:00	5:30	
F.	8:00	1:00	2:00	5:00	
S.					
RATE $8.35			TOTAL HOURS __		
WAGES _____					

b.

			Timecard		
No. _57_		NAME: ___PETER CHANG___			
WEEK ENDING _____					
DAY	IN	OUT	IN	OUT	HOURS
M.	9:00	12:00	12:30	5:00	
T.	8:30	11:30	12:00	4:00	
W.	8:30	12:00	1:30	6:00	
T.	8:00	12:30	1:00	4:00	
F.	8:30	12:00	1:00	5:30	
S.					
RATE $5.40			TOTAL HOURS __		
WAGES _____					

WEEKLY SALARIES, MONTHLY SALARIES, ANNUAL SALARIES

In other occupations the employed persons have fixed salaries on a weekly, semimonthly, monthly, or annual basis.

5. Find the hourly rate of pay when a person's weekly earnings are:
   a. $190 for 40 hours of work   **b.** $234 for 40 hours of work
   c. $150.10 for 38 hours of work   **d.** $193.45 for $36\frac{1}{2}$ hours of work

6. Find the annual salary of a person for the given weekly salary.
   a. $85   **b.** $425   **c.** $104.50   **d.** $257.50

7. Find the annual salary of a person for the given monthly salary.
   a. $700   **b.** $535   **c.** $1,000   **d.** $1,915

8. Find the annual salary for each.
   a. $172.50 per week   **b.** $965 per month
   c. $592.75 semimonthly   **d.** $5.60 per hour, 40-hour week

9. Find the weekly salary for the given annual salary.

   a. $9,360   **b.** $18,200   **c.** $14,560   **d.** $26,942

10. Find the monthly salary for the given annual salary.

   a. $8,100   **b.** $23,180   **c.** $13,320   **d.** $30,500

11. Find the weekly salary for the given monthly salary ($4\frac{1}{3}$ weeks per month).

   a. $936   **b.** $1,383   **c.** $1,750   **d.** $1,100

12. Which is a better salary?

   a. $180 per week or $750 per month   **b.** $15,400 per year or $1,250 per month
   c. $7,700 per year or $145 per week   **d.** $510 semimonthly or $240 per week

PIECE WORK

Some people work where their earnings are based on the number of pieces
or units of work that a person completes at a given pay rate per piece or unit.

13. A farm worker picked 49 bushels of apples per day at $.68 a bushel. How much did he earn for the day?

14. A typist charges $.99 per page and types an average of 40 pages per day. How much does the typist earn during a week of 5 workdays?

15. Find each person's earnings for the week.

	Number of Units					Total Units	Unit Rate	Total Wages
	M	T	W	T	F			
Charlotte	28	25	33	29	36		$.70	
Marilyn	32	41	37	45	39		$.65	
Leon	147	159	162	154	167		$.30	
Ronald	119	107	99	98	125		$.38	

FEES, TIPS, PENSIONS, INTEREST, DIVIDENDS

Other ways people earn income are by charging a fee for their work,
by receiving tips, by collecting pensions from work and social security,
and by receiving interest and dividends from their investments.

16. A TV repair person charges a $20 service fee to come to a home to repair a television. How much money in service fees were made on a day when she had 7 service calls?

17. Becky's mother has $25,000 invested at the rate of $10\frac{3}{4}\%$ per year. What is her annual income from this investment?

18. If a doctor sees 4 patients an hour and charges $25 an office visit, how much does she earn during a day when her office hours are from 2 P.M. to 5 P.M. and 7 P.M. to 9 P.M. and her appointment schedule is full?

19. A retired worker receives a social security pension of $250 per month and a pension of $310 per month.

   a. How much is her total monthly pension?
   b. How much is her total annual pension?
   c. How much is her total weekly pension?

5-2 Commission

Many people who buy or sell goods for someone else are paid partly or completely by **commission,** which is a percent of the sale or purchase. This means of payment is usually considered to be an incentive for the salesperson or buyer.

Examples

1. Find the commission and net proceeds when the sales are $364.89 and the rate of commission is 4%.

To find the commission and the net proceeds when the amount of sales and rate of commission are given, follow these steps.

a. Multiply the sales by the decimal equivalent of the rate of commission to find the commission.

$$
\begin{array}{r}
\$364.89 \leftarrow \text{sales} \\
\times \quad 0.04 \leftarrow \text{rate of commission} \\
\hline
\$14.5956 \text{ or } \$14.60 \leftarrow \text{commission}
\end{array}
$$

b. Subtract the commission from the amount of sales to find the net proceeds.

$$
\begin{array}{r}
\$364.89 \leftarrow \text{sales} \\
- \quad \$14.60 \leftarrow \text{commission} \\
\hline
\$350.29 \leftarrow \text{net proceeds}
\end{array}
$$

You might use a calculator.

Key: 364.89 [M+] [×] 4 [%] 14.5956

[M−] [M R/C] 350.2944

ANSWER: $14.60 commission; $350.29 net proceeds

2. Find the rate if the commission on a sale of $230 is $18.40.

To find the rate of commission, divide the amount of the commission by the sales (or the cost of the goods purchased) and change the decimal to a percent.

$$
\begin{array}{l}
\text{commission} \rightarrow \dfrac{\$18.40}{\$230} \quad \underset{\$230\overline{)\$18.40}}{\overset{0.08}{}} = 8\% \leftarrow \text{rate of commission} \\
\text{sales} \longrightarrow \qquad\qquad \underline{18\ 40}
\end{array}
$$

You might use a calculator.

Key: 18.40 [÷] 230 [%] 8

ANSWER: 8% commission

3. Find the total sales if the commission is $320 and the rate of commission is 8%.

To find sales when the commission and the rate of commission are given, divide the commission by the rate of commission in decimal form.

commission ———→ $320
rate of commission → 8%

$$\overset{\$4{,}000 \;\leftarrow\; \text{total sales}}{0.08\overline{)\$320.00}}$$

You might use a calculator.

Key: 320 ÷ 8 % 4000

ANSWER: $4,000 total sales

Think About It

1. List two reasons why a commission might be considered more of an incentive than a salary.

2. In what situation would an employer prefer to pay an employee a commission rather than a salary?

3. Use a calculator to find the net proceeds a grower receives if a merchant, charging $7\frac{1}{2}\%$ commission, sells 290 boxes of melons at $5.65 a box and 475 boxes of plums at $6.40 a box for the grower.

Practice Exercises

1. A real estate agent sold a house for $83,500 and was paid a 7% commission. How much was the agent's commission? What were the net proceeds the owner received?

2. Find the commission and net proceeds of each of the following.

	a.	b.	c.	d.
Sales	$426	$93.40	$6,570	$39.15
Rate of commission	14%	5%	6%	20%

3. Find the rate of commission if the commission is $78 on sales amounting to $650.

4. Mike Marshall buys 1,000 discontinued sweaters from a manufacturer for $4,732. He then sells them at cost to a discount store and receives a commission of $189.28. What is the rate of his commission?

5. What is the rate of commission when sales are $6,300 and net proceeds are $6,111?

6. What must the sales be for a 6% rate of commission to bring a commission of $150?

7. Find the amount of sales of each of the following.

	a.	**b.**	**c.**	**d.**
Commission	$253	$65	$6.83	$54.60
Rate of commission	11%	$12\frac{1}{2}$%	5%	14%

8. Find the commission and the total cost of goods purchased.

	a.	**b.**	**c.**	**d.**
Cost of goods	$628	$879	$956.40	$1,297.50
Rate of commission	9%	$5\frac{1}{2}$%	$8\frac{1}{4}$%	10%

9. Alice works in a clothing store. She receives $105 per week and 1% commission on sales. How much did she earn when she sold $12,028 worth of merchandise?

10. A salesperson sold 9 window fans at $77.99 each and 7 air conditioners at $224.95 each. At 4% commission, how much did he earn?

11. A stockbroker purchases shares with a total value of $2,450. She receives a 1.5% commission on this purchase. Find her commission.

12. A lawyer collected a debt of $824 for a client, charging $123.60 for her services. What rate of commission did she charge?

13. Abe's brother receives $90 per week as salary and an additional 2% commission on the amount of his sales. If he earned $261.34 as his total income in a week, find the amount of his sales.

Withholding Tax

Income taxes are deducted from each paycheck before the employee receives it. This money is forwarded to the office of the Collector of Internal Revenue, where it is entered and recorded in each employee's account.

There are two ways to compute the amount of tax withheld, the **percentage method** and the income tax **withholding table method.**

PERCENTAGE METHOD

Example

Compute the withholding tax when a married person's weekly wages are $345 and 4 allowances are claimed.

To compute the amount of weekly withholding tax for a married person with 4 allowances, follow these steps.

a. List the total wage payment.

Total wage payment $345.00

b. Subtract the withholding allowance $146.16 ← **($36.54 × 4)**

                                     $198.84

c. Use the table to calculate the tax and add the two amounts.

If the amount of wages is:		The amount of income tax to be withheld shall be:		
Not over $36 0				
Over—	But not over—			of excess over—
$36	—$93 . . . . 11%			—$36
$93	—$574 . . . . $6.27 plus 15%			—$93
$574	—$901 . . . . $78.42 plus 28%			—$574
$901	—$1,767 . . . . . . $169.98 plus 35%			—$901
$1,767	. . . . $473.08 plus 38.5%			—$1,767

tax on first $93. . . . . $ 6.27
tax on remainder $105.84 at 15%. . + $15.88
                                $22.15
                ↑
        $198.84 − $93

ANSWER: $22.15 tax withheld.

You might use a calculator.

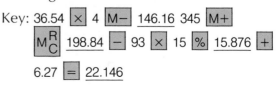

Key: 36.54 × 4 M− 146.16 345 M+
        M$_C^R$ 198.84 − 93 × 15 % 15.876 +
      6.27 = 22.146

Vocabulary

Withholding tax is the amount of income tax that employers are required by the federal government to withhold from their employees' earnings.

WITHHOLDING TABLE METHOD

Example

Compute the withholding tax when a married person's weekly wages are $278 and 2 allowances are claimed.

To compute the amount of weekly withholding tax for a married person with 2 allowances, follow these steps.

a. Locate the row indicating the total wage payment. Row reads: "At least 270 but less than 280."

b. Select amount of tax from column corresponding to number of allowances. Read down column headed "2" until you get to $23.00.

MARRIED Persons–WEEKLY Payroll Period

And the wages are–		And the number of withholding allowances claimed is–										
At least	But less than	0	1	2	3	4	5	6	7	8	9	10
		The amount of income tax to be withheld shall be–										
$0	$40	$0	$0	$0	$0	$0	$0	$0	$0	$0	$0	$0
40	42	1	0	0	0	0	0	0	0	0	0	0
42	44	1	0	0	0	0	0	0	0	0	0	0
44	46	1	0	0	0	0	0	0	0	0	0	0
46	48	1	0	0	0	0	0	0	0	0	0	0
48	50	1	0	0	0	0	0	0	0	0	0	0
50	52	2	0	0	0	0	0	0	0	0	0	0
52	54	2	0	0	0	0	0	0	0	0	0	0
54	56	2	0	0	0	0	0	0	0	0	0	0
56	58	2	0	0	0	0	0	0	0	0	0	0
58	60	3	0	0	0	0	0	0	0	0	0	0
60	62	3	0	0	0	0	0	0	0	0	0	0
62	64	3	0	0	0	0	0	0	0	0	0	0
64	66	3	0	0	0	0	0	0	0	0	0	0
66	68	3	0	0	0	0	0	0	0	0	0	0
68	70	4	0	0	0	0	0	0	0	0	0	0
70	72	4	0	0	0	0	0	0	0	0	0	0
72	74	4	0	0	0	0	0	0	0	0	0	0
74	76	4	0	0	0	0	0	0	0	0	0	0
76	78	5	1	0	0	0	0	0	0	0	0	0
78	80	5	1	0	0	0	0	0	0	0	0	0
80	82	5	1	0	0	0	0	0	0	0	0	0
82	84	5	1	0	0	0	0	0	0	0	0	0
84	86	5	1	0	0	0	0	0	0	0	0	0
86	88	6	2	0	0	0	0	0	0	0	0	0
88	90	6	2	0	0	0	0	0	0	0	0	0
90	92	6	2	0	0	0	0	0	0	0	0	0
92	94	6	2	0	0	0	0	0	0	0	0	0
94	96	7	2	0	0	0	0	0	0	0	0	0
96	98	7	3	0	0	0	0	0	0	0	0	0
98	100	7	3	0	0	0	0	0	0	0	0	0
100	105	8	3	0	0	0	0	0	0	0	0	0
105	110	8	4	0	0	0	0	0	0	0	0	0
110	115	9	4	0	0	0	0	0	0	0	0	0
115	120	10	5	1	0	0	0	0	0	0	0	0
120	125	11	6	2	0	0	0	0	0	0	0	0
125	130	11	6	2	0	0	0	0	0	0	0	0
130	135	12	7	3	0	0	0	0	0	0	0	0
135	140	13	7	3	0	0	0	0	0	0	0	0
140	145	14	8	4	0	0	0	0	0	0	0	0
145	150	14	9	4	0	0	0	0	0	0	0	0
150	160	16	10	5	1	0	0	0	0	0	0	0
160	170	17	12	6	2	0	0	0	0	0	0	0
170	180	19	13	8	3	0	0	0	0	0	0	0
180	190	20	15	9	4	0	0	0	0	0	0	0
190	200	22	16	11	5	1	0	0	0	0	0	0
200	210	23	18	12	7	3	0	0	0	0	0	0
210	220	25	19	14	8	4	0	0	0	0	0	0
220	230	26	21	15	10	5	1	0	0	0	0	0
230	240	28	22	17	11	6	2	0	0	0	0	0
240	250	29	24	18	13	7	3	0	0	0	0	0
250	260	31	25	20	14	9	4	0	0	0	0	0
260	270	32	27	21	16	10	5	1	0	0	0	0
270	280	34	28	23	17	12	6	2	0	0	0	0
280	290	35	30	24	19	13	8	3	0	0	0	0
290	300	37	31	26	20	15	9	4	0	0	0	0
300	310	38	33	27	22	16	11	6	1	0	0	0
310	320	40	34	29	23	18	12	7	3	0	0	0
320	330	41	36	30	25	19	14	8	4	0	0	0
330	340	43	37	32	26	21	15	10	5	1	0	0
340	350	44	39	33	28	22	17	11	6	2	0	0
350	360	46	40	35	29	24	18	13	7	3	0	0
360	370	47	42	36	31	25	20	14	9	4	0	0
370	380	49	43	38	32	27	21	16	10	5	1	0
380	390	50	45	39	34	28	23	17	12	6	2	0
390	400	52	46	41	35	30	24	19	13	8	3	0
400	410	53	48	42	37	31	26	20	15	9	4	0
410	420	55	49	44	38	33	27	22	16	11	6	2
420	430	56	51	45	40	34	29	23	18	12	7	3
430	440	58	52	47	41	36	30	25	19	14	8	4

ANSWER: $23.00

1. Name some ways in which the federal government spends the money collected through taxes.

2. If you had a choice, would you prefer to have your taxes withheld from every paycheck or to pay your taxes in one large sum once a year? Explain.

3. If two people earn the same amount, but one person is single and the other is married with one child, who would have the larger paycheck and why?

Practice Exercises

1. Use the percentage method to compute the amount of income tax to be withheld each week from the following married persons' weekly wages.
   a. $119 and 2 withholding allowances are claimed
   b. $97 and 1 withholding allowance is claimed
   c. $439.50 and 5 withholding allowances are claimed
   d. $247.50 and 3 withholding allowances are claimed

2. Use the table of withholding tax to compute the amount of income tax to be withheld each week from the following married persons' weekly wages.
   a. $93 and 1 withholding allowance is claimed
   b. $410 and 3 withholding allowances are claimed
   c. $242.50 and 5 withholding allowances are claimed
   d. $162.25 and 2 withholding allowances are claimed

3. Use the table of withholding tax to compute the amount of income tax withheld for the year from the following married persons' weekly wages.
   a. $98 and 1 withholding allowance is claimed
   b. $175 and 2 withholding allowances are claimed
   c. $312.50 and 5 withholding allowances are claimed
   d. $274.25 and 3 withholding allowances are claimed

4. Use the percentage method to compute the amount of income tax withheld for the year from the following married persons' weekly wages. (Round weekly withholding before calculating for a year).
   a. $165 and 2 withholding allowances are claimed
   b. $220 and 3 withholding allowances are claimed
   c. $445 and 1 withholding allowance is claimed
   d. $287.75 and 4 withholding allowances are claimed

Social Security Tax

Payments to retired, disabled, dependent, or unemployed persons are financed by a tax on a percentage of wages that is also matched by employer contributions. In a recent year, the amount withheld was 7.15% of the first $43,800 of a worker's annual earnings.

Vocabulary

Social security tax is the money deducted from employees' earnings to help pay old-age pensions, disability, etc.

Example

Find the amount deducted from an employee's weekly wage of $330 for social security tax.

To find the social security tax deducted from an employee's wages, follow these steps. Multiply the employee's wage by 7.15% to find the amount deducted from the employee's earnings.

$330 ← weekly wage
× 0.0715 ← tax rate
$23.5950 or $23.60 ← social security tax

ANSWER: $23.60 social security tax

You might use a calculator.
Key: 330 ⊠ 7.15 % 23.595

Think About It

1. If $100 is deducted from an employee's semimonthly paycheck for social security tax, what is the total contribution to the fund?

2. Would you prefer to have social security tax deducted only once a year instead of every paycheck? Why or why not?

Practice Exercises

1. Find the amount deducted from the given weekly wages for social security tax.
   a. $60   **b.** $725   **c.** $218.40   **d.** $608.25

2. Find the amount deducted for social security tax for the given work period.
   a. $395 semimonthly   **b.** $2,000 monthly   **c.** $32,400 annually   **d.** $687.50 monthly

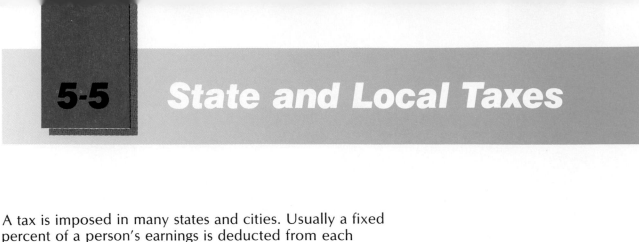

State and Local Taxes

A tax is imposed in many states and cities. Usually a fixed percent of a person's earnings is deducted from each paycheck for state and city programs.

Example

If the tax rate is 3% of the wages, find the tax owed for the week on weekly earnings of $429.

To find the state tax: Multiply the wages for the given pay period by the tax rate.

$429 ← wages
× 0.03 ← tax rate
$12.87 ← tax

You might use a calculator.
Key: 429 ⊠ 3 % 12.87

ANSWER: $12.87 state tax

Think About It

1. Find out whether your state and city have income taxes. If they do, find out the rate.

2. A person earns $400 a week. If the tax is $12 and the social security tax is $28.60, how much is left after taxes?

Practice Exercises

1. If the tax rate is 2% of the wages, find the weekly tax for the given week's earnings.
  a. $85   **b.** $116   **c.** $220   **d.** $174.25

2. If the tax rate is 3% of the wages, find the weekly tax for the given week's earnings.
  a. $68   **b.** $245   **c.** $407   **d.** $191.75

3. If the tax rate is $2\frac{1}{2}$% of the wages, find the semimonthly tax a person owes for the given semimonthly earnings.
  a. $182   **b.** $634   **c.** $570   **d.** $245.50

4. If the tax rate is $1\frac{3}{4}$% of the wages, find the monthly tax a person owes for the given monthly earnings.
  a. $900   **b.** $1,385   **c.** $1,025   **d.** $652.75

Take-Home Pay

In addition to the various taxes, many deductions such as union dues and health insurance costs are deducted from a paycheck before a person receives it. The amount a person takes home is sometimes called **net pay,** or **net wage.**

Example

Determine the take-home pay when a married person's weekly wages are $215, there is a state tax rate of 2%, and 2 allowances are claimed.

To determine the take-home pay, follow these steps.

a. Find the federal withholding tax.

According to the table on page 279, the tax is $14.00.

b. Find the social security tax.

$215 × 0.0715 = $15.37

c. Find the state tax, if any.

$215 × 0.02 = $4.30

d. Subtract the sum of the taxes from the amount of the person's wage.

(1) Federal withholding tax (by table) $14.00
(2) Social security tax ($215 × 7.15%). $15.37
(3) State tax ($215 × 2%) . . . . + $4.30
                                         sum of taxes $33.67
(4) Take-home pay (earnings − sum of taxes). $181.33

$$\uparrow$$
$215 − $33.67

You might use a calculator.

Key: 215 $\boxed{\times}$ 7.15 $\boxed{\%}$ 15.3725 $\boxed{M+}$ 215 $\boxed{\times}$ 2 $\boxed{\%}$

 4.3 $\boxed{M+}$ 14 $\boxed{M+}$ 14 215 $\boxed{-}$ $\boxed{\substack{M R \\ C}}$ $\boxed{=}$ 181.3275

ANSWER: $181.33 take-home pay

1. If you know the take-home pay and the total deductions, what else do you know and how do you find it?

2. How might a person's take-home pay change when he gets married? Why?

3. Name some reasons why money is withheld from a paycheck and what the money might be used for.

Practice Exercises

In the following problems, use the table of withholding tax (p. 279). The social security tax rate is 7.15%.

1. Mrs. Brown earns $182 each week and claims 3 withholding allowances. How much federal income tax is withheld each week? How much is deducted for social security? What is her take-home pay?

2. Find the take-home pay for each of the following weekly wages.

	Wage	Withholding Allowances		Wage	Withholding Allowances
a.	$94	4	c.	$191.60	3
b.	$105	2	d.	$420.25	1

3. Find the take-home pay for each of the following weekly wages when a local tax is also levied.

	Wage	Withholding Allowances	Local Tax		Wage	Withholding Allowances	Local Tax
a.	$96	1	2%	c.	$279.00	2	2%
b.	$193	4	$1\frac{1}{2}\%$	d.	$327.50	3	$2\frac{1}{2}\%$

4. Which take-home pay is greater, a weekly wage of $140.25 with 1 withholding allowance claimed or $129.75 with 5 withholding allowances claimed?

5. Mr. Wilson earns $217.50 each week and claims 2 withholding allowances. He works in a city where a tax of $1\frac{1}{2}\%$ is levied on wages. How much federal income tax is withheld each week? How much is deducted for social security? What is his weekly city tax? What is his take-home pay?

Federal Income Tax

The federal income tax forms describe different types of income and information to help you decide which tax form to use. Use Form 1040, 1040A, or 1040EZ to determine the amount of taxable income. Turn to the federal tax table on page 288 to determine the amount of tax based on taxable income. For a complete table, use the Internal Revenue Service publications. Income tax rates and the deductions allowed are subject to change.

Vocabulary

The **total income** includes wages, salaries, tips, interest income, business income, pensions, etc.

The **adjusted gross income** is the total income less adjustments, such as IRA payments, moving expenses, etc. It may be the same as the total income if these deductions are not allowed.

The **taxable income** is the adjusted gross income decreased by the sum of allowable exemptions, claimed credit, and itemized deductions.

At the end of each year, most people who have received income must file a **tax return.**

Employers must provide the employee with a **Form W-2,** a statement showing the income earned and the amount of tax withheld during the year.

Examples

1. Hilda is the head of a household whose taxable income is $10,382. Determine the amount of her federal income tax.

Read down the income column of the table on page 288 until you reach:

At least	But less than
10,350	10,400

Read across to the column headed "Head of a Household."

ANSWER: $1,456, amount of tax

2. If Hilda's withholding tax payments for the year amounted to $1,389.30, how much income tax does she still owe?

$$\begin{array}{r} \$1,456.00 \leftarrow \text{amount of tax} \\ -\ 1,389.30 \leftarrow \text{withholding tax payments} \\ \hline \$66.70 \leftarrow \text{owed} \end{array}$$

ANSWER: $66.70 owed

Department of the Treasury - Internal Revenue Service

Income Tax Return for
Single filers with no dependents (O)

OMB No. 1545-0675

Name & address

Use the IRS mailing label. If you don't have one, please print.

▶ Mary E. Smith
Print your name above (first, initial, last)

18 Elliot Street
Present home address (number and street). (If you have a P.O. box, see instructions.)

NC
City, town, or post office, state, and ZIP code

Please print your numbers like this:

| 0 | 1 | 2 | 3 | 4 | 5 | 6 | 7 | 8 | 9 |

Your social security number

| 0 | 7 | 4 | 3 | 6 | 5 | 3 | 7 | 6 |

Please read the instructions for this form on the reverse side.

Presidential Election Campaign Fund
Do you want $1 to go to this fund? ▶

Note: *Checking "Yes" will not change your tax or reduce your refund.*

Yes ✓ No

Report your income

			Dollars	Cents
1	Total wages, salaries, and tips. This should be shown in Box 10 of your W-2 form(s). (Attach your W-2 form(s).)	**1**	17,639	00
2	Taxable interest income of $400 or less. If the total is more than $400, you cannot use Form 1040EZ.	**2**	376	32
3	Add line 1 and line 2. This is your **adjusted gross income.**	**3**	18,015	32
4	Can you be claimed as a dependent on another person's return? ☐ Yes. Do worksheet on back; enter amount from line E here. ☒ No. Enter 2,540 as your standard deduction.	**4**	2,540	00
5	Subtract line 4 from line 3.	**5**	15,475	32
6	If you checked the "Yes" box on line 4, enter 0. If you checked the "No" box on line 4, enter 1,900. This is your **personal exemption.**	**6**	1,900	00
7	Subtract line 6 from line 5. If line 6 is larger than line 5, enter 0 on line 7. This is your **taxable income.**	**7**	13,575	32

Attach Copy B of Form(s) W-2 here

Figure your tax

8	Enter your Federal income tax withheld. This should be shown in Box 9 of your W-2 form(s).	**8**	2,158	00
9	Use the **single** column in the tax table on pages 32–37 of the Form 1040A instruction booklet to find the **tax** on the amount shown on **line 7** above. Enter the amount of tax.	**9**	1,964	00

Refund or amount you owe

Attach tax payment here

10	If line 8 is larger than line 9, subtract line 9 from line 8. Enter the **amount of your refund.**	**10**	194	00
11	If line 9 is larger than line 8, subtract line 8 from line 9. Enter the **amount you owe.** Attach check or money order for the full amount, payable to "Internal Revenue Service."	**11**		

Sign your return

I have read this return. Under penalties of perjury, I declare that to the best of my knowledge and belief, the return is true, correct, and complete.

Your signature   Date

For IRS Use Only—Please do not write in boxes below.

For Privacy Act and Paperwork Reduction Act Notice, see page 31.

Form **1040EZ**

Instructions for Form 1040EZ

Use this form if:	• Your filing status is single. • You are not 65 or over, OR blind. • You do not claim any dependents. • Your taxable income is less than $50,000. • You had **only** wages, salaries, and tips, and your taxable interest income was $400 or less. **Caution:** If you received tips (including allocated tips) that are not included in Box 14 of your W-2 form, you may not be able to use Form 1040EZ. See page 17 in the **Instructions for preparing 1040EZ and 1040A.** If you can't use this form, you must use Form 1040A or Form 1040. See pages 6 through 8 in the instruction booklet. If you are uncertain about your filing status, see page 9 of the booklet.
Completing your return	It will make it easier for us to process your return if you print your numbers (do not type) and keep them inside the boxes. Do not use dollar signs. You may find calculations easier if you round off cents to whole dollars. See page 15 of the instruction booklet for details.
Name & address	Use the mailing label we sent you. After you complete your return, carefully place the label in the name and address area. Mark through any errors on the label and print the correct information on the label. Use of the label saves processing time. If you don't have a label, print the information on the name and address lines. If your post office does not deliver mail to your street address and you have a P.O. box, enter your P.O. box number on the line for your present home address instead of your street address.
Presidential campaign fund	Congress set up this fund to help pay for Presidential election campaigns. You may have one of your tax dollars go to this fund by checking the "Yes" box. Checking the "Yes" box does not change the tax or refund shown on your return.
Report your income	**Line 1.** Enter on line 1 the total amount you received in wages, salaries, and tips. This should be shown in Box 10 of your 1987 wage statement(s), **Form W-2.** If you don't receive your W-2 form by February 15, contact your local IRS office. You must still report your earnings even if you don't get a Form W-2 from your employer. Attach the first copy or Copy B of your W-2 form(s) to your return. **Line 2.** Enter on line 2 the total taxable interest income you received from all sources, such as banks, savings and loans, and credit unions. You should receive a **Form 1099-INT** from each institution that paid you interest. You cannot use Form 1040EZ if your total taxable interest income is over $400. If you received tax-exempt interest, such as interest on municipal bonds, in the space to the left of line 2, write "TEI" and show the amount of your tax-exempt interest. DO NOT include tax-exempt interest in the total entered in the boxes on line 2. **Line 4.** If you checked the "Yes" box because you can be claimed as a dependent on another person's return (such as your parents'), complete the following worksheet to figure the amount to enter on line 4. For information on dependents, see page 12 of the instruction booklet.

Standard deduction worksheet for dependents	A. Enter the amount from line 1 on front.	A. _____
	B. Minimum amount.	B. ___500.00
	C. **Compare** the amounts on lines A and B above. Enter the LARGER of the two amounts here.	C. _____
	D. Maximum amount.	D. ___2,540.00
	E. **Compare** the amounts on lines C and D above. Enter the SMALLER of the two amounts here and on line 4 on front.	E. _____

	Line 6. Generally, you should enter 1,900 on line 6 as your personal exemption. However, if you can be claimed as a dependent on another person's return (such as your parents'), you cannot claim a personal exemption for yourself; enter 0 on line 6. If you are entitled to additional exemptions for your spouse, for your dependent children, or for other dependents, you cannot use Form 1040EZ.
Figure your tax	**Line 8.** Enter the amount of Federal income tax withheld. This should be shown in Box 9 of your W-2 form(s). If you had two or more employers and had total wages of over $43,800, see page 26 of the instruction booklet. If you cannot be claimed as a dependent and you want IRS to figure your tax for you, complete lines 1 through 8, sign and date your return. If you want to figure your own tax, continue with these instructions. **Line 9.** Use the amount on line 7 to find your tax in the tax table on pages 32–37 of the instruction booklet. Be sure to use the column in the tax table for **single** taxpayers. Enter the amount of tax on line 9. If your tax from the tax table is zero, enter 0.
Refund or amount you owe	**Line 10.** If line 8 is larger than line 9, you are entitled to a refund. Subtract line 9 from line 8, and enter the result on line 10. **Line 11.** If line 9 is larger than line 8, you owe more tax. Subtract line 8 from line 9, and enter the result on line 11. Attach your check or money order for the full amount. Write your social security number, daytime phone number, and "Form 1040EZ" on your payment.
Sign your return	You must sign and date your return. If you pay someone to prepare your return, that person must also sign it below the space for your signature and supply the other information required by IRS. See page 29.
Mailing your return	File your return by **April 15.** Mail it to us in the addressed envelope that came with the instruction booklet. If you don't have an addressed envelope, see page 3 for the address.

Left Panel

At least	But less than	Single (and 1040EZ filers)	Married filing jointly	Married filing separately	Head of a household
8,000					
8,000	8,050	1,132	1,084	1,144	1,104
8,050	8,100	1,139	1,091	1,151	1,111
8,100	8,150	1,147	1,099	1,159	1,119
8,150	8,200	1,154	1,106	1,166	1,126
8,200	8,250	1,162	1,114	1,174	1,134
8,250	8,300	1,169	1,121	1,181	1,141
8,300	8,350	1,177	1,129	1,189	1,149
8,350	8,400	1,184	1,136	1,196	1,156
8,400	8,450	1,192	1,144	1,204	1,164
8,450	8,500	1,199	1,151	1,211	1,171
8,500	8,550	1,207	1,159	1,219	1,179
8,550	8,600	1,214	1,166	1,226	1,186
8,600	8,650	1,222	1,174	1,234	1,194
8,650	8,700	1,229	1,181	1,241	1,201
8,700	8,750	1,237	1,189	1,249	1,209
8,750	8,800	1,244	1,196	1,256	1,216
8,800	8,850	1,252	1,204	1,264	1,224
8,850	8,900	1,259	1,211	1,271	1,231
8,900	8,950	1,267	1,219	1,279	1,239
8,950	9,000	1,274	1,226	1,286	1,246
9,000					
9,000	9,050	1,282	1,234	1,294	1,254
9,050	9,100	1,289	1,241	1,301	1,261
9,100	9,150	1,297	1,249	1,309	1,269
9,150	9,200	1,304	1,256	1,316	1,276
9,200	9,250	1,312	1,264	1,324	1,284
9,250	9,300	1,319	1,271	1,331	1,291
9,300	9,350	1,327	1,279	1,339	1,299
9,350	9,400	1,334	1,286	1,346	1,306
9,400	9,450	1,342	1,294	1,354	1,314
9,450	9,500	1,349	1,301	1,361	1,321
9,500	9,550	1,357	1,309	1,369	1,329
9,550	9,600	1,364	1,316	1,376	1,336
9,600	9,650	1,372	1,324	1,384	1,344
9,650	9,700	1,379	1,331	1,391	1,351
9,700	9,750	1,387	1,339	1,399	1,359
9,750	9,800	1,394	1,346	1,406	1,366
9,800	9,850	1,402	1,354	1,414	1,374
9,850	9,900	1,409	1,361	1,421	1,381
9,900	9,950	1,417	1,369	1,429	1,389
9,950	10,000	1,424	1,376	1,436	1,396
10,000					
10,000	10,050	1,432	1,384	1,444	1,404
10,050	10,100	1,439	1,391	1,451	1,411
10,100	10,150	1,447	1,399	1,459	1,419
10,150	10,200	1,454	1,406	1,466	1,426
10,200	10,250	1,462	1,414	1,474	1,434
10,250	10,300	1,469	1,421	1,481	1,441
10,300	10,350	1,477	1,429	1,489	1,449
10,350	10,400	1,484	1,436	1,496	1,456
10,400	10,450	1,492	1,444	1,504	1,464
10,450	10,500	1,499	1,451	1,511	1,471
10,500	10,550	1,507	1,459	1,519	1,479
10,550	10,600	1,514	1,466	1,526	1,486
10,600	10,650	1,522	1,474	1,534	1,494
10,650	10,700	1,529	1,481	1,541	1,501
10,700	10,750	1,537	1,489	1,549	1,509
10,750	10,800	1,544	1,496	1,556	1,516
10,800	10,850	1,552	1,504	1,564	1,524
10,850	10,900	1,559	1,511	1,571	1,531
10,900	10,950	1,567	1,519	1,579	1,539
10,950	11,000	1,574	1,526	1,586	1,546

Middle Panel

At least	But less than	Single (and 1040EZ filers)	Married filing jointly	Married filing separately	Head of a household
11,000					
11,000	11,050	1,582	1,534	1,594	1,554
11,050	11,100	1,589	1,541	1,601	1,561
11,100	11,150	1,597	1,549	1,609	1,569
11,150	11,200	1,604	1,556	1,616	1,576
11,200	11,250	1,612	1,564	1,624	1,584
11,250	11,300	1,619	1,571	1,631	1,591
11,300	11,350	1,627	1,579	1,639	1,599
11,350	11,400	1,634	1,586	1,646	1,606
11,400	11,450	1,642	1,594	1,654	1,614
11,450	11,500	1,649	1,601	1,661	1,621
11,500	11,550	1,657	1,609	1,669	1,629
11,550	11,600	1,664	1,616	1,676	1,636
11,600	11,650	1,672	1,624	1,684	1,644
11,650	11,700	1,679	1,631	1,691	1,651
11,700	11,750	1,687	1,639	1,699	1,659
11,750	11,800	1,694	1,646	1,706	1,666
11,800	11,850	1,702	1,654	1,714	1,674
11,850	11,900	1,709	1,661	1,721	1,681
11,900	11,950	1,717	1,669	1,729	1,689
11,950	12,000	1,724	1,676	1,736	1,696
12,000					
12,000	12,050	1,732	1,684	1,744	1,704
12,050	12,100	1,739	1,691	1,751	1,711
12,100	12,150	1,747	1,699	1,759	1,719
12,150	12,200	1,754	1,706	1,766	1,726
12,200	12,250	1,762	1,714	1,774	1,734
12,250	12,300	1,769	1,721	1,781	1,741
12,300	12,350	1,777	1,729	1,789	1,749
12,350	12,400	1,784	1,736	1,796	1,756
12,400	12,450	1,792	1,744	1,804	1,764
12,450	12,500	1,799	1,751	1,811	1,771
12,500	12,550	1,807	1,759	1,819	1,779
12,550	12,600	1,814	1,766	1,826	1,786
12,600	12,650	2,822	1,774	1,834	1,794
12,650	12,700	1,829	1,781	1,841	1,801
12,700	12,750	1,837	1,789	1,849	1,809
12,750	12,800	1,844	1,796	1,856	1,816
12,800	12,850	1,852	1,804	1,864	1,824
12,850	12,900	1,859	1,811	1,871	1,831
12,900	12,950	1,867	1,819	1,879	1,839
12,950	13,000	1,874	1,826	1,886	1,846
13,000					
13,000	13,050	1,882	1,834	1,894	1,854
13,050	13,100	1,889	1,841	1,901	1,861
13,100	13,150	1,897	1,849	1,909	1,869
13,150	13,200	1,904	1,856	1,916	1,876
13,200	13,250	1,912	1,864	1,924	1,884
13,250	13,300	1,919	1,871	1,931	1,891
13,300	13,350	1,927	1,879	1,939	1,899
13,350	13,400	1,934	1,886	1,946	1,906
13,400	13,450	1,942	1,894	1,954	1,914
13,450	13,500	1,949	1,901	1,961	1,921
13,500	13,550	1,957	1,909	1,969	1,929
13,550	13,600	1,964	1,916	1,976	1,936
13,600	13,650	1,972	1,924	1,984	1,944
13,650	13,700	1,979	1,931	1,991	1,951
13,700	13,750	1,987	1,939	1,999	1,959
13,750	13,800	1,994	1,946	2,006	1,966
13,800	13,850	2,002	1,954	2,014	1,974
13,850	13,900	2,009	1,961	2,021	1,981
13,900	13,950	2,017	1,969	2,029	1,989
13,950	14,000	2,024	1,976	2,036	1,996

Right Panel

At least	But less than	Single (and 1040EZ filers)	Married filing jointly	Married filing separately	Head of a household
14,000					
14,000	14,050	2,032	1,984	2,047	2,004
14,050	14,100	2,039	1,991	2,061	2,011
14,100	14,150	2,047	1,999	2,075	2,019
14,150	14,200	2,054	2,006	2,089	2,026
14,200	14,250	2,062	2,014	2,103	1,034
14,250	14,300	2,069	2,021	2,117	2,041
14,300	14,350	2,077	2,029	2,131	2,049
14,350	14,400	2,084	2,036	2,145	2,056
14,400	14,450	2,092	2,044	2,159	2,064
14,450	14,500	2,099	2,051	2,173	2,071
14,500	14,550	2,107	2,059	2,187	2,079
14,550	14,600	2,114	2,066	2,201	2,086
14,600	14,650	2,122	2,074	2,215	2,094
14,650	14,700	2,129	2,081	2,229	2,101
14,700	14,750	2,137	2,089	2,243	2,109
14,750	14,800	2,144	2,096	2,257	2,116
14,800	14,850	2,152	2,104	2,271	2,124
14,850	14,900	2,159	2,111	2,285	2,131
14,900	14,950	2,167	2,119	2,299	2,139
14,950	15,000	2,174	2,126	2,313	2,146
15,000					
15,000	15,050	2,182	2,134	2,327	2,154
15,050	15,100	2,189	2,141	2,341	2,161
15,100	15,150	2,197	2,149	2,355	2,169
15,150	15,200	2,204	2,156	2,369	2,176
15,200	15,250	2,212	2,164	2,383	2,184
15,250	15,300	2,219	2,171	2,397	2,191
15,300	15,350	2,227	2,179	2,411	2,199
15,350	15,400	2,234	2,186	2,425	2,206
15,400	15,450	2,242	2,194	2,439	2,214
15,450	15,500	2,249	2,201	2,453	2,221
15,500	15,550	2,257	2,209	2,467	2,229
15,550	15,600	2,264	2,216	2,481	2,236
15,600	15,650	2,272	2,224	2,495	2,244
15,650	15,700	2,279	2,231	2,509	2,251
15,700	15,750	2,287	2,239	2,523	2,259
15,750	15,800	2,294	2,246	2,537	2,266
15,800	15,850	2,302	2,254	2,551	2,274
15,850	15,900	2,309	2,261	2,565	2,281
15,900	15,950	2,317	2,269	2,579	2,289
15,950	16,000	2,324	2,276	2,593	2,296
16,000					
16,000	16,050	2,332	2,284	2,607	2,304
16,050	16,100	2,339	2,291	2,621	2,311
16,100	16,150	2,347	2,299	2,635	2,319
16,150	16,200	2,354	2,306	2,649	2,326
16,200	16,250	2,362	2,314	2,663	2,334
16,250	16,300	2,369	2,321	2,677	2,341
16,300	16,350	2,377	2,329	2,691	2,349
16,350	16,400	2,384	2,336	2,705	2,356
16,400	16,450	2,392	2,344	2,719	2,364
16,450	16,500	2,399	2,351	2,733	2,371
16,500	16,550	2,407	2,359	2,747	2,379
16,550	16,600	2,414	2,366	2,761	2,386
16,600	16,650	2,422	2,374	2,775	2,394
16,650	16,700	2,429	2,381	2,789	2,401
16,700	16,750	2,437	2,389	2,803	2,409
16,750	16,800	2,444	2,396	2,817	2,416
16,800	16,850	2,455	2,404	2,831	2,424
16,850	16,900	2,469	2,411	2,845	2,431
16,900	16,950	2,483	2,419	2,859	2,439
16,950	17,000	2,497	2,426	2,873	2,446

Practice Exercises

Use the tax table to find each of the following.

1. What is your federal income tax if you are single and your taxable income is $11,800?

2. How much federal income tax do you still owe if you are married and filing jointly? Your taxable income is $13,930 and your withholding tax payments are $1,690.

3. How much tax do you still owe or what is the amount of your refund if you are single, your taxable income is $13,425, and your withholding tax payments are $1,750?

4. Use copies of Form 1040EZ to prepare income tax returns for the following.
   a. Rosa Gomez of 2518 Green St., Chicago, IL 36941 earned $13,548 in wages and $279.93 interest. Her federal withholding tax payments for the year were $1,125. Rosa's social security number is 555-12-1200.
   b. Tom Wong of 25 So. Pine St., Los Angeles, CA 94216 has the social security number 808-08-0808. His wages for the year were $14,629 and the interest on his savings account was $316.85. His federal withholding tax payments for the year were $1,430.
   c. Hal Turner's wages for the year were $17,500. His federal withholding tax payments were $1,975. He resides at 8900 West Ave., Dallas, TX 46952. His social security number is 444-22-6666.
   d. Mary Williams of 125 Main St., New York, NY 10000 has the social security number 999-99-0000. Last year she earned $18,200 in salary and $389 in interest. She paid $2,138 federal withholding tax for the year.

Refresh Your Skills

1. 1–8
$$
\begin{array}{r}
4,396 \\
5,287 \\
8,962 \\
+\ 6,435 \\
\end{array}
$$

2. 1–9
$$
\begin{array}{r}
80,000 \\
-\ 9,607 \\
\end{array}
$$

3. 1–11
$$
\begin{array}{r}
409 \\
\times\ 506 \\
\end{array}
$$

4. 1–13
$$48\overline{)46,080}$$

5. 2–5
$12.46 + $9.07 + $.59

6. 2–6
$10 − $5.26

7. 2–7
24 × $.69

8. 2–8
$1.25\overline{)$20}

9. Find $16\frac{2}{3}$% of $24
4–6

10. What percent of $400 is $25?
4–7

11. 3% of what amount is $12?
4–8

Problem Solving Strategy: Solve Another Way

Mario earns $475 per week. Almost 26% of his salary is withheld each week. How much does Mario receive each week?

Strategy: Solve Another Way
Sometimes a second way of solving a problem makes the solution easier to find or gives a better understanding of the problem.

1. **READ** the problem carefully.

a. Find the question asked.
   How much does Mario receive each week?
   Think: The amount Mario receives is his salary less the amount of the deductions.

b. Find the given facts.
   $475 salary
   26% is withheld

---

2. **PLAN** how to solve the problem.

a. Choose a strategy.

Solution 1	**Solution 2**
Find the amount withheld and then subtract.	Find the percent received and multiply.

b. Think or write out your plan.

Solution 1	**Solution 2**
Salary × % withheld = amount withheld	100% − % withheld = % received
Salary − amount withheld = amount received	% received × salary = amount received

---

3. **SOLVE** the problem.

Solution 1	**Solution 2**
Estimate the answer.	Estimate the answer.

Solution 1	**Solution 2**
$25\% \times 480 = n$	$100\% - 25\% = 75\%$
$\frac{1}{4} \times \$480 = \$120$	$\$480 \times 75\% = n$
$\$480 - \$120 = \$360$	$\$480 \times \frac{3}{4} = \$360$

Solution 1	**Solution 2**
Solve.	Solve.
$26\% \times \$475 = \$123.50$	$100\% - 26\% = 74\%$
$\$475 - \$123.50 = \$315.50$	$\$475 \times 74\% = \$315.50$

You might use a calculator.

Solution 1

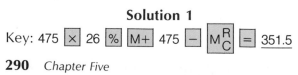

Key: 475 ⊠ 26 % M+ 475 ⊟ M℞C = 351.5

Solution 2

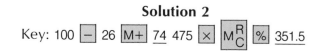

Key: 100 ⊟ 26 M+ 74 475 ⊠ M℞C % 351.5

4. CHECK

a. Check the accuracy of your arithmetic. Work backward to arrive at the given facts.

Solution 1	**Solution 2**
$351.50 + $123.50 = $475	$351.50 ÷ 74% = $475
$123.50 ÷ $475 = 26%	100% − 74% = 26%

b. Compare the answer to the estimate. The answer $351.50 compares favorably with the estimate $360 in both solutions.

ANSWER: Mario receives $351.50.

Practice Exercises

Work with a partner to solve each problem in two ways.

1. The regular price of a steel-belted radial tire is $65. What is its sale price if it is reduced 20%?

2. Of the 18 students in the home economics class, $\frac{2}{3}$ are girls. How many boys are enrolled in the class?

Think of two ways to solve each problem. Use the easier way.

3. Elena got a $4\frac{1}{2}$% cost-of-living raise. If her salary was $375 per week, what is her new salary?

4. Sun Li spent $3,638 on a used car. The registration fee was 8% of the cost. How much did she spend for the car and the registration fee?

5. Sondra spent $11.20 for dinner. She left a 15% tip. How much did she spend in all?

6. Roberto's team won $\frac{8}{10}$ of their 60 games last season. How many games did they lose?

7. At Boardman School, 56% of the 425 students take the bus to school. How many students do not take the bus to school?

8. Curtis got 24 problems correct out of 40. What percent did he miss?

9. Kathi earns a commission of 6% on each airline ticket she books. During 1 week she booked tickets costing $395, $410, $230, and $788. How much commission did she earn?

10. Carrie bought a computer system for $1,895. She received a 5% discount for paying cash. How much did the system cost?

11. Rob's stockbroker charges a 3% commission on all sales. If Rob bought $4,210 worth of stock, how much did he pay in all?

Career Application
TAX PREPARER

Many people cannot prepare their own income tax forms. Others worry that they will make a mistake or pay more tax than is necessary. *Tax preparers* help people fill out income tax forms. A tax preparer must have a good understanding of tax laws. Tax preparers who prepare tax returns incorrectly may be subject to penalties from the federal government.

Tax preparers may work for tax preparation companies or on their own. The federal government gives examinations on tax laws. Some tax preparation companies have training courses.

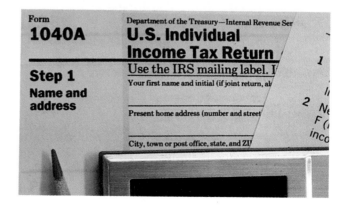

1. Taxpayers who use Form 1040 can itemize deductions. A taxpayer has the following deductions: $138.44 in medical expenses, $2,954.86 in state income taxes and real estate taxes, $3,129.57 in interest on mortgages, $827.41 in contributions, and $50 for the tax preparer's fee. What is the total deduction?

2. Medical expenses in excess of 7.5% of your gross income are deductible. For example, if your gross income was $20,000, any medical expenses over $1,500 are deductible. If the Costas had income of $42,500, and medical expenses of $3,900, how much is their deduction?

3. A taxpayer paid $4,236.81 in withholding tax during the year. Her total tax is $4,189.52. Does she owe more tax, or will she get a refund? How much?

Chapter 5 Review

Page
272 **1.** Find the weekly salary for a person who works 40 hours per week at $8.35 per hour.

275 **3.** If a realtor charges 5% commission on the sale of a house, how much will the owners receive on the sale of their $90,000 house?

278 **5.** Using the table of withholding tax (p. 278), compute the amount of income tax to be withheld when a married person's weekly wages are $405.62 and 2 allowances are claimed.

282 **7.** If the state tax rate is $2\frac{1}{4}$% of the wages, find the tax a person owes for the week when she earns $410.58.

Use the table of withholding tax (p. 278) and the social security tax rate of 7.15% in each of the problems.

283 **8.** Find the take-home pay when a married person's weekly salary is $388 with 5 allowances claimed.

Use the federal tax table (p. 288) to find the following.

285 **10.** What is your federal income tax if you are head of a household and your taxable income is $15,892?

285 **12.** What is the amount of your refund if you are married, filing jointly with a taxable income of $14,235, and withholding payments are $2,383?

Page
2. Which is the better wage, $24,275 per year or $1,945 per month?

4. Marie earns $215 plus 3% commission on sales per week. If she sold $5,075 in merchandise, what was her weekly income?

281 **6.** How much social security tax is deducted weekly from a wage of $524 if the tax rate is 7.15%?

9. Find the take-home pay when a married person's weekly salary is $438 with 1 allowance claimed and a $2\frac{1}{2}$% state tax.

11. How much federal income tax do you still owe if you are married, filing separately, taxable income is $13,110, and your withholding payments are $1,758?

272 **13.** A fixed sum of money paid for working a definite period of time is called ___?___.

Chapter 5 Test

Page
272
1. Will works 38 hours per week. He earns $218.50. What is his wage per hour?

Page
2. Ramona earns $8.20 per hour plus time-and-a-half over 40 hours. How much did she make if she worked 47 hours?

3. Tad receives a monthly salary of $2,325 and a bonus of $880. What is his total income for the year?

4. In 1984 in Oregon the average annual wage was $17,474. What was the average weekly wage?

275
5. An automobile salesman receives 4% commission on new cars. If a car sold for $13,250, what was the amount of the commission?

6. A real estate agent sold a house for $125,000 and received a 6% commission. How much was his commission, and how much did the seller receive?

278
7. If you use the percentage method to compute the withholding tax for wages over $574 but not over $901, the amount of tax is $78.42 plus 28% over $574. What is the withholding tax for a person earning $712 per week?

281
8. When the tax rate is 7.15%, how much is the social security tax on a weekly salary of $388.50?

282
9. At the tax rate of $3\frac{1}{4}$%, what is the state tax on a weekly salary of $588.90?

283
10. How much is the take-home pay if the withholding tax on a salary of $451 is $66.90 and the social security tax rate is 7.15%?

283
11. If the withholding tax on a salary of $332 is $42.12, the social security tax is $23.74, and the state tax is $11.26, what is the take-home pay?

285
12. If your federal income tax for the year is $1,834 and your withholding tax payments are $2,241, what is the amount of your refund?

13. If your federal income tax for the year is $2,109 and your withholding tax payments are $1,892, how much do you still owe?

6-1 Buying Food

When the purchase price of a food item is computed and the result contains any fractional part of a cent, the fraction is always rounded up to the nearest cent.

Example

If 3 bars of soap sell for 67¢, what is the price of 1 bar?

a. Divide the price for 3 bars of soap by 3 to find the price of 1 bar.

$$22\frac{1}{3}¢$$
$$3\overline{)67¢}$$

b. Round the price up to the nearest cent.

$$22\frac{1}{3}¢ \rightarrow 23¢$$

You might use a calculator. Key: 0.67 ÷ 3 =
0.223333

ANSWER: The purchase price of 1 bar of soap is 23¢, or $.23.

Think About It

1. What words are the division indicator in the example above?

3. Why don't stores just round the cost up or down according to the rules for rounding?

2. Give a keying sequence you might use in finding the total cost of 4 cans of peaches, 10 pounds of flour, 6 cans of tuna, and 8 cans of soup. (Use the Grocery Price List on page 297 for prices.)

Practice Exercises

Use the price list on the next page to find the cost.

1. a. 1 can of corn
b. 8 bars of soap
c. 1 can of soup
d. 1 can of peaches

2. a. 10 cans of corn
b. 3 cans of peas
c. 2 bars of soap
d. 12 cans of soup

3. a. 1 loaf of bread
b. 3 boxes cereal
c. 1 roll
d. $\frac{3}{4}$ lb granola bar

4. a. 1 qt milk
b. $2\frac{3}{4}$ lb flounder
c. 6 oz cheese
d. $1\frac{1}{2}$ dozen eggs

5. a. 2 lb string beans
b. 1 lb onions
c. $\frac{3}{4}$ lb pears
d. 1 bunch carrots

6. a. 3 lemons
b. $2\frac{1}{2}$ doz oranges
c. 6 grapefruit
d. $1\frac{1}{4}$ lb grapes

SUPERMARKET WEEKLY SPECIALS

FRUIT AND PRODUCE PRICE LIST

Apples, 49¢ a pound
Grapes, 89¢ pound
Peaches, 39¢ a pound
Celery, 59¢ a bunch
Peppers, 3 for 85¢
Bananas, 32¢ a pound
Lemons, 6 for 84¢
Pears, 48¢ a pound

Lettuce, 2 heads for 95¢
Potatoes, 10 pounds for $1.49
Grapefruit, 3 for 69¢
Oranges, 98¢ a dozen
Carrots, 2 bunches for 97¢
Onions, 3 pounds for 81¢
String beans, 40¢ a pound

MEATS, FISH, AND DAIRY PRICE LIST

Chicken, $1.25 a pound
Bluefish, $3.79 a pound
Rib roast, $3.99 a pound
Margarine, 96¢ a pound
Eggs, 89¢ a dozen
Lamb, $2.24 a pound
Cheese, $2.56 a pound
Flounder, $2.92 a pound
Milk, 2 quarts for $1.25
Smoked ham, $1.76 a pound

BREAD AND CEREAL PRICE LIST

Assorted Muffins, $1.92 a dozen
Bread, 2 loaves for $1.89
Cereal, $1.89 a box

Granola bars, $3.20 a pound
Rolls, 6 for 78¢

GROCERY PRICE LIST

Beverages, 2 bottles for 95¢
Flour, 10 pounds for $2.99
Peas, 6 cans for $2.16
Sugar, 5 pounds for $1.89

Coffee, $2.69 a pound
Detergent, $1.45 a package
Soap, 4 bars for 95¢
Tomato juice, 3 cans for $1.49

Corn, 5 cans for $1.79
Peaches, 2 cans for $1.29
Soup, 2 cans for 79¢
Tuna, 2 cans for $2.25

Use the price list to find the cost of each of the following orders.

7. a. 1 loaf bread, 1 dozen eggs, $\frac{1}{2}$ lb margarine, 10 lb flour, 1 lb coffee

b. $1\frac{1}{2}$ dozen oranges, 8 oz margarine, 2 lb 3 oz flounder, $4\frac{1}{2}$ lb chicken, 3 peppers, 1 lemon, 1 lb string beans, 1 lb onions

c. 4 cans tuna, $1\frac{1}{2}$ dozen rolls, 1 can tomato juice, 1 bar soap, 1 pkg. detergent, 1 can peaches, 2 grapefruit, 2 lb 4 oz smoked ham

d. 5 lb sugar, 1 can peaches, 3 cans corn, 8 rolls, 2 lemons, $1\frac{1}{2}$ dozen eggs, 12 cans soup, 4 bottles beverage, $4\frac{3}{4}$ lb rib roast, 2 lb 3 oz flounder, 6 peppers

e. 2 bunches carrots, 3 heads of lettuce, $4\frac{1}{2}$ lb chicken, $1\frac{1}{2}$ lb string beans, 4 cans corn, 10 rolls, 1 loaf bread, $1\frac{1}{4}$ lb peaches, 4 grapefruit, 12 oz margarine, 1 lb 8 oz lamb, 2 lb 15 oz bananas

f. 1 qt milk, $3\frac{1}{2}$ lb bluefish, 4 oz cheese, 1 lb 4 oz margarine, 2 lb 14 oz apples, 1 can peaches, 2 pkgs. detergent, 12 bars soap, 3 bunches carrots, 12 peppers, 3 dozen eggs, 12 cans corn, 18 cans soup, 20 rolls

Unit Pricing—Determining the Better Buy

When buying food or other items packaged in two different sizes or quantities, a wise shopper likes to know which is the better buy. Although the larger size or larger quantity is generally marked at the lower unit price, sometimes the smaller size or smaller quantity can be the better buy.

Example

Which is the better buy: 10 oranges for 85¢ or 12 oranges for 99¢?

To determine the better buy when two different quantities of the same item are priced at two rates, follow these steps.

a. Find the unit price (cost per unit) of the item at each rate.

$$
\begin{array}{ll}
0.085 & \text{10 for 85¢} \\
10\overline{)0.850} & \text{cost 8.5¢ each.} \\
\underline{80} \\
50
\end{array}
\qquad
\begin{array}{ll}
0.0825 & \text{12 for 99¢} \\
12\overline{)0.9900} & \text{cost 8.25¢ each.} \\
\underline{96} \\
30 \\
\underline{24} \\
60 \\
\underline{60}
\end{array}
$$

b. Select the lower unit price as the better buy.
   8.25¢ is less than 8.5¢

ANSWER: The better buy is 12 oranges for 99¢.

You might use a calculator.

Key: 0.85 ÷ 10 = 0.085 = 8.5¢

Key: 0.99 ÷ 12 = 0.0825 = 8.25¢

Vocabulary

Unit price is the cost per unit of measure. This may be per item, per ounce, per pound, per quart, per liter, per kilogram, or per any other unit of measure. @ is a symbol that means "at" and indicates the unit price.
3 shirts @ $14.95 is read "three shirts at $14.95 each." Sometimes unit prices are written in thousandths. For example, $.467 means 46.7¢.

1. How does knowing the unit price of items help you to make better buying decisions?

2. Give some reasons why you might choose a smaller package, even though a larger package is a better buy.

Practice Exercises

1. Write each of the following, using a dollar sign ($) and a decimal point.

   a. 95¢   b. 4¢   c. 23.9¢   d. $30\frac{3}{4}$¢   e. 43.57¢

2. Write each of the following, using the ¢ symbol.

   a. $.03   b. $.017   c. $2.81   d. $.9402   e. $.0075

3. Find the unit price (cost per pound) to the nearest tenth of a cent.

   a. 2-lb package bacon costing $2.98   b. 8-oz package sliced cheese costing $1.69
   c. 10-oz jar jelly costing 72¢   d. 1-lb 6-oz pie costing $1.49

4. Find the unit price (cost per quart) to the nearest tenth of a cent.

   a. $\frac{1}{2}$ qt vinegar costing 43¢   b. 4 qt bleach costing 96¢
   c. 1.5 qt milk costing $1.28   d. 1.75 qt can juice costing 83¢

5. Find the unit price (cost per ounce) to the nearest tenth of a cent.

   a. 3-oz jar olives costing 89¢   b. 12-oz box cereal costing $1.29
   c. 6.5-oz can tuna costing $1.09   d. $7\frac{3}{4}$-oz can salmon costing $1.95

6. Find the unit price (cost per 50 pieces) for each package, to the nearest tenth of a cent.

   a. 100 tea bags costing $1.80   b. 75 storage bags costing $1.50
   c. 10 soap pads costing 63¢   d. 160 paper napkins costing 81¢

7. Find the unit price (cost per quart) for each bottle of vegetable oil and determine the difference between the highest and lowest unit prices.

   a. 4 qt at $5.49   b. $\frac{1}{2}$ qt at 81¢   c. 0.75 qt at $1.13

8. Find the unit price (cost per pound) for each package of cereal and determine the difference between the highest and lowest unit prices.

   a. 12 oz at $1.39   b. 8 oz at 97¢   c. 15 oz at $1.68

9. Find the unit price (cost per single item) in each of the following.

   a. 5 jars of baby food costing $1.45   c. 10 oranges costing 95¢
   b. 2 shirts costing $37   d. 4 tires costing $165

10. In each, the same item is priced at two rates. Which is a better buy?

    a. Pears: 5 for 49¢ or 2 for 25¢   b. Grass seed: 5 lb for $6.50 or 25 lb for $27.75
    c. Doughnuts: 12 for $1.89 or 3 for 49¢   d. Soup: 7 cans for $2 or 3 cans for 95¢

11. How much do you save by buying the larger size of each item instead of an equivalent quantity in the smaller size?

   a. Pretzels: 4-oz bag at 33¢; 12-oz bag at 89¢   **b.** Bleach: 1-qt bottle at 45¢; $\frac{1}{2}$-qt bottle at 27¢

   c. Paint: 1-qt can at $7.49; 4-qt can at $12.95   **d.** Spaghetti: 8-oz box at 37¢; 1-lb box at 69¢

12. How much do you save per ounce by buying the larger size?

   a. Peanut butter: 6-oz jar at 85¢; 12-oz jar at $1.59
   b. Mayonnaise: 32-oz jar at $1.69; 16-oz jar at 95¢
   c. Mouthwash: 14-oz bottle at $1.09; 20-oz bottle at $1.39
   d. Tomato juice: 18-oz can at 40¢; 46-oz can at 83¢

13. Which is the better buy per unit price (cost per quart or cost per pound)?

   a. Milk: 1 qt at 62¢; a half-gallon at $1.12   **b.** Laundry detergent: 1-lb 4-oz pkg at 99¢; 5-lb 4-oz pkg. at $3.34

   c. Baked beans: 1-lb can at 41¢; 1-lb 10-oz can at 59¢   **d.** Beverage: 6 16-oz bottles at $2.39; 8 10-oz bottles at $2.49

14. Find how much you save on each can when buying in quantity.

   a. 1 dozen cans of peas for $4.20 or 41¢ each   **b.** 7 cans of soup for $1.75 or 31¢ each
   c. 5 cans of beans for $1.75 or 37¢ each   **d.** 12 cans of beverage for $2.99 or 29¢ each

15. Find the unit price (cost per kilogram) to the nearest tenth of a cent.

   a. 5 kg of potatoes costing $1.19   **b.** 2.27-kg bag of sugar costing $1.65
   c. 283-g jar of jelly costing 69¢   **d.** 964-g jar of applesauce costing 73¢

16. Find the unit price (cost per liter) to the nearest tenth of a cent.

   a. 2-L, bottle of beverage costing $1.25   **b.** 1.36-L can of fruit punch costing 64¢
   c. 532-mL can of tomato juice costing 45¢   **d.** 946-mL carton of milk costing 68¢

17. Which is the better buy?

   a. Flour: 2.26-kg bag at 95¢ or 907-g bag at 51¢   **b.** Bleach: 950-mL bottle at 45¢ or 1.89 L bottle at 67¢

   c. Crackers: 454-g box at $1.05 or 340-g box at 93¢   **d.** Pickles: 946-mL jar at $1.09 or 473-mL jar at 65¢

18. How much can you save buying the larger size of each item instead of an equivalent quantity in the smaller size?

   a. Fabric softener: 0.94-L bottle at 33¢; 3.76-L bottle at 87¢   **b.** Grapefruit sections: 227-g can at 31¢; 1.362-kg can at $1.29

   c. Vinegar: 1 L at 73¢; 0.5 L at 43¢   **d.** Salad dressing: 250 g at $.67; 500 g at $1.20

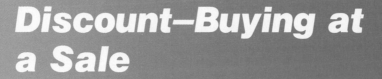

6-3 Discount—Buying at a Sale

As a means of attracting customers, many merchants offer periodic discounts on items in their stores. Sale prices are usually stated as a percent or dollar amount taken off the regular list price.

Vocabulary

Discount is the amount an article is reduced in price.
The **list price**, or **marked price**, is the regular or full price of an article.

The **net price**, or **sale price**, is the reduced price.
The **rate of reduction**, or **rate of discount**, is the percent taken off the list price.

Examples

1. Find the discount and the net price of a portable TV when the list price is $290 and the rate of discount is 15%.

To find the discount and net price when the list price and rate of discount are given, follow these steps.

a. Multiply the list price by the rate of discount to find the discount:

$$
\begin{array}{r}
\$290 \\
\times\ 0.15 \\
\hline
1450 \\
290 \\
\hline
\$43.50\ \textbf{discount}
\end{array}
$$

b. Subtract the discount from the list price to find the net price:

$$
\begin{array}{r}
\$290.00\ \textbf{list price} \\
-\quad 43.50\ \textbf{discount} \\
\hline
\$246.50\ \textbf{net price}
\end{array}
$$

ANSWER: $43.50 discount; $246.50 net price

You might use a calculator. Key: 290 M+ × 15 % 43.5 − MRC = 246.5

2. What is the rate of discount when the list price of a calculator is $48 and the discount is $12?

Divide the discount by the list price to find the rate of discount:
$$\frac{\$12}{\$48} = \frac{1}{4} = 25\%$$

ANSWER: 25%

You might use a calculator. Key: 12 ÷ 48 % 25

3. Find the list price of a dress that sold for $60 at a 20% discount sale.

To find the list price when the sale price and rate of discount are given, follow these steps.

a. Subtract the given rate from 100%. 100% − 20% = 80% = 0.8

b. Divide the sale price by the answer found in Step **a.** $60 ÷ 0.8 = $75

ANSWER: $75 list price

You might use a calculator.   Key: 100 − 20 M+ 60 ÷ M^R_C % 75

Think About It

1. On very expensive items, such as appliances or automobiles, why do you think a dealer may decide to offer a dollar amount discount rather than advertise a percent reduction of the original price?

2. If a store is selling picture frames at a "Buy Two and Get One Free" sale, what is the overall percent reduction in price for three picture frames?

Practice Exercises

CORY'S DEPARTMENT STORE

FURNITURE—Reduced 30%	CLOTHING—Reduced $33\frac{1}{3}$%	APPLIANCES—Reduced 20%
Tables, $99	Dresses, $69	Refrigerators, $595
Lamps, $54.50	Coats, $90	Toaster-ovens, $39.95
Sofas, $425	Shoes, $49.95	Hair dryers, $36.75
Bedroom sets, $899	Suits, $117.75	Color TVs $445
Bookcases, $72.75	Sweaters, $18.98	Blenders, $29.95
Mirrors, $37.98	Skirts, $21.45	Can openers, $14.69

1. Use the advertisement to find the sale price.
a. furniture items   **b.** items of clothing
c. appliances

2. How much is a football reduced in price at a 25% reduction sale if its regular price is $15? What is its sale price?

3. At a year-end clearance sale an automobile, regularly selling for $6,850, can now be purchased at a reduction of 17%. What is its sale price?

4. Find the rate of reduction when a lawn mower that regularly sells for $185 was purchased at $148.

5. Find the sale price of each of the following articles.
   a. Baseball; regular price, $3.50; reduced 40%
   b. Camera; regular price, $79.95; reduced 20%
   c. Stereo system; regular praice, $225; $\frac{1}{3}$ off.
   d. Sewing machine; regular price, $204.50; reduced 15%.

6. Find the rate of discount on each of the following articles.

   a. Typewriter; regular price, $240; sale price, $180
   b. Kitchen clock; regular price, $18; sale price, $15
   c. Tire; regular price, $48; sale price, $42
   d. Vacuum sweeper; regular price $87.50; sale price, $75

7. Find the list price of a desk that sold for $149.85 at a 25% reduction sale.

8. Find the regular price of each of the following.

   a. Battery, $28 sale price when reduced 30%
   b. Tablecloth, $20.40 sale price when reduced 15%
   c. Sprinkler, $9.60 sale price when reduced $37\frac{1}{2}$%
   d. Freezer, $246.66 sale price when reduced $33\frac{1}{3}$%

9. A calculator that regularly sells for $49.88 is now on sale at $\frac{1}{4}$ off. What is its sale price?

10. What is the marked price of a pair of slacks that sold for $23.60 at a 20%–off sale?

11. Mr. Warner bought a camera at a 15% discount. If the regular price was $56, how much discount was he allowed? What net price did he pay?

12. Find the trade discount and the net price of each.

	a.	b.	c.	d.
Catalog list price	$183	$13.49	$93.75	$514
Rate of trade discount	27%	12%	20%	$16\frac{2}{3}$%

13. What is the net price of a drill press listed at $342.95 with a trade discount of 12% and an additional cash discount of 5%?

14. Find the net price of each.

	a.	b.	c.	d.
List price	$160	$539.80	$316.50	$875
Rates of discount	25%, 4%	30%, 2%	15%, 10%	20%, 8%

15. Find the rate of discount when a radio, listed for $125, is sold for $95.

16. Find the list price of a stereo when the net price is $280 and the rate of discount is $12\frac{1}{2}$%.

17. Find the list price of each.

	a.	b.	c.	d.
Net price	$42	$368	$73.80	$31.32
Rate of discount	25%	8%	10%	$16\frac{2}{3}$%

Taxes on Purchases

Many cities and states have a sales tax, which is a percent of the total price. The Federal government and most state governments levy excise taxes on many goods and services such as air fare, gasoline, and telephone calls. Excise taxes are sometimes called "hidden," because they are not usually shown separately. Thus, you may not be aware that you are paying such a tax.

When figuring tax, any fractional part of a cent is rounded up to the nearest cent.

> ## Vocabulary
>
> **Sales tax** is a tax on the selling price of articles that consumers purchase.
>
> **Excise tax** is a tax on the manufacture, sale, or consumption of certain goods and services.

Examples

1. Find the sales tax on a purchase of $15.50 when the tax rate is 7%.

a. Multiply the tax rate and the amount of purchase.

7% = 0.07   Write the percent as a decimal.
0.07 × $15.50 = $1.085   Multiply.

b. Round the fractional part of a cent up to the nearest cent.

$1.085→$1.09

ANSWER: The sales tax is $1.09

2. Find the total cost of an airplane ticket, including federal excise tax of 8%, when the base fare (excluding tax) is $269.79.

a. Multiply the tax rate and the base fare.

8% = 0.08
0.08 × $269.79 = $21.5832

b. Add the tax to the base fare.

$269.79 + $21.5832 = $291.3732 → $291.37   Round up to the nearest cent.

ANSWER: The total cost is $291.37.

3. Use a calculator to find the total cost, including 3% excise tax, for telephone service when the basic charge is $17.50.

Key: 17.50 $\boxed{+}$ 3 $\boxed{\%}$ $\underline{18.025}$ → 18.03   Round up to the nearest cent.

ANSWER: The total cost is $18.03

Think About It

1. Name some goods and services that you think have an excise tax. How might you determine if you are paying a "hidden" tax?

2. What are some factors you should consider in deciding whether traveling to a state with no sales tax is really a bargain?

Practice Exercises

1. Find the sales tax on each of the following purchases with the given tax rate.
   a. $25.00; 4% tax   **b.** $93.50; 8% tax   **c.** $305.86; 6%   **d.** $258.75; 5%

2. Find the total cost, including sales tax, of each of the following articles.
   a. Bag. $38.98; tax 8%
   c. Electric dryer. $269.95; tax $4\frac{1}{2}$%
   b. Calculator. $24.99; tax 3%
   d. Safe. $799; tax 5%

3. Find the total cost of airplane fare inlcuding excise tax of 8% when the base fare (excluding tax) is given.
   a. $82   **b.** $63.25   **c.** $257.60   **d.** $329.15

4. Find the full amount of the bill including federal excise tax at a rate of 3%, and, where given, state tax on the following telephone services.
   a. $12.80   **b.** $33.75   **c.** $27.15, 5% state tax   **d.** $46.30, 4% state tax

5. A motorist paid 94.9¢ per gallon for 10 gallons of gasoline. If the excise tax is 9¢ per gallon and the state tax is 8¢ per gallon, find the total tax paid for the gasoline purchased.

6. Use a calculator to check your answer to Exercise 2c above. Write the keying sequence.

Most people use natural gas, electricity, and water in their apartment or home. A meter is installed to measure the amount used by each household. By reading the meter, the utility company knows how much to charge each customer.

Examples

1. Jason's house has an electric meter that he wants to read. What does the meter read?

To read a gas, electric, or water meter, follow these steps.

a. Select on each dial the number that was just passed by the pointer as it goes from 0 to 9.
Note: Some pointers move clockwise and others move counterclockwise.

The dials read, 4-6-2-7

b. Then put the dial readings together and read from left to right.

ANSWER: 4627

We read the meter in the same way for each of the services, but different units are used to express their consumption.

Gas is expressed in hundred cubic feet.
The reading 5486 represents 5,486 × 100 cubic feet or 548,600 cubic feet (ft³).

Electricity is expressed in kilowatt hours (kWh).
The reading 8291 represents 8,291 kilowatt hours.

Water is expressed in tens of gallons (gal).
The reading 25736 represents 25,736 × 10, or 257,360 gallons.

---

2. The gas meter for Judy's house now reads 4737, and it read 4276 last month. How much natural gas has been used?

To find the amount used during a given period, follow these steps.

a. 4737 − 4276 = 461

Subtract the reading from the beginning of the period from the reading at the end of the period.

b. 461 × 100 cu. ft = 46,100 cu. ft

Then express the difference in the appropriate units.

ANSWER: 46,100 cu. ft

Think About It

1. What types of utility meters do you have where you live? Why are they located where they are?

2. Think of another system to replace the "meter reading system" for electricity consumption.

Practice Exercises

1. a. What is the reading of the gas meter on April 1? On May 1?

Cubic Feet   Cubic Feet

April 1   May 1

b. How many cubic feet of gas were consumed during the month of April?
c. Find the cost of the gas consumed at the following rates.

First 100 cu. ft $3.30   Next 5,000 cu. ft $.61 per 100 cu. ft
Next 900 cu. ft $.77 per 100 cu. ft   Next 12,500 cu. ft $.59 per 100 cu. ft
Next 1,500 cu. ft $.67 per 100 cu. ft   Over 20,000 cu. ft $.58 per 100 cu. ft

2. Given the reading at the beginning and at the end of the period, use the above rates to find the cost of the gas consumed during the period.

	READING				READING	
	At Beginning	At End			At Beginning	At End
a.	2418	2487		**b.**	5523	5541
c.	8594	8691		**d.**	9905	0298

3. a. What is the reading of the electric meter on November 1? on December 1?

Kilowatt Hours   Kilowatt Hours

November 1   December 1

b. How many kilowatt hours of electricity were used during the month of November?
c. Find the cost of the electricity used at the following rates.

First 12 kWh or less $4.20   Next 46 kWh @ $.08
Next 42 kWh @ $.09   Over 100 kWh @ $.069

4. Given the reading at the beginning and at the end of the period, use the above rates to find the cost of the electricity used during the period.

	READING				READING	
	At Beginning	At End			At Beginning	At End
a.	4726	4773		**b.**	9887	0069
c.	6175	6266		**d.**	2261	2345

5. At the average cost of 8.6¢ per kilowatt hour (1 kilowatt hour = 1,000 watts/hour), find the hourly cost of operating each of the following.
a. Toaster, 1,000 watts   **b.** Dryer, 4,000 watts
c. 100-watt light bulb   **d.** Heater, 1,500 watts

In many communities water bills are issued every 3 months. The reading on a certain water meter on February 1 was 43582; May 1, 46106; August 1, 51263; November 1, 54091.

6. How many gallons of water were used during the 3-month period?
a. February 1–April 30   **b.** May 1–July 31   **c.** August 1–October 31

7. Find the cost of the water consumed during each of these periods using the rates: First 2,500 gallons or less, $7.58; Over 2,500 gallons, $1.77 for every 1,000 gallons.

8. During one year Mr. Ross bought 210 gallons of fuel oil at $1.249 per gallon, 185 gallons at $1.219 each, 235 gallons at $1.226 each, 220 gallons at $1.235 each, and 190 gallons at $1.253 each. How many gallons of fuel oil did he buy? What was the total cost of his fuel oil for the year?

6-6

Owning an Automobile

An automobile is one of the largest purchases a person makes. In addition to the cost of the car, many other expenses are involved in operating an automobile such as the registration fees, insurance, gasoline, and repairs. Depreciation, which is the decrease in value of an automobile on a yearly basis, is another cost.

Example

During the first year that Mary owned her car she drove it 9,425 miles and had the following operating expenses: gasoline, 620 gallons at $1.29 per gal; oil, 30 quarts at $1.30 per qt; insurance, $585 per yr; license fees, $36 per yr; miscellaneous, $125. How much did it cost Mary to operate her car for the year?

To find the cost per year, follow these steps.

a. Find the total cost for gasoline and oil.

620 gallons × $1.29 per gal = $799.80
30 quarts × $1.30 per qt = $39

b. Add all the costs.

$799.80 + $39 + $585 + $36 + $125
= $1,584.80

ANSWER: The total cost per year was $1,584.80

Think About It

1. What things should you consider before you purchase a car?

2. What factors determine automobile insurance rates?

Practice Exercises

1. Mary bought the car for $8,250. After using it for 4 years, she sold it for $1,725. What was the average yearly depreciation of the car?

2. Using the figures obtained from the Example and Problem 1 above, determine the total cost of using the car for a year.

3. Mr. Potter insures his car costing $9,000 against fire and theft for 80% of its value at the annual rate of 85¢ per $100. If the charge per year for property damage insurance is $81.75, for liability insurance is $68.50, and for collision insurance is $76, what is his total annual premium?

4. How long will it take a person, driving at an average speed of 50 miles per hour, to travel from Washington to Boston, a distance of 460 miles? If the car can average 24 miles on a gallon of gasoline, how many gallons are required to make the trip?

Installment Buying

Costly purchases such as major home appliances, furniture, automobiles, and so on, are often made on the installment plan. Many stores offer deferred, or postponed, payment plans as a convenience to their customers.

Vocabulary

Installment buying is a credit plan that allows for payment in equal amounts, called installments, over a period of time.

A **down payment** is a partial payment made at the time of purchase.

A **finance charge**, or **carrying charge**, is the interest paid for installment buying. It is the difference between the cash price and the total amount paid.

Deferred payment purchase plans are plans in which a specified monthly payment is made to reduce the balance owed. Sometimes these payments are fixed amounts. Other times these payments decrease as the bracket of the balance owed falls. However, a finance charge, usually $1\frac{1}{4}$% or $1\frac{1}{2}$% per month, is applied to the previous monthly unpaid balance and is added to this previous balance.

Example

Mr. Comstock bought an automatic dishwasher with a list price of $375. He made a down payment of $25. If the carrying charge interest rate is 6% for 8 months, how much is each monthly payment?

To find the monthly payment when cash price, down payment, carrying charge interest rate, and number of months to pay are given, follow these steps.

a. Subtract the down payment from the cash price to find the amount financed:

$375 − $25 = $350 financed

c. Add the finance charge to the amount financed to find the balance owed:

$350 + $21 = $371 owed

b. Multiply the amount financed by the finance charge interest rate to find the carrying charge:

$350 × $0.06 = $21 carrying charge

d. Divide the balance owed by the number of monthly payments to find the amount of each monthly payment:

$371 ÷ 8 = $46.38 owed each month

ANSWER: The monthly payment is $46.38.

You might use a calculator.

Key: 375 − 25 M+ × 6 % + MRC ÷ 8 = 46.375

1. It always costs more to buy on the installment plan than to use cash. List several reasons why someone might choose installment buying.

2. Would you expect to pay more if you spread installment payments over a longer period? Why might someone choose to do this?

Practice Exercises

1. Lisa's mother can purchase a refrigerator for the cash price of $489 or $39 down and 10 equal monthly payments of $48.38 each. How much is the carrying charge? How much can be saved by paying cash?

2. A TV set can be purchased for the cash price of $395 or $35 down, and the balance and finance charge to be paid in 12 equal monthly payments. If 6% was charged on the full balance, how much is the monthly payment?

3. Find the monthly payment when each of the following articles is purchased on the installment plan and the finance-charge interest rate is on the full balance.

	Cash Price	Down Payment	Finance Charge Interest Rate	Number of Monthly Payments
a. Camera	$118	$10	6%	12
b. VCR	$595	$55	6%	12
c. Bedroom set	$869	$59	8%	18
d. Sewing machine	$275	$35	$8\frac{1}{2}$%	6

4. A department store uses the following deferred payment schedule.

Monthly Balance	Monthly Payment	Monthly Balance	Monthly Payment
$0 to $10	Full Balance	$300.01 to $350	$25
$10.01 to $200	$10	$350.01 to $400	$30
$200.01 to $250	$15	$400.01 to $500	$40
$250.01 to $300	$20	Over $500	$\frac{1}{10}$ of balance

If finance charges are computed at the monthly rate of $1\frac{1}{4}$% on the previous month's balance, find the monthly payment due on each of the following previous month's balances. First compute the finance charge, add it to the month's balance, and then find in the above schedule the monthly payment due.

a. $79   **b.** $198   **c.** $600   **d.** $265

Life Insurance

People who are responsible for the welfare of others (food, clothing, shelter, and education) often purchase life insurance. When they die, the life insurance benefits can continue to provide money for family and friends.

Annual Life Insurance Premiums Per $1,000

Age	5–Year Term		Ordinary Life		20–Payment Life		20–Year Endowment	
	Male	Female	Male	Female	Male	Female	Male	Female
18	3.77	3.74	13.48	12.95	23.97	23.22	45.17	45.17
20	4.01	4.01	14.20	13.62	24.91	24.11	45.30	45.30
22	4.20	4.20	15.00	14.35	25.89	25.05	45.46	45.46
24	4.40	4.40	15.86	15.15	26.94	26.03	45.64	45.64
25	4.50	4.50	16.34	15.59	27.49	26.55	45.75	45.75
26	4.60	4.60	16.83	16.04	28.06	27.09	45.86	45.85
28	4.83	4.83	17.90	17.02	29.26	28.21	46.14	46.10
30	5.09	5.05	19.09	18.11	30.54	29.42	46.48	46.40
35	6.08	5.86	22.71	21.43	34.24	32.88	47.72	47.49
40	8.03	7.44	27.20	25.56	38.49	36.85	49.45	49.00
45	11.10	10.13	32.79	30.73	43.34	41,40	51.81	51.06
50	16.00	14.45	40.07	37.50	49.28	46.94	55.40	54.18

Example

Ms. Higgins, who is 35 years old, wants to purchase $90,000 worth of ordinary life insurance. What will be her annual premium?

To find the annual premium when the age of the insured and the type of insurance policy are given, follow these steps.

a. Select from the table the type of insurance (ordinary life), and whether male or female.

b. Read down the column until you reach the appropriate age, 35 years old.

c. Find out how many units of $1,000 are in the policy needed, and multiply this by the stated rate.

$90,000 ÷ $1,000 = 90   90 × $21.43 = $1,928.70

ANSWER: The annual premium is $1,928.70.

Think About It

1. At what point in your life would you consider purchasing life insurance? Who would be your beneficiary?

2. Why, do you think, is the premium for a 20-payment policy more for a 35-year-old than for a 20-year-old?

Vocabulary

Life insurance offers financial protection to the dependents of an insured person in the event of the insured person's death. The **beneficiary** is the person who receives the money when the insured dies.

Term insurance—The person is insured for a specified period of time. Premiums are paid only during that time. The beneficiary receives face value of policy if the insured person dies within the specified time.

Ordinary life insurance—The person is insured until death. Premiums are paid until death of the insured person. The beneficiary receives face value of policy when the insured person dies.

Limited-payment life insurance—The person is insured until death. Premiums are paid for a specified period of time, generally 20 years. The beneficiary receives face value of policy when the insured person dies.

Endowment insurance—The person is insured for a specified period of time. Premiums are paid only during that time. The beneficiary receives face value of policy if the insured person dies within the specified time. The insured person, or policyholder, receives face value of policy if alive at the end of the specified time.

The **premium** is the amount of money the insured person pays for the insurance.

The **policy** is the written contract between the insured person and the insurance company. This specifies the conditions of the insurance.

The **value of the policy** is the amount of insurance specified in that policy.

The **term of the policy** is the length of time that the insurance contract is in force.

Practice Exercises

1. Find the annual premium on each of the following policies.
   a. 20-payment life for $10,000 issued to a woman at age 30.
   b. 5-year term for $50,000 issued to a man at age 40.
   c. 20-year endowment for $5,000 issued to a woman at age 25.
   d. Ordinary life for $60,000 issued to a man at age 20.

2. Ms. Myers bought a 20-payment life insurance policy for $20,000 when she was 35 years old. What is the total amount of premium she must pay before the policy is fully paid up?

3. Mrs. Miller purchased a 20-year endowment policy for $5,000 when she was 30 years old. She is now 50 years old.
   a. How much will she receive?   **b.** What was the total premium she paid?

4. When Mr. Walker was 40 years old he purchased a 5-year term policy for $25,000. He died after paying three annual premiums. How much more did the beneficiary receive than Mr. Walker paid?

5. Rosa's brother, now 25 years old, wishes to buy either a 20-payment life policy or an ordinary life policy. What will be the difference in total premiums paid per $1,000 insurance for these policies when he becomes 45 years of age? 65 years of age?

Property Tax

All consumers pay property taxes, either directly or indirectly. The owner of property or a building pays the tax directly. The owner of an apartment building includes the amount of taxes in the rental charges.

Example

Find the taxes on a house assessed for $19,800 when the tax rate is $2.70 per $100.

To find the amount of taxes on property:

a. Express the assessed value so that the appropriate tax rate can be applied.

$$\$19,800 \div \$100 = 198$$

b. Multiply by the tax rate.

number of $100s × tax rate per $100 = amount of taxes
198 × $2.70 = $534.60

ANSWER: $534.60 taxes

Vocabulary

A tax on buildings and land is called a **real property**, or **real estate** tax. The tax is paid by the owners.

The value that tax officers place on property for tax purposes is the **assessed value** of the property.

The **tax rate** on property may be expressed as: cents per dollar, mills per dollar, dollars and cents per $100, dollars and cents per $1,000, or percent.

1 mill = one-thousandth of a dollar = $.001
10 mills = one cent = $.01

Think About It

1. One town has a property tax rate of 5 mills per $1. Another town has a property tax rate of $4 per $1,000. Describe how you would determine which town has the higher rate.

2. List in order the keys you would press on a calculator (assume it has a percent key) to do the following Practice Exercises.
a. Exercise 13c. **b.** Exercise 14.

Would you choose to use a calculator for exercises like these? Why or why not?

Practice Exercises

1. Write as a decimal part of a dollar.
  a. 7 mills   **b.** 19 mills
  c. 175 mills   **d.** 92 mills

2. Write as cents.
  a. 9 mills   **b.** 35 mills
  c. 146 mills   **d.** 58 mills

3. Write as mills.
  a. $.08   **b.** $.005
  c. $.105   **d.** $.095

4. Write as cents per dollar.
  a. $4 per $100   **b.** 28 mills per $1
  c. $43 per $1,000   **d.** 3.4%

5. Write as mills per dollar.
  a. $5 per $100   **b.** $33 per $1,000
  c. 4.5%   **d.** $.19 per $1

6. Write as dollars and cents per $100.
  a. $.08 per $1   **b.** 5.3%
  c. 36 mills per $1   **d.** $19 per $1,000

7. Write as dollars and cents per $1,000.
  a. $4.50 per $100   **b.** $2.9%
  c. 42 mills per $1   **d.** $.25 per $1

8. Write as a percent.
  a. $4 per $100   **b.** $24.50 per $1,000
  c. $.07 per $1   **d.** 61 mills per $1

Find the amount of taxes on property having the following assessed valuations and tax rates.

9.

	a.	b.	c.	d.
Assessed valuation	$16,000	$9,500	$54,000	$41,700
Tax rate per $100	$2.90	$4.25	$3.80	$5.45

10.

	a.	b.	c.	d.
Assessed valuation	$14,800	$8,600	$45,000	$79,500
Tax rate per dollar	$.04	$.10	$.058	$.07$\frac{1}{2}$

11.

	a.	b.	c.	d.
Assessed valuation	$20,000	$7,500	$43,200	$59,800
Tax rate per dollar	26 mills	41 mills	75 mills	69 mills

12.

	a.	b.	c.	d.
Assessed valuation	$8,000	$32,000	$15,600	$70,500
Tax rate per $1,000	$25	$53	$91	$33

13.

	a.	b.	c.	d.
Assessed valuation	$5,000	$28,300	$66,500	$85,000
Tax rate	3%	5%	4.6%	2.8%

14. Mrs. Moore's house is assessed for 75% of its cost. The tax rate is $2.45 per $100. How much property tax does she pay if the house cost $42,400?

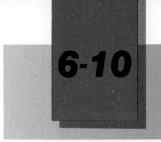

Buying and Renting a Home

Most people either rent or own the places where they live. In order to buy a house, condominium, or shares in a cooperative, people usually borrow some of the money from a bank.

The banks charge interest on the mortgage for the use of their money. This amount of interest depends on the period of the loan, the amount of the loan, and the present interest rate for that bank. The interest rate can vary from bank to bank.

Monthly Payments on a $1,000 Loan						
Period of Loan	Yearly Interest Rate					
	10%	11%	$11\frac{1}{2}\%$	12%	13%	14%
20 years	9.66	10.33	10.67	11.01	11.72	12.44
25 years	9.09	9.81	10.17	10.53	11.28	12.04
30 years	8.78	9.53	9.91	10.29	11.07	11.85
35 years	8.60	9.37	9.77	10.16	10.96	11.76
40 years	8.50	9.29	9.69	10.09	10.90	11.72

Example

The Tungs bought a house for $81,500 with a down payment of $16,300 and a loan for the balance on a 30-year mortgage at a yearly rate of 12%. How much were their monthly payments for the loan?

To find the monthly payment for a loan, follow these steps.

a. Find the amount of the mortgage by subtracting the down payment from the selling price.

Amount of mortgage: $81,500 − $16,300 = $65,200

b. Express the amount of the mortgage in $1,000's by dividing by $1,000.

$65,200 ÷ $1,000 = 65.2 thousands

c. Find the monthly payment per $1,000 at the given rate of interest for the period of the loan:

From the chart: $10.29   monthly payment
                              per $1,000

d. Multiply this amount by the number of $1,000s:

$10.29 × 65.2 = $670.91

ANSWER: The monthly payment is $670.91.

> ### Vocabulary
>
> A **mortgage** is the written claim that the bank has to a house or property until all of the money has been repaid.

1. Would you rather have a longer period of loan with smaller monthly payments OR a shorter period of a loan with larger monthly payments? Explain.

2. If you did have enough money to buy a house without borrowing from a bank, would that be a good idea? Why or why not?

Practice Exercises

1. If no more than 25% of a person's income should be spent for the rent of a house or apartment, find the highest monthly rent someone earning each of the following incomes can afford to pay:
  a. $18,500 per year
  b. $750 per month
  c. $440 per week

2. If a family should not buy a house costing more than $2\frac{1}{2}$ times its annual income, what is the highest price a family can afford to pay if its income is:
  a. $27,250 per year?
  b. $1,250 per month?
  c. $250 per week?

3. What is the monthly payment on each of the following mortgage loans?
  a. A $50,000 loan for a period of 20 years at 14% yearly interest
  b. A $68,500 loan for a period of 35 years at $11\frac{1}{2}$% yearly interest

4. The Mugridges bought a house for $107,000 with a down payment of $21,400 and a loan for the balance on a 25-year mortgage at a yearly rate of 13%. How much were their monthly payments for the loan?

5. Todd's parents bought a condominium for $47,500, paying $7,500 down. They are charged 13% interest on a loan for the balance. The condominium is assessed for $20,600 and the tax rate is $5.20 per $100. The condominium is insured for 80% of its cost at $.28 per $100. The mortgage is to be paid off in 20 years in monthly installments. The down payment of $7,500 was formerly invested at 8.5% interest.
  a. How much interest is due on the mortgage for the first year?
  b. What is the property tax?
  c. What does the insurance cost?
  d. How much interest do Todd's parents lose on money used as down payment?
  e. How much does the ownership of the condominium cost Todd's parents for the first year? What is the monthly cost?
  f. How much is the annual installment due on the mortgage?
  g. How much is the monthly payment on the loan?

6. Compare the costs of renting and buying a house for the first year when the house can be rented for $385 per month or can be purchased for $39,900 with a down payment of $4,900. It is assessed for $22,500 with a tax rate of $3.85 per $100. Insurance costs $325 for a year. The principal used as a down payment could be invested at 9% interest. The interest on the mortgage is 14%. Repairs average $400 per year.

Problem Solving Strategy: Work Backward

Mr. Jones spent $79.40 at the garden store. He bought 50 tree seedlings, a shovel for $19.95, and a bag of fertilizer for $9.95. How much did each seedling cost?

> **Strategy: Work Backward**
> Some problems are stated in such a way that working backward becomes an easier and faster way than starting from the beginning. Use the known values to turn a problem into a simple arithmetic computation.

1. **READ** the problem carefully.

a. Find the **question asked**.
How much did each seedling cost?

b. Find the **given facts.**
Total bill: $79.40   Cost of fertilizer: $9.95
Cost of shovel: $19.95   Number of seedlings: 50

2. **PLAN** how to solve the problem.

a. Choose a strategy: Work backward

b. Think or write out your plan relating the given facts to the question asked.
Step 1. Add to find the total amount Mr. Jones spent for the shovel and the fertilizer.
Step 2. Subtract to find the amount Mr. Jones spent for the seedlings.
Step 3. Divide to find the cost of each seedling.

3. **SOLVE** the problem.

		Estimate	Solve
Step 1.	Cost of shovel:	$20.00	$19.95
	Cost of fertilizer:	+ 10.00	+ 9.95
		$30.00	$29.90
Step 2.	Total cost:	$80.00	$79.40
	Cost of shovel and fertilizer:	− 30.00	− 29.90
	Cost of seedlings:	$50.00	$49.50

Step 3.
$$\frac{\$1.00 \text{ per seedling}}{50 \text{ seedlings})\$50.00} \text{ cost of seedlings}$$ $$\frac{\$0.99 \text{ per seedling}}{50)\$49.50} \text{ cost of seedlings}$$

You might use a calculator.   Key: 79.40 − 19.95 − 9.95 ÷ 50 = 0.99

4. **CHECK** .

a. **Check the accuracy** of your arithmetic.

$0.99
× 50
─────
$49.50
9.95
+ 19.95
─────
$79.40 ✔

b. **Compare the answer to the estimate.** The answer $.99 compares reasonably with the estimate $1.00.

ANSWER: Each seedling cost $.99.

Practice Problems

1. If you add 8 years to 3 times my age, you get 50 years. What is my age?

2. If you subtract 4 from 5 times a number, you get 36. Find the number.

3. If you divide the sum of 6 and a number by 2, the quotient is 5. What is the number?

4. If you multiply a number minus 8 by 7, the product is 28. Find the number.

5. If 3 is added to 4 times the number, the answer is 51. What is the number?

6. You need to go to the airport to catch a flight that leaves at 5:30 P.M. It takes 1 hour 14 minutes to get to the airport and you want to be there 45 minutes early. What time should you leave home?

7. Ira needed to earn $4,000 to pay his bills. He sells cars and earns a 10% commission on his sales. If the average cost of a car is $10,000, how many cars does he need to sell?

8. Kelley wants to buy a new bicycle for $145. She sold her old bicycle for $35 and has $40 saved. If Kelley earns $2.50 an hour for babysitting, how many hours will she have to babysit before she has enough money to buy the new bicycle?

9. Paula planned to send 12 pounds of books and magazines to a friend in Africa. She had to allow 1.5 pounds for the packaging. Of the remaining weight, she wanted $\frac{2}{3}$ to be books. How many pounds of magazines can she send?

10. Andy paid $102 for a new tire for his car. The price included a $20 labor charge and $7 for sales tax. The store owner marked each tire up 150%. What was the original cost of a tire to the store owner?

Career Application

INSURANCE AGENT

Spending too much for insurance could ruin your budget, but not having enough insurance coverage could be much worse. *Insurance agents* sell insurance. An insurance agent can analyze a customer's needs and plan a package with the appropriate insurance on the customer's life, health, car, and home or apartment.

Most insurance agencies will hire high-school graduates who have sales ability. An agent must pass a state examination on insurance laws.

1. A unit of life insurance has a face value of $1,000. A customer purchases 50 units. The annual premium per unit is $14.30. What is the total annual premium?

2. A customer's employer pays 85% of the cost of medical insurance. If the total monthly premium is $64.80, how much does the employer pay?

3. An insurance agent explains that a customer can pay quarterly premiums (four times a year) instead of one large annual premium. The quarterly premium is 25.5% of the annual premium. If the quarterly premium is $191.25, what is the annual premium?

4. To find the annual premium for automobile insurance, an agent multiplies the base premium by the customer's driver-rating factor. One customer's driver-rating factor is 1.85. She chooses insurance with a base premium of $115.40. What is her annual premium?

5. A homeowner decides to insure his house for $83,700. The replacement value of the house is $93,000. What percent of the replacement value is covered by insurance?

6. Because an apartment house has smoke detectors, the annual premium for renter's insurance is 8% less than the base premium, $110. How much does the tenant save?

Chapter 6 Review

Vocabulary

assessed value p. 314
beneficiary p. 313
carrying charge p. 310
deferred payment p. 310
discount p. 301
down payment p. 310
endowment insurance p. 313
finance charge p. 310
installment buying p. 310
life insurance p. 313

limited-payment life insurance p. 313
list price p. 301
marked price p. 301
mortgage p. 316
net price p. 301
ordinary life insurance p. 313
premium p. 313
policy p. 313
rate of discount p. 301
rate of reduction p. 301

real estate tax p. 314
real property tax p. 314
sale price p. 301
sales tax p. 305
tax rate p. 314
term insurance p. 313
term of the policy p. 313
unit price p. 298
value of the policy p. 313

Page
296   **1.** What is the cost of $6\frac{1}{2}$ pounds of apples at 69¢ a pound?

Page
298   **2.** What is the unit cost (cost per quart) of a 48-ounce container of juice costing $1.59?

298   **3.** Which is a better buy: 1 pound of cereal for $2.59 or 12 ounces for $1.99?

301   **4.** A refrigerator marked $885 was sold at a reduction of 25%. Find the sale price.

301   **5.** Find the regular price of a record that sold for $6.40 at a 20% reduction sale.

304   **6.** Amanda purchased some food items for $23.56 and some nonfood items for $5.23. How much change should she get from $30?

305   **7.** If the sales tax rate is 5%, what would be the cost of a motorcycle that sells for $795?

306   **8.** The meter reading at the beginning of the month was 3962 and at the end of the month was 4401. At the average cost of $.085 per kilowatt hour, find the cost of the electricity.

309   **9.** Nicole took her car in for service. She was charged $54.80 for parts and $109.30 for labor. If there was a sales tax of 6% on parts only, how much was the bill for repairs?

310   **10.** Sam's mother purchased a used car for a cash price of $3,100 or $400 down and $132.34 per month for 24 months. How much does she save by paying cash?

312   **11.** How much will a $45,000 life insurance policy cost per year if the premium is $8.95 per $1,000?

314   **12.** If the tax rate is $22.50 per $1,000, how much must you pay in taxes on a house assessed at $84,200?

316   **13.** Paula decided to buy a house. She did not want to pay more than $2\frac{1}{2}$ times her annual salary for a house. If she earns $590 per week, what is the most she could pay for a house?

313   **14.** The amount of money an insured person pays for a policy is called the _____?_____.

Chapter 6 Test

Page
296 **1.** Find the cost of 1.74 pounds of steak selling at $3.99 per pound.

298 **2.** Find the unit cost (cost per pound) of a $1\frac{1}{2}$ pound jar of coffee costing $3.81.

298 **3.** When you buy 4 gallons of milk at $1.88 each instead of 8 half-gallons costing $.99 each, how much do you save?

301 **4.** A tennis racquet that regularly sells for $54 is reduced 40%. What is the sale price?

301 **5.** What is the rate of discount when a printer with a list price of $880 is sold for $748?

304 **6.** Maury buys a notebook for $1.59 and two pens that sell for $.89 each. How much change should he receive from $10?

305 **7.** If the rate is 6.5%, what is the sales tax on a $395 coat?

309 **8.** What is the cost of 800 gallons of fuel oil at $1.235 per gallon?

309 **9.** Laura's car had a full tank of gas when the odometer read 8854.9. When the odometer read 8977.3, it took 8 gallons of gas to fill the tank. How many miles per gallon did the car travel?

310 **10.** John buys a bicycle for $10 down and 10 payments of $15.20. He could have bought the bicycle for $140 if he had paid cash. If he had paid cash, how much would he have saved?

312 **11.** At $12.66 per $1,000, how much would the premium for a $75,000 life insurance policy be?

314 **12.** If the tax rate is $39.40 per $1,000, what is the property tax on a house costing $85,000 and assessed at $54,000?

316 **13.** Mary bought a house for $153,000. She paid 15% down. What was the amount of the mortgage?

Chapter 7 Managing Money

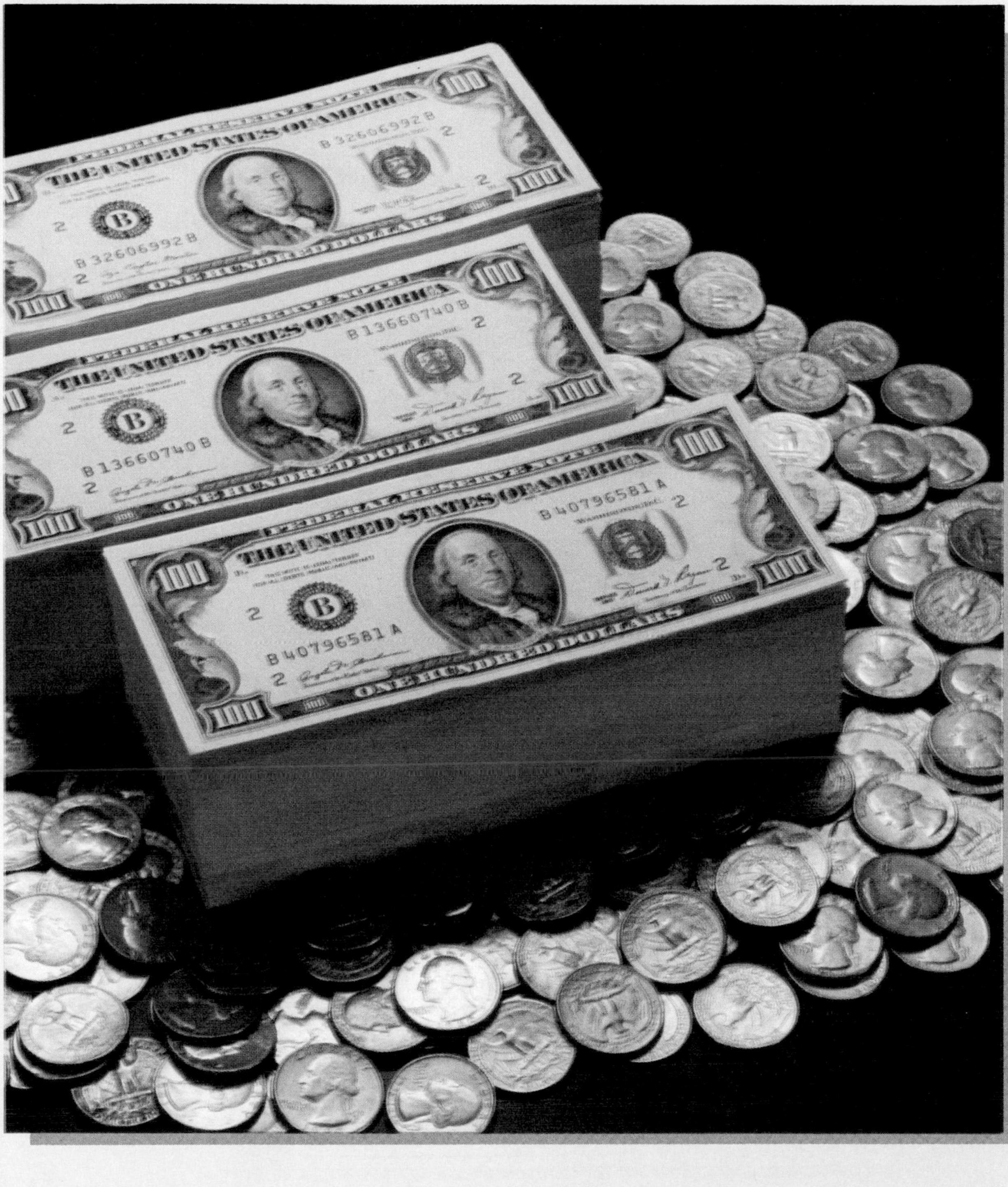

When you keep your money in a bank you use different forms to withdraw or deposit money, to write checks, and to record your transactions.

Vocabulary

A **deposit slip** is used to deposit money in a savings account or checking account.

A **check** is used to transfer funds from a checking account to persons or business establishments. When a check is deposited, it must be signed (endorsed) on its back by the payee.

A **check register** is used in place of check stubs. When a check is written or a deposit or withdrawal is made, the transaction is recorded in the check register.

A **withdrawal slip** is used to withdraw cash from a savings account.

The **check stub** is used if attached to the check to keep a record of the checks written and the deposits and withdrawals made.

When you **reconcile** your checkbook, you bring the bank statement and your records into agreement.

The following forms are used in banking.

Deposit Slip

Example

On December 15, 1988, Nancy Mugridge made a deposit to her account, numbered 999-123456, consisting of two $10 bills, three $5 bills, a check #85-24/260 in the amount of $128.63, and a check #23-847/105 in the amount of $79.82.

CENTRAL BANK — December 15, 19 88		DOLLARS	CENTS
Pondfield, CA 98765 DATE	CASH ▶	35	00
ACCOUNT NUMBER 9 9 9 · 1 2 3 4 5 6	CHECKS BY BANK NO		
	¹85-24/260	128	63
	²23-847/105	79	82
DEPOSIT TO THE ACCOUNT OF:	3		
Name *Nancy Mugridge*	4		
	5		
	6		
Teller Validation	TOTAL ▶	243	45

DEPOSIT

Withdrawal Slip

Example

On January 18, 1988, Stephen Scott wrote a withdrawal slip for the sum of $125 to be withdrawn from his savings account, numbered 129-345786.

CENTRAL BANK
Pondfield, CA 98765

JANUARY 18 19 88   DATE | 1 2 9 • 3 4 5 7 8 6 | ACCOUNT NUMBER

PASSBOOK WITHDRAWAL (Non-Negotiable)

PAY TO STEPHEN SCOTT   OR BEARER, THE SUM OF $ 125 | 00

ONE HUNDRED, TWENTY-FIVE AND   00/100 DOLLARS

If this is a Joint Account, I represent that both depositors are now living:

Stephen Scott

DEPOSITOR SIGNS HERE—DO NOT PRINT   # _____ ☐ M.O. $ _____
PASSBOOK MUST ACCOMPANY   # _____ ☐ Check $ _____
THIS ORDER   # _____ ☐ Redeposit $ _____
INTEREST CREDIT   ☐ Cash $ _____
$ _____   DATE OF NOTICE _____ 19 _____   TELLER'S STAMP

PASSBOOK WITHDRAWAL (Non-Negotiable)

Check

Example

Lisa Todd wrote a check on June 18, 1988, payable to the Town Department Store in the amount of $87.35 for clothing. In order to prevent anyone from altering a check, write the numbers as close to the dollar sign as possible and draw a line from the amount written as words to the word *dollars*.

Amount written in words with cents expressed as a fraction of a dollar

Payee   Date

LISA TODD
194 MAIN STREET
PONDFIELD, CA 98765   June 18 19 88   25-60/001   101

Check number

PAY TO THE ORDER OF Town Department Store $ 87.35

Amount written as a number

Eighty-seven and 35/100   DOLLARS

CENTRAL BANK
Pondfield, CA 98765

FOR Clothing   Lisa Todd   Signature

⑈000⑈00605⑈ 0101 808 7989611⑈

Purpose

Check Stub

Example

Assume that Lisa Todd has a bank balance of $826.59 and wrote the check for $87.35. She would note the transaction on the corresponding check stub, as shown at the right.

101	BAL. BRO'T FOR'D	826 | 59
June 18 19 88		
TO Town Department Store		
FOR Clothing		
	TOTAL	826 | 59
	THIS CHECK	87 | 35
	BALANCE	739 | 24
102		
___ 19 ___		
TO ___		
FOR ___		
	TOTAL	
	THIS CHECK	
	BALANCE	

Check Register

Example

Assume that Lisa Todd has the same bank balance as above ($826.59). She would record the transaction in her check register as shown below. She would also record other transactions in the same manner.

CHECK NUMBER	DATE	ISSUED TO OR DESCRIPTION OF DEPOSIT	AMOUNT OF PAYMENT	OTHER DEDUCT.	AMOUNT OF DEPOSIT	BALANCE FORWARD 826 | 59
101	6-18-88	TO Town Department Store	87 | 35			− 87 | 35
		FOR Clothing				739 | 24
102	6-19-88	TO ABC Garage	52 | 45			− 52 | 45
		FOR repairs				686 | 79
	6-20-88	TO Deposit			50 | 00	+ 50 | 00
		FOR				736 | 79
103	6-20-88	TO John Knowlton	350 | 00			− 350 | 00
		FOR rent				386 | 79
104	6-20-88	TO Anderson Electric	49 | 95			− 49 | 95
		FOR lamp				336 | 84
105	6-23-88	TO Humbolt Gas & Electric	65 | 72			− 65 | 72
		FOR utilities				271 | 12
	6-23-88	TO Deposit			373 | 62	+ 373 | 62
		FOR				644 | 74
		TO				

RECORD OF ACTIVITY   Please be sure to deduct charges that affect your account.

Bank Statement

Banks send, usually at the end of each month, a statement showing the deposits, withdrawals, cleared checks, bank charges, previous monthly balance, and closing monthly balance of the account. Thus, check stubs or check register records may be **reconciled** with the bank records. Compare the statement below with the check register on page 326.

CENTRAL BANK
Pondfield, CA 98765

LISA TODD
194 MAIN STREET
PONDFIELD, CA 98765

ACCOUNT NUMBER	ACCOUNT	BEGINNING BALANCE	DEPOSITS, PAYMENTS AND CREDITS	CHECKS, WITHDRAWALS AND OTHER DEBITS	INTEREST PAID	MONTHLY CHARGES	ENDING BALANCE
808 7989611	Checking	826.59	50.00	489.80		.30	386.49
DATE LAST STATEMENT							
5-26-88							
DATE THIS STATEMENT							
6-26-88							

CHECKS AND OTHER CHARGES			DEPOSITS AND CREDITS		BALANCE
Date	**Number**	**Amount**	**Date**	**Amount**	
06-22	101	87.35	06-20	50.00	876.59
06-23	102	52.45			779.24
60-25	103	350.00			716.79
					386.79
	Service Charge	.30			386.49

HOW TO RECONCILE YOUR CHECKING BALANCE

1. Mark off in your checkbook each of the checks paid and each deposit recorded by the bank.

2. List the numbers and amounts of checks still outstanding in the space provided to the right.

3. Compare the checking portion of your statement with your checkbook. Any bank charges such as service charges or penalties for overdrawing the account should be subtracted from your checkbook balance.

CHECKS OUTSTANDING	
NUMBER	AMOUNT
104	49 95
105	65 72
TOTAL CHECKS OUTSTANDING	115 67

4. Enter the balance shown on this statement here. — 386 49

5. If you have made deposits not shown on this statement, enter total of these items here. — 373 62

6. Total lines 4 and 5 and enter here. — 760 11

7. Enter Total Checks Outstanding here. — 115 67

8. Subtract Line 7 from Line 6 and enter here. This adjusted balance should agree with your checkbook balance. Be sure you have subtracted any service charges from your checkbook balance. — 644 44

1. What is the purpose of a checking account?

2. State the steps you must follow to deposit a check.

3. When you write a deposit in a check register, do you add or subtract the amount?

Practice Exercises

Use deposit slips, checks, either check stubs or a check register, and a bank statement to record the following. If these forms are not available, copy them as illustrated.

1. On August 1, you have a bank balance of $1,593.07. Write this as the balance brought forward on the first check stub or balance forward in the check register.

2. On August 2, you deposit four $20 bills, six $5 bills, twelve $1 bills and a check numbered 83-91/218 drawn on the Valley Bank for $284.75. Fill in a deposit slip. Write the total deposit on the check stub or the check register.

3. On August 5, you send a check numbered 296 to Andrew Jaffe for $508.40 for purchases of appliances. Fill in the first check stub or use check register. Calculate the account balance. Write the check.

4. On August 12, you send a check numbered 297 to the Stephen Furniture Co. in the amount of $241.95 for the furniture purchased. Fill in the second check stub or use the check register. Calculate the account balance. Write the check.

5. On August 16, you deposit sixteen $10 bills, nine $5 bills, twenty-three $1 bills, a check numbered 85-106/295 drawn on the First City Bank for $305.50 and a check numbered 6-39/318 drawn on the Second National Bank for $193.84. Fill in the deposit slip. Write the total deposit on the third check stub or use the check register.

6. On August 22, you send a check numbered 298 to the Scott Department Store for $485.69 for merchandise purchased. Fill in the third check stub or use the check register. Calculate the account balance. Write the check.

7. On August 29, you send a check numbered 299 to Peter Todd, Inc. for $94.95 for a desk purchased. You also deposit six $20 bills, seven $5 bills, and a check numbered 72-105/302 drawn on the County Bank for $127.43. Fill in the fourth check stub or use the check register. Calculate the account balance. Write the check.

8. On August 30, you receive your bank statement. It shows a beginning balance of $1,593.07. It lists checks numbered 296, 297, and 298 and a service charge of $.30. It also lists deposits totaling $1,134.09 and shows an ending balance of $1,490.82. Fill in the bank statement to show the transactions recorded by the bank and then reconcile the statement with your checkbook.

Simple Interest

When you deposit money in a savings account, you are permitting the bank to use the money and the bank pays you interest for this privilege. You also receive interest when you invest in a money market certificate. You pay interest on a loan and that interest is included in the monthly payment.

Examples

1. Find the interest on $900 for 3 years 6 months at 12%.

interest (i) = principal (p) × rate (r) × time (t)
$i = prt$

To find the interest, multiply the principal by the rate of interest per year and the time expressed in years.

$p = \$900$

$r = 12\% = \dfrac{12}{100}$

$t = 3$ yr 6 mo or $3\frac{1}{2}$ yr

$i = ?$

$i = prt$

$i = \$900 \times \dfrac{12}{100} \times 3\frac{1}{2}$

$i = \dfrac{\overset{9}{\cancel{900}}}{1} \times \dfrac{\overset{6}{\cancel{12}}}{\underset{1}{\cancel{100}}} \times \dfrac{7}{\underset{1}{\cancel{2}}} = 378$

$i = \$378$

ANSWER: $378 interest

You might use a calculator.

Key: 900 ×️ 12 %️ ×️ 3.5 =️ <u>378</u>

---

2. Find the total amount due on $625 borrowed for 7 years at 5%.

a. Find the interest using the formula $i = prt$.

```
    $625    principal
×    0.05    rate
   $31.25   interest for 1 year
×        7   years
  $218.75   interest for 7 years
```

b. Add the principal and the interest.

```
    $625.00   principal
+    218.75   interest
    $843.75   amount
```

ANSWER: Total amount due is $843.75.

3. What is the annual (yearly) rate of interest if the annual interest on a principal of $250 is $20?

a. Divide the annual interest by the principal.

b. Write the decimal as a percent.

$\dfrac{\$20}{\$250} = \$20.00 \div \$250 = 0.08 = 8\%$

ANSWER: The annual rate of interest is 8%.

You might use a calculator.

Key: 625 M+ ×️ 5 %️ ×️ 7 + M$\frac{R}{C}$ =️ <u>843.75</u>

1. A person plans to put a sum of money in a bank for a year. Bank A pays $1\frac{1}{2}$% simple interest quarterly (four times a year) and bank B pays 6% simple interest annually (once a year). Which bank is preferable? Why?

2. If you have decided how much interest you want to earn in a year and you know the best simple interest rate you can receive, how can you determine how much you would have to invest to earn this return?

Practice Exercises

1. Find the interest on each amount.

   a. $200 for 1 y at 15%

   b. $2,800 for 1 y at 10.25%

   c. $50 for 1 y 6 mo at 15%

   d. $3,000 for 4 y 1 mo at $9\frac{1}{2}$%

2. Find the semiannual (6 mo) interest on each amount.

   a. $400 at 7%

   b. $11,000 at $6\frac{1}{4}$%

   c. $40,000 at 9.383%

   d. $24,000 at $7\frac{1}{2}$%

3. Find the quarterly (3 mo) interest on each amount.

   a. $900 at 20%

   b. $8,000 at $10\frac{3}{4}$%

   c. $1,600 at 8.64%

   d. $550 at $5\frac{1}{2}$%

4. Find the monthly interest on each amount.

   a. $1,200 at 8%

   b. $900 at $12\frac{1}{2}$%

   c. $12,000 at 9.791%

   d. $600 at $6\frac{3}{4}$%

5. Using 1 year = 360 days, 1 month = 30 days, find the interest on each amount.

   a. $800 for 30 days at 6%

   b. $675 for 90 days at 4%

   c. $2,400 for 60 days at 15%

   d. $120 for 15 days at 10%

6. Find the exact interest (1 year = 365 days) on each amount.

   a. $425 for 73 days at 7%

   b. $2,920 for 15 days at 12%

   c. $2,400 for 30 days at 9%

   d. $1,653 for 25 days at $6\frac{3}{4}$%

7. Find the interest and total amount on each loan.

   a. $300 for 1 y at 15%

   b. $8,000 for 6 y at $10\frac{1}{2}$%

   c. $2,700 for 9 y 2 mo at 8.85%

   d. $12,000 for 1 y 11 mo at 6%

8. Find the annual rate of interest.

   a. p = $150, t = 1 y, i = $18

   b. p = $2,000, t = 3 y, i = $600

   c. p = $1,600, t = 4 y, i = $288

   d. p = $7,500, t = 5 y, i = $3,375

Financing a Car

People who finance the purchase of automobiles usually pay in monthly installments.

Example

Find the total payment if you borrow $2,000 for 36 months.

To find the total payment given the amount borrowed and the number of months to pay:

a. Use the table to find the monthly payment.

   The monthly payment is $63.88.

b. Multiply the monthly payment by the number of months to find the total payment.

   $63.88 × 36 = $2,299.68

ANSWER: $2,299.68 total payment

Amount to be Borrowed	Amount of Each Monthly Payment		
	24 mo	30 mo	36 mo
$1,000	$45.83	$37.50	$31.94
1,200	55.00	45.00	38.33
1,500	68.75	56.25	47.91
1,800	82.50	67.50	57.50
2,000	91.66	75.00	63.88
2,500	114.58	93.75	79.86
3,000	137.50	112.50	95.83
3,500	160.41	131.25	111.80

Think About It

Would you prefer to borrow an amount of money for 30 months or 36 months? Why?

Vocabulary

The **finance charge** is the difference between the cash price and the total amount paid.

Practice Exercises

Use the table above in each of the following exercises.

1. What is the monthly payment to finance a car if you borrow:

   a. $1,200 for 30 months?
   c. $1,500 for 24 months?

   b. $2,000 for 36 months?
   d. $3,000 for 30 months?

2. Find the total payment if you borrow:

   a. $1,800 for 24 months.
   b. $2,500 for 30 months.
   c. $3,500 for 36 months
   d. $1,500 for 30 months

3. Find the total finance charge if you borrow:

   a. $3,500 for 24 months.
   b. $1,500 for 36 months.
   c. $2,000 for 30 months.
   d. $3,000 for 36 months.

4. How much more is the total finance charge on the 36-month loan of $2,500 than on the 24-month loan?

7-4 Amortizing a Loan

When money is borrowed to buy property, each monthly payment lowers the principal of the mortgage. Interest is paid on the *new reduced balance* of the principal owed each month.

Example

If a person borrows $15,000 at 13% annual interest, what is the monthly payment when the loan is to be amortized over 35 years?

The table indicates the monthly payments necessary to amortize a loan made at 13% annual interest.

Term	$5,000	$10,000	$15,000	$20,000	$25,000	$30,000	$35,000	$40,000	$45,000	$50,000
15 y	63.27	126.53	189.79	253.05	316.32	379.58	442.84	506.10	569.36	632.63
20 y	58.58	117.16	175.74	234.32	292.90	351.48	410.06	468.64	527.21	585.79
25 y	45.12	112.79	169.18	225.57	281.96	338.36	394.75	451.14	507.53	563.92
30 y	44.25	110.62	165.93	221.24	276.55	331.86	387.17	442.48	497.79	533.10
35 y	43.81	109.52	164.28	219.04	273.80	328.56	383.32	438.08	492.84	547.60
40 y	43.59	108.96	163.43	217.91	272.38	326.86	381.33	435.81	490.29	544.76

ANSWER: $164.28 monthly payment

Think About It

If the amount borrowed is increased, but the money is still borrowed for the same number of years at 13% annual interest rate, what happens to the monthly payments?

Practice Exercises

Use the table of monthly payments to solve the following problems.

1. If a person borrowed $40,000 at 13% annual interest, what is the monthly payment when the loan is to be amortized over each period?

   a. 25 years   b. 40 years

   c. 15 years   d. 30 years

2. If a loan at 13% annual interest is to be amortized in 25 years, what is the monthly payment for the amount of each loan?

   a. $45,000   b. $35,000

   c. $25,000   d. $40,000

3. Find the total interest that was paid on each of the following loans at 13% annual interest when the loans were amortized in the specified terms.

	Amount	Term
a.	$25,000	15 years
b.	$30,000	40 years
c.	$45,000	20 years
d.	$20,000	25 years

4. a. Find the total payment over the full term required on a loan of $45,000 at 13% annual interest when it is amortized in 30 years.
   b. When it is amortized in 40 years.
   c. For which term is the interest less?
   d. How much less?

7-5   Compound Interest

People put money into savings accounts to earn interest. Sometimes, rather than earning simple interest, the money earns interest compounded over a certain period.

The table on page 335 shows how much $1 will amount to based on the given rate and the given number of years.

Examples

1. Find the amount and the interest earned on $800 deposited for 3 years at 8%, when the interest is compounded annually.

To find how much a given principal will amount to at a given rate for a certain period of time if compounded annually, follow these steps.

a. Read down the column for the given rate until you reach the given period of time (years).

Rate, 8%; Periods, 3; 1.25971

b. Multiply the amount in Step **a** by the given principal.

$$\begin{array}{r} 1.25971 \\ \times \quad \$800 \\ \hline \end{array}$$
$1,007.768 → $1,007.77 amount

c. To find the compound interest only, subtract the principal from the amount.

$$\begin{array}{r} \$1,007.77 \text{ amount} \\ - \quad \$800.00 \text{ principal} \\ \hline \$200.77 \text{ interest} \end{array}$$

ANSWER: The amount is $1,007.77; the interest is $200.77.

---

2. Find the amount and the interest earned on $650 deposited for 4 years at 8% interest when the interest is compounded semiannually.

a. Read down the column for one-half the given rate until you reach two times as many periods as the given number of years.

Rate, 4%; Periods, 8; 1.36856

b. Multiply the amount found in Step **a** by the given principal.

$$\begin{array}{r} 1.36856 \\ \times \quad \$650 \\ \hline \end{array}$$
$889.564 → $889.56 amount

c. To find the compound interest, subtract the principal from the amount.

$889.56 amount
− $650.00 principal
$239.56 interest

ANSWER: The amount is $889.56; the interest earned is $239.56.

---

3. What is the total amount and the interest earned when $1,200 is deposited for 2 years at 8% interest if the interest is compounded quarterly?

a. Read down the column for one-fourth the given rate until you reach four times as many periods as the given number of years.

Rate, 2%; Periods, 8; 1.17165

b. Multiply the amount in Step **a** by the given principal.

1.17165
× $1,200
$1,405.98

c. Determine the compound interest.

$1,405.98 amount
− $1,200.00 principal
$205.98 interest

ANSWER: The amount is $1,405.98; the interest is $205.98.

4. How much will Jeffrey have after one year if he deposits $300 at the interest rate of 12% compounded monthly?

a. Read down the column for one-twelfth the given rate until you reach twelve times as many periods as the given number of years.

Rate, 1%; Periods, 12; 1.12682

b. Multiply the amount found in Step **a** by the given principal.

1.12682
× $300
$338.046 = $338.05 amount

ANSWER: Jeffrey will have $338.05.

Think About It

1. If the same amount of money is deposited and kept for the same time, would it be better if the interest rate were compounded quarterly or monthly? Why?

2. Name a situation when simple interest would be preferred to compound interest.

Compound Interest Table
Showing How Much $1 Will Amount to at Various Rates

Periods	1%	2%	3%	4%	5%	6%
1	1.01000	1.02000	1.03000	1.04000	1.05000	1.06000
2	1.02010	1.04040	1.06090	1.08160	1.10250	1.12360
3	1.03030	1.06120	1.09272	1.12486	1.15762	1.19101
4	1.04060	1.08243	1.12550	1.16985	1.21550	1.26247
5	1.05101	1.10408	1.15927	1.21665	1.27628	1.33822
6	1.06152	1.12616	1.19405	1.26531	1.34009	1.41851
7	1.07213	1.14868	1.22987	1.31593	1.40709	1.50362
8	1.08285	1.17165	1.26676	1.36856	1.47745	1.59384
9	1.09368	1.19509	1.30477	1.42330	1.55132	1.68947
10	1.10462	1.21899	1.34391	1.48024	1.62889	1.79084
11	1.11566	1.24337	1.38423	1.53945	1.71033	1.89829
12	1.12682	1.26823	1.42575	1.60102	1.79585	2.01219
13	1.13809	1.29360	1.46852	1.66506	1.88564	2.13292
14	1.14947	1.31947	1.51258	1.73167	1.97992	2.26089
15	1.16096	1.34586	1.55796	1.80093	2.07892	2.39654
16	1.17257	1.37277	1.60469	1.87297	2.18286	2.54034
17	1.18429	1.40023	1.65283	1.94789	2.29201	2.69276
18	1.19614	1.42823	1.70242	2.02580	2.40661	2.85432
19	1.20810	1.45680	1.75349	2.10683	2.52694	3.02558
20	1.22018	1.48593	1.80610	2.19111	2.65328	3.20712

Periods	8%	10%	12%	14%	16%	20%
1	1.08000	1.10000	1.12000	1.14000	1.16000	1.20000
2	1.16640	1.21000	1.25440	1.29960	1.34560	1.44000
3	1.25971	1.33100	1.40492	1.48154	1.56089	1.72800
4	1.36048	1.46410	1.57351	1.68896	1.81063	2.07360
5	1.46932	1.61051	1.76234	1.92541	2.10034	2.48832
6	1.58687	1.77156	1.97382	2.19497	2.43638	2.98598
7	1.71382	1.94871	2.21067	2.50226	2.82621	3.58318
8	1.85092	2.14358	2.47596	2.85258	3.27841	4.29981
9	1.99900	2.35794	2.77307	3.25194	3.80295	5.15977
10	2.15892	2.59374	3.10584	3.70721	4.41143	6.19173
11	2.33163	2.85311	3.47854	4.22622	5.11725	7.43008
12	2.51816	3.13842	3.89597	4.81789	5.93601	8.91609
13	2.71961	3.45226	4.36348	5.49240	6.88578	10.69930
14	2.93718	3.79749	4.88710	6.26133	7.98750	12.83910
15	3.17216	4.17724	5.47355	7.13792	9.26550	15.40700
16	3.42593	4.59496	6.13038	8.13723	10.74790	18.48840
17	3.70000	5.05446	6.86602	9.27644	12.46760	22.18600
18	3.99600	5.55990	7.68995	10.57510	14.46240	26.62320
19	4.31568	6.11589	8.61274	12.05560	16.77640	31.94790
20	4.66094	6.72748	9.64627	13.74340	19.46060	38.33750

Practice Exercises

1. Find the amount and the interest earned when the interest is compounded annually.
   a. $1,400 for 3 y at 16%
   b. $10,000 for 15 y at 5%
   c. $8,500 for 7 y at 8%
   d. $2,750 for 20 y at 10%

2. Find the amount and the interest earned when the interest is compounded semiannually.
   a. $2,300 for 7 y at 16%
   b. $15,000 for 10 y at 4%
   c. $7,450 for 8 y at 8%
   d. $20,000 for 5 y at 10%

3. Find the amount and the interest earned when the interest is compounded quarterly.
   a. $350 for 2 y at 12%
   b. $1,500 for 1 y at 8%
   c. $6,300 for 3 y at 8%
   d. $50,000 for 4 y at 16%

A bank advertises an 8% interest rate compounded quarterly on all deposits.

4. What would a deposit of $500 amount to at the end of each period?
   a. 9 months
   b. 15 months
   c. 30 months
   d. 5 years

5. What is the interest earned on a $1,000 deposit at the end of each period?
   a. 1 year
   b. 21 months
   c. 45 months
   d. 5 years

At the interest rate of 12% compounded monthly (12 times a year),

6. What will each of the following deposits amount to at the end of one year?
   a. $700
   b. $1,250
   c. $4,800
   d. $10,000

7. How much interest is earned at the end of one year on each of the following deposits?
   a. $250
   b. $980
   c. $3,600
   d. $5,000

8. What will each of the following deposits amount to?
   a. A deposit of $75 at the end of 8 months
   b. A deposit of $400 at the end of 3 months
   c. A deposit of $1,800 at the end of 1 month
   d. A deposit of $2,000 at the end of 6 months

9. How much interest is earned on each?
   a. A deposit of $60 at the end of 9 months
   b. A deposit of $840 at the end of 6 months
   c. A deposit of $3,000 at the end of 2 months
   d. A deposit of $4,200 at the end of 4 months

10. Find the interest earned on a $5,000 deposit at the end of one year at each of the following rates.
    a. 12% compounded quarterly
    b. 12% compounded annually
    c. 12% compounded semiannually
    d. 12% compounded monthly

Use your answers to Exercise 10 to answer the following questions.

11. Which rate brings the greatest amount of interest?

12. Which rate brings the smallest amount of interest?

7-6 Stocks, Bonds, and Mutual Funds

One way to earn money is to invest your money in stocks, bonds, mutual funds, or treasury bills and notes.

Stocks

Vocabulary

A person buying a **share of stock** becomes part owner of the company and sometimes receives **dividends,** or a share of the profits earned by the company.

Stock quotations are listed as a whole number or mixed numbers that represent dollars and cents.

> $36\frac{1}{2}$ means $36.50

The **P.E. (price to earnings) ratio** compares the closing price of a share of stock to the company's annual earnings per share.

Fees, or commissions, are paid to stockbrokers for the service of buying or selling stocks.

Examples

1. Find the annual earnings, if the P.E. ratio is 7 and the closing price is $24\frac{1}{2}$.

$24\frac{1}{2} \div 7 = \frac{49}{2} \times \frac{1}{7} = \frac{7}{2} = 3\frac{1}{2}$

To find the annual earnings, divide the closing price by the P.E. ratio.

ANSWER: $3.50 annual earnings for each share of stock.

---

2. Annette Hays purchased 75 shares of OpTech stock at $22\frac{7}{8}$ and sold it 2 years later at $31\frac{5}{8}$. She paid a 2% sales commission when she purchased the stock and a 2.75% commission when she sold. What was the profit or loss from the sale?

$75 \times 22\frac{7}{8} = 1715.625 = \$1,715.63$

Find the cost of 75 shares.

$\$1,715.63 \times 2\% = 34.3126 = \$34.31$

Find the commission.

$\$1,715.63 + \$34.31 = \$1,749.94$

Add the cost and the commission to find the total cost of purchase.

$75 \times 31\frac{5}{8} = 2371.875 = \$2,371.88$

Find the selling price of 75 shares.

$\$2,371.88 \times 2.75\% = 65.2267 = \$65.23$

Find the commission.

$\$2,371.88 - \$65.23 = \$2,306.65$

Subtract the commission from the selling price.

$\$2,306.65 - \$1,749.94 = \$556.71$

Subtract the purchase costs from the amount earned from the sale.

ANSWER: $556.71 profit.

Bonds

Vocabulary

A **bond** is a written promise of a private corporation or of a local, state, or the national government to pay a given rate of interest at stated times, and to repay the **face value** (original value) of the bond at a specified time (date of maturity).

A person buying a bond is lending money to the business or government and receives interest on the face value of the bond.

The United States Government borrows money by selling **savings bonds** and **Treasury bills, notes,** and **bonds.** The selling prices and interest rates vary.

The **rate of income,** or **yield,** of a bond is the ratio of the annual interest to the market price of the bond, expressed as a percent.

The value at which a bond sells at any given time is called **market value.**

The quoted price of a bond represents a percent.

$101\frac{1}{4}$ means $101\frac{1}{4}\%$

Example

Find the cost, annual interest, and the yield of a $1,000 NamCo bond if the market price is $80\frac{1}{4}$ and the rate of interest is $10\frac{5}{8}$.

$1,000 \times 80\frac{1}{4} = 802.50$

$1,000 \times 10\frac{5}{8} = 106.25$

$10\frac{5}{8} \div 80\frac{1}{4} = \frac{85}{8} \times \frac{4}{361} = \frac{85}{722} = 0.13239$

or $106.25 \div 802.50 = 0.13239$

To find the cost of a bond, multiply the face value by the quoted price.
To find the annual interest paid, multiply the face value by the interest rate.
To find the annual yield, divide the interest rate by the bond cost.

ANSWER: $802.50; $106.25; 13.24%

Mutual Funds

Vocabulary

A **mutual fund** is an investment company. It obtains money for investment by selling shares of stock of the company. Mutual fund quotations are expressed as a **Net Asset Value (NAV)** and an **offer price**. When we *buy* shares, we pay the offer price of each share. When we *sell* shares, we receive the Net Asset Value (NAV) of each share.

NL means **no load**, or no commission charged for buying shares. When NL is indicated, the offer price is the same as the NAV.

Example

Find the cost of buying 75 shares of a mutual fund if the offer price is $14.21. If you were to sell the shares at a NAV of $13.18, how much money would you receive?

$75 \times 14.21 = \$1,065.75$

$75 \times 13.18 = \$988.50$

To find the cost of buying shares, multiply the number of shares by the offer price.
To find out how much money you would receive if you sold the shares, multiply the number of shares by the NAV.

ANSWER: $1,065.75 to buy; $988.50 received from sale

1. What fractions are used for stocks? Why do you think not all fractions are used?

2. What decimal does $18\frac{3}{8}\%$ represent?

3. What is a difference between stocks and bonds?

Practice Exercises

1. Find the cost of 75 shares of stock in PHI Co. if the listed price is $26\frac{5}{8}$.

2. Find the total cost of 60 shares of Irb, Inc. if the listed price is $13\frac{7}{8}$ and the broker charges a 1.5% commission.

3. Find the cost of a bond with a face value of $1,000 if the highest quotation is $88\frac{1}{2}$.

4. Mrs. Berk owns three $10,000 Treasury notes that pay an annual rate of $10\frac{3}{4}\%$ interest. How much interest does she receive semiannually?

5. The interest rate on a 6-month Treasury bill was announced at 9.5%. If you purchase a $5,000 T bill at auction for $4,985 and the interest is also deducted, what is your final cost?

6. Andy purchased 85 shares of the Leo Fund at the offer price of $13.40. What was his total cost?

Using Mathematics

MONEY RATES—CERTIFICATES OF DEPOSIT

Prime Rate: The prime rate is the basic interest rate commercial banks charge their most valued customers. In recent years this rate has ranged from 6.5% to 10.75%.

Passbook Interest Rates: Passbook interest on savings is paid by savings institutions and commercial banks at the fixed annual rate of 5.5%.

Money Market Rates: Banks offer money-market accounts with a minimum deposit of $2,500, paying rates of interest that change weekly.

Certificates of Deposit: Banks and savings and loan associations issue savings certificates called certificates of deposit (CD), which pay varying rates of interest depending on the term of the certificate.

The interest rates on money-market accounts and on certificates of deposit change weekly. These rates can change as little as a basis point, which is one-hundredth of one percent (0.01%). It takes 100 basis points to make 1%.

Exercises

1. A money-market account pays 5.5% annually on an average monthly balance of $15,000. How much interest is earned for the month?

In the table below, the annual yield shows the actual interest rate being earned due to the daily compounding of the savings in a certificate of deposit.

Interest Rate		Annual Yield
6.25%	6-month certificate	6.43%
6.50%	1-year certificate	6.78%
6.75%	18 month certificate	7.08%

2. Using the annual yield rates above, find the total interest earned by a 1-year certificate of deposit for $10,000.

3. Find the total interest earned by an 18-month certificate of deposit for $25,000.

The chart shows typical certificate of deposit rates for a minimum investment of $100,000. These rates change daily.

Use the annual rates shown to find the interest earned on a $100,000 CD purchased for:

4. 30 days   **5.** 90 days

6. 75 days   **7.** 160 days

30– 59 days	4.85%
60– 89 days	5.10%
90–119 days	5.35%
120–179 days	5.75%
180–360 days	6.25%

8. If a 5.45% interest rate is increased by 25 basis points, what is the new rate?

Making Decisions

MAKING A BUDGET

You receive $10 a week for doing chores at home and usually can earn $20 per week doing odd jobs. At the end of the first week, you had spent all your money and didn't have any left for a movie with your friends. So, you decide to make a budget to help you plan your spending. You make a table to list the things on which you want or need to spend your money.

Weekly Expenses

Lunches	$5 – 10
Bus	$2
Clothes	?
Movies at least	$5
Savings at least	$2
Tapes	?
Other	?

You know the bus will be $2 per week and lunch will be $5–$10. You have an idea how much you want to budget for movies and savings, but you don't know how much is reasonable for clothes and records. Use the following questions to help you decide on a budget. Explain your answers.

1. What three categories are probably most important to budget for?
2. If you budget $5 for food, you will need to buy the school lunch. If you budget $10, you can eat at the snack bar.
   a. What are some factors you should consider in deciding how much to budget for food?
   b. If you budget $10 for food, what effect will it have on the rest of your budget?

Weekly Expense

Lunch   Savings
Bus   Tapes
Clothes   Other
Movies

Typical Teen Budget

Clothing 30%

Recreation 25%

Personal 20%

Other 15%

Savings 10%

3. Make a budget sheet like this. Enter the amounts you would like to budget for lunches and bus.

4. To help you decide how much to budget for other categories, you find a budget showing the typical budget amounts for teens.
   a. If you follow this budget, how much would you budget for clothing?
   b. Do you feel that 30% of your weekly income is reasonable for clothing? If your parents buy most of the clothes you need, but you have to buy "luxuries," how would that affect the amount you budget for clothing?
   c. Decide on an amount to budget for clothing and enter it on your budget sheet.

5. a. If you follow the budget, how much would you budget for savings?
   b. Do you feel that is enough? If you are saving up for a bike, how would that affect the amount you budget for savings?
   c. Decide on an amount to budget for savings and enter it on your budget sheet.

6. a. How much money do you have left for the budget categories of Movies, Tapes, and Others?
   b. Choose amounts to budget for each of these categories and enter them on your budget sheet.

7. Make a circle graph to illustrate your budget.

8. Compare your budget to others' budgets. Be prepared to defend your choices!

9. How would you decide if your budget is a good one? What can you do if you feel it is not good?

In Your Community

Survey your friends and find out how they spend their money. Make up a budget for yourself. What are some of the things you need to consider in budgeting your money? Why is having a budget worthwhile?

Career Application

BANK TELLER

Customers go to banks to make deposits and withdrawals, cash checks, buy savings bonds and travelers' checks, and apply for loans. *Bank tellers* handle the money and paperwork for all these transactions. In most banks they use computers to do calculations and keep records.

Banks hiring tellers look for high-school graduates who are careful, accurate, and courteous. Tellers learn their jobs in training programs run by banks, and by watching experienced tellers.

1. A teller receives a deposit of $23.50 in cash plus checks for $189.27 and $46.75. What is the total deposit?

2. A customer has $463.12 in her account. She withdraws $275. How much money is left in her account?

3. A teller explains that the cost of a savings bond is 50% of its face value. If the cost is $37.50, what is the face value?

4. A bank will grant a loan for 80% of the cost of a new car. The car costs $11,782.65. What is the amount of the loan?

5. A money order is a safe way to send money through the mail. A customer wants a money order for $119.95 to buy a camera. The bank charges a service fee of $1.55. What is the total cost of the money order?

6. A customer wants to buy $500 worth of travelers' checks. The bank charges a 1% service fee. What is the total cost of the travelers' checks?

Chapter 7 Review

Vocabulary

amortizing p. 332	finance charge p. 331	principal p. 329
bond p. 338	interest p. 329	rate of income p. 338
certificate of deposit (CD) p. 341	market value p. 338	rate of interest p. 329
check p. 324	money market rate p. 341	reconcile p. 324
check register p. 324	mutual fund p. 339	share of stock p. 337
check stub p. 324	net asset value (NAV) p. 339	simple interest p. 329
compound interest p. 333	no load (NL) p. 339	stock quotations p. 337
deposit slip p. 324	offer price p. 339	total amount p. 329
dividend p. 337	periods p. 333	withdrawal slip p. 324
face value p. 338	price to earnings ratio p. 337	yield p. 338

Page

324 **1.** On January 15, 1988, Maria Salez made a deposit to her account, number 005723-4, consisting of five $10 bills, five $5 bills, a check #110-22 in the amount of $423.88 and a check #101-34 in the amount of $38.92. Write the deposit slip.

324 **3.** On December 10, 1988, Sun Chow withdrew the cash sum of $301.44 from her account, #2091009. Write the withdrawal slip.

329 **5.** Janet Lloyd-Jones receives $320 semiannual interest on an investment of $4,000. What is the rate of interest?

331 **7.** Using the table on page 331, find the total finance charge if you borrow $3,000 for 30 months to finance your car.

333 **9.** Using the table on page 335, find the amount and interest earned on $4,800 deposited for 3 years at 12%, compounded quarterly.

338 **11.** Find the current yield of a bond whose annual interest is $125 and whose closing price is 108, if the face value of the bond is $1,000.

333 **13.** Interest paid on both the principal and the interest earned previously is called _____?_____ interest.

Page

324 **2.** On July 5, 1988, Bill Conway had a bank balance of $566.87 in his account #4-003-009. On the same day he wrote a check to Alpha Mart in the amount of $102.98 for groceries. Write the check and record the transaction on the check stub.

329 **4.** What is the amount due on $4,300 borrowed at 15% and repaid at the end of 2 years 6 months?

337 **6.** Rich owns two $10,000 Treasury notes bearing 11.25% interest. How much interest does he receive every 6 months?

332 **8.** Using the table on page 332, find the total interest that is paid on a loan of $45,000 at 13% interest, to be amortized over 35 years.

337 **10.** Excluding broker's fee, how much profit do you make if you buy 200 shares of stock at $16\frac{3}{4}$ and sell them at $19\frac{1}{8}$?

338 **12.** What is the cost of buying six bonds, excluding broker's fee, if the face value is $1,000 and they are quoted at $105\frac{1}{4}$?

Chapter 7 Test

```
                                                          145
MICHELE LEWIS
                                        _____19___

Pay to the
order of _____(2)_____  $_____

_____(3)_____ DOLLARS
SOUTHERN STATE BANK
for_____                          (1)
                                         _____
⑃021000089⑃ 030 953 28‖° 145
```

For questions 1–3 choose from the following.

a. $35.00   **b.** Michele Lewis (signature)   **c.** thirty-five and 00/100

d. Southern Electric Co.   **e.** electric bill Feb. 87   **f.** 10 Mar 1987

Page
324 **1.** What should go on line (1)?

3. What should go on line (3)?

329 **5.** What is the amount due on $6,600 borrowed at $15\frac{1}{2}$% and repaid at the end of 4 years?

332 **7.** If the monthly payment to amortize a loan of $45,000 over 25 years at 13% annual interest is $507.53, what is the total interest paid?

337 **9.** Tammy owns three $10,000 Treasury notes that pay annual interest at the rate of $11\frac{1}{4}$%. How much does she receive semiannually?

Page **2.** What should go on line (2)?

329 **4.** How much is the interest on $8,200 for $1\frac{1}{2}$ years at 14%?

331 **6.** If you borrow $3,500 for 36 months to buy a car and pay $111.80 per month, what is the total finance charge?

333 **8.** A sum of $1 compounded annually at 12% interest amounts to $3.10584 in 10 years, and compounded semiannually amounts to $3.20712. If $10,000 is invested at each rate, what is the difference in the interest earned?

337 **10.** If you purchase 150 shares of stock at $18\frac{1}{2}$ and sell at $21\frac{1}{8}$, what is your profit, excluding commissions?

Achievement Test

Page
275 **1.** Esther receives $185 per week plus 2% commission on sales. One week her sales totaled $8,509. What were her total earnings that week?

Page
283 **2.** When the withholding tax is $25.40, Social Security tax is 7.15% of earnings, and the city tax is $1\frac{1}{2}$% of earnings, what is the take-home pay on weekly wages of $275?

285 **3.** If your Federal taxes for the year are $2,059 and your withholding tax payments are $1,775, what is the amount of tax you still owe?

298 **4.** Which is the better buy of the same kind of dinner rolls, 12 for $1 or 20 for $1.69?

301 **5.** A camera which regularly sells for $62.50 is reduced 18%. What is its sale price?

305 **6.** If the rate is 6%, what is the sales tax on a radio selling for $49.50?

306 **7.** At the average cost of 6.5¢ per kilowatt hour, find the weekly cost of operating 4 hours each day a TV that uses 200 watts of power.

310 **8.** When the monthly finance charge rate is $1\frac{1}{4}$%, what is the annual rate?

312 **9.** At the annual rate of $22.71 per $1,000 for life insurance, what is the total annual premium for $75,000 insurance?

318 **10.** Lydia picked a number, added 3 to it, multiplied the sum by 5, then subtracted 8 and doubled the result. If the answer was 44, what number did Lydia start with?

324 **11.** Complete a check register showing the following transactions. Beginning balance, $463.29; check number 169 to City Gas & Electric for $55.69; check number 170 to Acme Department Store for $39.50 for a blanket; a payroll deposit of $225.81; and a cash withdrawal of $50.

329 **12.** Find the interest on $2,600 for 3 years at $7\frac{1}{2}$% annual rate.

329 **13.** What is the monthly payment on an add-on-interest loan, if $6,000 is borrowed at 10% for 12 months?

331 **14.** If you borrow $3,000 for 24 months to buy a car and pay $137.50 per month, what is the finance charge?

333 **15.** By compounding interest semiannually at the annual rate of 12%, each $1 is worth $2.01219 at the end of 6 years. Find the total interest earned on $1,000 compounded semiannually at a 12% annual rate at the end of 6 years.

337 **16.** If you purchase 100 shares of TEL at 24 and sell at $30\frac{1}{2}$, what is your profit, excluding commissions?

337 **17.** Ms. Costello owns a $10,000 Treasury Note that pays an annual interest rate of $9\frac{1}{4}$%. How much does she receive semiannually?

Unit 3

Measurement, Graphs, Statistics, and Probability

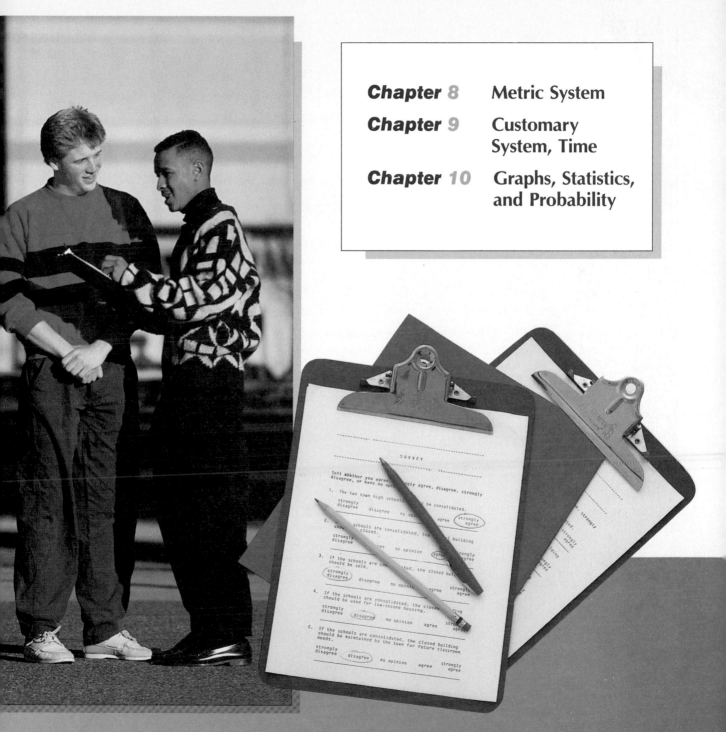

Inventory Test

394 **20.** Make a bar graph showing the elevations of the following mountain peaks in the United States.

Mt. Rainier, 4,395 meters   Mt. Washington, 1,918 meters   Guadalupe Peak, 2,669 meters
Mt. McKinley, 6,198 meters   Mt. Hood, 3,427 meters

396 **21.** Make a broken-line graph showing the average number of days per month that Seattle and New Orleans have precipitation.

	Jan.	Feb.	Mar.	Apr.	May	June	July	Aug.	Sept.	Oct.	Nov.	Dec.
Seat.	20	16	18	14	10	10	5	6	9	14	18	21
N.O.	10	10	9	7	7	10	15	13	10	6	7	10

398 **22.** Make a circle graph showing the average annual outlays of the Federal government in recent years.

Education, 5%; Health, 8%; Interest, 7%; Defense, 33%; Income Security, 28%; Other, 19%

400 **23.** Make a frequency distribution table for the following list of scores.

72, 66, 75, 67, 73, 74, 69, 72, 68, 71, 70, 65, 66, 71, 68, 74,
65, 71, 66, 74, 70, 67, 75, 68, 74, 69, 68, 70, 72, 69, 74

402 For the following scores—39, 42, 39, 48, 41, 36, 49—find each of the following.

24. Mean   **25.** Median

26. Mode   **27.** Range

409 **28.** What is the probability of drawing at random on the first draw a yellow ball from a box containing 12 white balls and 18 yellow balls?

412 **29.** What is the probability of drawing 2 yellow balls in succession?

Introduction to the Metric System

The United States monetary system is a decimal system, that is, based on ten, in which the dollar is the basic unit. In the metric system, also a decimal system, the meter (m) is the basic unit of length, the gram (g) is the basic unit of weight or mass, and the liter (L) is the basic unit of capacity (dry and liquid measures).

Other metric units of length, weight, or capacity are found by combining the following prefixes with the basic units of measure.

Prefix	Symbol	Value
mega-	M	million (1,000,000)
kilo-	k	thousand (1,000)
hecto-	h	hundred (100)
deka-	da	ten (10)
deci-	d	one-tenth (0.1 or 1/10)
centi-	c	one-hundredth (0.01 or 1/100)
milli-	m	one-thousandth (0.001 or 1/1,000)
micro-	μ	one-millionth (0.000001 or 1/1,000,000)
nano-	n	one-billionth (0.000000001 or 1/1,000,000,000)
pico-	p	one-trillionth (0.000000000001 or 1/1,000,000,000,000)

In the chart below, each unit of United States money, length, weight, or capacity is 10 times as great as the unit to the right. Therefore, 10 of any unit is equivalent to 1 unit of the next larger size.

It should be noted that just as 3 dollars 8 dimes 4 cents may be written as a single quantity, 3.84 *dollars* or $3.84, so 3 meters 8 decimeters 4 centimeters may also be written as a single quantity, 3.84 *meters*.

Abbreviations or symbols are written without periods after the last letter. The same symbol is used for both one or more quantites. Thus, "cm" is the symbol for centimeter *or* centimeters.

	1,000	100	10	1	0.1	0.01	0.001
Decimal Place Value	thousands	hundreds	tens	ones	tenths	hundredths	thousandths
United States Money	$1,000 bill	$100 bill	$10 bill	dollar	dime	cent	mill
Metric Length	kilometer km	hectometer hm	dekameter dam	meter m	decimeter dm	centimeter cm	millimeter mm
Metric Weight	kilogram kg	hectogram hg	dekagram dag	gram g	decigram dg	centigram cg	milligram mg
Metric Capacity	kiloliter kL	hectoliter hL	dekaliter daL	liter L	deciliter dL	centiliter cL	millimeter mL

The number of units of the smaller denomination that is equivalent to one unit of the larger denomination is called the *conversion factor*.

Each conversion factor is some power of 10, so that the short method of computation may be used.

Metric Measures of Length

The metric system is an international system of measurement in which units are related by powers of ten.

Vocabulary

The **millimeter, centimeter, meter,** and **kilometer** are the units commonly used to measure length in the metric system.

The **conversion factor** between two units of measure is the number you multiply or divide by to change from one unit to another.

Examine the following section of a centimeter ruler.

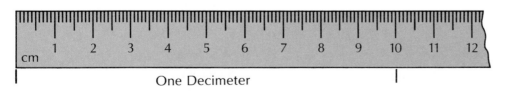

One Decimeter

Each subdivision represents 1 millimeter (mm).
Ten millimeters equal 1 centimeter (cm).
Ten centimeters equal 1 decimeter (dm).

Centimeter rulers and meter sticks are commonly used to measure length. The kilometer is used to measure longer distances, such as those between cities.

10 millimeters (mm) = 1 centimeter (cm) 10 centimeters (cm) = 1 decimeter (dm) 10 decimeters (dm) = 1 meter (m)	10 meters (m) = 1 dekameter (dam) 10 dekameters (dam) = 1 hectometer (hm) 10 hectometers (hm) = 1 kilometer (km)

1 meter (m) = 100 centimeters (cm) = 1,000 millimeters (mm)
1 kilometer (km) = 1,000 meters (m)

Examples

1. The distance between two points on a map is 23 centimeters. Fred is working with a computer program that requires all map distances to be entered in millimeters. How many millimeters are in 23 centimeters?

To change a given metric unit of length to another, follow these steps.

a. Find the conversion factor.

The conversion factor between centimeters and millimeters is 10.

1 cm = 10 mm

b. **Multiply** by the conversion factor when changing to a **smaller** unit.

Millimeters are smaller than centimeters.

23 cm = 23 × 10 = 230 mm

ANSWER: 230 millimeters

2. Change 9,000 meters to kilometers.
When changing to a **larger** unit, **divide** by the conversion factor.

Kilometers are larger than meters.

1 km = 1,000 m

The conversion factor is 1,000.

9,000 m = 9,000 ÷ 1,000 = 9 km

ANSWER: 9 kilometers

3. Express 9 km 3 hm 5 m 2 dm in meters.

9 km = 9 × 1,000 = 9,000   m
3 hm = 3 × 100 =   300   m
5 m   5   m
2 dm = 2 ÷ 10   =   0.2 m
                  9,305.2 m

Shortcut: Use place value.

9 km 3 hm 0 dam 5 m   2 dm
↓   ↓   ↓   ↓   ↓
9   3   0   5 · 2

ANSWER: 9,305.2 meters

1. What metric unit is best for measuring the length of your classroom?

2. What metric unit is best for measuring the length of your pencil?

3. List in order the keys you would press on a calculator to do each of the following.
   a. Practice Exercise 12d.

   b. Practice Exercise 18b.

4. Complete: Changing dekameters to meters is equivalent to changing decimeters to __?__.

Practice Exercises

1. What metric unit of length does each of these symbols represent?

   a. km   **b.** cm   **c.** m   **d.** mm   **e.** dm

2. What measurement is indicated by each labeled point?

   a. B   **b.** C   **c.** E   **d.** H   **e.** A

3. Give the distance between each pair of points.

   a. point B to point C   **b.** point D to point E   **c.** point A to point G

4. How many centimeters are there between the points?

   a. point B to point A   **b.** point C to point G   **c.** point E to point F

5. How many millimeters are there between the points?

   a. point E to point H   **b.** point D to point F   **c.** point F to point G

6. How many millimeters are in the length from 0 to each of the following?

   a. 8 cm   **b.** 6 cm 2 mm   **c.** 3 cm 4 mm   **d.** 1 dm 1 cm 6 mm

7. In each of the following, locate the marking for the first measurement on the ruler and add to this the second measurement. Simplify each sum as indicated.

   a. 8 mm + 6 mm = _____ mm = _____ cm _____ mm
   b. 2 cm 5 mm + 4 cm 2 mm = _____ cm _____ mm
   c. 5 cm 9 mm + 3 cm 7 mm = _____ cm _____ mm = _____ cm _____ mm
   d. 1 dm 2 mm + 3 cm 5 mm = _____ dm _____ cm _____ mm

8. Find the 108 mm mark on the metric ruler. Subtract 4.3 cm. What measurement does the mark you reach indicate?

9. Complete.

    a. __?__ mm = 1 cm   **b.** __?__ cm = 1 dm   **c.** __?__ dm = 1 m

    d. __?__ cm = 1 mm   **e.** __?__ dm = 1 cm   **f.** __?__ m = 1 dm

10. Complete.

    a. __?__ mm = 1 m   **b.** __?__ cm = 1 m   **c.** __?__ dm = 1 m

    d. 1 km = __?__ m = __?__ dm = __?__ cm = __?__ mm

11. Express each of the following in meters.

    a. 7 km 5 hm 8 dam 3 m   **b.** 8 m 2 dm 6 cm 5 mm

    c. 6 m 4 cm 9 mm   **d.** 3 km 5 m 1 cm 6 mm

12. Change each of the following to millimeters.

    a. 34 cm   **b.** 28 m

    c. 4 km   **d.** 8.5 cm

13. Change each of the following to centimeters.

    a. 4.7 dm   **b.** 6.4 m

    c. 851 mm   **d.** 9.3 m

14. Change each of the following to decimeters.

    a. 439 m   **b.** 760 cm

    c. 6,875 mm   **d.** 2.6 km

15. Change each of the following to meters.

    a. 78 dam   **b.** 4,206 cm

    c. 569 mm   **d.** 12,600 mm

16. Change each of the following to kilometers.

    a. 81 hm   **b.** 5,328 m   **c.** 794,000 cm   **d.** 9,270,000 mm

17. Give the missing equivalent measurements.

	km	hm	dam	m	dm	cm	mm
a.				700			
b.	3						
c.						5,000	
d.							6,800,000

18. Solve.

    a. If each curtain panel requires 125 cm of fabric and you need 8 panels, how many meters of fabric should you buy?

    b. Sound travels at a speed of 1,450 m/s in water. How many kilometers does it travel in water in 15 s?

    c. How much longer is a metal strip that measures 1.2 m than one that measures 894 mm?

    d. A car is traveling at 50 km/h. How many meters does it travel in 30 seconds?

19. Arrange the following measurements in order of size (longest first).
6,800 m   526,000 cm   6.7 km   692,400 mm

20. Arrange the following measurements in order of size (shortest first).
8.45 km   84.5 mm   845 m   8,450,000 cm

8-2 Metric Measures of Mass or Weight

When you compare two packages to determine which is heavier, you are concerned about the mass of the packages.

In everyday use, the word "weight" almost always means "mass." Therefore, for our purposes we will use kilogram and gram as units of weight.

10 milligrams (mg) = 1 centigram (cg)	10 grams (g) = 1 dekagram (dag)
10 centigrams (cg) = 1 decigram (dg)	10 dekagrams (dag) = 1 hectogram (hg)
10 decigrams (dg) = 1 gram (g)	10 hectograms (hg) = 1 kilogram (kg)
1,000 kilograms (kg) = 1 metric ton (t)	

Mass, or weight, can be measured using a balance scale. The scale at the right shows that 10 mg is the same as 1 cg. Use a balance scale to find the weight in grams of some handy objects (your pencil, watch, notebook, or keys).

1 cg   10 mg

Examples

1. A pharmacist has 2 grams of medicine in a bottle. A doctor prescribed 600 milligrams. Does the pharmacist have enough to fill the order? Change 2 grams to milligrams and compare.

a. Find the conversion factor.

1 g = 1,000 mg, so the conversion factor is 1,000.

b. Multiply by the conversion factor to change to a **smaller** unit.
Divide by the conversion factor to change to a **larger** unit.

2 g = 2 × 1,000 = 2,000 mg

ANSWER: The pharmacist has enough medicine to fill a prescription for 600 mg.

2. Change 58 milligrams to centigrams.

1 cg = 10 mg
The conversion factor is 10.
58 mg = 58 ÷ 10 = 5.8 cg

ANSWER: 5.8 centigrams

3. Change 1 cg 4 mg to centigrams.

1 cg 4 mg = 1 cg + 0.4 cg = 1.4 cg

ANSWER: 1.4 centigrams

1. What does it mean to be weightless in space?

2. If you multiply dekagrams by 100, what unit is your answer?

Practice Exercises

1. What metric unit of weight does each symbol represent?

   a. mg   **b.** kg   **c.** g   **d.** cg   **e.** hg   **f.** dag

2. Complete each of the following.

   a. _____ mg = 1 cg   **b.** _____ cg = 1 dg   **c.** _____ g = 1 dag   **d.** _____ g = 1 dg

3. Find each weight.

   a. paper clip in grams   **b.** shoe in grams
   c. dictionary in kilograms **d.** rock in kilograms

4. Change each of the following to milligrams.

   a. 9 cg   **b.** 16 g
   c. 2.36 kg   **d.** 0.841 g

5. Change each of the following to centigrams.

   a. 39 mg   **b.** 62 g
   c. 1.4 kg   **d.** 28.9 g

6. Change each of the following to grams.

   a. 530 dg   **b.** 9 kg
   c. 25.8 cg   **d.** 750 mg

7. Change each of the following to kilograms.

   a. 7,864 g   **b.** 85 hg
   c. 961,000 cg   **d.** 7,400,000 mg

8. Change each of the following to metric tons.

   a. 3,000 kg   **b.** 570 kg
   c. 6,300,000 g   **d.** 82,600 kg

9. Give the missing equivalent weights.

	kg	g	cg	mg
a.		5,000		
b.	7.3			
c.			2,900	
d.				800,000

10. Complete each of the following.

    a. 5 cg 6 mg = _____ mg   **b.** 8 kg 130 g = _____ g
    c. 2 g 84 mg = _____ mg   **d.** 3 cg 9 mg = _____ mg

11. Solve.

    a. A shipment of 595 dag of topsoil has a delivery charge of $.25 per kilogram. Find the delivery charge.

    b. A color technician needs 0.4 g of cobalt blue. She has 555 mg of that color. How many milligrams will be left over?

    c. A rabbit is to receive 20 g of food a day. How many kilograms of food are needed for 40 days?

    d. One rock specimen weighs 1.2 kg. Another weighs 750 g. How much heavier is the first rock?

8-3 Metric Measures of Capacity

To measure capacity, measuring cups and beakers are used. Estimate, and then determine the number of liters of water your sink would hold.

10 milliliters (mL) = 1 centiliter (cL)	10 liters (L) = 1 dekaliter (daL)
10 centiliters (cL) = 1 deciliter (dL)	10 dekaliters (daL) = 1 hectoliter (hL)
10 deciliters (dL) = 1 liter (L)	10 hectoliters (hL) = 1 kiloliter (kL)

The units that are used to measure capacity are also used to measure volume.

The difference between volume and capacity is that volume is the amount of space within a three-dimensional figure and capacity is the amount of substance that the figure will hold.

Vocabulary

The liter and milliliter are units commonly used to measure capacity in the metric system.

Examples

1. The class determined that they needed 7,250 milliliters of milk for lunch. How many liters of milk is this?

a. Find the conversion factor.

1 L = 1,000 mL, so 1,000 is the conversion factor.

b. **Divide** by the conversion factor when changing to a **larger** unit.

7,250 mL = 7,250 ÷ 1,000 = 7.250 L

ANSWER: 7.25 L of milk are needed.

---

2. Change 4 liters to centiliters.

1 L = 100 cL   100 is the conversion factor.

Multiply by the conversion factor to change to a **smaller** unit.

4 L = 4 × 100 = 400 cL

ANSWER: 400 cL

---

3. Change 1 liter 8 centiliters to liters.

1 L 8 cL = 1 L + 0.08 L
       = 1.08 L

Shortcut: Use place value.

1 L	0 dL	8 cL
↓	↓	↓
1	0	8

ANSWER: 1.08 L

1. What unit of capacity is appropriate for measuring the amount of gasoline that will fill a car's gas tank?

2. List in order the keys you would press on a calculator to do each of the following.

   a. Exercise 7c below.

   b. Exercise 10c below.

3. When you divide kiloliters by 10,000, what is the unit of your answer?

Practice Exercises

1. What metric unit of capacity does each symbol represent?

   a. kL   b. L   c. cL   d. cL   e. hL   f. dL

2. Complete each of the following.

   a. _____ mL = 1 cL   b. _____ dL = 1 L   c. _____ cL = 1 mL   d. _____ dL = 1 cL

3. Write the most likely measure for each.

   a. bathtub of water   100 mL   100 L   1 kL
   b. pitcher of juice   750 mL   7 L   0.7 kL
   c. eyedropper of medicine   3 mL   0.3 L   0.3 kL

4. Change each of the following to milliliters.

   a. 7 cL   b. 4.6 L   c. 375 L   d. 80.4 cL

5. Change each of the following to centiliters.

   a. 85 L   b. 18 mL   c. 6.7 L   d. 44.2 mL

6. Change each of the following to deciliters.

   a. 31 L   b. 68 cL   c. 2.25 L   d. 520 mL

7. Change each of the following to liters.

   a. 9 dL   b. 60 hL   c. 21.6 kL   d. 8,405 mL

8. Change each of the following to kiloliters.

   a. 6,582 L   b. 875 L   c. 79.5 hL   d. 200 daL

9. Complete each of the following.

   a. 7 cL 4 mL = _____ mL   b. 1 L 6 cL = _____ cL
   c. 8 kL 600 L = _____ L   d. 6 L 1 cL 9 mL = _____ mL

10. Solve.

    a. Frank mixed 15 centiliters of a weedkiller with 485 centiliters of water. How many liters of the mixture did he have?

    b. A chemist mixed 4.7 liters of distilled water with 8 milliliters of acid. How many milliliters of the mixture does she have?

c. A dairy produces 7,483 kiloliters of milk a day. How many two-liter containers can it fill?

d. A laboratory produces 20,000 kiloliters of hydrogen peroxide in a week. How many deciliter bottles can be filled?

11. Which capacity is less?

a. 40 cL or 65 mL   **b.** 6.8 L or 2,000 cL   **c.** 54.8 mL or 1 L

d. 2.6 cL or 3 mL   **e.** 65 dL or 65 daL   **f.** 8.1 L or 9,700 mL

Computer Know-How

The computer program below converts any number of *kilometers* to the *equivalent* number of: hectometers, dekameters, or meters.

```
1Ø PRINT "HOW MANY KILOMETERS";
2Ø INPUT K
3Ø PRINT K; "KILOMETERS IS EQUAL TO:"
4Ø PRINT K*1Ø; "HECTOMETERS. . .OR"
5Ø PRINT K*1ØØ; "DEKAMETERS. . .OR"
6Ø PRINT K*1ØØØ; "METERS"
```

a. ENTER the program and then RUN it to make sure it works.

b. Modify the program so that it converts kilograms to hectograms, dekagrams, or grams.

c. Modify the program so that it converts kiloliters to hectoliters, dekaliters, or liters.

d. Write a program that converts meters to decimeters, centimeters, or millimeters.

e. Modify the program so that it converts grams to decigrams, centigrams, or milligrams.

f. Modify the program so that it converts liters to deciliters, centiliters, or milliliters.

Refresh Your Skills

1.   93,514 [1–8]
     6,648
    87,529
      896
  +   9,388

2.   109,625 [1–9]
  −   87,592

3.   1,658 [1–11]
  ×   973

4. 5,280)237,600 [1–13]

5. Add. [2–5]
0.008 + 0.8 + 0.08

6. Subtract. [2–6]
0.97 − 0.3

7. Multiply. [2–7]
0.09 × 0.09

8. Divide. [2–8]
$4 ÷ $.05

9. Add. [3–8]
$1\frac{7}{12} + 2\frac{1}{6} + 4\frac{2}{3}$

10. Subtract. [3–9]
$1\frac{3}{4} - \frac{5}{6}$

11. Multiply. [3–11]
$\frac{2}{3} \times 1\frac{4}{5}$

12. Divide. [3–12]
$5\frac{1}{4} \div \frac{3}{10}$

13. $1.25 is what percent of $10? [4–7]

14. Find $13\frac{1}{4}$% of $5,000. [4–6]

15. $4.90 is 10% of what amount? [4–9]

Using Mathematics

SIGNIFICANT DIGITS

Digits in an approximate number are significant when they indicate the precision (closeness to the true measurement), which is determined by the value of the place of the last significant digit on the right.

SIGNIFICANT DIGITS
  I All nonzero digits.
 II All zeros located between significant digits.
III All zeros at the end of a decimal.
IV All digits of the first factor when a number is expressed in scientific notation.
 V Underscored or specified zeros of a whole number ending in zeros.

DIGITS THAT ARE NOT SIGNIFICANT
  I Zeros at the end of a whole number (unless specified to be significant).
 II Zeros following the decimal point in a number between 0 and 1.

Examples

684 has three significant digits: 6, 8, and 4.

5,006 has four significant digits: 5, 0, 0, and 6.

86.190 has five significant digits: 8, 6, 1, 9, and 0.

7.32×10^9 has three significant digits: 7, 3, and 2.

51,000 has two significant digits: 5 and 1.

51,0_0_0 has four significant digits: 5, 1, 0, and 0.

0.0070 has two significant digits: 7 and 0.

2.00960 has six significant digits: 2, 0, 0, 9, 6, and 0.

0.003 has one significant digit: 3.

In 64.75, the significant digit 5 indicates precision to the nearest hundredth.

In 83,0_0_0, the underlined zero indicates precision to the nearest ten.

Exercises

In each of the following, determine the number of significant digits and name them.

1. 28	**2.** 600	**3.** 4,271	**4.** 37,000	**5.** 9
6. 0.08	**7.** 0.0095	**8.** 0.013	**9.** 0.0130	**10.** 9,050,000
11. 23.05	**12.** 0.0001000	**13.** 87,0_0_0	**14.** 350,000	**15.** 6.3×10^5
16. 8.49×10^7	**17.** 2,500	**18.** 0.0007	**19.** 6,810	**20.** 51.0
21. 3.1416	**22.** 7,060	**23.** 95,280	**24.** 16,003	**25.** 40,0_0_0
26. 0.000000015	**27.** 9.00000027	**28.** 8,259,078	**29.** 4.0060	**30.** 203,000

8-4 Metric Measures of Area

To measure area, we use square units. In square measure, 100 of any metric unit is equivalent to 1 of the next higher unit. The exponent ^{2} is used to replace the word *square*.

1 cm

1 cm

Area: 1 cm^2

One way to find area is to place a transparent grid of square centimeters on the surface of an object such as your textbook. Then you can count to find the area in square centimeters.

Vocabulary

The **square centimeter**, **square meter**, and **square kilometer** are the units commonly used to measure area in the metric system.

Centare may be used in place of square meter and **hectare** in place of square hectometer.

100 square millimeters (mm^2) = 1 square centimeter (cm^2)
100 square centimeters (cm^2) = 1 square decimeter (dm^2)
100 square decimeters (dm^2) = 1 square meter (m^2)
100 square meters (m^2) = 1 square dekameter (dam^2)
100 square dekameters (dam^2) = 1 square hectometer (hm^2)
100 square hectometers (hm^2) = 1 square kilometer (km^2)
100 centares = 1 are (a)
100 ares = 1 hectare (ha)
100 hectares = 1 square kilometer

Examples

1. The Bruins' farm measures 9 km^2. They are ordering bags of fertilizer that cover 10,000 m^2. How many bags do they need for the entire farm?

Change 9 km^2 to square meters.

1 km^2 = 1,000,000 m^2
9 km^2 = 9 × 1,000,000 = 9,000,000 m^2

Since each bag covers 10,000 m^2, divide to find the number of bags needed.

9,000,000 m^2 ÷ 10,000 m^2 per bag = 900 bags

ANSWER: 900 bags of fertilizer are needed

2. Change 670 square millimeters to square centimeters.

1 cm^2 = 100 mm^2
670 mm^2 = 670 ÷ 100 = 6.70 cm^2

ANSWER: 6.7 square centimeters

3. Change 900 hectares to square kilometers.

1 km^2 = 100 ha
900 ha = 900 ÷ 100 = 9 km^2

ANSWER: 9 square kilometers

1. Name three things that it would be useful to measure in square millimeters.

2. What is the conversion factor between hectares and square meters?

3. Area is measured in square units. Why, do you think, the rectangle, the triangle, or the circle are not used?

Practice Exercises

1. What metric unit of square measure does each of the symbols represent?

   a. cm^2 b. km^2 c. m^2 d. mm^2 e. ha

2. How many square centimeters are in 1 square meter?

3. How many square millimeters are in 1 square meter?

4. How many square meters are in 1 square kilometer?

5. How many square meters are in 1 hectare?

6. How many hectares are in 1 square kilometer?

7. Change each of the following to square millimeters.

   a. $6 cm^2$ b. $2.15 m^2$ c. $1.7 dm^2$ d. $0.85 m^2$

8. Change each of the following to square centimeters.

   a. $37 dm^2$ b. $700 mm^2$ c. $0.66 m^2$ d. $5,300 mm^2$

9. Change each of the following to square decimeters.

   a. $819 cm^2$ b. $30 m^2$ c. $600,000 mm^2$ d. $7.5 m^2$

10. Change each of the following to square meters (or centares).

    a. $45 km^2$ b. $19 dm^2$ c. $20,000 cm^2$ d. $0.259 km^2$

11. Change each of the following to square dekameters (or ares).

    a. $8,000 m^2$ b. $2.77 m^2$ c. 6.8 hectares   d. $12,300 m^2$

12. Change each of the following to hectares (or square hectometers).

    a. $75,000 m^2$ b. $14 km^2$ c. $387.5 m^2$ d. $4.3 km^2$

13. Change each of the following to square kilometers.

    a. 500 hectares   b. 72.6 hectares

    c. $3,825,000 m^2$ d. $609,220 m^2$

14. A farm contains 40 hectares of land. How many square meters does it measure? What part of a square kilometer is it?

8-5 Metric Measures of Volume

When you want to know how much space you have for juice in a bottle or for water in a swimming pool, you find the volume of the container. Fill a small box with cubic centimeter cubes to find its volume. How many blocks fit in the box?

Volume = 1 cm^3

Vocabulary

Cubic centimeters and cubic meters are the units commonly used to measure volume in the metric system. Volume is related to capacity, so milliliters and liters can also be used to measure volume.

To measure volume, we use cubic units. In cubic measure, 1,000 of any metric unit is equivalent to 1 of the next higher unit. The exponent ^{3} is used to replace the word *cubic*.

One cubic decimeter holds one liter.

One cubic centimeter holds one milliliter.

$$1,000 \text{ cubic millimeters (mm}^3\text{)} = 1 \text{ cubic centimeter (cm}^3\text{)}$$
$$1,000 \text{ cubic centimeters (cm}^3\text{)} = 1 \text{ cubic decimeter (dm}^3\text{)}$$
$$1,000 \text{ cubic decimeters (dm}^3\text{)} = 1 \text{ cubic meter (m}^3\text{)}$$

$$1 \text{ L} = 1 \text{ dm}^3 = 1,000 \text{ cm}^3$$
$$1 \text{ mL} = 1 \text{ cm}^3$$

Metric units of weight and volume (or capacity) are related as follows:

1,000 cubic centimeters (or 1 liter) of water weighs 1 kilogram at 4° Celsius.
1 cubic centimeter (or 1 milliliter) of water weighs 1 gram at 4° Celsius.

Examples

1. A camper fills a small tank with 7 cubic decimeters of water. How many cubic centimeters does the tank hold?

$1 \text{ dm}^3 = 1,000 \text{ cm}^3$ Change 7 cubic decimeters to cubic
$7 \text{ dm}^3 = 7 \times 1,000 = 7,000 \text{ cm}^3$ centimeters.

ANSWER: 7,000 cubic centimeters

2. Two liters of liquid occupy how many cubic centimeters?

$1 L = 1,000 cm^3$
$2 L = 2 \times 1,000 = 2,000 cm^3$

ANSWER: 2,000 cubic centimeters

3. At 4° Celsius, 8 liters of water weigh about how many kilograms?

1 L weighs 1 kg
8 L weighs $8 \times 1 = 8$ kg

ANSWER: 8 kilograms

Think About It

1. Name three things that it would be useful to measure in cubic meters.

3. Explain why 1 cm = 10 mm but $1 cm^3 = 1,000 mm^3$.

2. Why do you think that the temperature of the water must be given when relating the volume and the weight of the water?

Practice Exercises

1. What metric unit of cubic measure does each of the following symbols represent?
  a. m^3 **b.** mm^3 **c.** km^3 **d.** cm^3 **e.** hm^3

2. How many cubic millimeters are in 1 cubic decimeter?

3. How many cubic millimeters are in 1 cubic meter?

4. How many cubic centimeters are in 1 cubic meter?

5. Change each of the following to cubic millimeters.
  a. $8 cm^3$ **b.** $15 m^3$ **c.** $25.8 cm^3$ **d.** $53.6 m^3$

6. Change each of the following to cubic centimeters.
  a. $27 m^3$ **b.** $6,226 mm^3$ **c.** $8.5 m^3$ **d.** $740 mm^3$

7. Change each of the following to cubic decimeters.
  a. $63 m^3$ **b.** $17.9 m^3$ **c.** $73,250 mm^3$ **d.** $86.4 cm^3$

8. Change each of the following to cubic meters.
  a. $17,000 dm^3$ **b.** $36,800 cm^3$ **c.** $9,400,000 mm^3$ **d.** $600,000 cm^3$

9. 21 milliliters of liquid will fill how many cubic centimeters?

10. 9 liters of liquid will fill how many cubic centimeters of space?

11. A space of $430 cm^3$ will hold how many milliliters of liquid?

12. A space of $18.6 dm^3$ will hold how many liters of liquid?

13. A space of $7,900 cm^3$ will hold how many liters of liquid?

14. Assume that the water is at a temperature of 4° Celsius.
    a. 5 liters of water weigh approximately how many kilograms?
    b. 67 milliliters of water weigh how many grams?
    c. 82 centiliters of water weigh how many grams?
    d. 9.1 liters of water weigh how many kilograms?

How much space is occupied by each of the following capacities?

15. In cubic centimeters:
    a. 7 mL   *b.* 43.6 mL   *c.* 61 cL   *d.* 0.75 L

16. In cubic decimeters:
    a. 6 L   *b.* 84 dL   *c.* 257 cL   *d.* 14.2 L

Find the capacity that will fill each of the following volumes.

17. In milliliters:
    a. 13 cm^3 *b.* 2.5 cm^3 *c.* 800 mm^3 *d.* 0.09 m^3

18. In centiliters:
    a. 8 cm^3 *b.* 3.18 m^3 *c.* 5.07 cm^3 *d.* 9,100 mm^3

19. In liters:
    a. 7 dm^3 *b.* 5.9 m^3 *c.* 6,000 cm^3 *d.* 43,000 mm^3

Find the weight of each of the following volumes or capacities of water at 4°C.

20. In kilograms:
    a. 79 L   *b.* 196 cm^3 *c.* 4.81 L   *d.* 6,300 cm^3

21. In grams:
    a. 11 cm^3 *b.* 0.225 L   *c.* 8.7 mL   *d.* 5,600 mm^3

Find the volume or capacity occupied by each of the following weights of water at 4°C.

22. In cubic centimeters:
    a. 6 kg   *b.* 0.6 kg   *c.* 950 mg   *d.* 8.74 g

23. In liters:
    a. 15 kg   *b.* 4,100 g   *c.* 520 g   *d.* 0.004 kg

24. In milliliters:
    a. 7 g   *b.* 138.51 g   *c.* 1.8 kg   *d.* 6,270 mg

25. In centiliters:
    a. 26 g   *b.* 5.7 kg   *c.* 17,400 cg   *d.* 9,000 mg

26. Solve.
    a. A tank holds 800 liters of oil. How many cubic decimeters of space are in the tank?

    b. It takes 190,000 liters of water to fill a swimming pool. What is the weight of the water when the pool is 0.9 full?

    c. How many liters of water will fill an aquarium if it has a volume of 6,600 cubic centimeters? What is the weight of the water when the aquarium is full?

    d. A water tank has a volume of 7.54 cubic meters. How many liters of water can the tank hold? What is the weight of the water when the tank is full?

Career Application
MEDICAL LABORATORY ASSISTANT

L aboratory tests are important in the diagnosis and treatment of many diseases. A large laboratory may employ a variety of workers. *Medical laboratory assistants* help other workers and do some work on their own. A medical laboratory assistant may weigh and measure samples, use a microscope to look for abnormal cells, do blood tests, and prepare solutions for tests.

After high-school graduation, medical laboratory assistants take a one-year training program at a college or hospital.

1. A medical laboratory assistant needs to measure the weight of a sample. The weight of the empty container is 126.83 g. The weight of the container and the sample is 174.36 g. What is the weight of the sample?

2. A laboratory assistant looks at a parasite egg under a microscope. The microscope field is divided into units 0.0023 mm long. The egg is 8 units long. How long is the egg in millimeters?

3. A medical laboratory assistant plans to test 23 5-mL blood samples. What is the total volume of the samples?

4. To test blood for sickle cell anemia, a medical laboratory assistant uses a solution of 2 g of sodium metabisulfate in 100 mL of water. Sodium metabisulfate is available as 0.2 g tablets. How much water should be used for one tablet?

5. A solution is to be 3.8% sodium citrate by weight. How much sodium citrate is needed for 500 g of solution?

Chapter 8 Review

Vocabulary

centare p. 363
centimeter p. 353
conversion factor p. 353
cubic centimeter p. 365
cubic meter p. 365
gram p. 357

hectare p. 363
kilogram p. 357
kilometer p. 353
liter p. 359
mass p. 357
meter p. 353

milliliter p. 359
millimeter p. 353
square centimeter p. 363
square kilometer p. 363
square meter p. 363

Page

353

1. 48 cm = _____ mm

2. 23,900 m = _____ km

3. 0.97 km = _____ cm

4. 16.7 hm = _____ m

5. Which is longer, 4.8 m or 0.01 km?

357

6. 821 g = _____ cg

7. 34,000 mg = _____ kg

8. 5.5 cg = _____ mg

9. 6 g 5 cg = _____ cg

10. Which is heavier, 0.7 kg or 698,000 cg?

359

11. 3 cL = _____ mL

12. 2,309 L = _____ kL

13. 0.635 L = _____ cL

14. 7 L 5 cL = _____ L

15. Which holds more liquid, a 1.5-L bottle or six 250-mL bottles?

16. A large bottle holds 48-L of vinegar. How many 400-mL bottles can be filled from the large bottle?

363

17. 52 mm² = _____ cm²

18. 8 m² = _____ cm²

19. Jim sold 4.8 km² of land. How many hectares does it measure?

365

20. 4,500 mm³ = _____ cm³

21. 5.2 dm³ = _____ cm³

22. 1.7 L of a liquid will occupy how many cubic centimeters of space?

23. How many grams does 85 cm³ of water weigh?

24. A space of 75 cm³ will hold how many millimeters of a liquid?

25. How many kilograms do 14,000 mL of water weigh?

Chapter 8 Test

1. 45 mm = _____ cm

2. 1.68 = _____ cm

3. Which is longer, 0.09 m, 91 cm, or 9 mm?

4. 35 m 6 cm 8 mm = _____ mm

5. Which is longer, four 1-meter pieces of wood or seven 60-cm pieces of wood?

6. If a space of 50 cm is needed for one person to sit down, how many people can fit on a bench 1 dekameter long?

357 **7.** 82.1 g = _____ mg

8. 910 cg = _____ dag

9. Which is heavier, 0.0045 kg, 45 g, or 550 cg?

10. 7 kg 25 g = _____ kg

11. One bag of fertilizer weighs 5 kg and another weighs 4,300 g. What is the total weight of the bags?

359 **12.** 420 ml = _____ dL

13. 5.33 L = _____ kL

14. Which has the least capacity, 0.04 kL, 30 L, or 4,500 dL?

15. 3 L 4 cL = _____ L

16. Which holds more liquid, a 3-liter bottle or four 700-cL bottles?

363 **17.** $12mm^2$ = _____ cm^2

18. $2.4 m^2$ = _____ cm^2

19. How many hectares are in 3 km^2?

365 **20.** Which is larger, a hectare or 50,000 m^2?

21. $4 m^3$ = _____ dm^3

22. 2,700 mm^3 = _____ cm^3

23. How many grams does 5.3 cL of water weigh?

24. A tank that holds 1,000 L of water has how many cubic decimeters of space?

25. If a quantity of water weighs 450 grams, how many liters is it?

9-1 Customary Measures of Length

The system of measurement that has been used in the United States for more than two hundred years is called the U.S. customary system.

The line segment below measures $2\frac{1}{4}$ inches to the nearest quarter-inch. Measure the width of your textbook to the nearest half-inch.

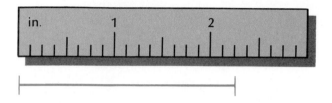

In the customary system, there are many different conversion factors. Once the particular conversion factor is known, the same procedures that you used in the metric system are used to change from one unit to another.

1 foot (ft) = 12 inches (in.) 1 yard (yd) = 3 feet = 36 inches	1 statute mile (mi) = 0.8684 nautical mile = 1,760 yards = 5,280 feet = 1.1515 statute miles 1 nautical mile = 6,080 feet

Examples

1. Joe is 6 feet tall. He wants to record his height in inches on the class height graph. How many inches are 6 feet? To change a given customary unit of measure to another, follow these steps.

  a. Find the conversion factor.

    1 ft = 12 in.

  b. **Multiply** by the conversion factor when changing to a **smaller** unit.

    6 ft = 6 × 12 = 72 in.

ANSWER: Joe's height in inches is 72.

2. Change 2,464 yards to miles. When changing to a **larger** unit, **divide** by the conversion factor.

1 mi = 1,760 yd

2,464 yd = 2,464 ÷ 1,760 = 1.4 mi

ANSWER: 2,464 yd = 1.4 mi

To add or subtract denominate numbers:

a. Arrange the numbers in columns.
b. Add or subtract each column.
c. Simplify the answer.

To multiply or divide denominate numbers:

a. Multiply or divide each unit.
b. In multiplication, simplify the product. In division, if there is a remainder with any unit, change the remainder to the next smaller unit and add to the number of these units.

---

3. Add: 6 ft 7 in. + 1 ft 9 in.

```
    6 ft   7 in.
  + 1 ft   9 in.
    7 ft 16 in. = 8 ft 4 in.
```
16 in. = 1 ft 4 in.

ANSWER: 8 ft 4 in.

4. Subtract: 3 mi 200 yd − 1 mi   650 yd

```
1 mi = 1,760 yd → 3 mi 200 yd =     2 mi 1,960 yd
                  1 mi 650 yd = −   1 mi   650 yd
                                    1 mi 1,310 yd
```

ANSWER: 1 mi 1,310 yd

---

5. Multiply: 5 yd 11 in. × 4

```
    5 yd 11 in.
  ×         4
   20 yd 44 in. = 21 yd 8 in.
```
44 in. = 1 yd 8 in.

ANSWER: 21 yd 8 in.

6. Divide: 6 yd 1 ft 6 in. by 3

```
        2 yd 0 ft 6 in.
   3)6 yd 1 ft 6 in.
      6 yd 0 ft
              1 ft 6 in. = 18 in.
                           18 in.
```

ANSWER: 2 yd 6 in.

Think About It

1. From earliest times, parts of the body were used for measuring. For example, a cubit was defined to be the length of the forearm from the elbow to the tip of the middle finger. List advantages and disadvantages of using this kind of unit.

2. In the mythical land of Quar, there are 25 quods to a quell. How would you change quells to quods?

3. List in order the keys you would press on a calculator to do Practice Exercises 10b and 12c. Assume that the calculator has a memory key.

Practice Exercises

1. Measure each line segment to the nearest quarter-inch.

a. _____ **b.** _____ **c.** _____

2. Measure the length of your classroom in feet.

3. Find the number of inches in each.

  a. 8 ft   **b.** 6.25 ft

  c. $4\frac{2}{3}$ yd   **d.** 5 yd 21 in.

4. Find the number of feet in each.

  a. 6 yd   **b.** 6.5 mi

  c. $5\frac{3}{4}$ yd   **d.** 588 in.

5. Find the number of yards in each.

  a. 7 mi   **b.** 4.5 ft

  c. 50.4 in.   **d.** 89 ft

6. Find the number of miles in each.

  a. 15,840 ft   **b.** 3,300 ft

  c. 4,840 yd   **d.** 6,000 yd

7. Give the fractional part of a foot.

  a. 4 in.   **b.** 9 in.

  c. 11 in.   **d.** 10 in.

8. Give the fractional part of a yard.

  a. 30 in.   **b.** 7 in.

  c. 1 ft   **d.** $1\frac{1}{2}$ ft

9. Give the fractional part of a mile.

  a. 440 yd   **b.** 1,100 yd   **c.** 280 yd   **d.** 1,320 yd

10. Add and simplify.

  a.   2 ft 9 in.
     + 11 ft 2 in.

  b.   3 yd 27 in.
     + 5 yd 19 in.

11. Subtract.

  a.   9 yd 7 in.
    − 4 yd 5 in.

  b.   10 ft
    − 7 ft 8 in.

12. Multiply and simplify.

  a.   5 ft 2 in.
    ×   4

  b.   4 mi 300 yd
    ×   9

  c.   2 yd 9 in.
    ×   8

  d.   4 yd 1 ft 11 in.
    ×   6

13. Divide.

  a. $4\overline{)12\text{ ft 8 in.}}$ **b.** $7\overline{)17\text{ yd 4 in.}}$ **c.** $5\overline{)6\text{ mi 100 ft}}$ **d.** $3\overline{)7\text{ yd 2 ft 6 in.}}$

14. a. Sally attached a wire 4 feet 7 inches long to another wire 5 yards 1 foot 6 inches long. The wires overlapped for 1 inch. Find the length of the new piece.

  b. Len has a rope that measures 5 yards long. How many pieces, each 26 inches long, can be cut from it?

  c. Jane threw a ball 52 yards 4 inches. Ann threw it 49 yards 2 feet 7 inches. How much farther was Jane's throw than Ann's?

  d. Ron needs three boards each 4 yards 9 inches long. He has a board that is 41 feet long. Does he have enough wood?

Customary Measures of Weight

Weight is actually a measure of gravitational force, and so it depends on the amount of substance (mass) and the force of gravity. Therefore, your weight on earth is different from what your weight would be in a spaceship.

In the scale pictured at the right, the objects being weighed are pulled down by gravity and that pull is measured by the dial. Use a similar scale to find the weight in ounces of this textbook. In the customary system, the pound, ounce, and ton are units of weight, not mass. There are two weights commonly called tons, the short ton and the long ton.

> 1 pound (lb) = 16 ounces (oz)
> 1 short ton (T) = 2,000 pounds (lb)
> 1 long ton (l. ton) = 2,240 pounds (lb)

Vocabulary

The **pound, ounce,** and **ton** are basic units commonly used to measure the weight of an object.

Examples

1. A shipment of gravel weighs 2,800 pounds. How many short tons is the shipment?

Find the conversion factor:

1 T = 2,000 lb

Divide by the conversion factor when changing to a **larger** unit.

2,800 lb = 2,800 ÷ 2,000 = 1.4 T

ANSWER: The gravel weighs 1.4 short tons.

2. Change $3\frac{1}{2}$ pounds to ounces.

1 lb = 16 oz

Multiply by the conversion factor when changing to a **smaller** unit.

$3\frac{1}{2}$ lb = $3\frac{1}{2}$ × 16 = $\frac{7}{2}$ × 16 = 56 oz

ANSWER: 56 ounces

Think About It

1. If the force of gravity is decreased, how would you expect your weight to change?

2. Suppose that 40 sceets = 1 scat. Change 3 scats 15 sceets to sceets.

Practice Exercises

Choose the most likely measure.

1. a watch   **a.** 3 ounces   **b.** $\frac{1}{2}$ pound   **c.** 30 ounces

2. a full-grown cat   **a.** 20 ounces   **b.** 4 pounds   **c.** 10 pounds

3. a desk   **a.** 60 ounces   **b.** 60 pounds   **c.** 6 tons

4. Find the weight in pounds or ounces of three things in the classroom.

5. Find the number of ounces in each.

     a. 6 lb   **b.** 2.2 lb   **c.** $5\frac{3}{4}$ lb   **d.** 18 lb 13 oz

6. Find the number of pounds in each.

     a. 8.75 T   **b.** $17\frac{3}{4}$ l. ton   **c.** 20T 1,200 lb   **d.** 368 oz

7. Find the number of short tons in each.   **8.** Find the number of long tons in each.

     a. 6,000 lb   **b.** 24,000 lb   **c.** 7,000 lb   **a.** 6,720 lb   **b.** 17,920 lb   **c.** 4,000 lb

9. Give the part of a pound.   **10.** Give the part of a short ton.   **11.** Give the part of a long ton.

     a. 12 oz   **b.** 8 oz   **a.** 1,200 lb   **b.** 500 lb   **a.** 560 lb   **b.** 1,120 lb

     c. 14 oz   **d.** 5 oz   **c.** 750 lb   **d.** 1,800 lb   **c.** 840 lb   **d.** 1,960 lb

12.

a.	**b.**	**c.**	**d.**
3 lb 6 oz	2 T 1,000 lb	4 lb 10 oz	2 l. ton 1,200 lb
+ 1 lb 8 oz	+ 5 T 1,200 lb	7 lb 9 oz	1 l. ton 850 lb
		+ 6 lb 13 oz	+ 5 l. ton 1,000 lb

13.

a.	**b.**	**c.**	**d.**
6 lb 13 oz	9 lb 5 oz	4 T 700 lb	5 lb
− 3 lb 4 oz	− 8 lb 14 oz	− 2 T 1,900 lb	− 3 lb 6 oz

14.

a.	**b.**	**c.**	**d.**
3 lb 2 oz	7 T 400 lb	1 lb 8 oz	6 lb 4 oz
× 6	× 5	× 9	× 8

15. a. 5)‾10 lb 15 oz   **b.** 4)‾5 l. ton 160 lb   **c.** 6)‾14 lb 4 oz   **d.** 8)‾10 T 200 lb

16. There are 11 pounds of nails available for the carpentry class. How many packages containing 4 ounces of nails can be filled?

17. A metalsmith has 1 pound 4 ounces of 14 carat gold, and 12 ounces of 18 carat gold. How much does the metalsmith have in all?

18. How many ounces of cheese are in a 3-pound wheel?

19. How many short tons are in 8,400 pounds of gravel?

Customary Liquid Measures

The metric system uses one system of measure for liquid and dry measure. The customary system has two systems for measuring capacity. When you buy liquids like milk or juice, you find that they usually come packaged in customary units of liquid measure.

1 pint (pt) = 16 fluid ounces (oz)	1 gallon (gal) = 4 quarts (qt)
1 quart (qt) = 2 pints (pt)	= 8 pints (pt)
= 32 fluid ounces (oz)	= 128 ounces (oz)

Examples

1. The Wales are bringing 1 gallon 1 quart of juice on a picnic. How many 6-ounce cups can they fill with juice?

Change 1 gallon 1 quart to ounces.

Divide by 6 ounces to find how many cups can be filled.

1 gallon = 128 ounces   1 quart = 32 ounces

$128 + 32 = 160$ oz

$160 \div 6 = 26\frac{2}{3}$ cups

ANSWER: The Wales can fill 26 cups with juice and have a little left over.

2. Change 3 gallons to pints.

1 gal = 8 pt
3 gal = 3 × 8 = 24 pt

ANSWER: 24 pints

3. Change 224 fluid ounces to quarts.

1 qt = 32 oz
224 oz = 224 ÷ 32 = 7 qt

ANSWER: 7 quarts

Think About It

1. What do the conversion factors for liquid measure have in common?

2. List in order the keys you would press on a calculator to do each of the following Practice Exercises.

   a. Exercise 6b.   **b.** Exercise 9d.

Practice Exercises

1. Find four containers of various sizes (pitchers, cups, jars, etc.).
   Find the number of ounces each will hold.

2. Find the number of fluid ounces in each.

   a. 9 pt   **b.** 4.75 pt   **c.** $6\frac{3}{4}$ gal   **d.** 2 pt 8 oz

3. Find the number of pints in each.

   a. 3 qt   **b.** $7\frac{1}{2}$ gal   **c.** 9 qt 2 pt   **d.** 72 oz

4. Find the number of quarts in each.

   a. 8 gal   **b.** 1 gal 2 qt   **c.** 7 gal 3 qt   **d.** 700 oz

5. Find the number of gallons in each.

   a. 20 qt   **b.** 56 pt   **c.** 31 pt   **d.** 256 oz

6. Give the part of a pint.

   a. 7 oz   **b.** 4 oz   **c.** 6 oz   **d.** 10 oz

7. Give the part of a quart.

   a. 1 pt   **b.** $\frac{1}{2}$ pt   **c.** 24 oz   **d.** 4 oz

8. Give the part of a gallon.

   a. 3 qt   **b.** 1 pt   **c.** 96 oz   **d.** $3\frac{1}{2}$ qt

9. **a.**   5 gal 2 qt   **b.**   2 qt 14 oz   **c.**   5 qt 1 pt   **d.**   1 pt 8 oz
        + 3 gal 1 qt   + 1 qt 7 oz   − 2 qt 1 pt   −   14 oz

10. **a.**   2 qt 1 pt   **b.**   3 qt 8 oz   **c.**   1 pt 7 oz   **d.**   5 gal 3 qt
         ×   5   ×   8   ×   4   ×   6

11. **a.** 2)8 gal 2 qt   **b.** 3)9 pt 12 oz   **c.** 6)8 qt 2 oz   **d.** 9)20 gal 1 qt

Solve.

12. How many pints of milk are in 2 gallons?

13. How many ounces of lemon juice are in $2\frac{1}{2}$ pints?

14. The grocery store had 15 cans of tomato juice left. If each can
    contains 12 ounces, how many pints of the juice were on hand?

15. A punch was made of 6 quarts of apple juice, 5 quarts of
    cranberry juice, and 1 pint of lime juice. How many 4-ounce
    cups of punch can be served?

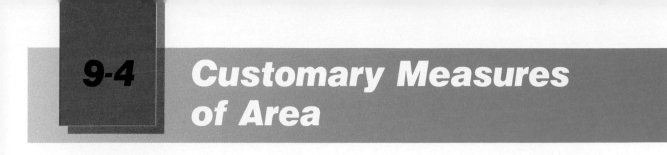

9-4 Customary Measures of Area

Square units are used to measure surfaces like walls, floors, and land. The number of square units is the **area** of the surface.

1 square foot (ft²) = 144 square inches (in.²)
1 square yard (yd²) = 9 square feet
= 1,296 square inches
1 square mile (mi²) = 640 acres
1 acre = 4,840 square yards
= 43,560 square feet

Vocabulary

The **square inch, square foot, square yard, square mile,** and **acre** are basic units commonly used to measure area.

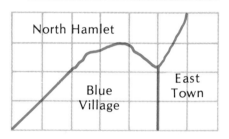

On the map at the right, the square units represent acres. You can count the units to estimate the area of Blue Village. The area of Blue Village is about 10 acres.

Examples

1. Julia figured that her quilt measured 20 square feet. How many square yards is this?

$1 \text{ yd}^2 = 9 \text{ ft}^2$ Find the conversion factor.

$20 \text{ ft}^2 = 20 \div 9 = 2\frac{2}{9} \text{ yd}^2$ **Divide** to change to a **larger** unit.

ANSWER: Julie's quilt measures $2\frac{2}{9}$ yd².

2. Change 10 square feet to square inches.

$1 \text{ ft}^2 = 144 \text{ in.}^2$
$10 \text{ ft}^2 = 10 \times 144 = 1,440 \text{ in.}^2$

ANSWER: 1,440 square inches

3. Change 96 square feet to square yards.

$1 \text{ yd}^2 = 9 \text{ ft}^2$
$96 \text{ ft}^2 = 96 \div 9 = 10\frac{2}{3} \text{ yd}^2$

ANSWER: $10\frac{2}{3}$ square yards

Think About It

1. Name three things that are appropriately measured in square inches.

2. List in order the keys you would press on a calculator to do Practice Exercises 5d and 9c.

Practice Exercises

1. Find the number of square yards of the classroom floor.

2. Find the number of square feet of one wall of your classroom.

Change each measure.

3. To square inches: **a.** 16 ft²   **b.** 5 yd²   **c.** 14.75 ft²   **d.** $\frac{2}{3}$ yd²

4. To square feet: **a.** 1,584 in.²   **b.** 12 yd²   **c.** 40 yd²   **d.** 5 acres

5. To square yards: **a.** 45 ft²   **b.** 9,072 in.²   **c.** 7 acres   **d.** 3.25 mi²

6. To acres: **a.** 80,000 ft²   **b.** 3.6 mi²   **c.** 24,200 yd²   **d.** 13,068 ft²

7. To square miles: **a.** 10,880 acres   **b.** 18,585,600 yd²   **c.** 14,720 acres

8. Give the part of a square yard.   **a.** 8 ft²   **b.** 6 ft²   **c.** 648 in.²

9. Give the part of a square mile.   **a.** 240 acres   **b.** 61,952 yd²   **c.** 278,784 ft²

10. Give the part of an acre.   **a.** 12,000 ft²   **b.** 1,210 yd²   **c.** 27,225 ft²

Solve.

11. A kitchen has 108 square feet of floor space. How many square yards of linoleum are needed to cover the entire floor? If linoleum cost $4.90 per square yard, how much would it cost in all?

12. The air pressure at sea level is 14.7 pounds per square inch. Find the force on 2 square feet of surface at sea level.

Computer Know-How

The computer program below converts any number of <u>gallons</u> into the <u>equivalent</u> number of quarts, pints, or fluid ounces.

```
1Ø PRINT "HOW MANY GALLONS";
2Ø INPUT G
3Ø PRINT G; "GALLONS IS EQUAL TO:"
4Ø PRINT G*4; "QUARTS. . .OR"  ←——— 4 quarts in a gallon
5Ø PRINT G*4*2; "PINTS. . .OR"  ←——— 2 pints in a quart
6Ø PRINT G*4*2*16; "OUNCES"  ←——— 16 ounces in a pint
```

a. ENTER the program and then RUN it to make sure it works.

b. Modify the program so that it converts any number of *miles* to yards, feet, or inches.

c. Modify the program so that it converts any number of *ounces* to pints, quarts, or gallons.

d. Modify the program so that it converts any number of *inches* to feet, yards, or miles.

Problem Solving Strategy: Make a Drawing

Amy made a drawing of the trails around her camping area. She wanted to hike from her tent to Rock Ledge by way of Webb Falls and Mingus Mountain. How far will she have to hike in all to get to Rock Ledge?

> **Strategy: Make a Drawing**
> Sometimes making or using a drawing helps clarify the given information and the problem to be solved.

1. **READ** the problem carefully.

a. Find the **question asked.**
How far does Amy have to hike in all to get to Rock Ledge?

b. Find the **given facts.**
Amy wants to hike from her tent to Rock Ledge by way of Mingus Mountain.

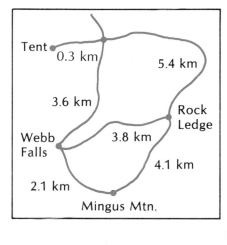

2. **PLAN** how to solve the problem.

a. Choose a strategy: Make a drawing.
Use Amy's drawing to find information.

b. Think or write out your plan relating the given facts to the question asked.
Read the distances from the map. The words *in all* with *unequal* distances indicate addition.
Equation: $0.3 + 3.6 + 2.1 + 4.1 = d$

3. SOLVE the problem.

Estimate and calculate:

0.3 ⟶ 0	0.3
3.6 ⟶ 4	3.6
2.1 ⟶ 2	2.1
+ 4.1 ⟶ + 4	+ 4.1
10 km	10.1 km

---

4. CHECK

a. **Check the accuracy** of your arithmetic.

0.3 ⎫
3.6 ⎪ add up
2.1 ⎪
+ 4.1 ⎭
10.1

b. Compare the answer to the estimate. The answer 10.1 kilometers compares reasonably to the estimate 10 kilometers. The unit is correct.

ANSWER: Amy will have to hike 10.1 km.

Practice Exercises

1. Frank planned to fence a plot for a garden that is 10 feet by 10 feet. He plans to put a stake at each corner and every 1 foot along the edges of the plot to hold the fence. How many stakes will he need?

2. If Frank (Problem 1) has 16 stakes, how far apart will he have to put them so they are evenly spaced around the plot?

3. Train A is traveling due east at 45 miles per hour. Train B is traveling due west at 60 miles per hour. How far apart will the trains be $1\frac{1}{2}$ hours after they pass each other?

4. Kevin drew a map of his neighborhood. He decided to drop some books off at the library on his way to school. How much farther did he have to walk than if he had gone the direct route?

5. At the end of a year, a tree splits into two branches. At the end of each succeeding year, each branch splits into two more branches. How many branches are there at the end of four years?

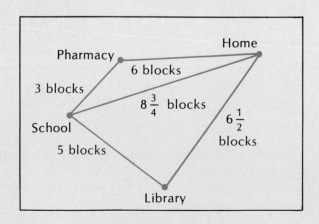

Customary Measures of Volume

Cubic units are used to measure **volume** The volume of a container tells you how much space is available to hold a given substance.

> 1 cubic foot (ft^3) = 1,728 cubic inches ($in.^3$)
> 1 cubic yard (yd^3) = 27 cubic feet = 46,656 cubic inches

Examples

1. A truck delivered 4.2 cubic yards of cement. Change this measure to cubic feet.

To change one unit to another, find the conversion factor.

$1 \ yd^3 = 27 \ ft^3$

Multiply by the conversion factor to change to a **smaller** unit.

$4.2 \ yd^3 = 4.2 \times 27 = 113.4 \ ft^3$

ANSWER: The truck delivered 113.4 ft^3 of cement.

2. Change 55,987.2 cubic inches to cubic yards.

$1 \ yd^3 = 46,656 \ in.^3$

Divide by the conversion factor to change to a **larger** unit.

$55,987.2 \ in.^3 = 55,987.2 \div 46,656 = 1.2 \ yd^3$

ANSWER: 1.2 yd^3

Think About It

1. Describe the difference between area and volume.

2. Choose three exercises from the Practice Exercises section that you would prefer to do with a calculator. For each, list in order the keys you would press on the calculator.

Practice Exercises

1. Determine the volume of a large box using your textbook as the unit of volume.

2. Change to cubic inches. **a.** 38 ft³ **b.** $\frac{3}{4}$ yd³ **c.** 7.25 ft³ **d.** $2\frac{1}{2}$ ft³

3. Change to cubic feet. **a.** 40 yd³ **b.** 5,184 in.³ **c.** 432 in.³ **d.** $7\frac{2}{3}$ yd³

4. Change to cubic yards. **a.** 135 ft³ **b.** 93,312 in.³ **c.** 233,280 in.³ **d.** 688.5 ft³

5. Give the part of a cubic foot. **a.** 432 in.³ **b.** 756 in.³ **c.** 1,440 in.³ **d.** 1,152 in.³

6. Give the part of a cubic yard. **a.** 9 ft³ **b.** $6\frac{3}{4}$ ft³ **c.** 21 ft³ **d.** $13\frac{1}{2}$ ft³

Applications

Solve each problem and select the letter corresponding to your answer.

1. Lead weighs 0.41 lb/in.³ Find the weight of one cubic foot of lead.
   a. 59.04 lb **b.** 4.92 lb
   c. 4,100 lb **d.** 708.48 lb

2. A certain truck can hold 2 yd³ of dirt. If 1 ft³ of dirt weighs 100 lb, what is the weight of a truckload of dirt?
   a. 50 lb **b.** 2,700 lb
   c. 5,400 lb **d.** Answer not given

3. The community center determined that they need 525 ft³ of concrete for building repairs. The town supplier will only sell concrete in whole cubic yards. How many cubic yards must be ordered?

4. A sculptor ordered 38 ft³ of clay. The cost is $10.50 per yd³. Find the cost of the clay.

5. Mrs. Garcia ordered $7\frac{1}{2}$ yd³ of pine bark mulch. The mulch costs $1.10 a cubic foot. Find the total cost.

6. A chemist wants to have 200 metal samples, each containing 10 in.³ of iron. Is 1 ft³ of iron enough to do all the experiments she wants?

Refresh Your Skills

1. 2–5
   6.2 + 0.47

2. 2–6
   0.076 − 0.0105

3. 2–7
   1.8 × 0.009

4. 2–8
   $0.5\overline{)60}$

5. 3–8
   $5\frac{7}{8} + 1\frac{2}{3}$

6. 3–9
   $8\frac{5}{12} - 4\frac{3}{4}$

7. 3–11
   $6\frac{1}{4} \times \frac{4}{5}$

8. 3–12
   $24 \div 2\frac{1}{4}$

9. Find 5% of 927.
   4–6

10. What percent of 40 is 32?
    4–7

11. 54 is 18% of what number?
    4–9

9-7 **Measuring Time**

The units of time are the same in both the metric and customary systems. Since you deal with time every day, these units and conversion factors are probably very familiar to you.

1 hour (h) = 60 minutes (min)	1 week (wk) = 7 days (da)
= 3,600 seconds (s)	1 day = 24 hours (h)
1 minute = 60 seconds	

1 year (y) = 12 months (mo)
= 52 weeks
= 365 days

Every four years, there is a leap year in which one day is added to the year (February 29). The reason for leap year is that the earth actually takes $365\frac{1}{4}$ days to complete its orbit of the sun.

Examples

1. Linda has a hospital insurance plan that pays for 120 days of hospital stay. If she is hospitalized for 5 weeks, will her insurance cover that many days?

Change 5 weeks to days.

a. Find the conversion factor.

1 wk = 7 da

b. Multiply to change to a smaller unit.

5 wk = 5 × 7 = 35 da

ANSWER: Her insurance should pay for a 35-day hospital stay.

2. Change 6 years to days.

1 y = 365 da

Add 1 day for every fourth year to account for leap year.

6 y = 6 × 365 + 1 = 2,191 da

ANSWER: 2,191 days

3. Change 4,500 seconds to hours.

1 h = 3,600 s
4,500 s = 4,500 ÷ 3,600 = $1\frac{1}{4}$ h

ANSWER: $1\frac{1}{4}$ hours

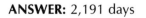

Think About It

1. There are 52 weeks in a year. Are there always exactly 52 Mondays in a year? Explain your answer.

2. If you add 91 days to today's date, will it be the same day of the week? Can you determine the answer without using a calendar?

Practice Exercises

Find the time that it takes.

1. To write your name (in seconds).

2. To get from home to school (in minutes).

3. Give the number of seconds.

   a. 24 min   **b.** 6.25 h   **c.** $1\frac{1}{2}$ h

4. Give the number of minutes.

   a. $\frac{3}{4}$ h   **b.** 13 h 52 min   **c.** 1,800 s

5. Give the number of hours.

   a. 30 da   **b.** 5 da 13 h   **c.** 3,180 min

6. Give the number of days.

   a. $6\frac{3}{7}$ wk   **b.** 3 y 200 da   **c.** 192 h

7. Give the number of months.

   a. $2\frac{1}{2}$ y   **b.** 1 y 6 mo   **c.** 12 y 5 mo

8. Give the number of weeks.

   a. 3 y   **b.** $5\frac{3}{4}$ y   **c.** 2 y 8 wk

9. Give the number of years.

   a. 300 mo   **b.** 260 wk   **c.** 1,095 da

10 Give the part of a minute.

   a. 45 s   **b.** 18 s   **c.** 5 s   **d.** 36 s

11. Give the part of an hour.

   a. 40 min   **b.** 15 min   **c.** 30 min   **d.** 54 s

12. Give the part of a day.

   a. 4 h   **b.** 16 h   **c.** 20 h   **d.** 15 h

13. Give the part of a week.

   a. 2 da   **b.** 6 da   **c.** $3\frac{1}{2}$ da   **d.** 5 da

14. Give the part of a year.

   a. 9 mo   **b.** 13 wk   **c.** 146 da   **d.** 4 mo

15. Add and simplify.
```
  17 h 22 min
+  5 h 38 min
```

16. Subtract.
```
  10 y 4 mo
-      7 mo
```

17. Multiply.
```
  5 y 3 mo
×      3
```

18. Divide.
```
6)10 da 2 h
```

19. On a clock, how long does it take the minute hand to move from:

   a. 4 to 5?   **b.** 2 to 7?   **c.** 1 to 9?   **d.** 8 to 3?

20. Find the length of time. **a.** From 2 A.M. to 6 P.M. the same day.
   b. From 10:45 A.M. to 7:15 P.M. the same day.
   c. From 12:26 P.M. one day to 3:10 A.M. the following day.

21. Look at your calendar. Give the date this month for each.

   a. The second Monday   **b.** The fourth Thursday   **c.** The third Saturday

22. Find the exact number of days from:

   a. March 15 to May 15.
   b. April 3 to October 16.
   c. July 16 to December 9
   d. June 12 of one year to January 8 of the following year.

23. A 4-day course on safety will meet daily from 9:00 A.M. to 2:30 P.M. with a half-hour off for lunch. How many hours is the course?

24. A satellite has been in orbit exactly 2 years. How many hours is this?

25. How old will Lisa be on her next birthday if she was born on June 3, 1972?

Using Mathematics

TIME ZONES

The four standard time belts in the Continental United States are Eastern (EST), Central (CST), Mountain (MST), and Pacific (PST). Central time is one hour earlier than Eastern, Mountain is one hour earlier than Central, and Pacific is one hour earlier than Mountain. Alaska has four time zones, Pacific, Yukon, Alaska-Hawaii, and Bering. Hawaii uses the Alaska-Hawaii time zone which is two hours earlier than Pacific. Parts of Canada are in the Atlantic Standard time zone, which is one hour later than Eastern.

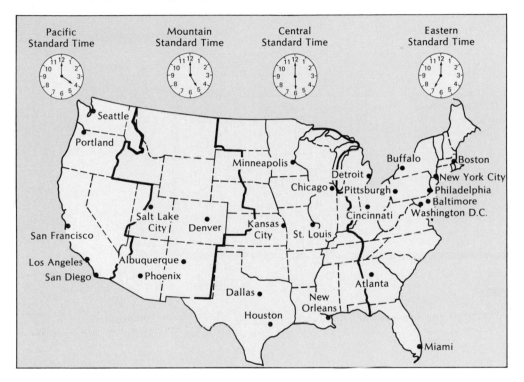

Exercises

1. In what time zone is your city located?

At 11 A.M. in the Mountain Standard zone, what time is it in each zone?

2. Pacific   **3.** Eastern   **4.** Central

When it is 8 P.M. in Dallas, what time is it in each city?

5. Phoenix   **6.** Washington, D.C.   **7.** St. Louis   **8.** San Francisco

9. Atlanta   **10.** Kansas City   **11.** Denver   **12.** Seattle

13. Detroit   **14.** New York   **15.** Houston   **16.** Your city

A program is telecast at 10 P.M. from Los Angeles. What time is it seen in each city?

17. Pittsburgh   **18.** Albuquerque   **19.** Minneapolis

20. Miami   **21.** Salt Lake City   **22.** Baltimore

Reading a Timetable

Traveling from place to place can be exciting and exhausting. Planning your time and budget are necessary for a successful trip. Public transportation companies offer information that helps you make your plans.

The train timetable lists in columns the times of arrival or departure of the train at each stop.

Vocabulary

A **timetable** is a schedule that gives arrivals and departures for trains, buses, or airplanes.

		Train Number		123	177	179	193	195
km	mi.	Operation		Daily	Daily	Daily	Daily	Daily
0	0	Boston, MA	Dp		1 45 P	4 00 P	5 30 P	7 00 P
19	12	Route 128, MA			2 03 P	4 18 P	5 48 P	7 18 P
71	44	Providence, RI			2 39 P	4 54 P	6 25 P	7 54 P
114	71	Kingston, RI				5 19 P		8 20 P
171	106	New London, CT			3 44 P	6 02 P	7 25 P	9 01 P
253	157	New Haven, CT	Ar		4 48 P	7 07 P	8 17 P	10 00 P
			Dp		4 58 P	7 17 P	8 27 P	10 10 P
280	174	Bridgeport, CT			5 18 P	7 37 P		10 30 P
315	196	Stamford, CT			5 41 P	8 00 P	9 10 P	10 55 P
373	232	New York, NY	Ar		6 40 P	8 50 P	10 00 P	11 47 P
			Dp	5 30 P	7 00 P	9 00 P	10 10 P	12 01 A
389	242	Newark, NJ		5 45 P	7 15 P	9 17 P	10 25 P	12 17 A

The letter A indicates A.M.
The letter P indicates P.M.

Practice Exercises

1. At what time does Train 195 leave Boston? At what time does it arrive in New Haven? In Bridgeport?

2. At what time does Train 193 leave Providence? At what time does it arrive in New Haven?

3. How far (in kilometers) is it by rail from Boston to Bridgeport? How long does it take Train 177 to go from Boston to Bridgeport? What is the average rate of speed during this trip?

4. Compare the time it takes Train 177 to travel from Providence to Bridgeport to that of Train 179. What is the difference in time?

The airline schedule gives separate listings for each city. Notice that all arrival and departure times are given in local time.

The letter a indicates A.M.
The letter p indicates P.M.
M—Meal   S—Snack

Leave	Arrive	Flight	Stops	Meals
From: DALLAS-FORT WORTH (CST)				
TO: CHICAGO (CST)				
6 30a	9 15a	242	One-Stop	M
8 20a	10 15a	54	Non-Stop	M
10 20a	2 10p	248	Two-Stops	S
2 40p	5 30p	140	One-Stop	S
3 20p	5 20p	44	Non-Stop	S
5 20p	7 20p	36	Non-Stop	M
7 15p	10 45p	120/156	Kansas City	M
TO: DENVER (MST)				
9 25a	10 05a	62	Non-Stop	M
11 25a	12 05p	68	Non-Stop	M
4 25p	5 30p	66	One-Stop	S
6 25p	7 05p	78	Non-Stop	M

Leave	Arrive	Flight	Stops	Meals
From: DALLAS-FORT WORTH (CST)				
TO: MIAMI (EST)				
8 35a	12 00n	79	Non-Stop	M
12 05p	3 25p	63	Non-Stop	M
5 00p	9 10p	169	One-Stop	M
8 10p	12 15a	405	One-Stop	S
TO: NEW ORLEANS (CST)				
8 35a	9 45a	157	Non-Stop	M
11 30a	12 40p	235	Non-Stop	M
4 40p	5 50p	133	Non-Stop	S
TO: SEATTLE-TACOMA (PST)				
11 30a	1 10p	95	Non-Stop	M
6 40p	8 20p	182	Non-Stop	M
7 45p	10 20p	184	One-Stop	M

Practice Exercises

5. At what time does Flight 54 leave Dallas-Fort Worth for Chicago? At what time does it arrive in Chicago? How long does the flight take? Are there any stops?

6. What is the number of the flight leaving Dallas-Fort Worth for New Orleans at 11:30 A.M.? When does it arrive in New Orleans? Is a meal or a snack served?

7. How long does it take Flight 95 to fly from Dallas-Fort Worth to Seattle-Tacoma? How much faster is it than Flight 184?

8. How much longer does it take Flight 66 to fly from Dallas-Fort Worth to Denver than Flight 78?

The bus timetable lists the times of arrival and departure for buses at each town. Read down on the left and up on the right. Notice that not all the towns are in the same time zone.

Times are given in local time.
Red type indicates A.M. time.
Blue type indicates P.M. time.

Read Down				Bus Number			Read Up		
824	822	820					823	825	827
3 30	9 30	3 15	Lv	Atlanta, GA	EST	Ar	7 00	10 55	6 35
6 00	11 59	5 05	Ar	Birmingham, AL	CST	Lv	1 45	6 40	2 45
6 30	12 45	6 00	Lv	Birmingham		Ar	1 05	6 00	2 20
10 00	4 35	9 20		Tupelo, MS			9 35	2 15	10 50
12 30	7 20	12 15	Ar	Memphis, TN		Lv	7 00	11 30	8 20
1 10	8 30	1 30	Lv	Memphis		Ar	6 00	10 50	7 00
4 55	12 15	5 15	Ar	Little Rock, AR		Lv	2 15	7 05	3 15
5 15	12 30	6 00	Lv	Little Rock		Ar	1 55	6 40	2 40
9 00	3 40	9 10	Ar	Fort Smith, AR		Lv	10 45	3 20	10 55
9 10	3 55	10 10	Lv	Fort Smith		Ar	10 30	2 55	10 30
3 30	9 30	3 45	Ar	Oklahoma City		Lv	4 15	11 10	5 45
5 00	11 40	4 45	Lv	Oklahoma City		Ar	2 40	10 45	4 40
11 25	6 10	10 35	Ar	Amarillo, TX		Lv	8 15	4 30	11 25
11 55	6 50	11 00	Lv	Amarillo	CST	Ar	7 30	4 15	11 10
4 50	11 45	4 10	Ar	Albuquerque, NM	MST	Lv	12 35	9 00	3 45

Practice Exercises

9. At what time does Bus 822 leave Atlanta? At what time does it arrive in Birmingham? In Memphis? In Little Rock?

10. At what time does Bus 827 leave? At what time does it arrive in Fort Smith? In Tupelo? In Atlanta?

11. At what time does Bus 824 leave Oklahoma City? At what time does it arrive in Albuquerque, NM? How long does it take to make this trip?

12. Compare the time it takes Bus 820 to travel from Atlanta to Amarillo with the time it takes Bus 824 to make the same trip.

Think About It

1. You can drive a car from Boston to New Haven. List some advantages of a car trip over a train ride. Then list some advantages of the train over a car.

2. Suppose that you live in Dallas and are invited to a cousin's wedding in New Orleans. The wedding is at 6:00 P.M. Would you take Flight 133 the day of the wedding? Explain your answer.

Career Application
JEWELER

You have probably admired the beautiful rings and necklaces in a jewelry store. A *jeweler* knows much more about jewelry than its price. Jewelers may work in factories and repair shops as well as stores. They may make jewelry by casting metal in molds or shaping it with hand tools. Some jewelers are gemologists, experts on diamonds and other precious stones. A jeweler needs good eyesight and skillful hands. Most jewelers learn their skills as apprentices or in technical schools.

1. Pure gold is very soft. Gold used in jewelry is an alloy, a mixture of gold and other metals. 24-carat gold is 100% pure gold. What percent gold does 18-carat gold contain?

2. The melting point of pure gold is 1064.43°C. The melting point of pure silver is 961.93°C. How much higher is the melting point of gold?

3. Gold can be made into sheets $\frac{1}{250,000}$ of an inch thick. How thick would a pile of 400,000 of these sheets be?

4. After repairing a silver cuff link with silver solder, a jeweler cleans it in pickling solution. To make pickling solution, the jeweler dissolves a tablespoon of pickling powder in a pint of water. How much pickling powder would be needed for a gallon of water?

Chapter 9 Review

Vocabulary

acre p. 379	fluid ounces p. 377	month p. 385	square inch p. 379
cubic foot p. 383	foot p. 372	ounce p. 375	square mile p. 377
cubic inch p. 383	gallon p. 377	pint p. 377	square yard p. 379
cubic yard p. 383	hour p. 385	pound p. 375	timetable p. 388
day p. 385	inch p. 372	quart p. 377	ton p. 375
denominate	mile p. 372	second p. 385	yard p. 372
numbers p. 372	minute p. 385	square foot p. 377	year p. 385

Page
372   **1.** How many inches is 15 feet?

Page
372   **2.** What part of a yard is $1\frac{1}{2}$ feet?

372   **3.**   7 ft 6 in.
              + 2 ft 9 in.

372   **4.** 26 miles 385 yards = _____ yards
            (marathon)

375   **5.** How many pounds equals 52 ounces?

375   **6.**   4 lb   7 oz
              − 1 lb 10 oz

375   **7.** What part of a pound is 4 ounces?

377   **8.** How many pints are in 6 quarts?

377   **9.**   4 qt 6 oz
              ×   6

377   **10.** How many gallons are in 14 quarts?

379   **11.** Change 468 ft² to square yards.

379   **12.** Change $6\frac{1}{2}$ ft² to square inches.

379   **13.** What part of a square yard is
            3 square feet?

383   **14.** Change 135 yd³ to cubic feet.

383   **15.** 5 cubic feet is how many cubic inches?

385   **16.** Find the number of hours in seven days.

385   **17.** Find the number of months in $3\frac{1}{3}$ years.

385   **18.**   4 h 18 min 33 s
              + 9 h 46 min 18 s

388   **19.** Use the train schedule on page 388 to find
            how long it takes Train 195 to go from
            Boston to Newark.

388   **20.** Use the plane schedule on page 388 to
            find how long it takes Flight 62 to go from
            Dallas to Denver.

Chapter 9 Test

Page
372 **1.** How many feet is $3\frac{1}{2}$ yards?

Page
372 **2.** What part of a yard is 20 inches?

372 **3.** The summit of Mt. Everest is 29,028 feet. About how many miles is that?

372 **4.**
$$\begin{array}{r} 3 \text{ yd } 2 \text{ ft } 6 \text{ in.} \\ -\ 1 \text{ yd } 2 \text{ ft } 8 \text{ in.} \\ \hline \end{array}$$

375 **5.** 6 lbs 8 oz = _____ oz

375 **6.** 96 oz = _____ lb

375 **7.**
$$\begin{array}{r} 4 \text{ lb } 9 \text{ oz} \\ +\ 3 \text{ lb } 8 \text{ oz} \\ \hline \end{array}$$

375 **8.** How many ounces of cashews must be added to 38 ounces of peanuts to make 4 pounds of mixed nuts?

377 **9.** $2\frac{1}{2}$ gallons is equal to how many pints?

377 **10.** A can of apple juice holds 12 ounces. A case of 24 cans would be how many quarts of juice?

377 **11.** $4\overline{)6 \text{ qt } 8 \text{ oz}}$

379 **12.** 360 in.^{2} = _____ ft^2

379 **13.** 23 yd^2 = _____ ft^2

383 **14.** 12 yd^3 = _____ ft^3

381 **15.** Judy rode her bike 4 blocks east, 8 blocks south, 9 blocks west, and 4 blocks north. How many blocks west and south of her starting point was she?

385 **16.** What part of a minute is 50 seconds?

385 **17.** Change 16 days 2 hours to hours.

385 **18.** Find the number of weeks in $2\frac{1}{2}$ years.

385 **19.**
$$\begin{array}{r} 3 \text{ da } 8 \text{ h } 15 \text{ min} \\ \times \qquad\qquad 5 \\ \hline \end{array}$$

388 **20.** Using the train schedule on page 388, find how long it takes Train 177 to go from Boston to New York.

388 **21.** Using the plane schedule on page 388, find how much faster Dallas-to-Miami Flight 79 is than Flight 169.

A bar graph presents categorical data quickly and clearly.

The bar graph at the right shows how many people rode a bus during the first six months of one year in one town. In this graph the categories are the months January through June, and the data points are the number of people who rode buses in that time period.

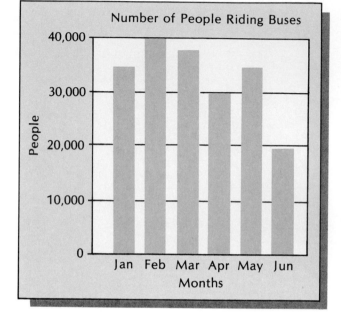

To make a bar graph, follow these steps.

a. On a sheet of graph paper, draw a horizontal guide line (horizontal axis) that will be the bottom of the graph. Draw a vertical guide-line (vertical axis) on the left.

b. Select a convenient scale for the numbers in each category that are being compared, first rounding large numbers. Use tick marks to indicate points on the scale. For a vertical bar graph, write the number scale along the vertical axis; for a horizontal bar graph, place the scale on the horizontal axis.

c. Label the remaining axis with the names of the categories that are being graphed. Also, provide an overall label for each axis, such as *People* and *Months* in the graph above.

d. Mark off the height (or length) corresponding to the given size of each category. Draw lines to complete the bars. All bars should have the same width and be equidistant from each other (or equally spaced) to emphasize the distinctness of each category.

e. Select an appropriate title and write it above the graph.

Vocabulary

Categorical data are data that can be placed into categories or groups. The data in each group are then counted. Each data point must belong to only one category.

A bar graph uses vertical or horizontal bars to compare the number of items in each category.

Examples

1. In the graph above, in which month did the most people ride the bus?

ANSWER: February

2. In the graph above, about how many people rode the bus in January?

ANSWER: 35,000 people

3. Make a bar graph showing the following information. Of the first ten presidents of the United States, 6 were born in Virginia, 2 in Massachusetts, 1 in South Carolina, and 1 in New York.

ANSWER:

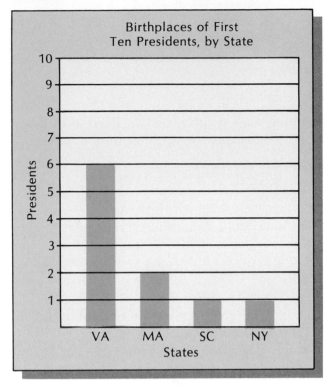

Think About It

1. In a bar graph, what do the bars represent?

2. Why may it be more meaningful to look at data presented in a graph rather than in a table of data values?

Practice Exercises

1. Use the graph on the right to answer the following questions regarding the lengths of some rivers of the world.

    a. How many miles does the side of a small square represent in the vertical scale of the bar graph? $\frac{1}{4}$ of the side of the square?

    b. Which river is the longest? The shortest?

    c. Find the approximate length of each river.

2. Make a bar graph showing the lengths of the channel spans of the following suspension bridges in the United States.

Golden Gate, 4,200 ft
Bear Mountain, 1,632 ft
George Washington, 3,500 ft
Delaware River, 1,750 ft
Brooklyn, 1,595 ft

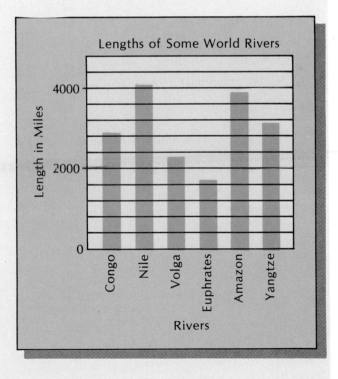

Broken-Line Graphs

A broken-line graph is usually used to display data that are measured rather than counted.

Vocabulary

Metric data are data that are measured such as distances, temperatures, or weights.

A **broken-line graph** uses line segments to show changes and relationships between quantities.

The broken-line graph at the right shows daily high temperatures in Tucson, Arizona during the first five days of July one year. Note the jagged portion of the vertical axis. This is used to indicate that part of the scale (0° to 94°) has been omitted.

To make a **broken-line graph**, follow these steps.

a. On a sheet of graph paper draw a horizontal axis and a vertical axis.

b. Select a convenient scale for the related numbers, first rounding large numbers. The measurement scale is usually placed on the vertical axis. Label the scale.

c. Below the horizontal axis print the names of the items being measured. Label the items.

d. Above each item, place a dot on the graph to show the measured value corresponding to the given item.

e. Connect the dots in order with straight lines.

f. Select an appropriate title and write it above the graph.

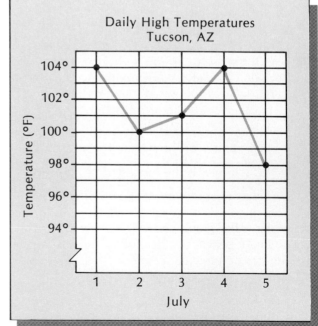

Examples

1. What was the temperature on July 3?

ANSWER: 101° F

2. On which days was it 104°?

ANSWER: July 1 and July 4

3. Make a broken-line graph showing this data:
Five laboratory rats (labeled A, B, C, D, and
E) completed a maze in 93 seconds, 112
seconds, 175 seconds, 128 seconds, and 140
seconds respectively.

ANSWER:

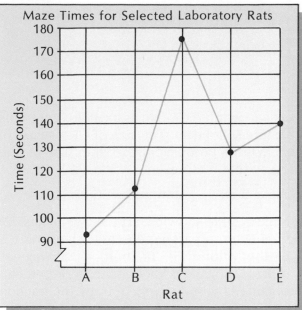

Maze Times for Selected Laboratory Rats

Think About It

1. In your own words, explain the difference
between categorical data and metric data.

2. The graphs in this lesson have used the
jagged vertical axis, indicating that only a
portion of the measurement scale large
enough to contain the data points was used.
How do you think your impression of these
graphs would change if the entire scales
(0°–105° and 0–180 seconds) were used?

Practice Exercises

1. Use the broken-line graph at the right
to answer the following questions about
the market value of goods produced by
Melmac Industries, Inc.

   a. How many dollars does the side of a
square represent in the vertical scale of
the graph? $\frac{1}{2}$ of the side of a square?

   b. What was the market value of goods
produced in each month?

   c. Between which two months was the
change in production the greatest? the
least?

2. Make a broken-line graph showing the
data below.

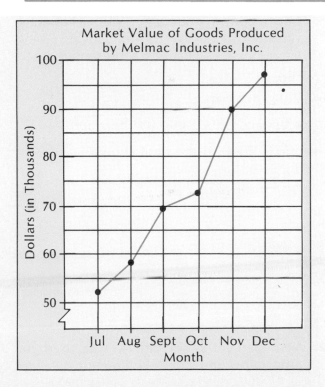

Market Value of Goods Produced
by Melmac Industries, Inc.

Monthly Normal Temperatures (°F) in Some U.S. Cities

	Jan	Feb	Mar	Apr	May	June	July	Aug	Sept	Oct	Nov	Dec
Chicago	24	26	35	47	58	67	72	72	65	54	40	29
Houston	51	55	61	69	75	81	83	83	78	70	60	54
Los Angeles	55	56	58	59	62	66	70	71	69	65	61	57
St. Paul	13	16	29	46	58	67	72	69	61	49	32	19

Circle Graphs

In a circle graph the circle as a whole represents the entire set of data. The size of each sector of the circle is proportional to the size of the category that it represents.

To make a *circle graph*, follow these steps.

a. Make a table showing: (1) the given facts, (2) the fractional part or percent each quantity is of the whole, (3) the number of degrees representing each fractional part or percent, obtained by multiplying 360° by the fraction or percent.

b. Draw a convenient circle. With a protractor construct successive central angles, using the number of degrees representing each part.

c. Label each part.

d. Select and print a title for the graph.

The circle graph at the right shows the uses of electricity in the home.

Vocabulary

A **circle graph** is used to show the relationship of the parts to a whole and to each other.

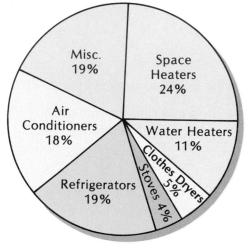

Uses of Electricity in the Home

Examples

1. Which use of electricity was most common?

ANSWER: for space heaters

2. What percent of electricity was used for refrigerators and stoves combined?

19% + 4% = 23%

ANSWER: 23%

3. Make a circle graph to show Heather's budget.

Food, 25%; Clothing, 18%; Bus, 10%; Rent, 33%; Savings, 10%; Misc, 4%

ANSWER:

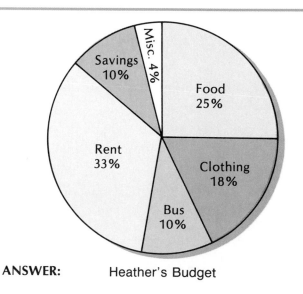

Heather's Budget

1. In a circle graph, how many degrees are needed to represent 1% of data? Is this true for any circle graph, regardless of how large or small the circle is? Why or why not?

2. Do you think that circle graphs are used to show categorical (counted) data or metric (measured) data? Explain.

Practice Exercises

1. Use the graph on the right to answer the following questions regarding the family budget.

   a. What fractional part of its entire income does the Harris family plan to spend for food? for shelter?

   b. Compare the percent to be spent for shelter to the percent of savings.

   c. If the family income is $18,600 per year, what are the amounts to be spent for each item per year? per month? per week?

2. Make circle graphs showing the following data.

   a. A certain large city planned to spend its income as follows: schools, 30%; interest on debt, 25%; safety, 15%; health and welfare, 12%; public works, 10%; other services, 8%.

The Harris Family Budget

Savings 10%
Clothing 12%
Family Car 10%
Food 25%
Operating Expenses 8%
Other Expenses 15%
Shelter 20%

   b. Charles spends his time as follows: 9 hours for sleep, 6 hours for school, 2 hours for study, 3 hours for recreation, and 4 hours for miscellaneous activities.

Refresh Your Skills

1. 3,586 [1-8]
2,975
4,598
+ 6,284

2. 305,000 [1-9]
− 214,069

3. 690 [1-11]
× 504

4. 365)75,920 [1-13]

5. Add. [2-5]

$.54 + $9.89 + $24.06

6. Subtract. [2-6]

$30 − $1.98

7. Multiply. [2-7]

0.006 × 0.02

8. Divide. [2-8]

0.1)0.586

9. Add. [3-8]

$8\frac{1}{4} + 5\frac{1}{3} + 2\frac{5}{12}$

10. Subtract. [3-9]

$\frac{9}{10} - \frac{3}{5}$

11. Multiply. [3-11]

$1\frac{3}{4} \times 1\frac{3}{5}$

12. Divide. [3-12]

$9\frac{3}{4} \div 13$

13. Find $11\frac{1}{2}$% of $9,500.
[4-6]

14. What percent of 50 is 75?
[4-7]

15. $1.62 is 60% of what amount?
[4-9]

10-4 Organizing Data

Histograms and frequency polygons are much like bar graphs and broken-line graphs. Such graphs are used a great deal by statisticians to organize and display grouped data.

The data studied in statistics may be scores, measurements, or other numerical facts. Histograms and frequency polygons are used to display metric data. Data are recorded by tallying.

To make a **frequency distribution**, follow these steps.

a. List the classes of data.

b. Tally the data. Each number should fall into a single class.

c. Count the tally marks to list the frequency of each class.

To make a **histogram**, follow these steps.

a. Make a frequency distribution.

b. Make a bar graph showing the classes and frequencies. Note that in a histogram the bars are *not* separated by spaces.

To make a **frequency polygon**, follow these steps.

a. Make a frequency distribution.

b. Make a broken-line graph showing the classes and frequencies. (Note: In a frequency polygon, a class with zero frequency should be shown at either end only if it makes sense to do so.)

Vocabulary

Statistics is the study of collecting, organizing, analyzing, and interpreting data. The following terms are used:

Frequency: The number of times a score or datum occurs.

Frequency Distribution: Arrangement of data in table form.

Histogram: Frequency distribution shown in a bar graph.

Frequency Polygon: Frequency distribution shown in a broken-line graph.

Example

Make a frequency distribution table, a histogram, and a frequency polygon for the given data.

A class of 34 pupils made the following scores on a test:

80, 95, 70, 80, 85, 65, 90, 75, 60, 90, 100, 65, 85, 75, 80, 95, 85, 65, 80, 70, 90, 70, 80, 85, 80, 60, 75, 80, 85, 95, 80, 70, 90, 85

ANSWER:

Frequency Distribution Table		
Score	Tally	Frequency
100	I	1
95	III	3
90	IIII	4
85	IIII I	6
80	IIII III	8
75	III	3
70	IIII	4
65	III	3
60	II	2
		Total 34

ANSWER:

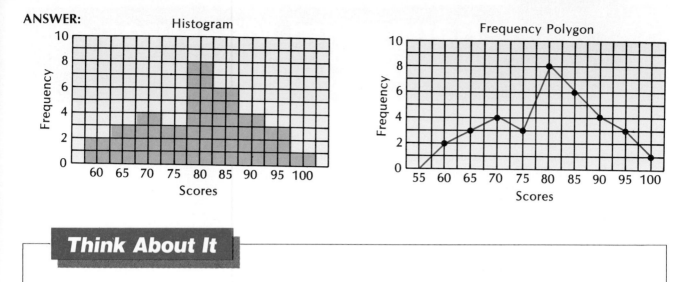

Think About It

What is the difference between a bar graph and a histogram?

Practice Exercises

Using the lists of scores given below, make for each:

1. A frequency distribution table   **2.** A histogram   **3.** A frequency polygon

a. 85 60 75 40 85 70 90 100 95 85 75 80 70 70 85 90 60 65 75 80
90 75 90 80 95

b. 9 2 7 6 4 5 8 8 6 9 7 10 5 7 2 8 3 7 4 6 5 9 1 8 7 6 9 7 5 8

c. 45 43 44 48 45 52 47 48 46 50 45 49 49 47 51 49 52 50 43 45
48 44 47 49 45 51 50 43 45 51 46 49 50 45 44

Refresh Your Skills

1. 1–8
6,109 + 56,274 + 398

2. 1–9
81,040 − 2,951

3. 1–11
9,008 × 406

4. 1–13
$6,080\overline{)54,720}$

5. 2–5
0.724 + 5.66 + 20.8

6. 2–6
45 − 3.9

7. 2–7
0.03 × 0.0008

8. 2–8
2.8 ÷ 0.007

9. 3–8
$3\frac{2}{3} + 4\frac{1}{2} + 2\frac{3}{5}$

10. 3–9
$8\frac{1}{6} - 5\frac{1}{3}$

11. 3–11
$12 \times 4\frac{5}{8}$

12. 3–12
$7\frac{1}{2} \div 2\frac{1}{2}$

13. Find $8\frac{3}{4}$% of $20,000
4–6

14. What percent of 60 is 48?
4–7

15. 9 is 36% of what number?
4–9

10-5 Range and Measures of Central Tendency

Statisticians use several types of measures to describe sets of data. Measures of central tendency are used to describe data in an "average" sense. Range is a measure of the extent to which data in a set vary.

Vocabulary

The **mean, median,** and **mode** are three commonly used **measures of central tendency.**

The **arithmetic mean** (or simply **mean**) is the average of a set of data.

The **median** is the middle data point when an odd number of data are arranged in order of size. It is the average of the two middle data points when an even number of data are arranged.

The **mode** is the number that occurs most frequently in a set of data. A set of data is **bimodal** if two numbers appear with equal greatest frequency.

The **range** is a measure of variance. It is the difference between the highest and lowest numbers in a set of data.

Using mean, median, mode, and range to describe a set of data tells us: the average, which one is in the middle, which one occurs most frequently, and how widely the data vary.

Examples

1. Find the arithmetic mean of 12.3, 18.6, 15.4, 16.7, and 19.5.

To find the mean, follow these steps.

a. Find the sum of the numbers. (To find the sum using a frequency distribution table, multiply each score by its frequency. Add these products.)

$12.3 + 18.6 + 15.4 + 16.7 + 19.5 = 82.5$

b. Divide the sum by the number of items.

$82.5 \div 5 = 16.5$

ANSWER: The mean is 16.5.

2. Find the mean of the data in the table.

Frequency Distribution Table		
Score	*Tally*	*Frequency*
80	‖	2
85	‖|	3
90			1
95	‖	2
		8 Total

$$80 \times 2 = 160$$
$$85 \times 3 = 255$$
$$90 \times 1 = 90$$
$$95 \times 2 = +\ 190$$
$$\text{Sum} = 695 \qquad 695 \div 8 = 86.875$$

ANSWER: The mean is 86.9.

---

3. Find the median of the scores 78, 85, 83, 81, 92, 86, and 90.

To find the median, follow these steps.

a. Arrange the numbers in order.

78   81   83   85   86   90   92

b. If the number of data is odd, count from either end to find the middle number. If the number of data is even, find the mean of the two middle numbers.

78   81   83   (85)   86   90   92

ANSWER: The median is 85.

---

4. Find the median of the data 18, 15, 24, 19, 22, 26, 29, 24, 18, and 16.

15   16   18   18   (19)   (22)   24   24   26   29

$$\frac{19 + 22}{2} = \frac{41}{2} = 20.5$$

ANSWER: The median is 20.5.

---

5. Find the mode of the numbers 80, 93, 75, 81, 93, 57, 80, 98, 82, and 93.

To find the mode, find the number that occurs most frequently. There may be more than one mode or no mode.

93 appears 3 times. All other numbers appear only once or twice.

ANSWER: The mode is 93.

---

6. Find the mode of the numbers 17, 24, 12, 28, 19, 16, 25, 28, 17, 21, and 15.

17 and 28 each appear twice. All other numbers appear only once.

ANSWER: The distribution is bimodal. The modes are 17 and 28.

7. Find the range of the scores 85, 91, 78, 83, 86, 94, and 79.

To find the range, follow these steps.

a. Identify the highest and lowest numbers in the set.

85   91   (78)   83   86   (94)   79

b. Subtract to find the range.

$94 - 78 = 16$

ANSWER: The range is 16.

Think About It

1. Which measure of central tendency (mean, median, or mode) can be most greatly affected by the range of a set of data? Explain.

2. Which measure(s) of central tendency would best represent each of the following sets of data? Explain.
   a. 51, 56, 48, 51, 97, 52, 123, 51
   b. 78, 84, 93, 87, 82, 85, 91, 79
   c. 23, 25, 22, 23, 48, 45, 46, 44

Practice Exercises

1. Find the mean for each of the following sets of data.
   a. 9, 5, 8, 4, 10, 3, 7, 8, 6, 7, 9, 9, 7, 3, 8, 9
   b. 83.4, 89.2, 80.9, 81.7, 84.2, 86.1, 85.3
   c. 47, 43, 50, 48, 42, 47, 49, 43, 45, 47, 48, 46, 47, 44, 40, 44, 47, 49, 42, 44, 50, 47, 46, 49

2. Find the median for each of the following sets of data.
   a. 8, 6, 7, 5, 9, 10, 3, 0, 5
   b. 4.2, 5.6, 3.8, 4.7, 6.1, 4.5
   c. 65, 85, 70, 95, 75, 80, 70, 85, 80, 85, 95, 60, 75, 80, 75, 95, 85, 60, 75, 90, 85

3. Find the mode(s) for each of the following sets of data.
   a. 11, 15, 10, 19, 15, 12, 11, 17, 14, 15, 13, 12
   b. 57, 49, 64, 53, 58, 55, 53, 60, 58, 52, 53, 64, 58, 49, 60
   c. 5, 4, 9, 7, 6, 8, 6, 4, 5, 9, 7, 7, 5, 6, 9, 8, 3, 2, 5, 7, 4, 9, 3, 6, 2, 7, 8

4. Find the range for each of the following sets of data.
   a. 31, 26, 29, 43, 34, 39, 47, 28, 45
   b. 85, 90, 75, 65, 35, 80, 55, 85, 60, 75, 40, 80
   c. 23, 41, 37, 16, 27, 32, 19, 49, 36, 52, 28, 46, 59, 44, 27, 30

5. Arrange each of the following sets of data in a frequency distribution table; then find the mean, median, mode, and range.
   a. 48, 47, 50, 43, 42, 47, 49, 43, 45, 47, 48, 46, 47, 44, 47, 44, 40, 49, 48
   b. 18, 15, 12, 19, 14, 16, 14, 18, 13, 17, 16, 19, 16, 12, 14, 17, 15, 16, 20, 18, 11, 14, 19, 15, 17, 20, 19, 17, 18, 13, 14
   c. 58, 62, 64, 59, 57, 61, 60, 59, 63, 58, 64, 60, 63, 59, 59, 56, 64, 61, 58, 62, 64, 60, 57, 59, 63, 55, 61, 56, 60, 59, 61, 58, 62, 59, 57, 62

Knowing something about probability helps the statistician understand the significance of the data being collected. The first step in learning about probability is to learn about possible outcomes of activities or experiments.

Vocabulary

An **outcome** is the result of an activity or experiment, such as getting "heads" or "tails" when a coin is tossed.

A **sample space** consists of all possible outcomes of a given activity or experiment.

An **event** is an outcome or group of outcomes of the sample space.

A **simple event** is an event that consists of only a single outcome.

A **tree diagram** is a drawing listing all possible outcomes for an activity.

Examples

1. What is the sample space of a coin being tossed?

When a coin is tossed, there are two possible **outcomes**— heads (**H**) and tails (**T**).

ANSWER: The sample space is (**H**), (**T**) .

2. What is the sample space when two coins are tossed?

The possible outcomes are:

heads, heads (H, H); heads, tails (H, T); tails, heads (T, H); tails, tails (T, T)

ANSWER: The sample space is (H, H), (H, T), (T, H), (T, T) .

3. When tossing two coins, what is the event that the two coins are the same?

The coins must both be heads or both be tails.

ANSWER: The event that both coins are the same is (H, H), (T, T) .

Tossing two coins can be illustrated by the following **tree diagram**:

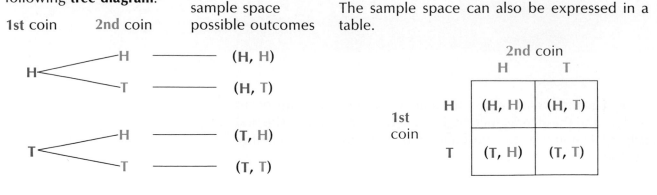

The sample space can also be expressed in a table.

	2nd coin	
	H	T
1st coin H	(H, H)	(H, T)
1st coin T	(T, H)	(T, T)

Multiplication can also be used to determine the total *number* of possible outcomes.

Examples

1. What is the number of possible outcomes when tossing two coins?

1st coin   2nd coin
  2   ×   2   = 4
outcomes   outcomes
(H or T)   (H or T)

ANSWER: There are 4 possible outcomes.

2. What is the number of possible outcomes when tossing three coins?

1st coin   2nd coin   3rd coin
  2   ×   2   ×   2   = 8

ANSWER: There are 8 possible outcomes.

Generally, the number of outcomes possible when performing two or more activities (such as tossing two or more coins) is the product of the number of outcomes for each activity.

Think About It

1. Describe in your own words the differences between *outcome*, *event*, and *sample space*.

2. Suppose you are performing a particular experiment that has four possible outcomes and you perform the experiment five times. How many total outcomes are possible if you know that all outcomes can repeat, but no two consecutive trials will result in the same outcome?

Practice Exercises

1. Write the sample space listing all the possible outcomes.

   a. Drawing a card from a set containing five cards numbered 1 to 5 inclusive.

   b. Picking a ball from a box containing 3 white balls, 2 yellow balls, and 1 orange ball.

   c. Picking a marble from a bag containing 1 red and 1 green marble **and** a button from another bag containing 1 black, 1 yellow, and 1 blue button.

2. For each of the following, draw a tree diagram listing all the possible outcomes.

   a. Drawing a bead from a bag containing 1 red bead, 1 purple bead, and 1 white bead and a button from a box containing 1 white button and 1 brown button.

   b. Selecting a white, blue, or tan shirt with a striped, blue, or tan tie.

   c. Spinning each spinner. Use numerals and letters where the pointers may stop.

   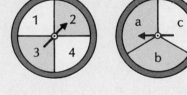

3. In each of the following, use multiplication to determine the total number of possible outcomes.

   a. Choosing apple, blueberry, or cherry pie with chocolate or vanilla ice cream.

   b. Pairing a white, blue, or pink blouse with a blue, white, or gray skirt.

   c. Pairing a blue or red sweater with black, white, or gray slacks.

   d. Choosing a hot dog, hamburger, or fish cake with relish, onion, or pickle with soda pop or orange drink.

4. In each of the following use multiplication to find the total number of outcomes, draw a tree diagram illustrating all these possible outcomes, then list the sample space.

   a. Painting your bedroom walls white, light blue, or gray with blue or gray trim.

   b. Tossing four coins. Use H for heads and T for tails.

   c. Choosing cake, pie, ice cream, or rice pudding for dessert and tea, coffee, or milk as a beverage.

   d. Selecting a brown, gray, or blue sports coat with a white or blue shirt and blue, white, or tan slacks.

Using Mathematics

FACTORIALS AND PERMUTATIONS

Factorials

The product of all natural numbers up to and including a given number is called the **factorial** of the given number.

Five **factorial**, written as **5!**, means $1 \times 2 \times 3 \times 4 \times 5$, or the product 120.

One **factorial**, written as **1!**, is equal to 1.

Exercises

Find the product represented by each of the following.

1. 4!　　**2.** 7!　　**3.** 9!　　**4.** 10!　　**5.** 3! 6!　　**6.** 2! 3! 8!

Permutations

A **permutation** is an arrangement of things in a specific order. The group of letters, R, S, and T may be arranged in any of 6 different ways:

RST; RTS; SRT; STR; TRS; TSR

The symbol $_nP_n$ represents the number of permutations (or arrangements) of **n** things taken **n** at a time. For this, the permutation formula is $_nP_n = n!$.

Example

Find the number of permutations of 3 things taken 3 at a time.

$$_nP_n = n!$$
$$_3P_3 = 3!$$
$$_3P_3 = 1 \times 2 \times 3$$
$$_3P_3 = 6$$

ANSWER: 6 permutations or arrangements

Exercises

Find the number of permutations.

1. 8 things taken 8 at a time.

2. 12 things taken 12 at a time.

3. How many arrangements are there for 7 books on a shelf?

4. How many ways can 15 people be seated in 15 seats?

5. How many ways can the letters of the word STAR be arranged, using all the letters each time? Write these arrangements of letters.

10-7 Probability

In studying probability, we are interested in the likelihood that a particular outcome, or "favorable" outcome, will occur when several different outcomes are possible.

Vocabulary

Probability is a numerical measure indicating the chance, or likelihood, that an event will occur. It is expressed as a ratio in the form of a fraction, a decimal, or a percent.

Mutually exclusive events are events that cannot occur at the same time. A tossed coin will fall either heads or tails, but not heads and tails at the same time.

Two events are **complementary** if the sum of their probabilities is equal to 1.

$$\text{Probability} = \frac{\text{number of favorable outcomes}}{\text{total number of outcomes}}$$

If no outcome is favorable, the probability is 0.
If all outcomes are favorable, the probability is 1.

Scale of Probability

```
        0                0.5               1
        |        low      |      high       |
        |    probability  |   probability   |
    impossible           even            certain
                        chance
```

Example

There are 3 blue marbles in a bag containing 12 marbles. What is the probability of drawing a blue marble if one is drawn at random?

To find the probability of an outcome, follow these steps.

Write the ratio of the number of ways this outcome can occur (favorable) to the total number of outcomes.

a. $\dfrac{\text{number of favorable outcomes}}{\text{total number of outcomes}} = \dfrac{3}{12} = \dfrac{1}{4}$

b. Write: P(blue marble) $= \dfrac{1}{4}$ Read: The probability of a blue marble is $\dfrac{1}{4}$.

ANSWER: P(blue marble) $= \dfrac{1}{4}$

When two events are mutually exclusive, the probability of one event *or* the other event happening is equal to the *sum* of their separate probabilities. If A and B are mutually exclusive events, we can express this as:

$$P(A \text{ or } B) = P(A) + P(B)$$

Example

A bag contains 11 red, 6 yellow, and 4 green marbles. What is the probability of drawing a red or a green marble at random? To find the probability of one event or another event happening:

Find the sum of the two separate probabilities.

$$P(\text{red}) = \frac{\text{number of red marbles}}{\text{total number of marbles}} = \frac{11}{21}$$

$$P(\text{green}) = \frac{\text{number of green marbles}}{\text{total number of marbles}} = \frac{4}{21}$$

$$P(\text{red or green}) = P(\text{red}) + P(\text{green}) = \frac{11}{21} + \frac{4}{21} = \frac{15}{21} = \frac{5}{7}$$

ANSWER: P (red or green) $= \frac{5}{7}$

If event **A** is the **complement** of event **B**, then we can write $P(A) + P(B) = 1$.

It makes sense that event **A happening** and event **A not happening** are complementary events, since **A** must either *happen* or *not happen*. You can express this as:

$$P(A) + P(\text{not } A) = 1$$

Example

A box contains 8 red checkers and 10 black checkers.

The probability of drawing a red checker is $P(\text{red}) = \frac{8}{18} = \frac{4}{9}$.

The probability of **not** drawing a red checker is

$P(\text{not red}) = P(\text{black}) = \frac{10}{18} = \frac{5}{9}$.

Then,

P (red) + P (not red) = P (red) + P (black) $= \frac{4}{9} + \frac{5}{9} = \frac{9}{9} = 1$

1. In terms of expressing probability as the ratio of favorable outcomes to total outcomes, explain why the probability of any event cannot be greater than 1.

2. Give an example of an event for which the probability is 0.

Practice Exercises

1. What is the probability of drawing each of the following at random on the first draw?

   a. A green ball from the box containing 9 balls of which 4 are green.
   b. A marked slip of paper from a hat containing 15 slips of which 9 are marked.
   c. A red checker from a box containing 20 checkers of which 8 are red.
   d. A blue bead from a bag containing 48 beads of which 30 are blue.
   e. A yellow marble from a bag containing 54 marbles of which 36 are yellow.

2. A bag contains 10 brown beads, 6 blue beads, 8 white beads, and 18 black beads. What is the probability of selecting each of the following at random on the first draw?

   a. A white bead or a black bead. b. A brown bead or a blue bead. c. A green bead or a yellow bead.
   d. A white, black, or blue bead. e. A brown, white, or purple bead.

3. A hat contains 12 cards that are numbered 1 through 12. What is the probability of selecting each of the following at random on the first draw?

   a. The number 5 or 10.
   b. An even number.
   c. A prime number.
   d. A number divisible by 3.
   e. A number greater than 6 and less than 9.

4. What is the probability of *not* drawing each of the following at random on the first draw?

   a. A red pencil from a box containing 40 pencils, of which 6 are red.
   b. A white marble from a bag containing 72 marbles, of which 10 are white.
   c. A green book from a shelf containing 8 red books and 4 green books.
   d. A purple bead from a bag containing 16 purple beads and 20 white beads.
   e. A yellow ball from a box containing 3 red, 8 yellow, and 7 white balls.

10-8 Independent and Dependent Events

In determining the probability of an event occurring, it is sometimes important to consider how that event may be related to some other event or outcome.

When we toss two coins, the outcome of the first toss does not affect the outcome of the second toss. The two events are independent.

In general, if events **A** and **B** are independent, the probability that both events will occur is equal to the product of the two separate probabilities. You can symbolize this as:

$$P(\textbf{A and B}) = P(\textbf{A}) \times P(\textbf{B}).$$

Example

When two coins are tossed, what is the probability that both coins will fall heads up?

The probability of heads on the first coin is $\frac{1}{2}$. Thus, $P(\textbf{H}_1) = \frac{1}{2}$.

The probability of heads on the second coin is also $\frac{1}{2}$. Thus $P(\textbf{H}_2) = \frac{1}{2}$.

The probability of both coins falling heads up is

$$P(\textbf{H}_1 \text{ and } \textbf{H}_2) = P(\textbf{H}_1) \times P(\textbf{H}_2) = \frac{1}{2} \times \frac{1}{2} = \frac{1}{4}$$

ANSWER: $P(H_1 \text{ and } H_2) = \frac{1}{4}$.

When two events are dependent, the outcome of the first event will change the size of the sample space, thus affecting the probability of the second event.

Example

What is the probability of drawing 2 blue marbles from a bag that contains 4 blue and 3 green marbles?

To find the probability of these two dependent events occurring, follow these steps.

a. The probability of drawing a blue marble on the first draw is

$$P(\textbf{B}_1) = \frac{4}{7}.$$

b. Reduce the sample space by 1, since one drawing has been made. Also reduce the number of possible favorable outcomes by 1, since a favorable outcome is assumed on the first draw.

Of the six remaining marbles, 3 will be blue if the first marble drawn is blue; thus, the probability of drawing a blue marble on the second draw is

$$P(B_2) = \frac{3}{6} = \frac{1}{2}.$$

c. Multiply the separate probabilities.

$$P(B_1 \text{ and } B_2) = \frac{4}{7} \times \frac{1}{2} = \frac{4}{14} = \frac{2}{7}$$

ANSWER: $P(B_1 \text{ and } B_2) = \frac{2}{7}$.

Think About It

Explain the difference between mutually exclusive events and independent events. Can two events be both mutually exclusive and independent at the same time? Why or why not?

Practice Exercises

1. Spinner A has five sectors, each of the same size, numbered 1 through 5. Spinner B has eight sectors, each of the same size, numbered 1 through 8. What is the probability of each of the following?

    a. Spinner A stops at 2, and Spinner B at 7.
    b. Both spinners stop at 3.
    c. Both spinners stop at 6.
    d. Both spinners stop at a prime number.

2. A bowl contains 6 blue beads, 10 white beads, and 8 yellow beads. Jeff draws two beads, replacing the first before the second is drawn. What is the probability that he draws each of the following?

    a. A blue bead, then a white bead.
    b. A yellow bead, then a blue bead.
    c. Two yellow beads.
    d. Two blue beads.

3. A bag contains 4 green, 5 black, and 6 red marbles. Naomi draws two marbles without replacement. What is the probability that she draws each of the following?

    a. A black marble, then a green marble.
    b. A red marble, then a black marble.
    c. Two green marbles.
    d. A green marble, then a yellow marble.

4. A hat contains eight cards numbered 1 through 8. Marian draws two cards without replacement. What is the probability that she draws each of the following?

    a. The 3-card, then the 7-card.
    b. An even numbered card, then an odd numbered card.
    c. Two prime-numbered cards.
    d. A card between 1 and 3, then a card greater than 5.

Making Predictions

Sometimes statisticians want to study the likelihood of certain events even when the size of the sample space is unknown. In these cases, it is necessary to observe a selected portion of the sample space in order to make predictions about the entire sample space.

Suppose a school librarian wants to know the likelihood that any given student will return a book past its due date. He has no idea of how many students will be returning books (the *population* size), but he can make a *prediction* or find the empirical probability that a borrower will be overdue by observing a sample of students.

- The librarian takes a sample of 100 students returning books.
- Of the 100, he finds that 18 students are returning overdue books.

Vocabulary

Empirical probability is a probability value that is found by observing a sample from an entire population.

A **population** is the sample space in empirical probability.

A **sample** is a part of the population. The characteristics of the sample are assumed to represent the entire population.

Examples

1. What is the empirical probability that any given student will return an overdue book?

Assuming that the sample is representative of the entire population:

$$P\,(\text{overdue}) = \frac{\text{number of favorable outcomes}}{\text{size of the sample}} = \frac{18}{100} = 18\%$$

ANSWER: The empirical probability is 18%.

2. If 250 students will return books on a given day, how many students would you predict to be returning overdue books?

Predicting that 18% of 250 students will be returning overdue books:

$$0.18 \times 250 = 45$$

ANSWER: 45 students

1. Look up "empirical" in the dictionary, and explain its meaning in your own words.

2. When the librarian is taking a random sample of 100 students, does it matter whether he observes 100 consecutive students, every other student, every third student, and so on? Explain why or why not.

3. Suppose you are sampling a population. How do you think the size of your sample might affect the accuracy of your predictions?

Practice Exercises

1. A quality control inspector is sampling lightbulbs. Give the empirical probability that any given bulb is defective.

   a. 1000 bulbs are sampled and 23 are defective.
   b. 500 bulbs are sampled and 18 are defective.
   c. 750 bulbs are sampled and 36 are defective.
   d. 600 bulbs are sampled and 28 are defective.
   e. 800 bulbs are sampled and 45 are defective.

2. A public opinion poll was conducted in a particular town to see how many residents would be in favor of building a new highway through the town. Give the number of people responding favorably.

   a. The empirical probability was 42% and 300 people were sampled.
   b. The empirical probability was 55% and 460 people were sampled.
   c. The empirical probability was 34% and 250 people were sampled.
   d. The empirical probability was 63% and 500 people were sampled.
   e. The empirical probability was 28% and 350 people were sampled.

Computer Know-How

The computer program below computes the *arithmetic mean* of any 5 numbers.

```
1 LET T = Ø
1Ø PRINT "PLEASE TYPE A NUMBER AFTER
EACH ?-MARK"
2Ø FOR I = 1 TO 5
3Ø INPUT X
4Ø LET T = T + X
5Ø NEXT I
6Ø PRINT "THE ARITHMETIC MEAN IS: "; T/5
```

a. Enter the program and then RUN it to make sure it works.
b. Modify the program to compute the arithmetic mean of any *10* numbers.

Making Decisions

USING AND MISUSING STATISTICS

USING AND MISUSING STATISTICS

You collected some data about how the students in your class did in selling tickets to the school play.

Mickey	21	Larry	27
Carol	28	Roberto	10
Bob	16	Jan	18

You made three different bar graphs to show the results.

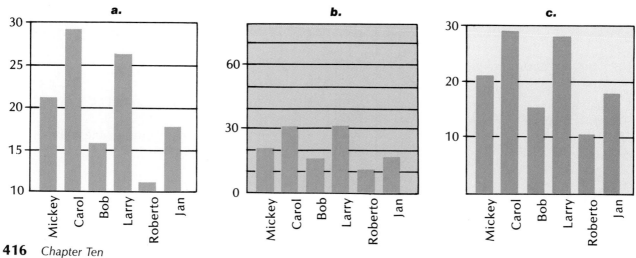

1. What is the difference between the graphs?

2. _a._ What impression does graph (a) give you?

   _b._ Does it look as though Larry sold twice as many tickets as Jan? Did he?

3. _a._ What impression does graph (b) give you?

   _b._ Did all the students sell about the same number of tickets?

4. What impression does graph (c) give you?

5. Which graph do you think shows the results most fairly?

6. Which graph would you use if you wanted to give the impression that some students sold many more tickets than others?

Your friend collects some data about tickets sold by another class.

Gayle	8	Maria	3
Sun	4	Allegra	7
Sean	12	Brad	6

7. Make a circle graph for each group of 6 students to show the part of their group's tickets each student sold.

8. Compare your graphs in Question 7. Is that a good way to show how the two classes did? Explain.

9. Make a graph that you feel fairly shows the sales by the two classes. Be prepared to defend your answer!

10. Collect some statistics about your class. See if you can use graphs to give different impressions about the data.

In Your Community

Watch your newspaper or news magazines for graphs or uses of statistics that you feel are misleading and bring them to class.

Career Application
FORESTRY TECHNICIAN

K eeping a forest healthy, beautiful, and productive requires a lot of work. *Forestry technicians* inspect trees for disease, prevent fires and flood damage, and plant new trees. They work in national and state parks to help maintain forest areas for hiking and camping, and teach visitors about park rules.

Most forestry technicians study forestry for one or two years after high school. Others learn their skills by working on-the-job training.

1. A forestry technician counted the trees in a certain area and made the graph at the right. How many more birch trees than spruce trees were counted?

2. The circle graph shows the distribution of campers at a park. If there were 5,394 tent campers this season, how many campers were there in all?

3. A forestry technician counted the diseased trees in fifteen plots. The numbers were: 53, 64, 52, 44, 39, 41, 60, 58, 49, 42, 44, 55, 59, 36, and 63. What are the mean, median, and mode of the data?

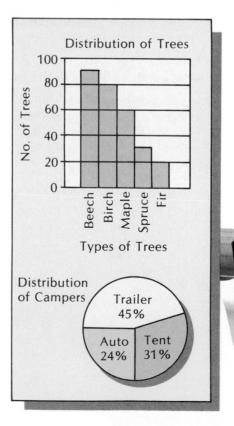

Distribution of Trees

No. of Trees: 100, 80, 60, 40, 20, 0

Types of Trees: Beech, Birch, Maple, Spruce, Fir

Distribution of Campers

Trailer 45%
Auto 24%
Tent 31%

Chapter 10 Review

Vocabulary

arithmetic mean p. 402	frequency p. 400	mutually exclusive events p. 409
bar graph p. 394	frequency distribution p. 400	outcome p. 405
bimodal p. 402	frequency polygon p. 400	permutation p. 408
broken-line graph p. 396	histogram p. 400	population p. 414
categorical data p. 394	independent events p. 412	probability p. 409
circle graph p. 398	mean p. 402	range p. 402
complementary events p. 409	measures of central	sample p. 414
dependent events p. 412	tendency p. 402	sample space p. 405
empirical probability p. 414	median p. 402	simple event p. 405
event p. 405	metric data p. 396	statistics p. 400
factorial p. 408	mode p. 402	tree diagram p. 405

Page
394 **1.** Make a bar graph showing the amounts of sodium in a 235 mL glass of water in the following cities.
Galveston, TX: 82 mg Tucson, AZ: 10 mg
Houston, TX: 19 mg Miami, FL: 5 mg
San Diego, CA: 30 mg

Page
396 **2.** Make a broken-line graph showing average hospital bills in recent years.
1977, $1,584 1978, $1,776 1979, $1,992
1980, $2,248 1981, $2,660 1982, $3,016

398 **3.** Make a circle graph showing how a local government spent its funds.
Utilities, 3.8% Highways, 9.1% Insurance Trust, 14.6%
Public Welfare, 12.4% Education, 36.3% Other, 23.8%

400 For the list of scores below, make each of the following.

11, 15, 14, 12, 19, 12, 16, 15, 17, 18, 19, 12, 10, 11,
15, 17, 14, 14, 15, 12, 12, 13, 16, 19, 12, 10, 14

4. A frequency distribution table **5.** A histogram **6.** A frequency polygon

402 For the list of scores below, determine each of the following.

10, 8, 9, 10, 3, 4, 10, 3, 5, 3, 8, 10, 10, 9, 3, 4, 7, 2, 8, 4

7. Range **8.** Mean **9.** Mode **10.** Median

405 **11.** What is the sample space for rolling a cube?

405 **12.** Use multiplication to determine the total number of possible outcomes in rolling two number cubes.

409 **13.** What is the probability of rolling a number cube and getting a 4 or less?

409 **14.** If a box contains 6 blue, 7 red, and 3 white stars, what is the probability of picking a blue star?

412 **15.** If you roll 2 number cubes, what is the probability of both being 6?

414 **16.** If 4 CDs out of 200 did not pass quality control, what is the empirical probability of finding a defective CD?

Chapter 10 Test

Page
394 **1.** Make a bar graph showing the height of the tall buildings listed below.

> John Hancock Tower: 790 ft
> Detroit Plaza Hotel: 720 ft
> Empire State Building: 1,250 ft
> World Trade Center: 1,350 ft
> Sears Tower, Chicago: 1,454 ft
> Texas Commerce Tower: 1,002 ft

Page
396 **2.** Make a broken-line graph showing the average high temperature in degrees Fahrenheit for each month in Madison, Wisconsin.

Jan	Feb	Mar	Apr	May	June
28	30	40	56	68	78

July	Aug	Sept	Oct	Nov	Dec
83	81	73	62	44	32

398 **3.** Make a circle graph showing the average costs of operating a car for one year in which 15,000 miles were driven.

> Running costs: $1,170   Depreciation: $1,298   License and Taxes: $97
> Insurance: $470   Finance Charges: $529

400 Use the scores below to make each of the following.

180, 183, 192, 199, 183, 180, 182, 195, 199, 191, 199, 180, 191, 199, 182, 185, 188, 191, 180, 185, 194, 195, 182, 188, 190, 189, 188, 199, 190, 180

4. A frequency distribution   **5.** A histogram   **6.** A frequency polygon

402 For the scores below, find each of the following.

78, 83, 83, 72, 83, 61, 75, 91, 95, 72

7. The mean   **8.** The median   **9.** The mode   **10.** The range

405 **11.** A _____?_____ consists of all the possible outcomes of a given activity or experiment.

405 **12.** Use multiplication to determine the total number of possible outcomes when spinning a spinner with the numbers 1–8 and a spinner with the letters A and B.

409 **13.** What is the probability of an event that is certain?

409 **14.** What is the probability of picking a multiple of 5 from cards numbered from 1 to 20?

409 A bag has 2 red, 7 yellow, and 4 blue marbles. Give each of the following probabilities.

15. Picking a blue marble   **16.** Not picking yellow   412 **17.** Picking 2 blue

412 **18.** When rolling 2 number cubes, what is the probability of getting a sum greater than 4?

414 **19.** If 3 radios out of 75 are found to be defective, what is the empirical probability of getting a defective radio?

414 **20.** If the empirical probability of a defective part is 1.5%, how many defective parts would you predict there would be in a batch of 20,000 pieces?

Achievement Test

Page

Change.

353 **1.** 63.8 m to cm

357 **3.** 12.4 kg to g

359 **5.** 759 mL to L

363 **7.** 26.45 cm²

365 **9.** 1,495 cm³ to dm³

365 **11.** 33.4 liters of water occupy _____ dm³ and weigh _____ kg.

372 **12.** How many yards are in $6\frac{2}{3}$ miles?

377 **14.** How many liquid ounces are in 3 quarts?

Page

353 **2.** 419 mm to cm

357 **4.** 35 mg to g

359 **6.** 12 L to cL

363 **8.** 31.7 km² to hectares

365 **10.** 83.9 m³ to dm³

375 **13.** Change $2\frac{5}{8}$ pounds to ounces.

379 **15.** What part of a square foot is 48 square inches?

381 **16.** If there are 16 teams playing in a single elimination soccer tournament, how many games will the championship team have to play? In a single-elimination tournament, a team is eliminated after one loss. Make a drawing to help you solve the problem.

383 **17.** Change 297 cubic feet to cubic yards.

385 **18.** What part of a minute is 45 seconds?

388 **19.** Using the bus schedule on p. 389, find at what times Bus 820 leaves Memphis and arrives in Oklahoma City.

394 **20.** Make a bar graph showing the heights of the world's tallest buildings:

World Trade Center, 1,350 ft; CNR Tower, 1,815 ft; Centerpoint Tower, 1,065 ft; Eiffel Tower, 984 ft; Sears Tower, 1,454 ft.

396 **21.** Make a broken-line graph showing the U.S. national income for five recent years: 1st yr, $1,760 billion; 2nd yr, $1,967 billion; 3rd yr, $2,117 billion; 4th yr, $2,373 billion; 5th yr, $2,450 billion.

398 **22.** Make a circle graph showing a family's budget:

Housing, 25%; Food, 30%; Transportation, 20%; Recreation, 10%; Savings, 5%; Other, 10%.

400 **23.** Make a frequency distribution table for the following list of scores.

5, 9, 6, 4, 5, 8, 7, 3, 9, 6, 2, 4, 5, 9, 7, 4, 8,
3, 5, 7, 4, 6, 8, 9, 7, 5, 8, 4, 6, 9, 4, 3, 8, 6, 7

402 Use the scores above to find each of the following.

24. Mean   **25.** Median   **26.** Mode   **27.** Range

409 A spinner has ten equal sectors numbered from 1 through 10.

28. What is the probability that the pointer on the spinner will stop on an even-numbered sector?

29. What is the probability that the pointer will stop on an odd number less than 4?

Unit 4

Geometry

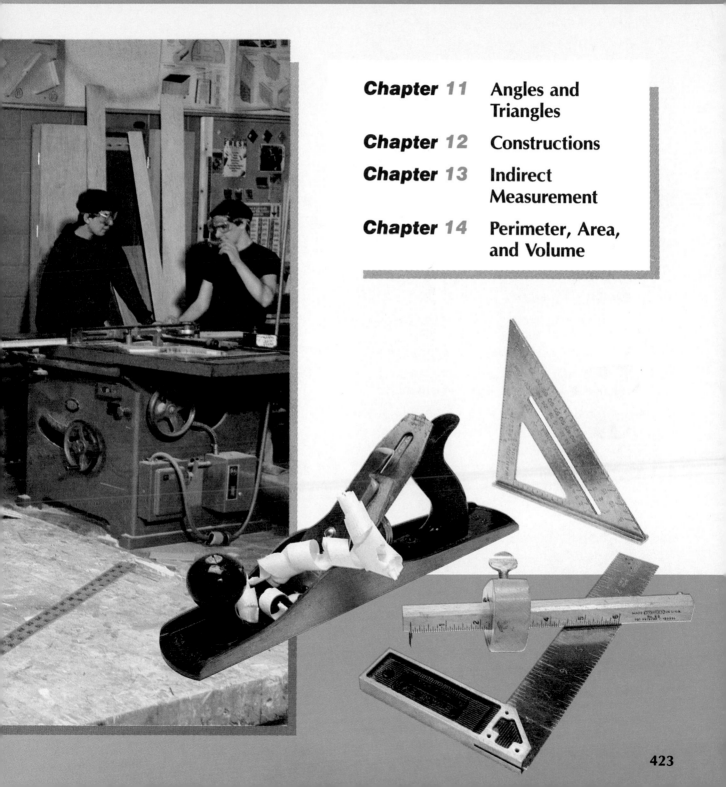

423

Inventory Test

428 Read, or write in words, each of the following.

1. $\overline{BC}$ **2.** $\overleftrightarrow{NR}$ **3.** $\overrightarrow{AT}$ **4.** $\overleftrightarrow{CF}$ **5.** $\overline{SY}$ **6.** $\overrightarrow{PQ}$

Name each of the following and express them symbolically.

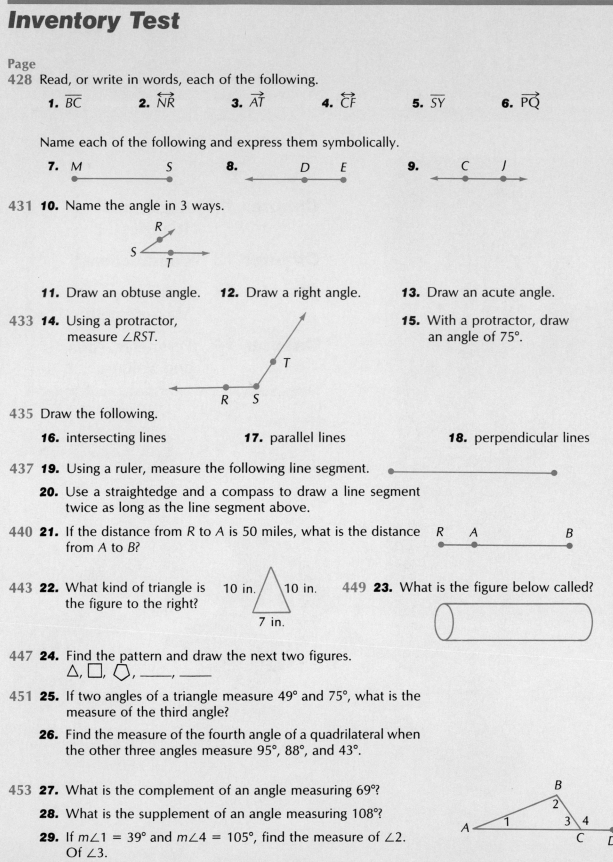

7. M   S

8. D   E

9. C   J

431 **10.** Name the angle in 3 ways.

11. Draw an obtuse angle.   **12.** Draw a right angle.   **13.** Draw an acute angle.

433 **14.** Using a protractor, measure ∠RST.

15. With a protractor, draw an angle of 75°.

435 Draw the following.

16. intersecting lines   **17.** parallel lines   **18.** perpendicular lines

437 **19.** Using a ruler, measure the following line segment.

20. Use a straightedge and a compass to draw a line segment twice as long as the line segment above.

440 **21.** If the distance from R to A is 50 miles, what is the distance from A to B?

443 **22.** What kind of triangle is the figure to the right?   10 in.   10 in.   7 in.

449 **23.** What is the figure below called?

447 **24.** Find the pattern and draw the next two figures.

△, □, ⬠, _____, _____

451 **25.** If two angles of a triangle measure 49° and 75°, what is the measure of the third angle?

26. Find the measure of the fourth angle of a quadrilateral when the other three angles measure 95°, 88°, and 43°.

453 **27.** What is the complement of an angle measuring 69°?

28. What is the supplement of an angle measuring 108°?

29. If $m\angle 1 = 39°$ and $m\angle 4 = 105°$, find the measure of ∠2. Of ∠3.

30. If $m\angle3 = 67°$, what is the measure of $\angle1$? $\angle2$? $\angle4$?

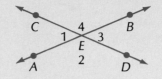

456 **31.** Parallel lines AB and CD are cut by transversal EF. If $m\angle6 = 95°$, what is the measure of $\angle1$? $\angle2$? $\angle3$? $\angle4$? $\angle5$? $\angle7$? $\angle8$?

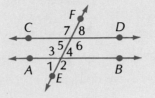

458 **32.** Which pair of triangles below are congruent?

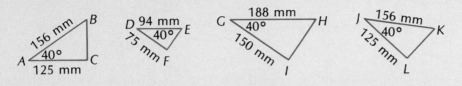

460 **33.** Which pair of triangles above are similar?

466 **34.** Make a copy of the following line segment: ●————————●

467 **35.** Draw an angle of 30° with a protractor, then copy the angle using a compass.

468 **36.** Construct a triangle with the sides measuring $1\frac{5}{8}$ in., $2\frac{1}{8}$ in., and $1\frac{3}{4}$ in.

37. Construct a triangle with sides measuring 5 cm and 4.5 cm and an included angle of 75°.

38. With a scale of $\frac{1}{4}$ in. = 1 ft, draw a triangle with angles measuring 30° and 60° and an included side measuring 9 ft.

471 **39.** Draw any line. Use a compass to construct a perpendicular at a point on the line.

472 **40.** Draw any line. Use a compass to construct a perpendicular from a point not on the line.

473 **41.** Twelve toothpicks were used to make the figure shown. How can you form 5 squares by moving only 3 toothpicks?

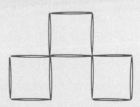

475 **42.** Draw any line segment. Use a compass to bisect it.

476 **43.** Draw any angle. Use a compass to bisect it.

477 **44.** Draw any line. Locate a point outside this line. Through this point draw a line parallel to the first line.

482 **45.** Find the square of 27.

483 **46.** Find $\sqrt{6,400}$.

484 **47.** Find $\sqrt{50}$ by the estimate, divide, and average method.

487 48. Find the base of a right triangle if the hypotenuse is 146 m and the altitude is 96 m.

49. Find the hypotenuse of a right triangle if the altitude is 112 ft and the base is 384 ft.

50. Find the altitude of a right triangle if the base is 45 cm and the hypotenuse is 117 cm.

489 51. Find the height of a flagpole that casts a shadow of 450 ft at a time when a girl, $5\frac{1}{2}$ ft tall, casts a shadow of 33 ft.

In right triangle *ABC:*

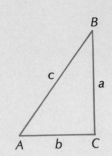

491 52. Find side *a* when angle $A = 89°$ and side $b = 250$ cm.

53. Find side *a* when angle $A = 71°$ and side $c = 800$ yd.

54. Find side *b* when angle $A = 47°$ and side $c = 500$ m.

498 55. Find the perimeter of a rectangle 138 m long and 57 m wide.

499 56. Find the perimeter of a square whose side measures 106 mm.

500 57. Find the perimeter of a triangle with sides measuring $8\frac{9}{16}$ in., $7\frac{5}{8}$ in., and $6\frac{3}{4}$ in.

Find the circumference of each of the following.

501 58. A circle whose radius is 49 ft.

59. A circle whose diameter is 64 m.

503 60. Find the area of a rectangle 293 m long and 185 m wide.

504 61. Find the area of a square whose side measures 78 km.

505 62. Find the area of a parallelogram with an altitude of 98 ft and a base of 107 ft.

506 63. Find the area of a triangle whose altitude is $10\frac{1}{2}$ in. and whose base is 9 in.

507 64. Find the area of a trapezoid with bases of 87 m and 41 m and a height of 62 m.

Find the area of each of the following.

508 65. A circle whose radius is 3.8 km.

66. A circle whose diameter is 56 in.

510 67. Find the total area of the outside surface of a rectangular solid $7\frac{1}{2}$ in. long, 6 in. wide, and $2\frac{3}{4}$ in. high.

511 68. Find the total area of the outside surface of a cube whose side measures 9.3 m.

Find the total area of the outside surface of each of the following.

512 69. A cylinder 20 cm in diameter and 34 cm high.

70. A sphere whose diameter is 84 mm.

515 71. Find the volume of a rectangular solid 109 cm long, 97 cm wide, and 63 cm high.

516 72. Find the volume of a cube whose edge measures 2 ft 8 in.

517 73. Find the volume of a cylinder with a radius of 8 cm and a height of 10 cm.

Find the volume of each of the following.

518 74. A sphere whose diameter is 4.5 m.

75. A cone 10 in. in diameter and 9 in. high.

76. A square pyramid 16 yd on each edge of the base and 13 yd high.

Points, Rays, Lines

The word geometry comes from ancient Greek words meaning "to measure the earth." Geometry is an important branch of mathematics, used by almost everyone.

Vocabulary

A **point** is a location in space.

A **line** is a set of points that continues in both directions, without end.

A **line segment**, or **segment**, is part of a line having two endpoints.

A **ray** is part of a line with one endpoint. It continues without end in one direction.

Collinear points are points on the same line.

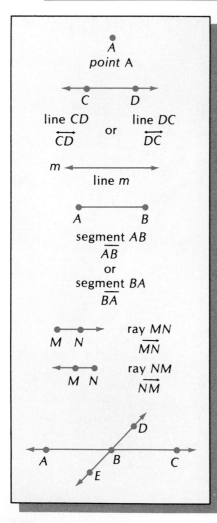

A point has no length or width and cannot be seen. A dot represents a point and is named by using one capital letter.

A line has no width but an infinite length. There are no endpoints on a line. We cannot actually see a geometric line, but we can draw figures that represent lines using arrows. A line can be named by using two points on the line or by a single lowercase letter.

A line segment has length but no width. A segment is named by its two endpoints.

A ray has no width but an infinite length even though it has an endpoint. A ray is named by first the endpoint and then any other point on the ray. The symbol above the two capital letters is always an arrow extending to the right.

Points A, B, and C are collinear. Points A, B, and D are not collinear.

Examples

Name and then express symbolically each of the following.

1.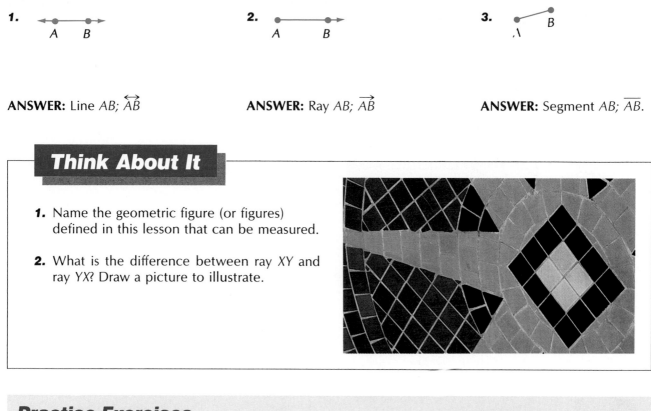
$$\underset{A \quad\ B}{\longleftrightarrow}$$

2.
$$\underset{A \qquad B}{\bullet\!\!-\!\!-\!\!-\!\!\longrightarrow}$$

3.
B

ANSWER: Line AB; $\overleftrightarrow{AB}$

ANSWER: Ray AB; $\overrightarrow{AB}$

ANSWER: Segment AB; $\overline{AB}$.

Think About It

1. Name the geometric figure (or figures) defined in this lesson that can be measured.

2. What is the difference between ray XY and ray YX? Draw a picture to illustrate.

Practice Exercises

Name and then express symbolically each of the following.

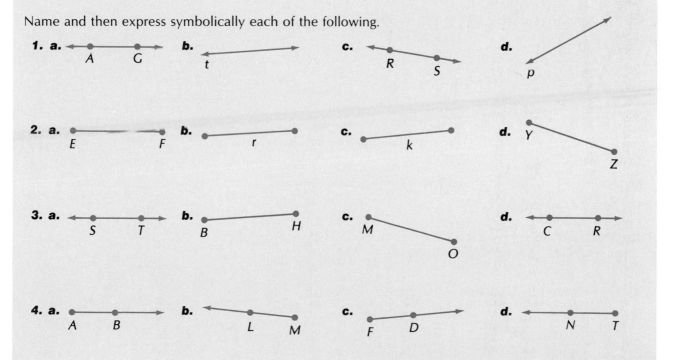

1. a. A G **b.** t **c.** R S **d.** p

2. a. E F **b.** r **c.** k **d.** Y Z

3. a. S T **b.** B H **c.** M O **d.** C R

4. a. A B **b.** L M **c.** F D **d.** N T

5. Name and then express symbolically
in three ways.

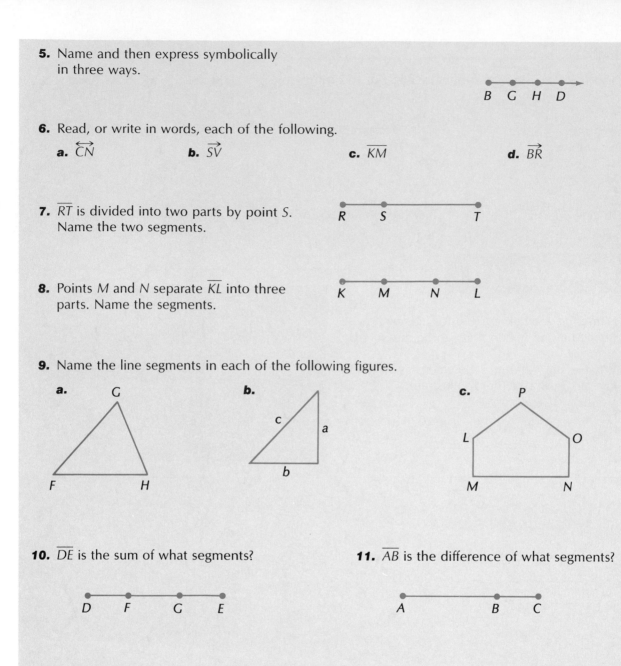

6. Read, or write in words, each of the following.

    a. $\overleftrightarrow{CN}$ **b.** $\overrightarrow{SV}$ **c.** $\overline{KM}$ **d.** $\overrightarrow{BR}$

7. $\overline{RT}$ is divided into two parts by point S.
Name the two segments.

8. Points M and N separate $\overline{KL}$ into three
parts. Name the segments.

9. Name the line segments in each of the following figures.

    a.   **b.**   **c.**

10. $\overline{DE}$ is the sum of what segments?

11. $\overline{AB}$ is the difference of what segments?

Make drawings to help you decide whether each of the following is
true or false.

12. An infinite number of straight lines can be drawn through
a point.

13. Two points determine a straight line and one and only one straight
line can pass through any two points.

14. The shortest path between two points is along a line.

11-2 Naming and Classifying Angles

Another basic concept in geometry is that of the angle.

Angles can be named in different ways.

1. They can be named by the vertex.

The angle shown can be named ∠A, read "angle A."

2. They can be numbered. The angle shown is ∠1, or angle 1.

3. They can be named by three points, the vertex and one on each side. The vertex is always in the middle. The angle shown can be named ∠ABC or ∠CBA.

Vocabulary

An **angle** is the figure formed by two rays with a common endpoint.

The **vertex** of an angle is the common endpoint.

The **sides** of an angle are the two rays.

Think About It

1. What is the difference between ∠XYZ and ∠YZX?

2. How many different angles are shown in the figure to the right? Name each angle in two ways.

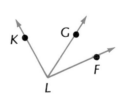

Angles are measured in degrees (symbol °). There are 360° in one complete rotation. Angles are classified by their number of degrees. The measure of an angle is written m.

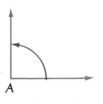

right angle

$\frac{1}{4}$ of a rotation
measures 90°
$m\angle A = 90°$

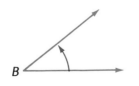

acute angle

measures more than 0°
but less than 90°
$0° < m\angle B < 90°$

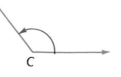

obtuse angle

measures more than 90°
but less than 180°
$90° < m\angle C < 180°$

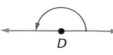

straight angle

$\frac{1}{2}$ of a rotation
measures 180°
$m\angle D = 180°$

Examples

Classify each angle as acute, right, obtuse, or straight.

1.

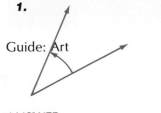

Guide: Art

ANSWER: acute

2.

ANSWER: straight

3.

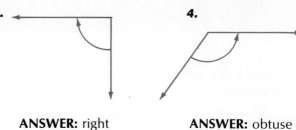

ANSWER: right

4.

ANSWER: obtuse

Practice Exercises

1. Name the sides and vertex of the following angle.

2. Name the following angle in two ways.

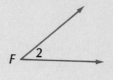

3. Name each of the following angles.

a.

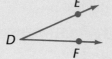

b.

c.

d.

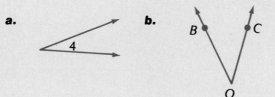

4. What unit is used to measure angles? What symbol is written for it?

5. Copy and complete the sentences.

    a. One complete rotation measures _____°.
    b. A right angle measures _____° and is _____ of a rotation.
    c. The measure of an acute angle is greater than _____° and less than _____°.
    d. The measure of an obtuse angle is greater than _____° and less than _____°.

6. Which of the following are measures of an acute angle?

    a. 67°   **b.** 98°   **c.** 89°   **d.** 155°

7. Which of the following are measures of an obtuse angle?
    a. 39°   **b.** 126°   **c.** 94°   **d.** 88°

8. For each angle, tell whether it is acute, right, obtuse, or straight.

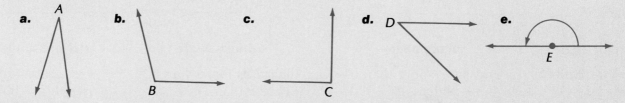

11-3 ## Using a Protractor

Many times it is important to measure or draw an angle accurately. A special instrument is used to do this.

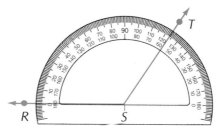

<table>
</table>

> **Vocabulary**
>
> A **protractor** is an instrument used to measure and draw angles.

Examples

1. Use a protractor to measure ∠RST.

To measure an angle, follow these steps.

a. Place the protractor on the angle so that its center is at the vertex and its zero line lies along a ray.

b. Read the measure of the angle on the scale that has zero on the side of the angle.

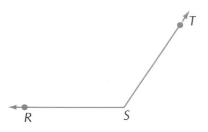

ANSWER: m∠RST = 124°

---

2. Draw an angle of 75°.

To draw an angle, follow these steps.

a. Draw a ray to represent one side of the angle.
b. Place the protractor so that its straight edge falls on this ray and its center mark is on the vertex of the angle.
c. Use the scale whose zero line lies along the ray and mark the correct number of degrees.
d. Remove the protractor and draw a ray from the vertex through the dot.

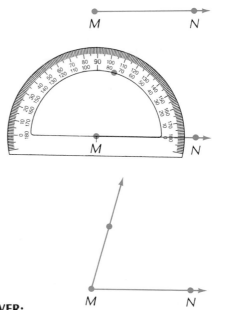

ANSWER:

Think About It

1. Does the length of the sides of an angle affect its degree measure? Draw a picture to illustrate your answer.

2. When you use a protractor to measure an angle, how do you decide which scale, the top or the bottom, to use?

Practice Exercises

1. Estimate the size of each angle. Then measure with a protractor.

   a. b. c. d.

2. a. Which angle is smaller, ∠r or ∠S?
   b. The size of an angle does not depend upon the length of the _____.

   r   s

3. How many degrees are in the angle formed by the hands of a clock at each time?

   a. 5 o'clock   b. 3 o'clock   c. 2 o'clock   d. 6 o'clock

4. Through how many degrees does the minute hand of a clock turn in each period?

   a. 30 min   b. 10 min   c. 25 min   d. 1 h

For each measure, first draw a ray with the endpoint either on the left or on the right as required. Then use a protractor to draw the angle.

5. Left endpoint as the vertex:

   a. 30°   b. 83°   c. 120°   d. 100°

6. Right endpoint as the vertex:

   a. 80°   b. 9°   c. 124°   d. 170°

7. With a protractor, measure each of the following angles, then draw an angle equal to it.

   a.   b. B   c. C

   A

8. Draw with a protractor.

   a. a right angle   b. a straight angle

Parallel, Intersecting, and Perpendicular Lines

You can think of a piece of paper as part of a plane. Lines in the same plane are either parallel or intersecting.

Vocabulary

A **plane** is a flat surface that goes on endlessly in all directions.

Intersecting lines are lines in the same plane that meet.

Parallel lines are lines in the same plane that never meet.

Perpendicular lines are lines in the same plane that intersect to form a right angle.

Examples

1.

Line AB ∥ Line CD $\overleftrightarrow{AB}$ ∥ $\overleftrightarrow{CD}$
("Line AB is parallel to line CD")

2.

$\overleftrightarrow{AB}$ intersects $\overleftrightarrow{EF}$ at point C

3.

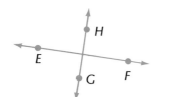

$\overleftrightarrow{EF} \perp \overleftrightarrow{GH}$
("Line EF is perpendicular to line GH")

1. When perpendicular lines intersect, how many right angles are formed?

2. Find an example of perpendicular segments and parallel segments in your classroom.

Practice Exercises

Tell which of the following figures show intersecting lines or rays, which show parallel lines or rays, and which show perpendicular lines or rays.

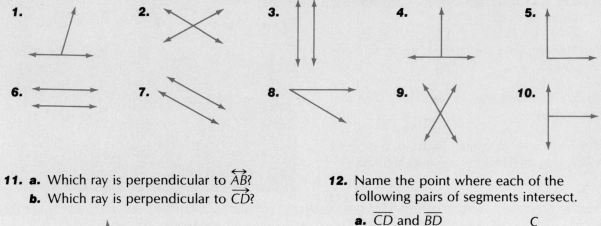

1. **2.** **3.** **4.** **5.**

6. **7.** **8.** **9.** **10.**

11. a. Which ray is perpendicular to $\overleftrightarrow{AB}$?
b. Which ray is perpendicular to $\overrightarrow{CD}$?

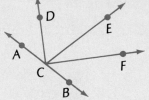

12. Name the point where each of the following pairs of segments intersect.

a. $\overline{CD}$ and $\overline{BD}$
b. $\overline{BC}$ and $\overline{CD}$
c. $\overline{BD}$ and $\overline{CB}$

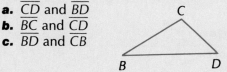

Points that lie on the *same* straight line are called collinear points.

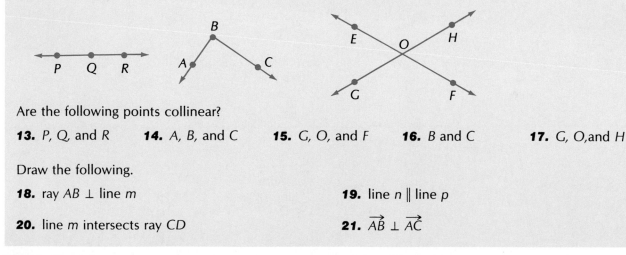

Are the following points collinear?

13. *P, Q,* and *R*   **14.** *A, B,* and *C*   **15.** *G, O,* and *F*   **16.** *B* and *C*   **17.** *G, O,* and *H*

Draw the following.

18. ray *AB* ⊥ line *m*

19. line *n* ∥ line *p*

20. line *m* intersects ray *CD*

21. $\overrightarrow{AB} \perp \overrightarrow{AC}$

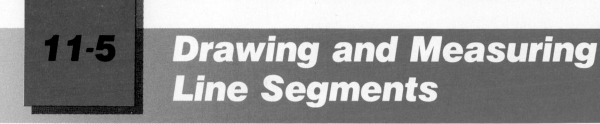

Drawing and Measuring Line Segments

A straightedge is used to draw line segments.
A ruler is a straightedge with calibrated measurements.
It is used to measure line segments and can be used to draw them also.

> ### Vocabulary
>
> A **straightedge** and a **ruler** are used to draw and measure segments.
>
> A **compass** is an instrument also used in measuring.

Examples

1. Draw a line segment $1\frac{3}{4}$ in. long.

ANSWER: _____

2. Measure the length of the segment.

ANSWER: $1\frac{7}{8}$ inches

3. Use a compass to draw a segment $1\frac{1}{2}$ in. long.

Draw a working line with a straightedge.

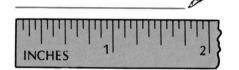

Open the compass to measure $1\frac{1}{2}$ in.

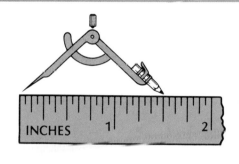

With the compass set at $1\frac{1}{2}$ in., mark the length on the working line.

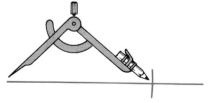

Think About It

1. What is the difference between a straightedge and a ruler?

2. Draw a line segment. Measure it using rulers with two different measurements (like eighths and sixteenths). Are the measurements the same? Why or why not?

Practice Exercises

1. Using a ruler, draw line segments having the following lengths.

   a. $4\frac{1}{2}$ in.　　　**b.** $2\frac{3}{4}$ in.　　　**c.** $3\frac{5}{8}$ in.　　　**d.** $4\frac{13}{16}$ in.

2. Measure the length in inches of each of the following line segments. Use a ruler for segments a and b, and a compass and ruler for segments c and d.

   a. ————————————————　　　**b.** ——————————

   c. ——————————————————　　　**d.** ———————————————

3. Using a metric ruler, draw line segments having the following lengths.

   a. 6 cm　　　**b.** 49 mm　　　**c.** 13 cm　　　**d.** 108 mm

4. Draw a line segment 15 centimeters long. Using a compass, mark in succession on this segment lengths of 2.8 centimeters, 4.9 centimeters, and 3.6 centimeters. How long is the remaining segment?

5. The symbol $m\overline{BR}$ is read "the measure of line segment BR" and represents the length of the segment.

   B————————————R　S————————————T

   A——————————D　N——————————H

   a. Find $m\overline{BR}$; $m\overline{AD}$; $m\overline{ST}$; $m\overline{NH}$　**b.** Does $m\overline{AD} = m\overline{NH}$?　　　**c.** Does $m\overline{ST} = m\overline{BR}$?

6. Draw any line segment EF. Label a point not on $\overline{EF}$ as A. Draw a line segment AG equal in length to $\overline{EF}$.

7. Draw a line segment twice as long as $\overline{BR}$ in Exercise 5.

Refresh Your Skills

1. $6 + 0.666 + 0.66$　　　**2.** $8.5 - 0.42$　　　**3.** 0.25×8.04　　　**4.** $1.5\overline{)0.75}$

　2–5　　　　　　　　　2–6　　　　　　　　　2–7　　　　　　　　　2–9

5. $1\frac{2}{3} + \frac{5}{6} + 1\frac{7}{8}$　　**6.** $16 - 3\frac{7}{10}$　　**7.** $3\frac{3}{5} \times 5\frac{1}{3}$　　**8.** $20 \div 1\frac{1}{4}$

　3–8　　　　　　　　　3–9　　　　　　　　　3–11　　　　　　　　3–12

9. Find 18.4% of $9,500　　　**10.** What percent of $15 is $.30?　　**11.** $10 is 5% of what amount?

　4–7　　　　　　　　　　　　　4–6　　　　　　　　　　　　4–9

Using Mathematics

PRECISION AND ACCURACY

Measurement is never exact—it is only approximate. Precision is the closeness of a measurement to the true measurement. *The smaller the unit of measure, the more precise the measurement.*
The same length measured on each of the following scales shows a measurement of:

 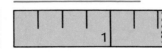  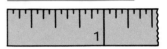

$1\frac{1}{2}$ inches
unit of measure: $\frac{1}{2}$ inch
least precise

$1\frac{1}{4}$ inches
unit of measure: $\frac{1}{4}$ inch

$1\frac{3}{8}$ inches
unit of measure: $\frac{1}{8}$ inch

$1\frac{5}{16}$ inches
unit of measure: $\frac{1}{16}$ inch
most precise

The greatest possible error for a measurement is $\frac{1}{2}$ the unit of measure being used. The relative error is the ratio of the greatest possible error to the measurement. *The smaller the relative error, the more accurate the measurement.*

Example

For the measurement $2\frac{3}{8}$ in., find the precision and accuracy.

ANSWER: The measurement is precise to the nearest $\frac{1}{8}$ in.   The greatest possible error is $\frac{1}{2} \times \frac{1}{8} = \frac{1}{16}$ in.
The relative error is the ratio:

$$\frac{1}{16} \text{ to } 2\frac{3}{8} = \frac{1}{16} \div 2\frac{3}{8} = \frac{1}{16} \times \frac{8}{19} = \frac{1}{38}$$

Exercises

To what unit of measure are these measurements precise?

1. 0.09 cm   **2.** 35.6 km   **3.** 0.0005 g   **4.** 16.92 kg   **5.** 8,006 L

6. $2\frac{13}{16}$ in.   **7.** $9\frac{3}{5}$ h   **8.** 6.8 s   **9.** 5 kg 7 g   **10.** 11 ft $8\frac{1}{4}$ in.

Which measurement in each of the following is more precise?

11. 9.6 m or 18.0 m   **12.** $7\frac{7}{16}$ in. or $3\frac{17}{32}$ in.   **13.** $8\frac{3}{4}$ mi or $5\frac{2}{3}$ yd   **14.** $2\frac{5}{6}$ h or 10 h 8 min

Find the greatest possible error and relative error.

15. 65 kg   **16.** 2.057 cm   **17.** 0.08 mm   **18.** 473 km   **19.** $9\frac{1}{4}$ h

20. $5\frac{7}{8}$ lb   **21.** 9,526.8 ft   **22.** 6 m 9 cm   **23.** 2 ft 11 in.   **24.** 2 h 19 min

Which measurement in each of the following is more accurate?

25. 28.1 cm or 6.52 cm   **26.** 30 mm or 0.003 mm   **27.** 0.06 m or 250 km

Scale

When large distances need to be represented on paper, ratios are used so that a scale drawing can be made.

A scale like 1 in. = 8 ft means that 1 scale inch represents 8 actual feet. The scale 1:96 means that 1 unit of scale represents 96 units in the actual dimensions. The scale 1 in. = 8 ft can also be written as $\frac{1}{8}$ in. = 1 ft or as 1:96, since 8 ft = 96 in.

Vocabulary

Scale shows the relationship between the dimensions of a drawing, plan, or map, and the actual dimensions.

Examples

1. If the scale is 1 cm = 30 km, what actual distance is represented by 4.7 cm?

To find the actual distance when the scale and scale distance are known:

Multiply the scale distance by the scale value of a unit.

Scale: 1 cm = 30 km   Scale distance: 4.7 cm = ? km
$$4.7 \times 30 = 141.0$$

ANSWER: 141 km

2. If the scale is 1 in. = 16 ft, how many inches represent 78 ft?

To find the scale distance when the scale and the actual distance are known:

Divide the actual distance by the scale value of a unit.

Scale: 1 in. = 16 ft   Scale distance: ? in. = 78 ft
$$78 \div 16 = 4\frac{7}{8}$$

ANSWER: $4\frac{7}{8}$ in.

3. Find the scale when the actual distance is 60 km and the scale distance is 15 mm.

To find the scale when the actual and scale distances are known:

Divide the actual distance by the scale distance.

$60 \div 15 = 4$ Each scale millimeter represents 4 actual kilometers.

ANSWER: 1 mm = 4 km, or 1:4,000,000

1. Name an advantage of scale drawings. Are there any disadvantages? Explain.

2. If the scale is 1 in. = 12 ft, represent the information two other ways.

3. If a scale drawing of 28 mm represents 119 km, what scale is used?

4. If you had a choice, would you prefer to solve Question 3 using a calculator, paper and pencil, mental arithmetic, or estimation? Why?

Practice Exercises

Find the actual distances using the given scale in each of the following.

1. Scale: 1 cm = 30 km   **a.** 6 cm   **b.** 23 cm   **c.** 8.5 cm   **d.** 17.1 cm

2. Scale: 1 in. = 48 mi   **a.** 5 in.   **b.** $6\frac{1}{2}$ in.   **c.** $2\frac{3}{4}$ in.   **d.** $3\frac{5}{8}$ in.

3. Scale: $\frac{1}{4}$ in. = 1 ft   **a.** 3 in.   **b.** $4\frac{1}{2}$ in.   **c.** $7\frac{3}{4}$ in.   **d.** $5\frac{9}{16}$ in.

4. Scale: 1:600   **a.** 4 m   **b.** 27 cm   **c.** 9.3 cm   **d.** 154 mm

Find the scale distances, using the given scale in each of the following.

5. Scale: 1 in. = 64 mi   **a.** 192 mi   **b.** 160 mi   **c.** 304 mi   **d.** 372 mi

6. Scale: $\frac{1}{8}$ in. = 1 ft   **a.** 48 ft   **b.** 126 ft   **c.** 29 ft   **d.** 167 ft

7. Scale: 1 cm = 25 m   **a.** 75 m   **b.** 325 m   **c.** 110 m   **d.** 42.5 m

8. Scale: 1:2,000,000   **a.** 8 km   **b.** 116 km   **c.** 98 km   **d.** 47 km

Find the scale.

9. The length 7 mm represents an actual distance of 420 km.

10. The length $6\frac{5}{8}$ in. represents an actual distance of 106 ft.

11. The length 10.7 cm represents an actual distance of 192.6 m.

12. The length $3\frac{7}{8}$ in. represents an actual distance of 186 ft.

13. Using the scale 1 cm = 100 km, draw line segments representing each of the following.

   a. 300 km   **b.** 60 km   **c.** 140 km   **d.** 290 km

14. Using the scale $\frac{1}{8}$ inch = 1 foot, draw line segments representing each of the following.

   a. 24 ft   **b.** 10 ft   **c.** 19 ft   **d.** $61\frac{1}{2}$ ft

15. What are the actual dimensions of a warehouse floor if plans drawn to the scale of 1:200 show dimensions of 12 cm by 9 cm?

16. Draw a floor plan of a schoolroom 33 feet long and 18 feet wide, using the scale 1 inch = 12 feet.

17. Using the scale of *miles*

0 12 24 36 48

find the distance represented by each of the following line segments.

a. _____
c. _____
b. _____
d. _____

18. Using the scale of *kilometers*

0   10   20   30   40   50

find the distance represented by each of the following line segments.

a. _____
c. _____
b. _____
d. _____

19. If the distance from *A* to *B* is 56 miles, what is the distance from *G* to *H*?

A _____ B   G _____ H

20. If the distance from *D* to *E* is 70 kilometers, what is the distance from *M* to *N*?

D _____ E   M _____ N

Computer Know-How

The computer program below computes the *actual distance* when the *scale* and *scale distance* are known.

```
1Ø PRINT "SCALE: 1 CM = ? KM"

2Ø PRINT "HOW MANY KILOMETERS (KM)";

3Ø INPUT K

4Ø PRINT "WHAT IS THE SCALE DISTANCE (IN
CM)";

5Ø INPUT D

6Ø PRINT "THE ACTUAL DISTANCE IS:"

7Ø PRINT K*D; "KILOMETERS"
```

a. ENTER the program and then RUN it to make sure it works

b. Modify the program so that
(1) the scale distance is expressed in *inches*.
(2) the actual distance is expressed in *feet*.

11-7 Geometric Figures—Plane

Triangles and circles are examples of simple closed figures.
Many simple closed figures have interesting properties.

Vocabulary

Coplanar points are points that are in the same plane.

Coplanar lines are lines that are in the same plane.

A **polygon** is a simple closed figure made up of line segments called *sides*.

A **vertex** is a point where a pair of intersecting sides of a polygon meet.

A **diagonal** is a line segment connecting two nonadjacent vertices of a polygon.

A **regular polygon** is a polygon with all sides and all angles equal in measure.

A **circle** is all the points in a plane, equidistant (the same distance) from a fixed point in the plane called the *center*.

A plane extends indefinitely far in any direction. We represent
and name a plane three ways:

plane *EFG*	By three capital letters which name three points not on the same line in a plane.
plane *RS*	By two letters written outside opposite corners.
plane *Q*	By one letter written inside a corner.

A, *B*, *C*, and *D* are coplanar points.
$\overleftrightarrow{AB}$ and $\overleftrightarrow{CD}$ are coplanar lines.

Types of Polygons			
Polygon	Number of Sides	Polygon	Number of Sides
Triangle	3	Heptagon	7
Quadrilateral	4	Octagon	8
Pentagon	5	Decagon	10
Hexagon	6	Dodecagon	12

A triangle can be classified by the length of its sides or by the measure of its angles.

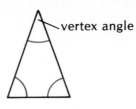
vertex angle

Equilateral Triangle
Three equal sides.

Isosceles Triangle
Two equal sides.

Scalene Triangle
No equal sides.

hypotenuse

legs

Equiangular Triangle
Three equal angles.
The **base** is generally the side on which a triangle rests.

Isosceles Triangle
Two equal base angles.
The **vertex angle** is the angle formed by the two equal sides.

Right Triangle
One right angle.
The **hypotenuse** is the side opposite the right angle. The other two sides are called legs.

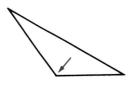

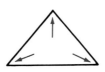

Obtuse Triangle
One obtuse angle.

Acute Triangle
All acute angles.

When working with triangles, three (3) line segments are commonly used: An **altitude** is a perpendicular segment from any vertex to the opposite side or extension of that side. A median is a segment connecting any vertex to the midpoint of the opposite side. An **angle bisector** is a segment that is drawn from any vertex to the opposite side and that bisects the angle at that vertex.

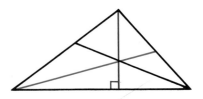

A quadrilateral can be described by its sides and their relation to each other, and by its angles.

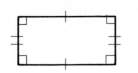

Rectangle
2 pairs of opposite sides are equal.
4 right angles.

Square
4 equal sides.
4 right angles.

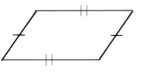

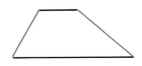

Parallelogram
2 pairs of opposite sides equal and parallel.

Trapezoid
1 pair of opposite sides parallel.

The square is a special rectangle. The rectangle and square are special parallelograms.

A circle has various line segments and curves associated with it.

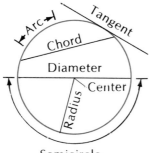

A **radius** is a segment with one endpoint on the circle and the other at the center of the circle.
A **chord** is a segment that has both its endpoints on the circle.
A **diameter** is a chord that passes through the center of the circle.
An **arc** is part of a circle.
A **semicircle** is an arc that is one-half a circle.
A **tangent** to a circle is a line that has one and only one point in common with the circle.

Think About It

1. When will an isosceles triangle also be equilateral? Is an equilateral triangle always isosceles? Explain.

2. Which type of triangle is a regular polygon? Which type of quadrilateral is a regular polygon?

3. State the relationship between a radius and a diameter of the same circle.

Practice Exercises

1. Name each of the following planes.

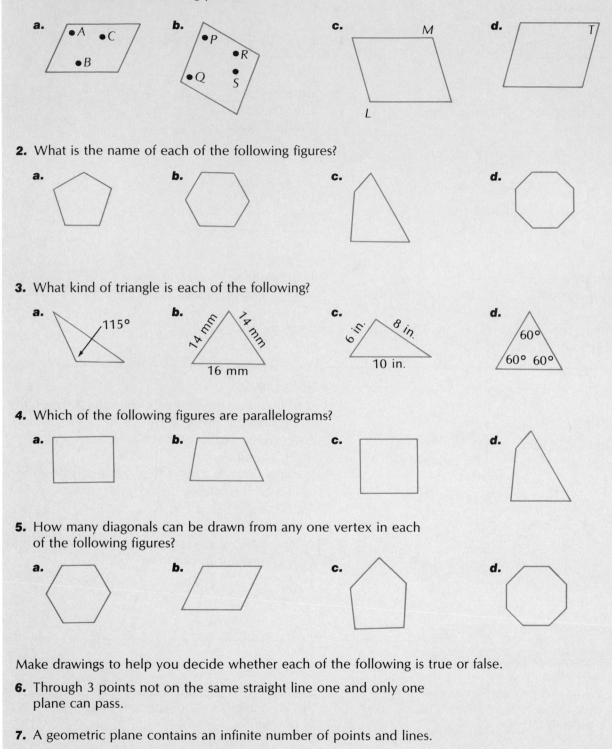

a.

b.

c.

d.

2. What is the name of each of the following figures?

a.

b.

c.

d.

3. What kind of triangle is each of the following?

a. 115°

b. 14 mm 14 mm 16 mm

c. 6 in. 8 in. 10 in.

d. 60° 60° 60°

4. Which of the following figures are parallelograms?

a.

b.

c.

d.

5. How many diagonals can be drawn from any one vertex in each of the following figures?

a.

b.

c.

d.

Make drawings to help you decide whether each of the following is true or false.

6. Through 3 points not on the same straight line one and only one plane can pass.

7. A geometric plane contains an infinite number of points and lines.

8. Two intersecting lines determine one and only one plane.

Problem Solving Strategy: Look for Patterns

Problem

Find the number of diagonals that can be drawn from any one vertex of a polygon with 7 sides; 8 sides; 10 sides; and 12 sides.

1. **READ** the problem carefully.

a. Find the question to be answered. What is the number of diagonals from one vertex of polygons with 7, 8, 10, and 12 sides?

b. Find the given facts. No facts are given.

> **Strategy: Look for Patterns**
>
> Finding a pattern is sometimes easier if you make a table or drawing. After you recognize the pattern, you can extend the table or diagram to obtain the missing information.

2. **PLAN** how to solve the problem.

a. Choose a strategy: Look for patterns.

b. Think or write out your plan relating the given facts to the questions asked:
Draw some polygons and see how many diagonals can be drawn. Organize the results in a table and look for a pattern.

3. **SOLVE** the problem.

Make a table:

Number of sides	3	4	5	6
Number of diagonals	0	1	2	3

Look for a pattern:

The number of diagonals is the number of sides minus 3.

Use the pattern to complete the table.

Number of sides	3	4	5	6	7	8	10	12
Number of diagonals	0	1	2	3	4	5	7	9

4. [CHECK] The answers are reasonable since a diagonal cannot be drawn from a vertex to itself or to the two adjoining vertices.

ANSWER: 7 sides, 4 vertices; 8 sides, 5 vertices; 10 sides, 7 vertices; 12 sides, 9 vertices.

Practice Exercises

1. Use the table to find how c was determined from a and b. Write a formula.

a	2	2	2	3	3	3
b	1	2	3	1	2	3
c	5	6	7	7	8	9

2. The sum of the first two consecutive odd numbers is 4, the sum of the first three consecutive odd numbers is 9, of the first four is 16, of the first 5 is 25.

$1 + 3 = 4$

$1 + 3 + 5 = 9$

$1 + 3 + 5 + 7 = 16$

$1 + 3 + 5 + 7 + 9 = 25$

What is the sum of the first six consecutive odd numbers? First 9? First 20? First 100?

3. The Fibonacci sequence is a sequence of natural numbers: 1, 1, 2, 3, 5, 8, 13, 21, . . .
Find the next five numbers in the sequence after 21.

4. Pascal's Triangle is a triangular arrangement of rows of numbers, each row increasing by one number. Each row, except the first, begins and ends in a 1, as shown.

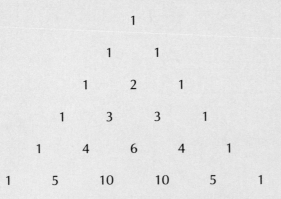

Find the numbers that belong in the next four rows. Hint: Look for an addition pattern.

Geometric Figures– Solid

Not all figures are contained in only one plane. Some figures lie in many different planes.

Vocabulary

Space is the infinite set of all points.

A **polyhedron** is a closed geometric figure consisting of four or more polygons and their interiors, all in different planes.

The **faces** of a polyhedron are the polygons and their interiors.

The **edges** of a polyhedron are line segments where the faces intersect.

The **vertices** of a polyhedron are points where the edges intersect.

A **sphere** has a curved surface in space on which every point is equidistant from a fixed point.

A **cylinder** has two equal and parallel circular bases and a lateral curved surface.

A **cone** has a circular base and a curved surface that comes to a point at the vertex.

The length, width, and height of space are endless.

Common polyhedra are the rectangular solid (right rectangular prism), the cube, and the pyramid.

Rectangular Solid
Six rectangular faces.

Cube
Six square faces.
All edges equal in length.

Pyramid
Base is a polygon.
Triangular faces meet in a common vertex.

The relationship among the number of faces, edges, and vertices of a polyhedron is expressed by Euler's formula, $F + V - E = 2$. This formula tells us that the number of faces (F) plus the number of vertices (V) minus the number of edges (E) is equal to two for *any* polyhedron.

Other solid geometric figures are the cylinder, sphere, and cone.

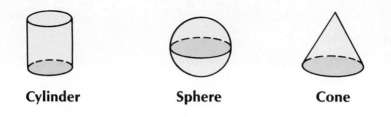

Cylinder   **Sphere**   **Cone**

Think About It

1. How are a cube and a square pyramid alike? (A square pyramid has a square base.) How are they different?

2. What is the difference between a sphere and a circle?

Practice Exercises

1. Classify each of the following figures.

a. 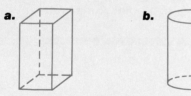   **b.**   **c.**   **d.**

2. a. How many faces (F) does a rectangular solid have?
   b. How many vertices (V)?
   c. How many edges (E)?
   d. Does $F + V - E = 2$?

3. a. How many faces (F) does a pyramid with a square base have?
   b. How many vertices (V)?
   c. How many edges (E)?
   d. Does $F + V - E = 2$?

Make drawings to help you decide whether each of the following is true or false.

4. When two different planes intersect, their intersection is a straight line.

5. The intersection of a plane and a line not in the plane is one and only one point.

6. A line that is perpendicular to each of two intersecting lines at their point of intersection is perpendicular to the plane in which these lines lie.

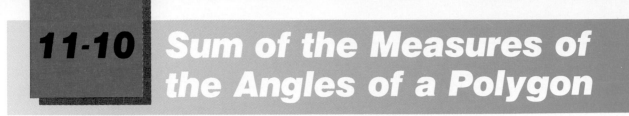

The sum of the three angles in a triangle is always the same.

Draw several triangles and measure the angles of each. Find the sum of the angles of each. You should find that the sum of the measures in each case is 180°.

> The sum of the measures of the angles of any triangle is 180°.

To find the measure of one angle of a triangle given the measures of the other two angles, follow these steps.

a. Add the given measures.
b. Subtract that sum from 180°.

Example

Find the measure of the third angle of a triangle if the other two angles have measures of 33° and 101°.

$$33° + 101° = 134° \qquad 180° - 134° = 46°$$

ANSWER: The measure is 46°.

The sum of the measures of the angles of a polygon with four or more sides, depends upon the number of triangles that can be formed by drawing diagonals from one vertex. For any given type of polygon—quadrilateral, pentagon, and so on—the sum of the measures of the angles is the same.

Draw several quadrilaterals of different sorts. Measure the angles and find their sum.

> The sum of the measures of the angles of any quadrilateral is 360°.

Draw a pentagon and a hexagon. Measure the angles and find the sum of the measures. To find the sum of the measures of a polygon with n sides, follow these steps.

a. Subtract 2 from the number of sides.
b. Multiply this difference by 180°.

$$180°(n - 2)$$

Example

Find the sum of the measures of the angles of an octagon.

$8 - 2 = 6$ $180°(6) = 1,080°$

ANSWER: The sum of the measures is 1,080°.

Think About It

1. What is the measure of each angle of an equiangular triangle?

2. If seven triangles can be formed by drawing all the diagonals from one vertex of a polygon, how many sides does it have? What is the sum of the measures of its angles?

Practice Exercises

1. In each of the following, find the measure of the third angle of a triangle when the other two angles have the given measures.

a. 29° and 87°   **b.** 63° and 72°   **c.** 106° and 51°   **d.** 90° and 45°

2. If each of the equal angles of an isosceles triangle measures 59°, find the measure of the third angle.

3. If the vertex angle of an isosceles triangle measures 70°, find the measure of each of the other two angles.

4. Find the measure of the third angle in each of the following.

a.

b.

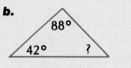

c.

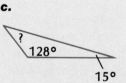

5. In each of the following, find the measure of the fourth angle of a quadrilateral when the other three angles have the given measures.

a. 45°, 85°, and 97°   **b.** 119°, 90°, and 90°   **c.** 76°, 102°, and 161°

6. The opposite angles of a parallelogram are equal. If one angle measures 65°, find the measures of the other three angles.

7. Three angles of a trapezoid measure 90°, 90°, and 108°. What does the fourth angle measure?

8. Find the number of degrees in each of the following polygons.

a. heptagon   **b.** decagon   **c.** dodecagon

11-11 Special Pairs of Angles

Some angles can be classified according to their relationship to each other.

Vocabulary

Complementary angles are two angles the sum of whose measures is 90°.

Supplementary angles are two angles the sum of whose measures is 180°.

Vertical angles are a pair of angles formed from two intersecting lines and are directly opposite each other.

An **exterior angle** is an angle formed by extending one side of a triangle.

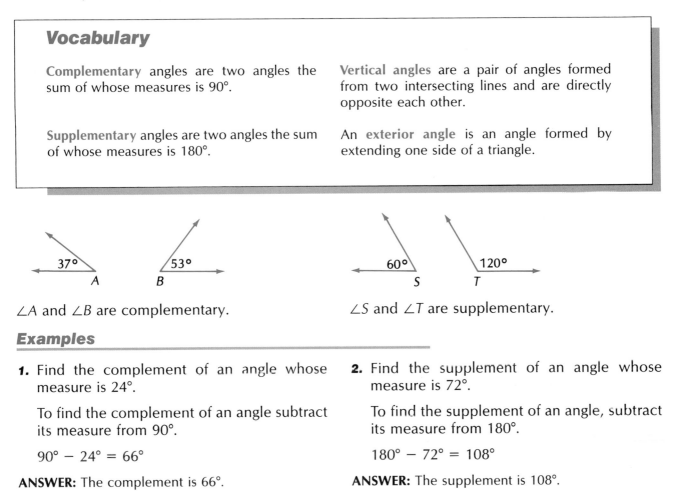

$\angle A$ and $\angle B$ are complementary.

$\angle S$ and $\angle T$ are supplementary.

Examples

1. Find the complement of an angle whose measure is 24°.

To find the complement of an angle subtract its measure from 90°.

$90° - 24° = 66°$

ANSWER: The complement is 66°.

2. Find the supplement of an angle whose measure is 72°.

To find the supplement of an angle, subtract its measure from 180°.

$180° - 72° = 108°$

ANSWER: The supplement is 108°.

Draw two intersecting lines, forming four angles. Measure a pair of angles that are directly opposite each other. Are the measures equal? Measure the other pair of opposite angles. Are the measures equal?

You should find that the angles opposite each other are equal in measure.

When two straight lines intersect, the opposite (vertical) angles are equal in measure.

Example

The measure of ∠1 is 115° and the measure of ∠2 is 65°. Find the measures of angles 3 and 4.

$m\angle 3 = m\angle 2 = 65°$ because angles 2 and 3 are vertical angles.
$m\angle 4 = m\angle 1 = 115°$ because angles 1 and 4 are vertical angles.

ANSWER: $m\angle 3 = 65°$, $m\angle 4 = 115°$

Angle 4 is an exterior angle of triangle *ABC* at the right.

Measure angles 1, 2, and 4.

> The measure of an **exterior angle** of a triangle is equal to the sum of the measures of the two opposite interior angles.

Example

Find $m\angle 4$ in the drawing above if $m\angle 1 = 35°$ and $m\angle 2 = 83°$.

$m\angle 4 = m\angle 1 + m\angle 2 = 35° + 83° = 118°$

ANSWER: The measure of ∠4 = 118°.

Think About It

1. If complementary angles are joined to form one angle, what kind of angle is it? If supplementary angles are joined to form one angle, what kind of angle is it?

2. If two lines intersect and one angle is acute, what type of angles are the other three?

3. Can an exterior angle of a triangle have the same measure as one of the angles of the triangle? As one of the opposite interior angles of the triangle? Explain your answers.

4. What is the sum of the three exterior angles formed by extending each side of the triangle?

Practice Exercises

1. Which of the following pairs of angles are complementary?

   a. $m\angle E = 50°$, $m\angle G = 40°$
   c. $m\angle 6 = 22°$, $m\angle 3 = 68°$
   b. $m\angle R = 37°$, $m\angle T = 53°$
   d. $m\angle a = 39°$, $m\angle b = 41°$

2. Find the measure of the angle that is the complement of each of the following angles.

   a. $m\angle 2 = 21°$ **b.** $m\angle L = 87°$ **c.** $m\angle B = 6°$ **d.** $m\angle 8 = 33°$

3. Angle *ABC* is a right angle.

    a. If $m\angle 2 = 26°$, find the measure of $\angle 1$.
    b. If $m\angle 1 = 53°$, find the measure of $\angle 2$.
    c. If $m\angle 2 = 14°$, find the measure of $\angle 1$.
    d. If $m\angle 1 = 67°$, find the measure of $\angle 1$.

4. In the figure, $\overrightarrow{FB} \perp \overrightarrow{FD}$ and $\overrightarrow{FC} \perp \overleftrightarrow{AE}$.

    a. If $m\angle 4 = 55°$, find the measure of $\angle 1$. Of $\angle 2$. Of $\angle 3$.
    b. If $m\angle 1 = 61°$, find the measure of $\angle 2$. Of $\angle 3$. Of $\angle 4$.
    c. If $m\angle 3 = 74°$, find the measure of $\angle 1$. Of $\angle 2$. Of $\angle 4$.
    d. If $m\angle 2 = 19°$, find the measure of $\angle 3$. Of $\angle 1$. Of $\angle 4$.

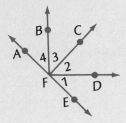

5. Why is the complement of an acute angle also an acute angle?

6. Which of the following pairs of angles are supplementary?

    a. $m\angle A = 134°$, $m\angle N = 46°$ **b.** $m\angle 3 = 85°$, $m\angle 4 = 105°$
    c. $m\angle B = 79°$, $m\angle D = 91°$ **d.** $m\angle c = 127°$, $m\angle e = 53°$

7. Find the measure of the angle that is the supplement of each of the following angles.

    a. $m\angle H = 44°$ **b.** $m\angle a = 117°$ **c.** $m\angle 5 = 89°$ **d.** $m\angle R = 8°$

8. Measure $\angle CDF$ and $\angle EDF$. Is the sum of their measures 180°? Do you see that *when one straight line meets another, the adjacent angles, which have the same vertex and a common side, are supplementary?*

    a. What is the measure of $\angle CDF$ when $m\angle EDF = 32°$?
    b. Find the measure of $\angle EDF$ when $m\angle CDF = 116°$.
    c. Find the measure of $\angle CDF$ when $m\angle EDF = 90°$.
    d. What is the measure of $\angle EDF$ when $m\angle CDF = 145°$?

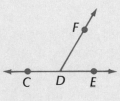

9. Is the supplement of an obtuse angle also an obtuse angle? Explain your answer.

10. In the drawing at the right, what angle is opposite to $\angle 1$? Does $\angle 1 = \angle 3$? What angle is opposite to $\angle 4$? Does $\angle 4 = \angle 2$?

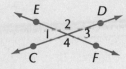

11. If $m\angle 3 = 68°$, what is the measure of each?

    a. $\angle 1$ **b.** $\angle 2$ **c.** $\angle 4$

12. If $m\angle 4 = 110°$, what is the measure of each?

    a. $\angle 2$ **b.** $\angle 1$ **c.** $\angle 3$

Use the figure at the right for Exercises 13–15.

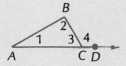

13. If $m\angle 1 = 42°$ and $m\angle 2 = 69°$, find the measure of $\angle 4$. Of $\angle 3$.

14. If $m\angle 3 = 53°$ and $m\angle 1 = 75°$, find the measure of $\angle 2$. Of $\angle 4$.

15. If $m\angle 2 = 38°$ and $m\angle 3 = 81°$, find the measure of $\angle 1$. Of $\angle 4$.

11-12 Parallel Lines and Angle Relationships

When a set of parallel lines is cut by another line, several pairs of angles are formed.

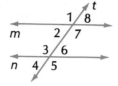

m is parallel to n, and m and n are cut by transversal t.
Angles 1 and 3 are corresponding angles.
Angles 2 and 6 are alternate interior angles.

Examples

1. Name other pairs of corresponding angles in the figure above.

ANSWER: ∠2 and ∠4; ∠6 and ∠8; ∠5 and ∠7

2. Name another pair of alternate interior angles in the figure above.

ANSWER: ∠3 and ∠7

When parallel lines are cut by a transversal, the measures of corresponding angles are equal, and the measures of alternate interior angles are equal.

When lines are cut by a transversal and the measures of a pair of corresponding or alternate interior angles are equal, the lines are parallel.

3. Are m and n parallel?

ANSWER: Yes, the measures of a pair of alternate interior angles are equal.

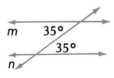

Think About It

1. A transversal cuts a pair of parallel lines and one of the angles formed is acute. What do you know about the other angles?

2. Of the eight angles formed by a transversal and a pair of parallel lines, how many are equal to each other?

3. When would all eight angles formed by a transversal and a pair of parallel lines be equal?

Practice Exercises

Use the figure shown for all exercises. Complete each statement as
required in Exercises 1 through 6.

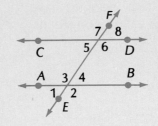

1. **a.** If lines *AB* and *CD* are cut by a third line *EF,* the line *EF* is
   called a _____.
   b. ∠6 and ∠3 are _____ angles.
   c. ∠5 and ∠4 are _____ angles.
   d. ∠5 and ∠1 are _____ angles.

2. If two parallel lines are cut by a transversal, the alternate
   interior angles are _____. Therefore, ∠3 = _____,
   ∠5 = _____.

3. If two parallel lines are cut by a transversal, the corresponding
   angles are _____.
   Therefore, ∠2 = _____, ∠7 = _____, ∠1 = _____, ∠8 = _____.

4. What is the sum of the measures of ∠5 and ∠7?
   They are a pair of _____ angles.

5. What is the sum of the measures of ∠1 and ∠2?
   They are a pair of _____ angles.

6. ∠5 and ∠8 are a pair of _____ angles.

7. If line *AB* is parallel to line *CD* in the figure above:

   a. Show that ∠2 = ∠7.
   b. Show that ∠1 and ∠6 are supplementary angles.
   c. Show that ∠8 = ∠1.
   d. Show that ∠3 and ∠5 are supplementary angles.

8. If line *AB* is parallel to line *CD* in the figure above, what is the
   measure of:

   a. ∠6 if $m\angle 2$ = 130?°   **b.** ∠7 if $m\angle 6$ = 106°?
   c. ∠5 if $m\angle 4$ = 75°?   **d.** ∠2 if $m\angle 7$ = 93°?

9. If two lines are cut by a transversal making a pair of
   corresponding angles or a pair of alternate interior angles equal,
   the lines are _____.

10. Are lines *AB* and *CD* parallel in the figure above when:

    a. $m\angle 1$ = 70° and $m\angle 5$ = 70°?   **b.** $m\angle 7$ = 120° and $m\angle 3$ = 120°?
    c. $m\angle 7$ = 135° and $m\angle 4$ = 45°?   **d.** $m\angle 2$ = 108° and $m\angle 8$ = 82°?

Congruent Triangles

Sometimes two triangles have the same size and shape. Such triangles are called **congruent.**

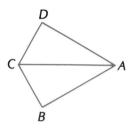

Triangle *ABC* is congruent to triangle *ADC*. The order of the letters indicates which vertices correspond to each other.

$$\triangle ABC \cong \triangle ADC$$

Vocabulary

Congruent triangles are triangles whose corresponding sides are equal in length and whose corresponding angles are equal in size.

The symbol ≅ means "is congruent to."

Conditions for Congruent Triangles

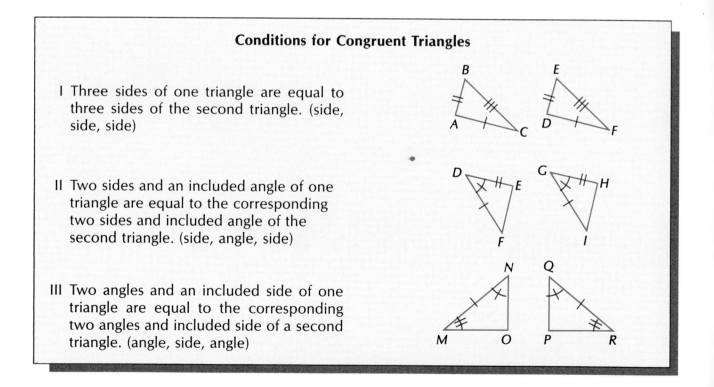

I Three sides of one triangle are equal to three sides of the second triangle. (side, side, side)

II Two sides and an included angle of one triangle are equal to the corresponding two sides and included angle of the second triangle. (side, angle, side)

III Two angles and an included side of one triangle are equal to the corresponding two angles and included side of a second triangle. (angle, side, angle)

To determine if two triangles are congruent, follow these steps.

a. Compare the sides and angles of one triangle to those of a second.

b. If one of the conditions above exists, the triangles are congruent.

Example

State why each pair of triangles is congruent.

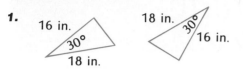

1. 16 in. / 30° / 18 in.   18 in. / 30° / 16 in.

ANSWER: side, angle, side

2. 7 cm / 7 cm / 6 cm   7 cm / 6 cm / 7 cm

ANSWER: side, side, side

Think About It

1. If $\triangle ABC \cong \triangle DEF$, name the corresponding sides and corresponding angles.

2. Draw two triangles that have congruent angles but are not congruent triangles.

Practice Exercises

Select in each of the following groups two triangles that are congruent and state the reason why.

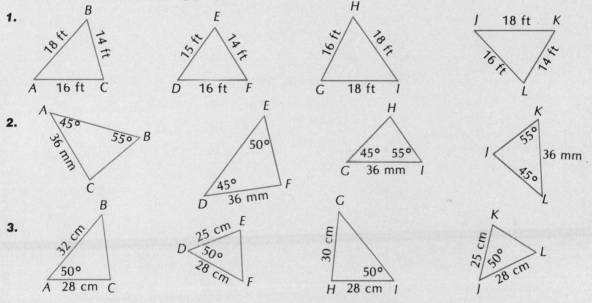

1.
- 18 ft, 14 ft, 16 ft (A, B, C)
- 15 ft, 14 ft, 16 ft (D, E, F)
- 16 ft, 18 ft, 18 ft (G, H, I)
- 18 ft, 16 ft, 14 ft (J, K, L)

2.
- 45°, 55°, 36 mm (A, B, C)
- 50°, 45°, 36 mm (D, E, F)
- 45°, 55°, 36 mm (G, H, I)
- 55°, 45°, 36 mm (I, K, L)

3.
- 32 cm, 50°, 28 cm (A, B, C)
- 25 cm, 50°, 28 cm (D, E, F)
- 30 cm, 50°, 28 cm (G, H, I)
- 25 cm, 50°, 28 cm (J, K, L)

Find the missing angles and sides in each of the following pairs of congruent triangles.

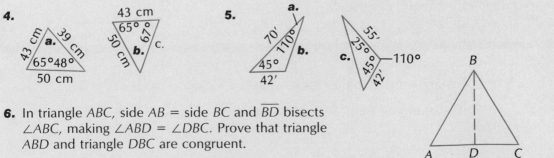

4. 43 cm, 39 cm, 65°, 48°, 50 cm, a., b., c., 43 cm, 65°, 67°, 50 cm

5. a., b., c., 70′, 110°, 45°, 42′, 55′, 25°, 45°, 42′, 110°

6. In triangle ABC, side AB = side BC and $\overline{BD}$ bisects $\angle ABC$, making $\angle ABD = \angle DBC$. Prove that triangle ABD and triangle DBC are congruent.

Triangles can have the same shape, but not be congruent.

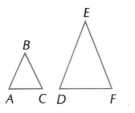

Triangle *ABC* is similar to △*DEF*. The order of the letters indicates the corresponding sides and angles.

△*ABC* ~ △*DEF*

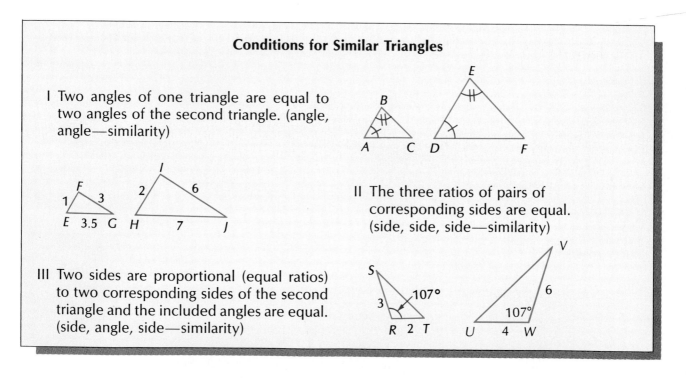

Conditions for Similar Triangles

I Two angles of one triangle are equal to two angles of the second triangle. (angle, angle—similarity)

II The three ratios of pairs of corresponding sides are equal. (side, side, side—similarity)

III Two sides are proportional (equal ratios) to two corresponding sides of the second triangle and the included angles are equal. (side, angle, side—similarity)

To determine if two triangles are similar, follow these steps.

a. Find the ratios of the corresponding sides and compare any angle measures known.

b. If one of the conditions above exists, the triangles are similar.

Examples

State why the triangles of each pair are similar.

1.

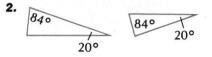

$\frac{10}{8} = \frac{5}{4}; \frac{5}{4} = \frac{5}{4}; \frac{7.5}{6} = \frac{5}{4}$

ANSWER: side, side, side—similarity

2.

ANSWER: angle, angle—similarity

Think About It

1. If $\triangle GHJ \sim \triangle KFT$, name the corresponding sides and angles.

2. What is true about pairs of corresponding angles for similar triangles?

Practice Exercises

Select in each of the following groups two triangles which are similar and state the reason why.

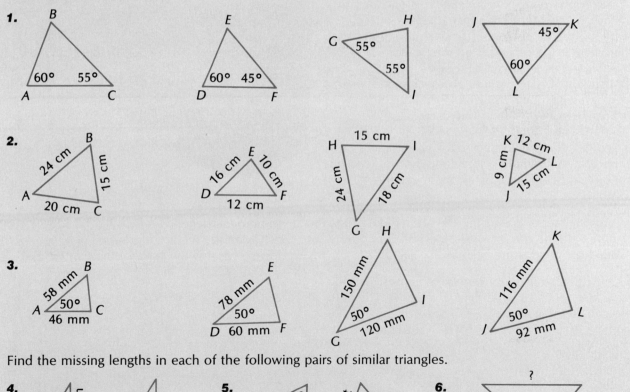

Find the missing lengths in each of the following pairs of similar triangles.

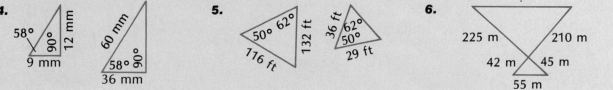

Career Application
PHOTOGRAPHER

L ook carefully at the excellent photographs that appear in advertisements and books. *Photographers'* knowledge of types of cameras and lenses and the amounts of light to use makes such high quality possible.

Many photographers specialize in one kind of picture. Portrait photographers take pictures of people, often for yearbooks or weddings. Industrial photographers photograph products and processes to be shown in catalogs and reports. Scientific photographers take pictures of research and new developments. They may use special equipment such as X-ray machines or microscopes.

Almost half of all photographers are self-employed. They usually have no formal training in photography. A person who wants to be a photographer should get as much experience as possible.

1. A photographer takes a picture of a building. One centimeter on the picture represents 5 m on the building. What is the scale of the picture?

2. A photographer takes a picture of a flower. Three inches on the picture represents one inch on the flower. What is the scale of the picture?

A photographer uses a variety of lenses. A wide angle lens "sees" a wide field of view. A telephoto lens can pick out a small detail of a scene. Measure the field of view (angle) of each of these lenses.

3. wide angle

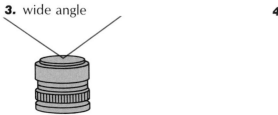

4. normal

5. telephoto

Chapter 11 Review

Vocabulary

acute angle p. 431	cylinder p. 449	obtuse triangle p. 444	right triangle p. 444
acute triangle p. 444	degree p. 431	parallel lines p. 435	scale p. 440
alternate interior angles p. 456	diagonal p. 439	parallelogram p. 445	sides p. 431
altitude p. 444	diameter p. 445	perpendicular lines p. 435	similar triangles p. 456
angle p. 431	edge p. 449	plane p. 435	space p. 449
angle bisector p. 444	equiangular triangle p. 444	point p. 428	sphere p. 449
arc p. 445	Euler's formula p. 449	polygon p. 443	square p. 445
base p. 444	exterior angle p. 453	polyhedron p. 445	straight angle p. 431
chord p. 445	face p. 449	precision p. 439	supplementary angles p. 453
circle p. 443	greatest possible error p. 439	protractor p. 433	tangent p. 445
collinear p. 428	hypotenuse p. 444	pyramid p. 449	transversal p. 456
compass p. 437	intersecting lines p. 435	radius p. 445	trapezoid p. 445
complementary angles p. 453	isosceles triangle p. 444	ray p. 428	vertex p. 431
cone p. 449	leg p. 444	rectangle p. 445	vertex angle p. 444
congruent triangles p. 458	line p. 428	rectangular solid p. 449	vertical angles p. 453
coplanar p. 443	line segment p. 428	regular polygon p. 443	
corresponding angles p. 456	median p. 444	relative error p. 439	
cube p. 449	obtuse angle p. 431	right angle p. 431	

Page
428 **1.** Read, or write in words each of the following.

   a. $\overline{BD}$ **b.** $\overleftrightarrow{HE}$ **c.** $\overrightarrow{SW}$

431 **3.** Point B is called the _____ of the angle.

443 **5.** If the scale is 1 cm = 40 km, what actual distance is represented by the scale distance of 12.3 cm?

451 **8.** If three angles of a quadrilateral are 75°, 82°, and 92° what is the fourth angle?

456 Given: $l \parallel m$.

   10. If $m\angle 2 = 53°$ then $m\angle 6 = ?$

   11. If $m\angle 2 = 74°$ then $m\angle 7 = ?$

458 12. Side RS corresponds to which side in the second triangle?

Page
431 **2.** Name the following angle in 3 ways.

431 **4.** Which measure is an obtuse angle?

   a. 24°   **b.** 90°   **c.** 95°

443 Name the figure.

   6.   **7.**

453 **9.** What is the complement of a 9° angle?

Chapter 11 Test

428 **1.** Name the figure.

2. Draw $\overleftrightarrow{AB}$.

431 **3.** An angle is formed by two different _____ with a common endpoint.

4. True or false? A 45° angle is an acute angle.

433 **5.** Draw an angle measuring 105°.

435 **6.** Draw a pair of perpendicular lines.

437 **7.** Measure using a ruler. _____

440 **8.** If the scale is 1:300,000, what is the actual number of kilometers represented by 25 mm?

443 Identify the figure.

9.

10.

447 **11.** Find the sum of the first 10 even numbers (0 + 2 + . . .). Look for a pattern.

449 **12.** Identify the figure.

451 **13.** If one of the angles in a right triangle measures 42°, what are the measures of the other two angles?

14. What is the sum of the measures of the angles in a parallelogram?

453 **15.** What is the supplement of a 53° angle?

16. What is the complement of a 25° angle?

456 Given: $l \parallel m$ and $m\angle a = 115°$.

17. What is the measure of $\angle b$?

18. What is the measure of $\angle f$?

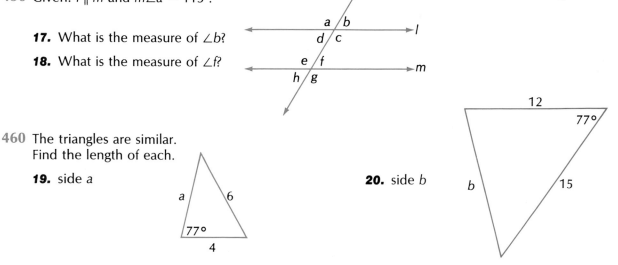

460 The triangles are similar. Find the length of each.

19. side a

20. side b

Copying a Line Segment

When copying a floor plan for a house, an architect's blueprint for a building, or an engineer's drawing of a machine, you need a way of making a copy of certain line segments. To copy a line segment means to draw a line segment equal in length to the given line segment.

Example

To copy a given line segment using a compass and straightedge, follow these steps.

A ———————— B

given line segment

a. Draw a line segment longer than the given line segment.

C ————————

b. Open the compass to the length of the given line segment.

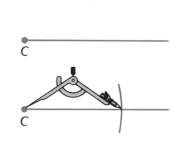

c. Transfer the compass to the new line segment, placing the metal point on one endpoint.

d. Draw an arc cutting the new line segment. The intersection is the second endpoint.

Segments *AB* and *CD* are congruent.　　$\overline{AB} \cong \overline{CD}$

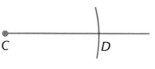

C　　　　　 D

Think About It

1. Describe another method of copying a line segment that does not require a compass.

2. Describe how you would construct a line segment that is twice the length of a given line segment using a compass and straightedge.

Practice Exercises

Make a copy of each of the following line segments, using a compass and a straightedge. Check with a ruler.

1. ————————————

2. ————————————

3. ——————

4. Draw any line segment and label the endpoints *D* and *E*. Label a point not on $\overline{DE}$ as *F*. Using a compass and a straightedge, draw a line segment *FG* equal in length to $\overline{DE}$.

12-2 Copying an Angle

You can see angles in the roof of a house, in the structure of a bridge, in the support cables for a flagpole, or in the sail of a boat. To copy an angle means to construct an angle congruent (equal in size) to the given angle.

Example

To copy a given angle using a compass and straightedge, follow these steps.

a. Draw a ray with endpoint *P*.

b. Open the compass and place the metal point on vertex *A* of the given angle. Draw an arc intersecting the sides of the given angle at points *B* and *C*.

c. With the same radius, place the metal point on endpoint *P* and draw an arc cutting the ray at *Q*.

d. Set the radius of the compass equal to *BC* and draw an arc with center *Q* intersecting at *R* the arc drawn in Step c.

e. Draw $\overrightarrow{PR}$. Angles *BAC* and *RPQ* are congruent.

$$m\angle BAC = m\angle RPQ \qquad \angle BAC \cong \angle RPQ$$

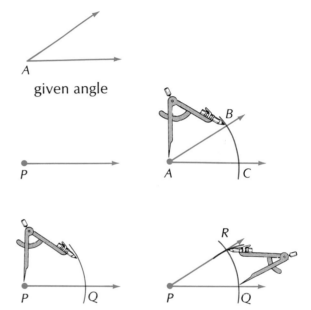

given angle

Think About It

1. Describe another method of copying an angle that does not require a compass.

2. Use the diagrams above. If you were to connect *B* and *C* with a line segment, and *R* and *Q* with a line segment, you would have two triangles, △*ABC*, △*PRQ*. Are these triangles congruent? Explain your answer.

Practice Exercises

Draw each of the following angles.
Construct with a compass an angle of equal size.

1. Acute angle   **2.** Obtuse angle   **3.** Right angle

12-3 Constructing Triangles

A triangle contains three sides and three angles. A triangle may be constructed when any of the following combinations of three parts are known:

I. Three sides   **II.** Two sides and an included angle.   **III.** Two angles and an included side.

Examples

I. To construct a triangle when three sides are given, follow these steps.

a ——————

b ——————

c ——————

given sides

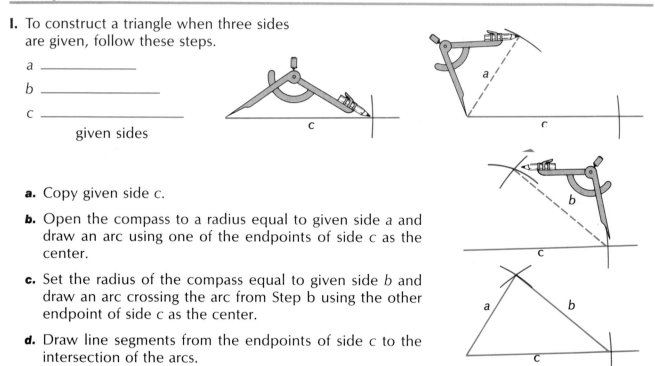

a. Copy given side c.

b. Open the compass to a radius equal to given side a and draw an arc using one of the endpoints of side c as the center.

c. Set the radius of the compass equal to given side b and draw an arc crossing the arc from Step b using the other endpoint of side c as the center.

d. Draw line segments from the endpoints of side c to the intersection of the arcs.

Practice Exercises

1. Construct triangles having sides of the following measures.

   a. 39 mm, 43 mm, 27 mm   **b.** $2\frac{3}{4}$ in., $2\frac{1}{4}$ in., 3 in.   **c.** 5.8 cm, 4.6 cm, 6.3 cm

2. Construct an equilateral triangle whose sides are each 3.1 cm long. Measure the three angles. Are their measures equal?

3. Construct an isosceles triangle whose base is $2\frac{3}{16}$ in. long and each of whose two equal sides is $1\frac{5}{8}$ in. long.

4. Using the scale 1 inch = 8 feet, construct triangles with sides of the following measures.

   a. 16 ft, 22 ft, 9 ft   **b.** 30 ft, 24 ft, 15 ft
   c. 12 ft, 12 ft, 12 ft   **d.** 18 ft, 10 ft, 10 ft

5. Using the scale 1:2,000,000, construct triangles with sides of the following measures.

a. 40 km, 70 km, 58 km   **b.** 30 km, 42 km, 38 km

c. 60 km, 54 km, 72 km   **d.** 26 km, 36 km, 48 km

II. To construct a triangle when two sides and an included angle are given, follow these steps.

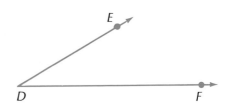

a —————————   given sides and
b —————————   included angle

a. Copy given angle *EDF*.

b. Open the compass to a radius equal to given side *a* and draw an arc crossing $\overrightarrow{D'E'}$ using *D'* as the center. Then, set the compass to a radius equal to given side *b* and draw an arc crossing $\overrightarrow{D'F'}$ using *D'* as the center.

c. Draw a line segment connecting the intersections of the arcs found above.

6. Construct triangles having the following sides and included angles.

a. 61 mm, 48 mm, 55°   **b.** 5.7 cm, 4.6 cm, 70°   **c.** $2\frac{7}{8}$ in., $2\frac{7}{8}$ in., 90°

7. Construct a right triangle with the following measures.

a. A base of 6.5 cm and an altitude of 5.9 cm

b. A base of $2\frac{15}{16}$ in. and an altitude of $3\frac{5}{8}$ in.

8. Construct an isosceles triangle in which the equal sides each measure $3\frac{1}{8}$ in. and the vertex angle formed by these sides measures 63°. Measure the angles opposite the equal sides. Are their measures equal?

9. Using the scale 1:500, construct a triangle in which two sides and the included angle are the following.

a. 15 m, 10 m, and 50°   **b.** 20 m, 35 m, and 75°   **c.** 18 m, 27 m, and 80°

III. To construct a triangle when two angles and an included side are given, follow these steps.

given angles and included side

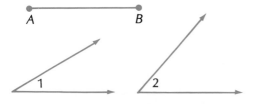

a. Copy side *AB*.

b. At *A'*, copy given ∠1 (∠1').

c. At *B'*, copy given ∠2 (∠2').

d. The intersection of the sides of these angles is the third vertex (*C'*) of the triangle.

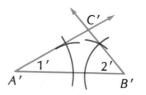

10. Construct triangles having the following angles and included side.

    a. 90°, 45°, 5 cm   **b.** 105°, 35°, $3\frac{1}{4}$ in.   **c.** 70°, 30°, 58 mm

11. Construct a triangle having two angles each measuring 50° and the included side measuring 46 mm. Measure the sides opposite the equal angles. Are their measures equal?

12. Construct a triangle having two angles each measuring 60° and the included side measuring $2\frac{11}{16}$ in. Measure the third angle. Measure the other two sides. Is the triangle equiangular? Equilateral?

13. Construct a triangle having two angles and an included side equal to these angles and line segment.

Think About It

1. Suppose you were given two sides of a triangle. How many different triangles could you construct?

2. Explain why the directions for Example II state that the given angle is an included angle.

Refresh Your Skills

1. $\boxed{1-8}$

$$\begin{array}{r} 485,296 \\ 246,583 \\ 627,416 \\ +\ 583,529 \\ \hline \end{array}$$

2. $\boxed{1-9}$

$$406,592 - 319,086$$

3. $\boxed{1-11}$

$$240 \times 509$$

4. $\boxed{1-13}$

$$3,600\overline{)86,400}$$

5. $\boxed{2-5}$

$$0.859 + 0.26 + 0.8$$

6. $\boxed{2-6}$

$$13.9 - 0.055$$

7. $\boxed{2-7}$

$$1.2 \times 0.06$$

8. $\boxed{2-9}$

$$0.4 \div 0.08$$

9. $\boxed{3-8}$

$$3\frac{11}{12} + 1\frac{5}{6} + 2\frac{3}{4}$$

10. $\boxed{3-9}$

$$1\frac{1}{2} - \frac{2}{3}$$

11. $\boxed{3-11}$

$$3\frac{1}{3} \times 1\frac{4}{5}$$

12. $\boxed{3-12}$

$$9\frac{1}{6} \div 2\frac{1}{5}$$

13. Find $9\frac{3}{4}$% of $10,000.

$\boxed{4-6}$

14. What percent of 144 is 96?

$\boxed{4-7}$

15. 21 is 10.5% of what number?

$\boxed{4-9}$

Constructing a Perpendicular I

Right angles are basic to the construction of buildings. On a floor plan, the corners of square or rectangular rooms are right angles. The angle that a wall makes with a floor is a right angle.

Example

To construct a line perpendicular to a given line at or through a given point on the given line, follow these steps.

given point on a given line

a. Use point C as the center and open the compass to any radius to draw an arc crossing $\overleftrightarrow{AB}$ at points D and E.

b. With D and E as centers, open the compass to a radius greater than $\overleftrightarrow{DC}$ and draw arcs intersecting at F.

c. Draw $\overleftrightarrow{CF}$.

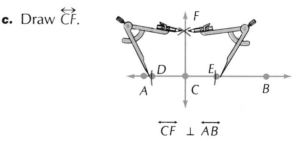

$$\overleftrightarrow{CF} \perp \overleftrightarrow{AB}$$

Think About It

1. When two lines are perpendicular, how many right angles are formed?

2. How can you determine whether two lines are perpendicular without using a protractor?

Practice Exercises

1. Draw any line. Select a point on this line. Construct a line perpendicular to the line you have drawn at your selected point.

2. Construct a rectangle 43 mm long and 35 mm wide.

3. Construct a square with each side $2\frac{1}{2}$ inches long.

Example

To construct a perpendicular to a given line from or through a given point not on the given line, follow these steps.

a. Use point C as the center and draw an arc crossing $\overleftrightarrow{AB}$ at points D and E.

b. With D and E as centers, open the compass to a radius greater than one-half the distance from D to E and draw arcs intersecting at F.

c. Draw $\overleftrightarrow{CF}$.

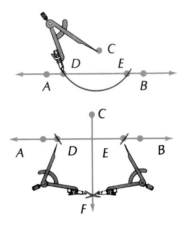

Think About It

1. In the construction above, suppose that line segments $\overline{DC}$, $\overline{EC}$, $\overline{DF}$, and $\overline{EF}$ are drawn. Are any of these line segments congruent to any others? Explain.

2. Can you use the example above to construct a perpendicular to a vertical line from a point to the left of the line? Explain.

Practice Exercises

1. Draw any line. Select a point not on this line. Construct a perpendicular from your selected point to the line you have drawn.

2. Construct an equilateral triangle with each side 6.2 cm long. From each vertex construct a perpendicular to the opposite side. Is each angle bisected? Is each side bisected?

3. Draw any acute triangle. From each vertex construct the altitude to the opposite side. Do the altitudes intersect?

4. Draw any right triangle. Construct the altitude to each side. Do they intersect? If so, what is the common point?

How high is a pyramid whose base is a square 8 centimeters by 8 centimeters and whose faces are equilateral triangles?

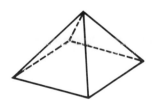

Strategy: Make a Model

Making a model often makes a problem clearer or makes it possible to solve.

1. **READ** the problem carefully.

a. Find the question asked.
How high is the pyramid? (The height is measured vertically from the center of the base to the point of the pyramid.)

b. Find the given facts.
Pyramid with base 8 centimeters square; sides are equilateral triangles.

2. **PLAN** how to solve the problem.

a. Choose a strategy: Make a model.

b. Think or write out your plan relating the given facts to the question asked. Use a ruler and compass to construct a pattern. Make a model from the pattern. Measure the height.

3. **SOLVE** the problem.

Construct a square 8 centimeters by 8 centimeters. Then construct an equilateral triangle on each side of the square.

Cut out your pattern. Fold along the sides of the square. Tape the edges. Use two rulers or a ruler and a straightedge to measure the height. The answer should be about 5.7 centimeters.

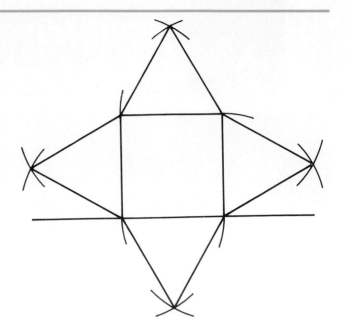

4. **CHECK**

Since the height is one leg of a right triangle whose hypotenuse is 8 centimeters (one edge of the pyramid), it should be less than 8 centimeters. So 5.7 centimeters is a reasonable answer.

Practice Exercises

1. Double the lengths of the sides of the base and of the sides of the faces of the preceding pyramid. How high is the pyramid now? Guess first and then check by making a model.

2. In a 12-hour period, how many times do the hands of a clock make a 90° angle? Guess first and then check by making a model.

3. A large cube is painted blue and then cut into eight small congruent cubes. How many sides of each small cube are painted?

4. A cube is painted red and then cut into 27 congruent cubes.

   a. How many of the smaller cubes have 3 red sides?
   b. How many of the smaller cubes have 2 red sides?
   c. How many of the smaller cubes have 1 red side?
   d. How many of the smaller cubes have 0 red sides?

5. Toothpicks were used to make an array of 9 small squares, as shown below. How can you move just 4 toothpicks and leave 5 small squares?

12-7 Bisecting a Line Segment

Example

To bisect a given line segment, follow these steps.

a. Open the compass so that the radius is more than half the length of $\overline{AB}$.

b. With A and B as centers, draw arcs which cross above and below the given line segment at points C and D.

c. Draw $\overleftrightarrow{CD}$ bisecting $\overline{AB}$ at point E. E is the midpoint of $\overline{AB}$.

given line segment

A   B

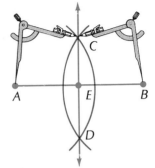

$\overleftrightarrow{CD} \perp \overline{AB}$
and bisects $\overline{AB}$ at E.

Vocabulary

The **midpoint** of a line segment is the point that separates the segment into two equal parts.

The **perpendicular bisector** of a line segment is a line that both bisects a line segment and is perpendicular to it.

Think About It

Is it possible for a line to be a bisector of a given line segment and *not* be perpendicular to it? Explain your answer, or draw a diagram to illustrate.

Practice Exercises

1. Draw a line segment of each length. Bisect each line segment, using a compass. Check with a ruler.

    a. 52 mm   **b.** $4\frac{3}{8}$ in.   **c.** 4.6 cm

2. Copy, then use a compass to divide the following segment into four equal parts. Check with a ruler.

_____

3. Draw any triangle.

    a. Bisect each side by constructing the perpendicular bisector of that side.
    b. Do these perpendicular bisectors intersect?
    c. Do they meet in a point equidistant from the vertices of the triangle? Check by measuring.
    d. Using this common point as the center and the distance from this point to any vertex of the triangle as the radius, draw a circle through the three vertices of the triangle.

    When each side of the triangle is a chord of the circle or each vertex is a point on the circle, we say that the circle is *circumscribed about the triangle* or that the triangle is *inscribed in a circle*.

Bisecting an Angle

Example

To bisect a given angle, follow these steps.

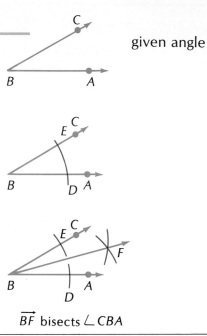

given angle

a. Use a compass with *B* as the center and any radius to draw an arc cutting $\vec{BA}$ at *D* and $\vec{BC}$ at *E*.

b. With *D* and *E* as centers and a radius of more than the distance from *D* to *E*, draw arcs crossing at *F*.

c. Draw $\vec{BF}$.

$\vec{BF}$ bisects ∠*CBA*

Think About It

1. If the measure of a given angle is 108°, what is the measure of each part of the angle after it is bisected?

2. Suppose that line segments *EF* and *DF* are drawn in the diagram above. What can you say about △*BEF* and △*BDF*?

Practice Exercises

1. Draw angles having the following measures. Bisect each angle using a compass. Check with a protractor.

   a. 70°   **b.** 54°   **c.** 145°   **d.** 137°

2. Draw any angle. Bisect it by using a compass only. Check by measuring your angle and each bisected angle.

3. Copy, then bisect each of the following angles.

   a.   **b.**   **c.**

4. Draw any angle. Using a compass only, divide the angle into four congruent angles. Check with a protractor.

12-9 Constructing a Line Parallel to a Given Line

Vocabulary

Parallel lines are lines in the same plane that do not meet.
The symbol ∥ means "is parallel to."

Example

To construct a line parallel to a given line through a
given point not on the given line, follow these steps.

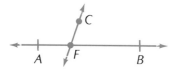

a. Draw a line through point C crossing $\overleftrightarrow{AB}$ at F.　　**b.** At point C, construct ∠2 equal to ∠1.

$$\overleftrightarrow{CD} \parallel \overleftrightarrow{AB}$$

Think About It

1. In the construction above, if ∠2 were greater
than ∠1, the lines would not be parallel and
would eventually meet. Would the lines meet
to the left or the right of $\overleftrightarrow{CF}$?

2. If a line is drawn perpendicular to $\overleftrightarrow{CD}$ will it
always, sometimes, or never be perpendicu-
lar to $\overleftrightarrow{AB}$? Explain your answer.

Practice Exercises

1. Draw any line. Select a point that is not on
this line. Through this point construct a line
parallel to the line you have drawn.

2. Construct a parallelogram with a base 75
millimeters long, a side 58 millimeters long,
and an included angle of 60°.

3. Draw any acute triangle. Bisect one of the sides. Draw a line segment from this
midpoint, parallel to one of the other sides, until it intersects the third side.

　a. Does this line segment bisect the third side? Check by measuring.
　b. Also check whether this line segment is one-half as long as the side to
　　　which it is parallel.
　c. Draw other triangles and check whether the above findings are true no
　　　matter which side is used first or which side is used as the parallel side.

Career Application

DRAFTER

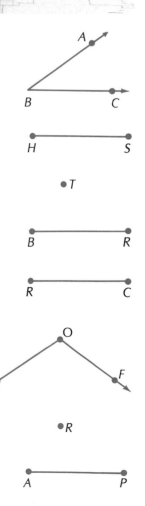

When workers are building a satellite, a house, or even a video recorder, they need detailed drawings that show the size and position of every part. The people who prepare these drawings are called *drafters*. The engineers or architects who design a product give drafters sketches and measurements. From these, drafters make accurate, detailed drawings. Drafters use rulers, compasses, protractors, and other drafting devices. Some drafters use computer-aided design systems. They also research or calculate numerical values when necessary.

People who are interested in drafting should have good eyesight and be able to picture objects in three dimensions. The work requires accuracy, attention to detail, and the ability to work with other people. Drafters can learn their skills in high school, by taking college courses, or on the job.

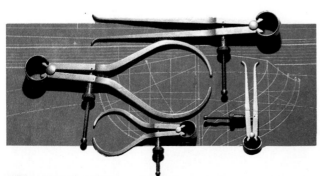

1. Drafters often have to copy part of a drawing. Copy angle *ABC*.

2. A structure is to be in the shape of an equilateral triangle. Construct an equilateral triangle with given base $\overline{HS}$.

3. A tower of a bridge is to be perpendicular to the roadway. Construct a perpendicular to $\overline{BR}$ through point *T*.

4. A knob is to be at the exact center of a radio cabinet. Bisect line segment *RC*.

5. Angle *ROF* represents the roof of a house. Bisect this angle.

6. The edges of an airport runway are to be parallel lines. Construct a line parallel to $\overleftrightarrow{AP}$ through point *R*.

Chapter 12 Review

Page

466   **1.** Use a metric ruler to copy the following line segment. ⎯⎯⎯⎯⎯⎯⎯⎯

466   **2.** Use a compass and an inch ruler to measure the following segment.

466   **3.** Use a compass and straightedge to copy the following segment.

467   **4.** Use a protractor to measure this angle.

467   **5.** Using a protractor, draw an angle of 120°. Then construct with a compass a second angle congruent to it. Check the measure of the second angle with a protractor.

468   **6.** Construct an isosceles triangle whose base is $3\frac{1}{4}$ inches long and each of whose base angles measure 30°.

468   **7.** Construct a triangle with two sides measuring 4.2 centimeters and 5.6 centimeters and an included angle of 56°.

468   **8.** Using a scale of 1:1000 construct a triangle with sides measuring 48 meters, 36 meters, and 60 meters.

471   **9.** Draw any line. Using a compass, construct a perpendicular to it at a point on the line.

472 **10.** Draw any line. Using a compass, construct a perpendicular to it from a point not on the line.

475 **11.** Draw a segment $1\frac{1}{8}$ inches long. Bisect it with a compass, check with a ruler.

476 **12.** Draw an angle measuring 56°. Bisect it with a compass, check with a protractor.

477 **13.** Draw a line. Using a compass, construct a line that is parallel to the first line and is 2 centimeters away.

477 **14.** Lines in the same plane that do not meet are called ⎯⎯**?**⎯⎯.

Chapter 12 Test

Page
466 **1.** Measure the given line segment using a ruler, and copy it.

---

466 **2.** Draw a line segment $1\frac{3}{4}$ inches long. Copy it using a compass and straightedge.

467 **3.** Use a protractor to measure and copy the given angle.

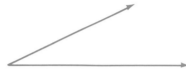

467 **4.** Draw an angle of 75°, using a protractor. Then, using a compass and straightedge, copy it.

468 **5.** Construct an equilateral triangle whose sides are $2\frac{1}{4}$ inches.

468 **6.** Construct an isosceles triangle whose base is $1\frac{5}{8}$ inches and whose congruent angles are 40°.

468 **7.** Construct, using a scale $\frac{1}{2}$ inch = 10 feet, a triangle whose three sides are 30 feet, 20 feet, and 25 feet.

471 **8.** Draw a line segment 2 inches long. At a point on the segment $\frac{1}{2}$ inch away from either endpoint, draw a perpendicular.

472 **9.** Draw any line. Using a compass and straightedge, construct a perpendicular to it from a point not on the line.

475 **10.** Draw a segment that measures $3\frac{1}{4}$ inches and bisect it with a compass.

476 **11.** Draw an angle that measures 64° and bisect it with a compass and straightedge.

477 **12.** Draw a line, then, using a compass and straightedge, construct a line that is parallel to the first line and 2.8 centimeters away.

13-1 Squares

Many formulas you will learn in math and science require you to find the square of a number. When you square a number you multiply it by itself. Is there a key on some calculators that will perform this operation for you?

Examples

To find the square of a number, multiply the number by itself or, if you are using a calculator with an x^2 key, enter the number and press the x^2 key.

1. Find the square of 15.
15×15, or $15^2 = 225$

ANSWER: 225

2. Find the square of 0.03 cm.
0.03 cm $\times$ 0.03 cm, or
$(0.03 \text{ cm})^2 = 0.0009 \text{ cm}^2$

ANSWER: 0.0009 cm^2

3. Find the square of $\frac{4}{5}$.
$\frac{4}{5} \times \frac{4}{5}$, or $\left(\frac{4}{5}\right)^2 = \frac{16}{25}$

ANSWER: $\frac{16}{25}$

Think About It

1. If a number is squared and the result is *smaller* than the original number, what can you deduce about the original number?

2. To evaluate an expression with more than one operation, work from left to right and square any numbers, then multiply or divide before you add or subtract. Explain how to find the value of the following expression:
$129 - 3 \times 5^2 + 42 \div 6$

Practice Exercises

Find the square of each of the following numbers.

1. a. 4   **b.** 1   **c.** 78   **d.** 100

2. a. 0.2   **b.** 0.7   **c.** 0.75   **d.** 0.18

3. a. $\frac{1}{2}$ **b.** $\frac{5}{6}$ **c.** $2\frac{3}{4}$ **d.** $3\frac{1}{3}$

4. Find the value of each of the following.

   a. 6^2 **b.** $(0.009)^2$ **c.** $(8.1)^2$ **d.** $\left(3\frac{1}{2}\right)^2$

5. To find the area of a square when the length of its side is known, square the given side. Find the area of a square with the given side.

   a. 7 in.   **b.** 13 mm **c.** $2\frac{1}{2}$ ft   **d.** 19 cm

6. The formula for finding the distance a body falls involves squaring the number of seconds it falls. Square the following numbers representing intervals of time.

   a. 5 s   **b.** 12 s   **c.** 30 s   **d.** 27 s

7. Find the value of each of the following expressions.

   a. $8 + 2 \times 3^2$ **b.** $9 \times 3^2 - 20$ **c.** $30 - 2^2 \times 5$ **d.** $9 \times 6^2 \div 12$

Square Roots of Perfect Squares

As you continue to learn more math and science, you will often have to find the square root of a number. In this lesson you will learn how to find the square root of a perfect square.

Examples

To find the square root of a perfect square, write all the factors of the number and give as your answer *one* of the two factors of a pair of *equal* factors. If you are using a calculator, enter the number and press the key labeled $\sqrt{}$.

Vocabulary

The **square root** (symbol $\sqrt{}$) of a number is that number which, when multiplied by itself, produces the given number.

A **perfect square** has a whole number square root.

1. Find $\sqrt{100}$.

$100 = 1 \times 100; 2 \times 50;$
$4 \times 25; 5 \times 20;$
10×10

ANSWER: 10

2. Find $\sqrt{144}$.

$144 = 1 \times 144; 2 \times 72;$
$3 \times 48; 4 \times 36;$
$6 \times 24; 8 \times 18;$
$9 \times 16; 12 \times 12$

ANSWER: 12

3. Find $\sqrt{\frac{25}{49}}$.

$25 = 1 \times 25; 5 \times 5$
$49 = 1 \times 49; 7 \times 7$
$\frac{25}{49} = \frac{5}{7} = \frac{5}{7}$

ANSWER: $\frac{5}{7}$

Think About It

A teacher explains another way to find the square root of 144. Factor it into primes: $3 \times 3 \times 2 \times 2 \times 2 \times 2$. Then take the square root of each pair of factors and multiply to find the answer: $3 \times 2 \times 2 = 12$. Explain how to use this technique to find the square root of 196.

Practice Exercises

Give each square root.

1. a. $\sqrt{4}$ **b.** $\sqrt{49}$ **c.** $\sqrt{1}$ **d.** $\sqrt{100}$

2. a. $\sqrt{400}$ **b.** $\sqrt{2,500}$ **c.** $\sqrt{6,400}$ **d.** $\sqrt{10,000}$

3. a. $\sqrt{\frac{9}{25}}$ **b.** $\sqrt{\frac{64}{225}}$ **c.** $\sqrt{\frac{144}{625}}$ **d.** $\sqrt{\frac{169}{2,500}}$

Finding Square Roots by Table and by Estimation

The table below can be used to find the square roots of whole numbers between 1 and 99 and the square roots of perfect squares up to 9,801.

Table of Squares and Square Roots

Number	Square	Square Root	Number	Square	Square Root	Number	Square	Square Root
1	1	1.000	34	1,156	5.831	67	4,489	8.185
2	4	1.414	35	1,225	5.916	68	4,624	8.246
3	9	1.732	36	1,296	6.000	69	4,761	8.307
4	16	2.000	37	1,369	6.083	70	4,900	8.367
5	25	2.236	38	1,444	6.164	71	5,041	8.426
6	36	2.449	39	1,521	6.245	72	5,184	8.485
7	49	2.646	40	1,600	6.325	73	5,329	8.544
8	64	2.828	41	1,681	6.403	74	5,476	8.602
9	81	3.000	42	1,764	6.481	75	5,625	8.660
10	100	3.162	43	1,849	6.557	76	5,776	8.718
11	121	3.317	44	1,936	6.633	77	5,929	8.775
12	144	3.464	45	2,025	6.708	78	6,084	8.832
13	169	3.606	46	2,116	6.782	79	6,241	8.888
14	196	3.742	47	2,209	6.856	80	6,400	8.944
15	225	3.873	48	2,304	6.928	81	6,561	9.000
16	256	4.000	49	2,401	7.000	82	6,724	9.055
17	289	4.123	50	2,500	7.071	83	6,889	9.110
18	324	4.243	51	2,601	7.141	84	7,056	9.165
19	361	4.359	52	2,704	7.211	85	7,225	9.220
20	400	4.472	53	2,809	7.280	86	7,396	9.274
21	441	4.583	54	2,916	7.348	87	7,569	9.327
22	484	4.690	55	3,025	7.416	88	7,744	9.381
23	529	4.796	56	3,136	7.483	89	7,921	9.434
24	576	4.899	57	3,249	7.550	90	8,100	9.487
25	625	5.000	58	3,364	7.616	91	8,281	9.539
26	676	5.099	59	3,481	7.681	92	8,464	9.592
27	729	5.196	60	3,600	7.746	93	8,649	9.644
28	784	5.292	61	3,721	7.810	94	8,836	9.695
29	841	5.385	62	3,844	7.874	95	9,025	9.747
30	900	5.477	63	3,969	7.937	96	9,216	9.798
31	961	5.568	64	4,096	8.000	97	9,409	9.849
32	1,024	5.657	65	4,225	8.062	98	9,604	9.899
33	1,089	5.745	66	4,356	8.124	99	9,801	9.950

Examples

To find the square root of a number from 1 to 99, locate the number in the Number column and look to the *right* to find the answer in the Square Root column.

1. Find $\sqrt{64}$. $\sqrt{64} = 8.000$

ANSWER: 8

2. Find $\sqrt{12}$. $\sqrt{12} = 3.464$

ANSWER: 3.464

To find a square root of a perfect square, locate the number in the Square column and look to the *left* to find the square root in the Number column.

3. Find $\sqrt{81}$. $\sqrt{81} = 9$

ANSWER: 9

4. Find $\sqrt{2,704}$. $\sqrt{2,704} = 52$

ANSWER: 52

Another way to find a square root is to use the estimate, divide, and average (E.D.A) method.

5. Find the square root of 12 using the E.D.A. method.

Since 12 is between 9 and 16, its square root will be between 3 and 4. Use 3.4 as the estimate.

a. Divide 12 by the estimate. Round the quotient.

$$3.4\overline{)12.00} \quad \frac{3.5}{}$$

b. Average the quotient and the divisor from step *a*.

$$\frac{3.4 + 3.5}{2} = 3.45$$

c. Divide the number by the average from step *b* and round.

$$3.45\overline{)12.00} \quad \frac{3.48}{}$$

d. Repeat steps *b* and *c* until the quotient and the divisor agree to the number of places you want. Add one decimal place each time.

$$\frac{3.45 + 3.48}{2} = 3.465$$

$$3.465\overline{)13.000} \quad \frac{3.463}{}$$

$$3.464\overline{)12.000} \quad \frac{3.464}{}$$

$$\frac{3.465 + 3.463}{2} = 3.464$$

ANSWER: The square root of 12, to the nearest thousandth, is 3.464.

Think About It

You can use the table to find the square root of 8,400 if you notice that $8,400 = 84 \times 100$. Since $\sqrt{84} = 9.165$ and $\sqrt{100} = 10$ then $\sqrt{8,400} = 9.165 \times 10$, or 91.65. Explain how to find the square root of 62.41. (Note that $62.41 = 6241 \times \frac{1}{100}$.)

Practice Exercises

1. Find the square root of each of the following numbers from the table.

a. 29　　**b.** 91　　**c.** 324　　**d.** 2,209

2. Find the square root of each number to the nearest thousandth by the E.D.A. method.

a. 6　　**b.** 21　　**c.** 59　　**d.** 105

Using Mathematics

PROBLEM SOLVING TECHNIQUES

The strategy below can help you apply problem solving techniques to set up and solve geometric problems.

1. **READ** the problem carefully to find:

   a. The fact, dimension, or value that is to be determined.

   b. The facts, dimensions, or values that are given.

2. **PLAN** how to solve the problem.

   Decide which formula or formulas can be used that relates the variable representing the value to be found with the variables representing the given values.

3. **SOLVE** by arranging the solution in three columns.

   a. In the left column:

   (1) Draw the geometric figure described in the problem.
   (2) Mark the given dimensions on the figure.
   (3) Below the figure write the variables with corresponding given values and the variable representing the unknown value.

   b. In the center column:

   (1) Write the formula that relates the variables.
   (2) Substitute the known given values for the corresponding variables in the formula.
   (3) Perform the necessary operations.
   (4) When necessary, solve the resulting equation.

   c. In the right column:

   Perform the necessary arithmetic operations.

4. **CHECK** the answer directly with the facts, dimensions, or values that are given.

For a sample solution see the completely worked-out problem on the next page.

It is not always possible to measure a distance directly. For example, it may be necessary to use an indirect method to measure the distance from a ship to the shore or the distance across a river.

The rule of Pythagoras is one means of finding distances indirectly. The rule of Pythagoras can be used to find the base, altitude, or hypotenuse of a right triangle if the other two sides are known.

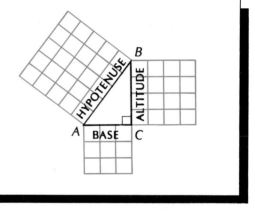

Rule of Pythagoras

The rule of Pythagoras states that the square of the hypotenuse of a right triangle is equal to the sum of the squares of the other two sides.

$$h^2 = a^2 + b^2$$

h is the length of the hypotenuse.
a is the length of the altitude.
b is the length of the base.

If any two sides of a right triangle are known, the third side can be found by using one of the following alternate forms of the rule of Pythagoras.

$$h = \sqrt{a^2 + b^2} \qquad a = \sqrt{h^2 - b^2} \qquad b = \sqrt{h^2 - a^2}$$

Examples

1. Find the hypotenuse of a right triangle if the altitude is 4 decimeters and the base is 3 decimeters.

Use the form of the rule of Pythagoras: $h = \sqrt{a^2 + b^2}$. $a = 4$ dm, $b = 3$ dm; $h = ?$

$$h = \sqrt{a^2 + b^2}$$
$$h = \sqrt{4^2 + 3^2}$$
$$h = \sqrt{16 + 9}$$
$$h = \sqrt{25}$$
$$h = 5$$

$4^2 = 4 \times 4 = 16$
$3^2 = 3 \times 3 = 9$

$$\begin{array}{r} 16 \\ + \ 9 \\ \hline 25 \end{array} \qquad \sqrt{25} = 5$$

ANSWER: $h = 5$ decimeters

2. Find the altitude of a right triangle if the hypotenuse is 20 centimeters and the base is 12 centimeters.

Use the form of the rule of Pythagoras: $a = \sqrt{h^2 - b^2}$.

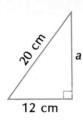

$$a = \sqrt{h^2 - b^2}$$
$$a = \sqrt{20^2 - 12^2}$$
$$a = \sqrt{400 - 144}$$
$$a = \sqrt{256}$$
$$a = 16 \text{ cm}$$

$$20^2 = 20 \times 20 = 400$$
$$12^2 = 12 \times 12 = 144$$

$$\begin{array}{r} 400 \\ - 144 \\ \hline 256 \end{array}$$

$\sqrt{256} = 16$
(from the table)

$h = 20$ cm, $b = 12$ cm; $a = ?$

ANSWER: $a = 16$ centimeters

Think About It

1. A student uses the rule of Pythagoras to find the hypotenuse of a right triangle. He finds the hypotenuse to be $\sqrt{50}$ inches long. Explain two techniques he could use to evaluate this to the nearest thousandth if he does not have a calculator.

2. In many of the next problems, you will find that all three sides of the right triangles are whole numbers. Make a list of these "Pythagorean triples."

Practice Exercises

1. Find the hypotenuse of each right triangle.

	Altitude	Base
a.	12 m	9 m
b.	8 cm	15 cm
c.	60 km	25 km
d.	33 ft	56 ft

2. Find the altitude of each right triangle.

	Hypotenuse	Base
a.	13 mm	5 mm
b.	35 cm	28 cm
c.	89 ft	80 ft
d.	53 km	45 km

3. Find the base of each right triangle.

	Hypotenuse	Altitude
a.	25 cm	24 cm
b.	87 m	63 m
c.	73 mm	48 mm
d.	91 yd	84 yd

4. Solve by the rule of Pythagoras and by actual measurement.

a. Find the length of the diagonal (line segment joining opposite corners) of a rectangle 8 cm long and 6 cm wide.

b. Find the length of the diagonal of a square whose side measures 7 cm.

5. A 15-foot wire is attached to a point in a tree 12 feet above the ground. The wire is held taut and attached to the ground. How far is the point on the ground from the base of the tree?

6. What is the shortest distance from first base to third base if the distance between bases is 90 feet?

7. How high up on a wall does a 25-foot ladder reach if the foot of the ladder is 7 feet from the wall?

13-5 Finding Distances by Using Similar Triangles

In the last section you used the rule of Pythagoras to find distances (in right triangles) that could not be measured directly. Similar triangles can also be used to find distances indirectly.

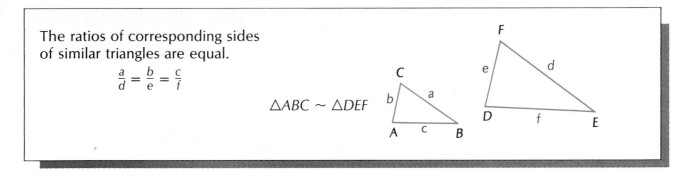

The ratios of corresponding sides of similar triangles are equal.

$$\frac{a}{d} = \frac{b}{e} = \frac{c}{f}$$

$\triangle ABC \sim \triangle DEF$

Example

A tree casts a shadow ($\overline{AB}$) 27 feet long, while a 4-foot pole ($\overline{EF}$) nearby casts a shadow ($\overline{DE}$) 3 feet long. What is the tree's height ($\overline{BC}$)?

To find a distance using similar triangles, follow these steps.

a. Use the given data to draw similar triangles.

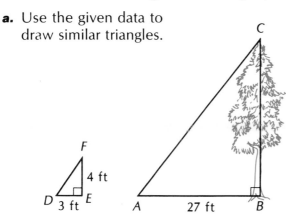

ANSWER: The tree is 36 feet high.

b. Write a proportion using the three known sides and the required distance as the fourth side.

$$\frac{\text{side } AB}{\text{side } DE} = \frac{\text{side } BC}{\text{side } EF}$$

Let n = the length of side BC.

$$\frac{27}{3} = \frac{n}{4}$$

c. Solve the proportion.

$$3 \times n = 27 \times 4$$
$$3n = 108$$
$$n = 36$$

$$\begin{array}{r} 27 \\ \times\ 4 \\ \hline 108 \end{array} \qquad \begin{array}{r} 36 \\ 3\overline{)108} \end{array}$$

Think About It

1. In the example above, if the pole were 4 yards long and the shadow were 3 yards long would the answer be different? Explain.

2. In the example above, if $DE = 6$ inches, $FE = 5$ feet, $AB = 3$ feet, explain how to set up the correct proportion to find BC.

Practice Exercises

1. Triangle *ABC* and triangle *DEF* are similar. Find the height (*BC*) of the flagpole.

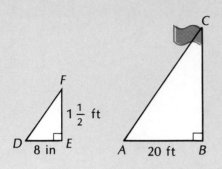

2. Triangles *ABC* and *CDE* are similar.
 a. Which side of triangle *ABC* corresponds to side *CE*?
 b. Which side corresponds to *DE*?
 c. What is the distance (*AB*) across the stream?

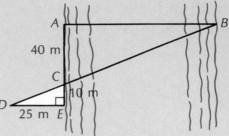

3. Triangles *ABE* and *CDE* are similar.
 a. Which side of triangle *ABE* corresponds to side *CE*?
 b. Which side corresponds to *DE*?
 c. Which side corresponds to *AB*?
 d. Find the length (*AB*) of the lake.

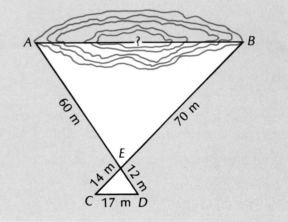

Enrichment

We can determine missing distances by scale drawings. First, find the scale. Then measure the scale lengths and apply the scale to find the actual lengths.

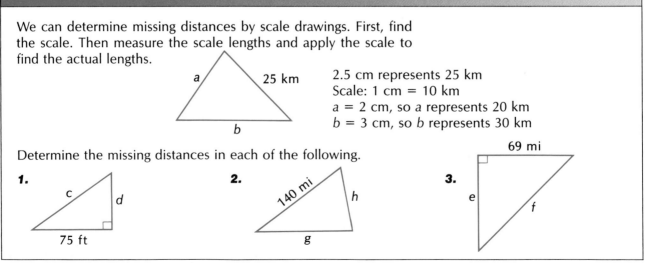

2.5 cm represents 25 km
Scale: 1 cm = 10 km
a = 2 cm, so *a* represents 20 km
b = 3 cm, so *b* represents 30 km

Determine the missing distances in each of the following.

1.
 c d
 75 ft

2.
 140 mi h
 g

3.
 69 mi
 e f

Numerical Trigonometry

If you know the measurements of two sides of a right triangle (or the measurements of a side and an angle) you can find the measurement of any other side or angle of the triangle by setting up a trigonometric ratio and using a table of trigonometric functions.

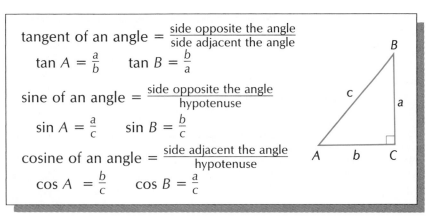

tangent of an angle $= \dfrac{\text{side opposite the angle}}{\text{side adjacent the angle}}$

$\tan A = \dfrac{a}{b}$ $\tan B = \dfrac{b}{a}$

sine of an angle $= \dfrac{\text{side opposite the angle}}{\text{hypotenuse}}$

$\sin A = \dfrac{a}{c}$ $\sin B = \dfrac{b}{c}$

cosine of an angle $= \dfrac{\text{side adjacent the angle}}{\text{hypotenuse}}$

$\cos A = \dfrac{b}{c}$ $\cos B = \dfrac{a}{c}$

Vocabulary

The ratio of the side opposite an acute angle in a right triangle to the adjacent side is the **tangent (tan)** of the angle.

The ratio of the side opposite an acute angle in a right triangle to the hypotenuse is the **sine (sin)** of the angle.

The ratio of the adjacent side of an acute angle in a right triangle to the hypotenuse is the **cosine (cos)** of the angle.

Examples

1. Find the value of cos 54°.

Use the table (page 607).

Find 54° in the angle column, then read to the *right* to find the corresponding cosine.

$\cos 54° = 0.5878$

ANSWER: $\cos 54° = 0.5878$

2. Find angle A if tan A = 0.7813.

Use the table (page 607).

Find 0.7813 in the tangent column, then read to the *left* to find the corresponding angle.

$\tan 38° = 0.7813$

ANSWER: $\angle A = 38°$

---

3. In right triangle *ABC*, find side *a* when angle $A = 53°$ and side $b = 100$ feet.

a. Draw a right triangle and label it with the given dimensions.

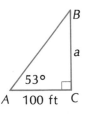

$b = 100$ ft, $\angle A = 53°$; $a = ?$

ANSWER: Side $a = 132.7$ feet.

b. Select the proper formula and substitute the given values. Use the table (page 607) for the tangent value.

$$\tan A = \frac{a}{b}$$
$$\tan 53° = \frac{a}{100}$$
$$1.3270 = \frac{a}{100}$$

c. Solve the resulting equation.

$100 \times 1.3270 = a$

$a = 132.7$ ft $1.3270 \times 100 = 132.7$

4. In right triangle *ABC,* find angle *B* when side *a* = 12 meters
and side *c* = 20 meters.

a. Draw a right triangle and label it with the
given dimensions.

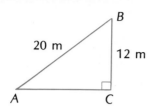

ANSWER: ∠*B* = 53°

b. Select the proper formula and substitute the
given values.

$$\cos B = \frac{a}{c}$$

$$\cos B = \frac{12}{20}$$

$$\cos B = 0.6 \qquad\qquad 12 \div 20 = 0.6$$

c. Use the table (page 607) to find the angle
whose cosine is closest to 0.6.

$$B = 53°$$

---

5. The angle of depression from a plane to an object is 13°. The
plane is 18 km above the ground. How far is the plane from
the object?

An angle measured vertically between the
horizontal line and the observer's line of sight
to an object is called the **angle of depression**
when the object is below the observer and
the **angle of elevation** when the object is
above the observer.

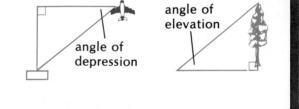

a. Draw a right triangle and label it.

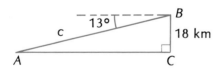

$$a = 18 \text{ km}$$
$$\angle ABC = 77° \text{ (complement of 13°)}$$
$$c = ?$$

ANSWER: The plane is 80 km from the object.

b. Select the proper formula and substitute the
given values. Use the table (page 607) for the
cosine value.

$$\cos B = \frac{a}{c}$$

$$\cos 77° = \frac{18}{c}$$

$$0.2250 = \frac{18}{c}$$

c. Solve the resulting equation.

$$0.2250c = 18 \qquad\qquad 18 \div 0.225 = 80$$

$$c = 80 \text{ km}$$

Practice Exercises

Find the value of the following.

1. a. tan 43°   **b.** tan 81°

2. a. sin 79°   **b.** sin 7°

3. a. cos 18°   **b.** cos 56°

Find angle *A.*

4. a. tan *A* = 0.3839   **b.** tan *A* = 3.0777

5. a. sin *A* = 0.9063   **b.** sin *A* = 0.3420

6. a. cos *A* = 0.8988   **b.** cos *A* = 0.6293

7. Use the tangent ratio to find the following.

    a. Side *b* if angle *B* = 65° and side *a* = 80 cm
    b. Side *b* if angle *A* = 22° and side *a* = 101 ft
    c. Side *a* if angle *B* = 70° and side *b* = 140 km
    d. Angle *B* if side *a* = 1,000 ft and side *b* = 1,804 ft

8. Find the indicated measure in each of the following right triangles, to the nearest tenth.

    a. Side *a*
    b. Side *b*
    c. ∠*A*

9. Use the sine ratio to find each of the following.

    a. Side *a* if angle *A* = 60° and side *c* = 75 m
    b. Side *b* if angle *A* = 86° and side *c* = 3.49 km
    c. Angle *A* if side *a* = 170.5 m and side *c* = 250 m
    d. Angle *B* if side *b* = 106 yd and side *c* = 125 yd

10. Find the indicated measure in each of the following right triangles, to the nearest tenth.

    a. Side *a*
    b. Side *b*
    c. ∠*B*

11. Use the cosine ratio to find each of the following.

    a. Side *b* if angle *A* = 46° and side *c* = 100 m
    b. Side *c* if angle *A* = 70° and side *b* = 17.1 km
    c. Side *c* if angle *B* = 59° and side *a* = 515 cm
    d. Angle *A* if side *b* = 23 mm and side *c* = 46 mm

12. Find each of the following.

    a. Side *a* if angle *A* = 61° and side *b* = 25 cm
    b. Side *a* if angle *A* = 40° and side *c* = 50 m
    c. Side *b* if angle *B* = 13° and side *c* = 80 km
    d. Side *b* if angle *A* = 30° and side *a* = 46 mm
    e. Angle *A* if side *a* = 451 ft and side *b* = 250 ft
    f. Angle *B* if side *a* = 511.5 mi and side *c* = 750 mi

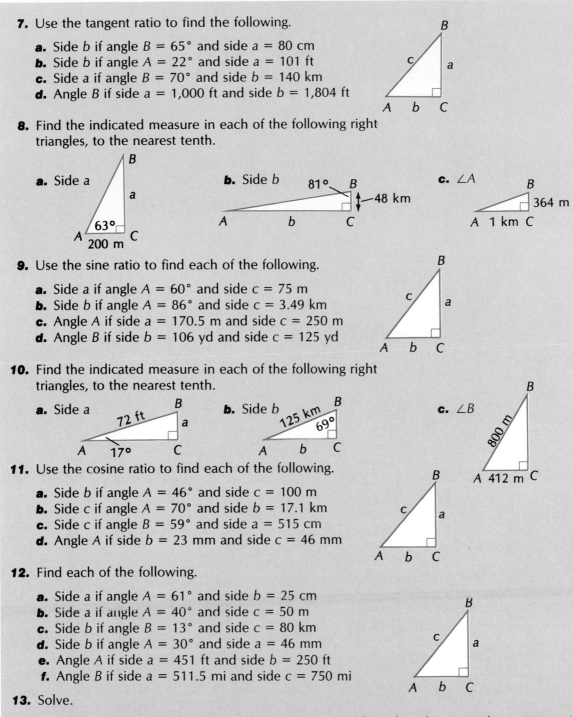

13. Solve.

    a. From a cliff 300 meters above the sea, the angle of depression of a boat is 87°. How far is the boat from the foot of the cliff?

    b. The sides of a rectangle are 35 centimeters and 50 centimeters. What angle is formed between one short side and the diagonal?

    c. The altitude of a right triangle is 220 millimeters and the angle opposite the base is 77°. How long is the hypotenuse?

    d. A guy wire 80 feet in length is attached to the top of a telephone pole and then to the ground. If the wire makes an angle of 75° with the ground, how tall is the telephone p

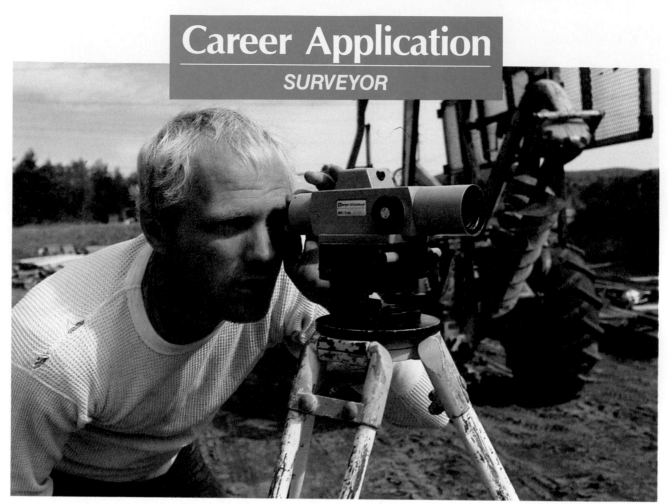

A contractor needs to know the exact size of a building site. A lawyer wants a precise description of the boundaries of a lot. An oil company needs a detailed map of the locations of its oil wells in an area. *Surveyors* are hired to do jobs like these. They usually work with a team of technicians. Together they establish the exact locations of points and measure distances and elevations. Their tools range from rods and levels to electronic devices that measure distances.

People interested in surveying should enjoy outdoor work and be able to make accurate measurements. Surveyors generally have college training. Surveyor technicians learn their skills on the job. Surveyors must have state licenses because they are legally responsible for the accuracy of their work.

1. A surveyor needs to find distance *JK,* the distance across a river. Triangle *JKL* and triangle *NML* are similar. What is the distance *JK*?

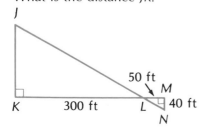

2. A surveyor needs to measure the distance *AB*. Use the rule of Pythagoras to help the surveyor find *AB* to the nearest tenth of a meter.

3. *PQRS* is a scale drawing of a building lot. The scale is 1 in. = 400 ft. What is the approximate length in feet of the actual side *QR*?

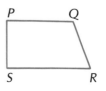

Vocabulary

angle of depression p. 492
angle of elevation p. 492
cosine p. 491

perfect square p. 483
rule of Pythagoras p. 487
sine p. 491

square root p. 483
squaring a number p. 482
tangent p. 491

Page
482 Find the value.

1. $(53)^2$ **2.** $\left(\frac{3}{7}\right)^2$ **3.** $(0.08)^2$

484 **6.** Between what two whole numbers is $\sqrt{172}$?

487 **8.** Is a triangle with sides of 90 m, 56 m, and 106 m a right triangle?

487 **10.** Find the base of a right triangle if the hypotenuse is 25 ft and the altitude is 7 ft.

489 **12.** Find the height of a tree that casts a shadow 42 ft long at the same time a 9-ft high sign casts a shadow 6 ft long.

489 **13.** Find the distance (PQ).

Page
483 Find the value.

4. $\sqrt{1600}$ **5.** $\sqrt{0.81}$

484 **7.** Find the square root of 56 to the nearest thousandth by the estimate, divide, and average method.

487 **9.** Find the hypotenuse of a right triangle if the base is 36 cm and the altitude is 77 cm.

487 **11.** Find the altitude of a right triangle if the base is 160 mm and the hypotenuse is 178 mm.

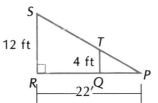

491 **14.** Find side b to the nearest hundredth.

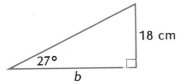

491 **15.** Find angle A.

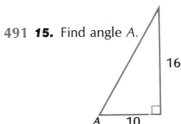

491 **16.** The angle of elevation from a sailboat to the top of an 800 ft cliff on the coast is 31°. How far from the coast at the base of the cliff is the boat, to the nearest foot?

487 **17.** The rule of Pythagoras states that the _____ of the hypotenuse of a right triangle is equal to the sum of the squares of the other two sides.

Chapter 13 Test

Find the value.

1. $(78)^2$ **2.** $\left(1\frac{3}{4}\right)^2$ **3.** $(0.03)^2$

7. Between which two whole numbers is the square root of 60?

9. Is a triangle with sides of 15 m, 18 m, and 23 m a right triangle?

11. What is the base of a right triangle if the altitude is 39 ft and the hypotenuse is 89 ft?

13. A child lets out 60 meters of string in flying a kite. The distance from a point directly under the kite to where the child stands is 36 meters. If the child holds the string 1.5 meters from the ground, how high is the kite? Disregard any sag.

Find the square root of each.

4. 1024 **5.** 2.25 **6.** $\frac{9}{4}$

8. Find $\sqrt{72}$ to the nearest thousandth by the estimate, divide, and average method.

10. What is the altitude of a right triangle if the hypotenuse is 29 m and the base is 21 m?

12. Joan left home and rode her bike 6 km west and then 8 km south. How far from home is she in a straight line?

14. An airplane, flying 252 kilometers from town A due west to town B, drifts off its course in a straight line and is 39 kilometers due south of town B. What distance did the airplane actually fly?

15. An office building casts a shadow 150 ft long at the same time a tree 40 ft tall casts a shadow 25 ft long. What is the height of the building?

16. Find the distance (AB).

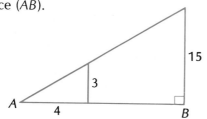

17. Find the length of side a.

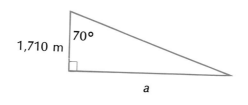

18. Find the length of side b.

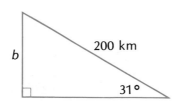

14-1 Perimeter of a Rectangle

Can you list several situations in which you might want to know the distance around a rectangle? It is often important to be able to find this distance, which is called the perimeter of the rectangle.

Vocabulary

The **perimeter** of a polygon is the sum of the lengths of its sides.

When computing perimeter, express all linear measurements with the same unit of measure. Since the pairs of opposite sides of a rectangle are equal in length, the perimeter of a rectangle is equal to twice its length (l) plus twice its width (w).

Perimeter $= l + w + l + w$
$ = 2l + 2w$
Formula: $P = 2l + 2w$ or $P = 2(l + w)$

w

l

Example

Find the perimeter of a rectangle 26 meters long and 17 meters wide.

17 m

26 m

$P = 2l + 2w$
$P = 2 \times 26 + 2 \times 17$
$P = 52 + 34$
$P = 86$ m

$\begin{array}{r} 26 \\ \times\ 2 \\ \hline 52 \end{array}$ $\qquad$ $\begin{array}{r} 17 \\ \times\ 2 \\ \hline 34 \end{array}$ $\qquad$ $\begin{array}{r} 52 \\ +\ 34 \\ \hline 86 \end{array}$

$l = 26$ m, $w = 17$ m; $P = ?$

ANSWER: $P = 86$ meters

Practice Exercises

1. What is the perimeter of a rectangle if its length is 47 mm and its width is 21 mm?

2. Find the perimeter of each rectangle.

   a. $l = 63$ ft $w = 49$ ft   **b.** $l = 7\frac{3}{8}$ in. $w = 4\frac{13}{16}$ in.   **c.** $l = 2$ ft 7 in. $w = 1$ ft 10 in

3. How many meters of fencing are required to enclose a rectangular garden 59 m long and 39 m wide? If each 2-m section costs $11.99, how much will the fencing cost?

4. How many meters of fringe are needed for a border on a bedspread 182 cm by 268 cm?

5. Fred wishes to make a frame for his class picture. The picture measures 25 in. by $12\frac{1}{2}$ in. If he allows $2\frac{1}{4}$ in. extra for each corner, how many feet of molding will he need?

6. A rectangle has a perimeter of 14 in. and a width of 3 in. What is the length of the rectangle? Construct the rectangle.

14-2 Perimeter of a Square

You could find the perimeter of a square using the formula for a rectangle but, since the sides of a square are all equal, there is an even easier way to find the perimeter.

Since the sides of a square are equal in length, the perimeter of a square is equal to four times the length of a side (s).

Perimeter $= s + s + s + s$
$\qquad\quad = 4s$

Formula: $P = 4s$

Example

Find the perimeter of a square whose side is 19 miles long.

19 mi

$s = 19$ mi
$P = ?$

$P = 4s$
$P = 4 \times 19$
$P = 76$ mi

$\begin{array}{r} 19 \\ \times\ 4 \\ \hline 76 \end{array}$

ANSWER: $P = 76$ miles

Think About It

1. A student wants to construct a square with a perimeter of 18 inches. Can he draw more than one square? Explain.

2. A student wishes to construct a rectangle with a perimeter of 18 inches. Can he draw more than one rectangle? Explain.

Practice Exercises

1. What is the perimeter of a square whose side is 8 km long?

2. Find the perimeters of the squares whose sides measure the following.

   a. 5,280 ft   **b.** 17.5 cm   **c.** $8\frac{3}{4}$ ft   **d.** 2 ft 9 in.

3. If the distance between bases is 90 ft, how many yards does a batter run when he hits a home run?

4. Find the cost of the wire needed to make a fence of 5 strands around a square lot 60 m by 60 m if a 400-m spool of wire costs $34.69.

5. Construct a square with a perimeter of 10 in. How long is one side of the square?

Perimeter of a Triangle

There are two formulas for the perimeter of a triangle. The first formula can be used for any triangle, the second can be used only for an equilateral triangle.

The perimeter of a triangle is equal to the sum of its sides.

Since all the sides of an equilateral triangle are equal, the perimeter is equal to 3 times the length of a side.

Perimeter = $a + b + c$
Formula: $P = a + b + c$

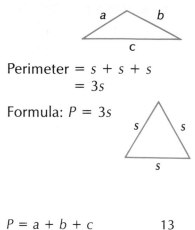

Perimeter = $s + s + s$
                    = $3s$

Formula: $P = 3s$

Example

Find the perimeter of a triangle with sides measuring 13 cm, 15 cm, and 19 cm.

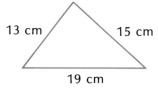

$a = 13$ cm, $b = 15$ cm,
$c = 19$ cm; $P = ?$

$P = a + b + c$
$P = 13 + 15 + 19$
$P = 47$ cm

```
  13
  15
+ 19
――――
  47
```

ANSWER: $P = 47$ centimeters

Think About It

1. If you know the perimeter of an equilateral triangle, how can you find the length of a side?

2. The sides of a triangle are 19 inches, $1\frac{1}{3}$ feet and $\frac{1}{4}$ yard in length. Explain how you would find the perimeter of the triangle in inches and convert your answer to feet or yards.

Practice Exercises

1. Find the perimeter of each triangle.

   a. 18 m, 9 m, 15 m

   b. 6.23 km, 4.7 km, 3.59 km

2. Find the perimeter of each equilateral triangle.

   a. $s = 63$ cm

   b. $s = 16\frac{5}{8}$ in.

3. Find the perimeter of an isosceles triangle if each of the equal sides is 15 cm and the third side is 8 cm.

4. How many meters of hedge are needed to enclose a triangular lot with sides measuring 196 m, 209 m, and 187 m?

5. How many feet of wire are needed to form a triangular shape with sides measuring 21 in., $2\frac{1}{4}$ ft, and $\frac{5}{6}$ yd?

Circumference of a Circle

You have learned how to find the perimeter of a polygon. In this lesson you will learn how to find the distance around a circle. This distance is called the circumference.

The diameter (d) is twice the radius (r).

Formula: $d = 2r$

The radius is one-half the diameter.

Formula: $r = \frac{d}{2}$

For computing, use $\pi = \frac{22}{7}$ or $\pi = 3.14$

The circumference (C) is equal to pi times the diameter.

Formula: $C = \pi d$

The circumference is equal to 2 times pi times the radius.

Formula: $C = 2\pi r$

Vocabulary

The distance around a circle is called the **circumference.**

π (read "pi") is the ratio of the circumference of a circle to its diameter.

Examples

1. Find the circumference of a circle when its diameter is 8 centimeters.

$d = 8$ cm, $\pi = 3.14$; $C = ?$

$C = \pi d$
$C = 3.14 \times 8$
$C = 25.12$ cm

$$\begin{array}{r} 3.14 \\ \times \quad 8 \\ \hline 25.12 \end{array}$$

ANSWER: $C = 25.12$ centimeters

2. Find the circumference of a circle when its radius is 21 inches.

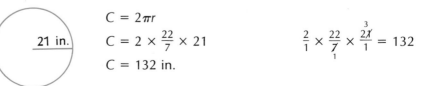

$r = 21$ in., $\pi = \frac{22}{7}$; $C = ?$

$C = 2\pi r$
$C = 2 \times \frac{22}{7} \times 21$
$C = 132$ in.

$\frac{2}{1} \times \frac{22}{7} \times \frac{\overset{3}{\cancel{21}}}{1} = 132$

ANSWER: $C = 132$ inches

1. If you know the circumference of a circle, explain how you can find the diameter of the circle.

2. If you want the circumference of a circle to be π inches, what length would you choose for the diameter?

Practice Exercises

1. Find the diameter if the radius is:

   a. 7 cm **b.** 6.5 km **c.** $10\frac{11}{16}$ in. **d.** 2 ft 3 in.

2. Find the radius if the diameter is:

   a. 38 m **b.** 8.9 km **c.** $10\frac{2}{3}$ ft **d.** 8 yd 2 ft

3. Find the circumference of a circle with the given diameter.

   a. 35 cm **b.** 260 m **c.** 49 mi **d.** 1.8 m

4. Find the circumference of a circle with the given radius.

   a. 7 mm **b.** 382 ft **c.** 8.4 km **d.** $5\frac{1}{2}$ mi

5. What is the diameter of a circle whose circumference is 286 cm?

6. Find the diameter of a circle with the given circumference.

   a. 176 m **b.** 330 yd **c.** 40 mm **d.** $6\frac{7}{8}$ in.

7. If the diameter of a circular table is 42 in., what is the circumference of the table?

8. What distance in feet does the tip of a propeller travel in one revolution if its length (diameter) is 7 ft?

9. How long a metal bar do you need to make a basketball hoop with a diameter of 48 cm?

10. The circumference of a circular table is 132 ft. What is the diameter of the table?

11. If the diameter of each wheel is 28 in., how far does a bicycle go when the wheels revolve once? How many times do the wheels revolve in a distance of 1 mi?

12. How much farther do you ride in one turn of a merry-go-round if you sit in the outside lane, 21 ft from the center, than if you sit in the inside lane, 14 ft from the center?

Enrichment

The Greek letter pi (written (π) stands for the number obtained when the circumference of any circle is divided by its diameter. $\pi = \frac{c}{d}$

The ancient Chinese used 3 as an approximation for pi. The Egyptians improved on this approximation around 1600 B.C. An astronomer, Ptolemy of Alexandria, calculated a value of pi to four decimal places. Later mathematicians attempted to calculate the exact value of pi. Their research showed that pi was an infinite decimal and impossible to calculate exactly.

Common values used for π are $\frac{22}{7}$, 3.14, 3.1416, and 3.14159.

Area of a Rectangle

A measurement given in square units represents the area of a surface.

When computing area, express all linear measurements in the same unit of measure.

The area of a rectangle is equal to its length (*l*) times its width (*w*).

Area = length × width
  = *l* × *w*

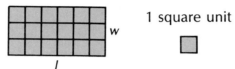

1 square unit

Formula: $A = lw$

Vocabulary

The **area** of any surface is the number of units of square measure contained in the surface. Area is measured in square units.

Example

Find the area of a rectangle 23 centimeters long and 16 centimeters wide.

16 cm

23 cm

$A = lw$
$A = 23 × 16$
$A = 368 \text{ cm}^2$

$$\begin{array}{r} 23 \\ \times\ 16 \\ \hline 138 \\ 23 \\ \hline 368 \end{array}$$

l = 23 cm, *w* = 16 cm; *A* = ?

ANSWER: 368 square centimeters

Think About It

1. The area of a rectangle is 15 square inches. Can more than one rectangle be drawn with this area? Explain.

2. The area of a rectangle is 96 square feet. How can this area be changed to square yards?

Practice Exercises

1. What is the area of a rectangle if its length is 14 m and its width is 9 m?

2. Find the area of each of the following rectangles.

   a. *l* = 23 mm, *w* = 17 mm

   b. *l* = 12 cm, *w* = 17 cm

   c. *l* = 6.8 m, *w* = 1.625 m

   d. $l = 1\frac{1}{3}$ yd, $w = 4\frac{1}{2}$ yd

3. Find the cost of resilvering a mirror 3 ft by 42 in., at $9.25 per square foot.

4. Find the area of the figure at the right.

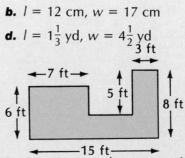

5. Find the area of a rectangular wing of an airplane if the span (length) is 47 ft 6 in. and chord (width) is 8 ft 3 in.

6. How many 4 cm by 6 cm tickets can be cut from 1.2 m by 1.5 m stock?

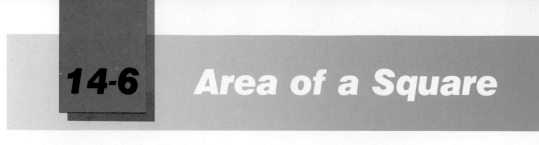

Since a square is a special kind of rectangle, you know how to find its area. However, since the length and width of a square are equal, the area formula can be expressed in a different way.

The area of a square is equal to the length of its side (s) squared.

Area = length × width
        = $s \times s$

Formula: $A = s^2$

Example

Find the area of a square whose side is 37 meters long.

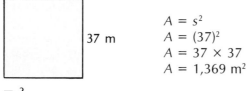

$s = 37$ m; $A = ?$

$A = s^2$
$A = (37)^2$
$A = 37 \times 37$
$A = 1{,}369$ m²

$$
\begin{array}{r}
37 \\
\times\ 37 \\
\hline
259 \\
1\ 11 \\
\hline
1{,}369
\end{array}
$$

ANSWER: 1,369 square meters

Think About It

1. When will the area of a square (in square units) be less than the length of a side (in linear units)?

2. Can you find a square whose perimeter (in linear units) is equal to its area (in square units)? Is there more than one such square? Explain.

Practice Exercises

1. What is the area of a square whose side is 23 ft long?

2. Find the area of each square.
   a. $s = 10$ cm   **b.** $s = 39.37$ in.
   c. $s = 18\frac{1}{2}$ ft   **d.** $s = 4\frac{3}{8}$ in.

3. What is the cross-sectional area of a square beam 14.5 cm on a side?

4. At \$19.95 per square meter, how much will a broadloom rug 5 m by 5 m cost?

5. The base of the Great Pyramid is a square whose sides measure 746 ft. How many acres (to nearest hundredth) does it cover? (43,560 ft² = 1 acre)

Area of a Parallelogram

If a right triangle is cut off one end of a parallelogram (as shown) and moved to the other end, the resulting figure is a rectangle. Thus, the area of a parallelogram can be found by multiplying the length of a base by the height.

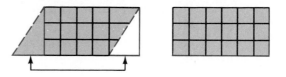

The area of a parallelogram is equal to the product of its base (*b*) and height (*h*).

Area = base × height
Area = $b \times h$

Formula: $A = bh$

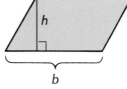

Vocabulary

The altitude of a parallelogram is a line segment drawn from one base perpendicular to the other base (or its extension).

The height is the measure of the altitude.

Example

Find the area of a parallelogram having a height of 25 inches and a base 32 inches long.

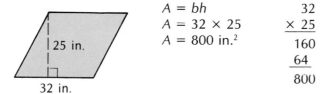

$A = bh$
$A = 32 \times 25$
$A = 800$ in.^{2}

```
   32
 × 25
  160
   64
  800
```

$b = 32$ in., $h = 25$ in.; $A = ?$

ANSWER: $A = 800$ in.^{2}

Think About It

In the example above, if the altitude had been drawn to the side to the right of base *b*, what information would you have needed to find the area?

Practice Exercises

1. What is the area of a parallelogram if its height is 6 m and its base is 8 m long?

2. Find the area of each of the following parallelograms.

   a. $b = 14$ in. $h = 26$ in.
   c. $b = 4.7$ km $h = 8.3$ km

   b. $b = 98$ cm $h = 75$ cm
   d. $b = 4$ yd 1 ft $h = 3$ yd 2 ft

3. Find the cost of seeding a lawn shaped like a parallelogram with a base length of 6 m and a height of 5 m. One kilogram of grass seed covers 30 m^2 and costs $7.49.

4. Draw a parallelogram with one pair of sides 4 cm in length, the other pair of sides 6 cm in length, and an angle of 30° (use a protractor). Construct the altitude to one base. What is the area of the parallelogram?

14-8 Area of a Triangle

Any triangle is half of a corresponding parallelogram with the same base length and height. Thus, the area of a triangle is one-half that of the corresponding parallelogram.

The area of a triangle is equal to one-half the base (b) times the height (h).

Area $= \frac{1}{2} \times$ base $\times$ height

$\qquad = \frac{1}{2} \times b \times h$

Formula: $A = \frac{1}{2}bh$

Vocabulary

The **altitude** of a triangle is a perpendicular segment from any vertex to the opposite side (or extension of that side).

The **height** of a triangle is the measure of the altitude.

Example

Find the area of a triangle with a height of 26 millimeters and a base 17 millimeters long.

$A = \frac{1}{2}bh$

$A = \frac{1}{2} \times 17 \times 26$ $\qquad \frac{1}{\overset{}{\underset{1}{2}}} \times \frac{17}{1} \times \frac{\overset{13}{\cancel{26}}}{1} = 221$

$A = 221$ mm²

26 mm

$b = 17$ mm, $h = 26$ mm; A = ?

ANSWER: $A = 221$ mm²

Think About It

If you are finding the area of a right triangle, what base and altitude are the easiest to use?

Practice Exercises

1. What is the area of a triangle if its height is 10 inches and the base is 8 inches in length?

2. Find the areas of the triangles whose dimensions are:

   a. $b = 12$ cm   **b.** $b = 4.8$ m   **c.** $b = 1\frac{7}{8}$ in.   **d.** $b = 16\frac{1}{2}$ ft   **e.** $b = 2$ ft

      $h = 18$ cm $h = 3.4$ m $h = 2\frac{1}{4}$ in. $h = 27$ ft $h = 1$ ft 4 in.

3. How many square feet of surface does one side of a triangular sail expose if it has a base length of 10 feet and a height of 12 feet 6 inches?

4. Find the area of a triangle with a base of 19 inches and a height of $1\frac{3}{4}$ feet.

14-9 Area of a Trapezoid

A diagonal separates a trapezoid into two triangles that have a common height but different bases. Thus, the area of the trapezoid is equal to the sum of the areas of the two triangles.

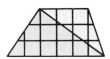

The area of a trapezoid is equal to one-half the height times the sum of the bases.

Area $= \frac{1}{2} \times$ height $\times$ sum of the two bases

$\qquad = \frac{1}{2} \times h \times (b_1 + b_2)$

Formula: $A = \frac{1}{2}h(b_1 + b_2)$ or $A = h \times \frac{b_1 + b_2}{2}$

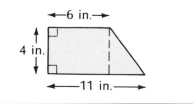

Example

Find the area of a trapezoid with bases 42 in. and 34 in. in length and a height of 29 in.

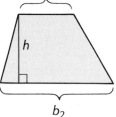

34 in.

29 in.

42 in.

$A = h \times \frac{b_1 + b_2}{2}$

$A = 29 \times \frac{42 + 34}{2}$

$A = 29 \times 38$

$A = 1{,}102 \text{ in.}^2$

$$
\begin{array}{r} 42 \\ + 34 \\ \hline 76 \end{array}
\qquad
\begin{array}{r} 29 \\ \times 38 \\ \hline 232 \\ 87 \\ \hline 1{,}102 \end{array}
$$

$\begin{array}{r} 38 \\ 2\overline{)76} \end{array}$

$h = 29$ in., $b_1 = 42$ in., $b_2 = 34$ in.; $A = ?$

ANSWER: $A = 1{,}102$ square inches

Think About It

Explain how to find the area of the trapezoid shown below by dividing it into two figures whose areas you can find.

←6 in.→

4 in.

←11 in.→

Practice Exercises

1. What is the area of a trapezoid if the height is 7 in. and the parallel sides are 8 in. and 14 in. long?

2. Find the area of each trapezoid.

 a. $h = 8$ cm   **b.** $h = 5$ mm   **c.** $h = 18$ m   **d.** $h = 6$ ft   **e.** $h = 10$ in.

  $b_1 = 4$ cm $b_1 = 9$ mm $b_1 = 29$ m $b_1 = 11\frac{3}{4}$ ft $b_1 = 1$ ft

  $b_2 = 10$ cm $b_2 = 13$ mm $b_2 = 36$ m $b_2 = 14\frac{1}{2}$ ft $b_2 = 1$ ft 4 in.

3. A section of a tapered airplane wing is shaped like a trapezoid. If the two parallel sides, measuring $3\frac{1}{2}$ feet and $5\frac{1}{2}$ feet, are 18 feet apart, find the area of the section.

4. Find the area of a trapezoid with bases 1 ft 3 in. and 2 ft 4 in., and height 10 inches.

Area of a Circle

If a circle is cut into sectors, the sectors can be arranged in the approximate shape of a parallelogram. The area of the parallelogram is found by multiplying the base (πr) and the height (r). Since the area of the circle and the parallelgram are equal, you have now found the formula for the area of a circle. ($A = \pi r^2$)

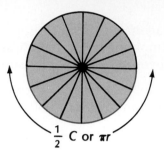

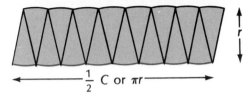

$\frac{1}{2}$ C or πr

You may use $\frac{22}{7}$, 3.14, or—for greater accuracy—3.1416 for π.

The area of a circle is equal to pi (π) times the radius squared.

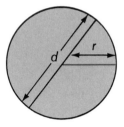

Area = π × radius × radius
   = $\pi \times r \times r = \pi r^2$

Formula: $A = \pi r^2$

or since $r = \frac{1}{2} d$

Area = $\pi \times \frac{1}{2}d \times \frac{1}{2}d = \frac{1}{4} \pi d^2$

or $A = \frac{1}{4} \pi d^2$

Examples

1. Find the area of a circle having a radius of 5 meters.

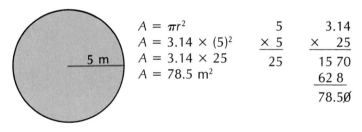

$A = \pi r^2$
$A = 3.14 \times (5)^2$
$A = 3.14 \times 25$
$A = 78.5 \text{ m}^2$

$$\begin{array}{r} 5 \\ \times\ 5 \\ \hline 25 \end{array}$$

$$\begin{array}{r} 3.14 \\ \times\ \ 25 \\ \hline 15\ 70 \\ 62\ 8\ \ \\ \hline 78.5\cancel{0} \end{array}$$

$r = 5$ m, $\pi = 3.14$; $A = ?$

ANSWER: $A = 78.5$ square meters

2. Find the area of a circle having a diameter of 14 feet.

$A = \frac{1}{4} \pi d^2$

$A = \frac{1}{4} \times \frac{22}{7} \times (14)^2$

$A = 154 \text{ ft}^2$

$\frac{1}{\underset{2}{\cancel{4}}} \times \frac{\overset{11}{\cancel{22}}}{\underset{1}{\cancel{7}}} \times \overset{2}{\cancel{14}} \times 14 = 154$

$d = 14$ ft, $\pi = \frac{22}{7}$; $A = ?$

ANSWER: $A = 154$ square feet

Think About It

1. Can you find a circle whose area (in square units) is equal to its circumference (in linear units)? Is there more than one such circle?

2. A circle is constructed inside a square so that it just touches all four sides of the square. If a side of the square is 8 inches, what is the area of the circle?

3. If you know the area of a circle, how can you find its radius?

Practice Exercises

1. What is the area of a circle whose radius is 6 cm?

2. Find the areas of circles having the following radii.

   a. 91 mm   **b.** 4.5 km   **c.** $5\frac{2}{3}$ yd   **d.** 7 ft 4 in.

3. What is the area of a circle whose diameter is 24 ft?

4. Find the areas of circles having the following diameters.

   a. 19 yd   **b.** 16.1 m   **c.** $8\frac{1}{6}$ ft   **d.** 5 yd 1 ft

5. If a forest ranger can see from her tower for a distance of 42 km in all directions, how many square kilometers can she watch?

6. The dial of one of the world's largest clocks has a diameter of 50 ft. What is the area of the dial?

7. A revolving sprinkler sprays a lawn for a distance of 7 m. How many square meters does the sprinkler water in 1 revolution?

8. If a station can televise programs for a distance of 91 km, over what area may the programs be received?

9. Which is larger: the area of a circle 6 cm in diameter or the area of a square whose side is 6 cm? How much larger?

10. What is the area of one side of a washer if its diameter is $\frac{3}{4}$ in. and the diameter of the hole is $\frac{1}{4}$ in.?

Refresh Your Skills

1. 39,627 + 46,858 + 90,377 [1-8]

2. 50,602 − 43,575 [1-9]

3. 806 × 450 [1-11]

4. $96\overline{)58{,}272}$ [1-13]

5. $8.79 + $13.63 + $.86 [2-5]

6. 20 − 9.57 [2-6]

7. 120 × 3.41 [2-7]

8. $0.75\overline{)30}$ [2-8]

9. $8\frac{1}{2} + 5\frac{7}{8}$ [3-8]

10. $1\frac{1}{4} - \frac{4}{5}$ [3-9]

11. $\frac{3}{4} \times 1\frac{1}{2}$ [3-11]

12. $2\frac{5}{8} \div 3$ [3-12]

13. Find 13% of $2,000 [4-6]

14. 25 is what percent of 20? [4-7]

15. $12\frac{1}{2}$% of what number is 5? [4-8]

Total Area of a Rectangular Solid

In the last six lessons you learned how to find the area of various plane figures. In this lesson and the next two, you will learn how to find the total surface areas of various three-dimensional solids.

Total area = 2 × length × width + 2 × length × height
+ 2 × width × height
= 2 × *l* × *w* + 2 × *l* × *h* + 2 × *w* × *h*
Formula: $A = 2lw + 2lh + 2wh$

Vocabulary

The **total area** of the surface of a rectangular solid is the sum of the areas of its six rectangular faces.

Example

Find the total area of a rectangular solid 9 meters long, 6 meters wide, and 7 meters high.

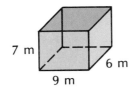

l = 9 m, *w* = 6 m, *h* = 7 m; *A* = ?

$A = 2lw + 2lh + 2wh$
$A = 2 \times 9 \times 6 + 2 \times 9 \times 7 + 2 \times 6 \times 7$
$A = 108 + 126 + 84$
$A = 318 \text{ m}^2$

ANSWER: *A* = 318 square meters

Think About It

1. Two parallel faces of a rectangular solid are squares with sides of 1 inch. If the total area of the solid is 30 square inches, explain how you can find the height of the solid.

2. The dimensions of a rectangular solid are 15 inches by 1 foot 4 inches by $\frac{5}{9}$ yard. What units would you use for total area? Explain your choice.

Practice Exercises

1. Find the total areas of rectangular solids with the following dimensions.

   a. *l* = 23 cm, *w* = 14 cm, *h* = 19 cm
   b. *l* = 8.1 m, *w* = 2.7 m, *h* = 5.9 m

2. A room is 18 ft long, 15 ft wide, and 9 ft high. Find the total area of the walls and ceiling, allowing a deduction of 64 ft² for the windows and doorway. How many gallons of paint are needed to cover the walls and ceiling with two coats if a gallon will cover 400 ft² with one coat? At $14.69 per gallon, how much will the paint cost?

3. How many square feet of plywood will be needed to make a packing box with the dimensions 2 ft 3 in. by 1 ft 4 in. by 1 ft 9 in.?

4. Make a paper model of a rectangular solid with the dimensions 6 in. by 5 in. by 4 in. What is the total area of the solid?

14-12 Total Area of a Cube

The total area of a cube could be found using the formula for the total area of a rectangular solid. However, since each face of a cube is a square, it is easier to find the total area by multiplying the area of one square face by 6.

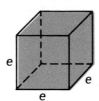

Total area = 6 × edge × edge
= 6 × e × e = 6e²

Formula: $A = 6e^2$ or
$A = 6s^2$ (where s is the side of a square face).

Example

Find the total area of a cube whose edges measure 15 in.

$e = 15$ in.; $A = ?$

$A = 6e^2$
$A = 6 \times (15)^2$
$A = 6 \times 225$
$A = 1,350$ in.²

$$
\begin{array}{r}
15 \\
\times\ 15 \\
\hline
75 \\
15 \\
\hline
225
\end{array}
\qquad
\begin{array}{r}
225 \\
\times\ \ \ 6 \\
\hline
1,350
\end{array}
$$

ANSWER: $A = 1,350$ square inches

Think About It

1. If you know the total area of a cube, how can you find the measure of an edge of the cube?

2. Can you find a cube whose total area (in square units) is equal to the total measurement of its edges (in linear units)? Is there more than one such cube?

Practice Exercises

1. Find the total area of each of the following cubes.

a. $e = 37$ mm   **b.** $e = 3\frac{7}{8}$ in.   **c.** $e = 1$ ft 11 in.   **d.** $e = 1\frac{2}{3}$ yd.

2. The total area of a cube is 294 in.². Find the measure of an edge of the cube.

3. A carton 1.5 m by 1.5 m by 1.5 m is made of cardboard. How many square meters of paper were used to make it if 10% extra was allowed for waste in cutting?

4. Make a paper cube with a total area of 150 in.². What is the measure of an edge of the cube?

14-13 Area of a Cylinder and a Sphere

In the last two lessons you learned how to find the areas of two solids with plane (flat) surfaces. In this lesson you will learn how to find the area of two solids with curved surfaces: the cylinder and the sphere.

If you cut open a cylinder and lay the curved surface flat, it looks like a rectangle. The area of the curved surface (**lateral area**) is equal to the circumference of the base times the height.

> ### Vocabulary
>
> The lateral area of a cylinder is the area of the curved surface.

Lateral Area = circumference × height
$$= \pi dh \text{ or } 2\pi rh$$

Formula: $A = \pi dh$ or $A = 2\pi rh$

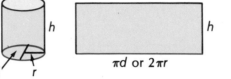

Examples

1. Find the lateral area of a cylinder having a diameter of 8 cm and a height of 9 cm.

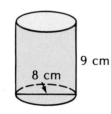

$A = \pi dh$
$A = 3.14 \times 8 \times 9$
$A = 226.08 \text{ cm}^2$

$$\begin{array}{r} 3.14 \\ \times \quad 8 \\ \hline 25.12 \end{array} \qquad \begin{array}{r} 25.12 \\ \times \quad 9 \\ \hline 226.08 \end{array}$$

$d = 8 \text{ cm} \qquad h = 9 \text{ cm}$
$\pi = 3.14 \qquad A = ?$

ANSWER: $A = 226.08$ square centimeters

To find the total area of a cylinder, add the lateral area to the area of the bases.

Total area = lateral area + area of bases
$$= 2\pi rh \qquad + \pi r^2 + \pi r^2 = 2\pi rh + 2\pi r^2$$

Formula: $A = 2\pi rh + 2\pi r^2$

    or $A = \pi dh + \frac{1}{2}\pi d^2$

    or $A = 2\pi r(h + r)$

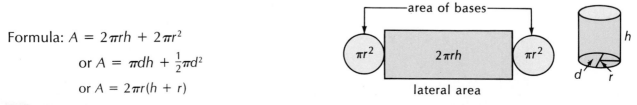

2. Find the total area of a cylinder whose radius is 21 ft and whose height is 30 ft.

30 ft

21 ft

$A = 2\pi rh + 2\pi r^2$

$A = 2 \times \frac{22}{7} \times 21 \times 30 + 2 \times \frac{22}{7} \times (21)^2$

$A = 3{,}960 + 2{,}772$

$A = 6{,}732 \text{ ft}^2$

$r = 21$ ft $\pi = \frac{22}{7}$

$h = 30$ ft $A = ?$

ANSWER: $A = 6{,}732$ square feet

The area of the surface of a sphere is equal to 4 times pi (π) times the radius squared.

Area $= 4 \times \pi \times r^2$
Formula: $A = 4\pi r^2$

r

---

3. Find the area of the surface of a sphere having a radius of 6 cm.

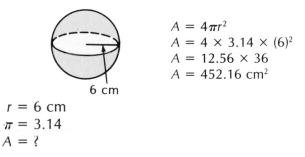
6 cm

$A = 4\pi r^2$
$A = 4 \times 3.14 \times (6)^2$
$A = 12.56 \times 36$
$A = 452.16 \text{ cm}^2$

$r = 6$ cm
$\pi = 3.14$
$A = ?$

ANSWER: $A = 452.16$ square centimeters

Think About It

1. A sphere is placed inside a cube so that it just touches all six faces of the cube. If an edge of the cube is 8 inches, how would you find the area of the sphere?

2. Explain how to find the radius of a sphere if you know its area.

3. Explain how you would find the total area of a cylinder in square inches if its radius is $1\frac{1}{3}$ feet and its height is $4\frac{1}{6}$ feet.

Practice Exercises

1. Find the lateral areas of cylinders with the following dimensions.

    a. $d = 4$ m   **b.** $d = 2$ ft 9 in.   **c.** $r = 14$ m   **d.** $r = 42$ cm
       $h = 10$ m $h = 4$ ft $h = 20$ m $h = 18$ cm

2. How many square inches of paper are needed to make a label on a can 4 inches in diameter and 5 inches high?

3. Find the total areas of cylinders with the following dimensions.

    a. $r = 7$ m
       $h = 12$ m

    b. $r = 4\frac{3}{8}$ in.
       $h = 20$ in.

    c. $d = 35$ cm
       $h = 40$ cm

    d. $d = 2$ ft 5 in.
       $h = 1$ ft 8 in.

4. How much insulation is need to enclose the curved surface and the two ends of a hot water storage tank 35 centimeters in diameter and 1.5 meters high?

5. Make a paper model of a cylinder with a radius of 5 inches and a height of 6 inches. What is the lateral area of the cylinder? What is the total area of the cylinder?

6. Find the surface areas of spheres having the following radii.

    a. 12 m   **b.** 6.3 cm   **c.** $9\frac{5}{8}$ in.   **d.** 5 ft 1 in.

7. Find the surface areas of spheres having the following diameters.

    a. 16 mm   **b.** 14.8 m   **c.** $6\frac{3}{4}$ yd   **d.** 4 ft 6 in.

8. What is the radius of a sphere whose area is 616 square inches?

9. What is the area of the earth's surface if its diameter is 7,900 miles?

Volume of a Rectangular Solid

If a measurement is given in cubic units it represents the volume of a solid. In this lesson and the next three, you will learn to find the volume of various solids.

When computing volume, express all linear measurements with the same unit of measure.

The volume of a rectangular solid is equal to the length (*l*) times the width (*w*) times the height (*h*).

Volume = length × width × height
$$= l \times w \times h$$
Formula: $V = lwh$
    or $V = Bh$, where *B* is the area of the base (*lw*)

Vocabulary

The **volume** of any solid is the number of units of cubic measure contained in the solid. Volume is measured in cubic units.

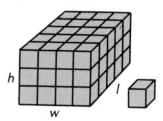

Example

Find the volume of a rectangular solid 8 meters long, 5 meters wide, and 7 meters high.

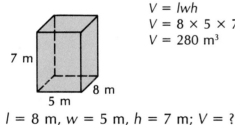

$$V = lwh$$
$$V = 8 \times 5 \times 7$$
$$V = 280 \text{ m}^3$$

$$
\begin{array}{r}
8 \\
\times\ 5 \\
\hline
40
\end{array}
\qquad
\begin{array}{r}
40 \\
\times\ 7 \\
\hline
280
\end{array}
$$

$l = 8$ m, $w = 5$ m, $h = 7$ m; $V = ?$

ANSWER: $V = 280$ cubic meters

Think About It

The volume of a rectangular solid is 5 cubic feet. How would you express this volume in cubic inches?

Practice Exercises

1. What is the volume of a rectangular solid if it is 7 inches long, 4 inches wide, and 9 inches high?

2. Find the volumes of rectangular solids having the following dimensions.

   a. $l = 8$ mm   **b.** $l = 17$ in.   **c.** $l = 36$ m   **d.** $l = 62$ cm
        $w = 3$ mm $w = 18$ in. $w = 36$ m $w = 40$ cm
        $h = 6$ mm $h = 14$ in. $h = 13$ m $h = 19$ cm

3. Find the volume of a rectangular solid $2\frac{1}{4}$ feet long, 15 inches wide, and $1\frac{1}{3}$ feet high.

Volume of a Cube

Since a cube is a rectangular solid, you could find its volume by multiplying length times width times height. However, since the lengths of all edges of a cube are the same, the formula for volume can be expressed as one edge cubed.

The volume of a cube is equal to the length of the edge (e) times itself, times itself, or the edge cubed.

Volume = length × width × height
$$= e \times e \times e$$
Formula: $V = e^3$

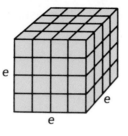

Example

Find the volume of a cube whose edge measures 17 inches.

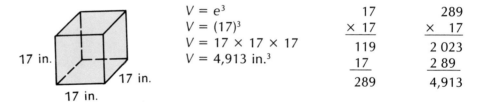

17 in.
17 in.
17 in.

$V = e^3$
$V = (17)^3$
$V = 17 \times 17 \times 17$
$V = 4{,}913$ in.^{3}

	17		289
	× 17		× 17
	119		2 023
	17		2 89
	289		4,913

e = 17 in.   V = ?

ANSWER: $V = 4{,}913$ cubic inches

Think About It

Can you find a cube whose total area (in square units) is equal to its volume (in cubic units)? Is there more than one such cube?

Practice Exercises

1. What is the volume of a cube whose edge is 25 cm?

2. Find the volume of each cube.

   a. e = 27 cm   **b.** e = 4.5 cm   **c.** $e = 4\frac{3}{4}$ ft   **d.** e = 5 yd 2 ft

3. How many cubic feet of space are in a bin 6 ft by 6 ft by 6 ft?

4. The volume of an 8-cm cube is how many times as large as the volume of a 2-cm cube?

Volume of a Cylinder

You found the area of a rectangular solid by multiplying the area of the base times the height. You can also find the volume of a cylinder by multiplying the area of the base (a circle whose area is πr^2) by the height (h).

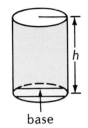

The volume of a cylinder is equal to the area of the base times the height.

Volume = (area of the base) × height
       = $B \times h$
       = $\pi r^2 \times h$

Remember, $\frac{22}{7}$ or 3.14 can be used for π.

Formula: $V = \pi r^2 h$ or $V = \frac{1}{4}\pi d^2 h$

Example

Find the volume of a cylinder 75 cm high whose base has a radius of 30 cm.

75 cm
30 cm

$r = 30$ cm, $h = 75$ cm,
$\pi = 3.14$; $V = ?$

$V = \pi r^2 h$
$V = 3.14 \times (30)^2 \times 75$
$V = 3.14 \times 900 \times 75$
$V = 211{,}950$ cm³

```
   30          3.14         2,826
×  30        × 900        ×   75
  900       2,826.00      14 130
                          192 82
                          211,950
```

ANSWER: $V = 211{,}950$ cubic centimeters

Think About It

Can you find a cylinder whose volume is equal to π cubic inches? Is there more than one such cylinder? Explain.

Practice Exercises

1. What is the volume of a cylinder if the radius of its base is 3 m and the height is 6 m?

2. What is the volume of a cylinder if the diameter of its base is 10 ft and the height is 16 ft?

3. Find the volumes of cylinders having the following dimensions.

   a. $r = 5$ mm $h = 8$ mm

   b. $r = 2\frac{5}{8}$ in. $h = 7\frac{1}{2}$ in.

   c. $r = 1$ ft 6 in. $h = 9$ ft 2 in.

4. Which cylinder holds more, one that is 3 cm in diameter and 4 cm high, or one that is 4 cm in diameter and 3 cm high? How many cubic centimeters more?

5. How many cubic yards of dirt must be dug to make a well 4 ft 6 in. in diameter and 42 ft deep?

6. What is the weight of a round steel rod 8 ft long and $\frac{3}{4}$ in. in diameter? A cubic foot of steel weighs 490 lb.

Volume of a Sphere, a Cone, and a Pyramid

The volume of a sphere is equal to $\frac{4}{3}$ times pi (π) times the radius cubed.

Volume $= \frac{4}{3} \times \pi \times$ radius $\times$ radius $\times$ radius

$\qquad = \frac{4}{3} \times \pi \times r^3$

Formula: $V = \frac{4}{3}\pi r^3$ or $V = \frac{\pi d^3}{6}$

Examples

1. Find the volume of a sphere with radius 1.2 cm.

$V = \frac{4}{3}\pi r^3$

$V = \frac{4}{3} \times 3.14 \times (1.2)^3$

$V = \frac{4}{3} \times 5.42592$

$V = 7.23456 \text{ cm}^3$

$\pi = 3.14$, $r = 1.2$ cm; $V = ?$

1.2	1.44	1.728
$\times 1.2$	$\times 1.2$	$\times 3.14$
1.44	1.728	5.42592

$\frac{4}{3} \times 5.42592 = \frac{21.70368}{3}$

$\begin{array}{r} 7.23456 \\ 3\overline{)21.70368} \end{array}$

ANSWER: $V = 7.23456 \text{ cm}^3$

The volume of a cone is $\frac{1}{3}$ that of a cylinder with the same base and height.

Volume $= \frac{1}{3} \times$ area of the base $\times$ height

$\qquad = \frac{1}{3} \times \pi r^2 \times h$

Formula: $V = \frac{1}{3}\pi r^2 h$

2. Find the volume of a cone with radius 5 m and height 7 m.

$V = \frac{1}{3}\pi r^2 h$

$V = \frac{1}{3} \times \frac{22}{\underset{1}{7}} \times 5^2 \times \overset{1}{7}$

$V = \frac{1}{3} \times 550$

$V = 183\frac{1}{3} \text{ m}^3$

5	25
$\times 5$	$\times 22$
25	550

$\begin{array}{r} 183\frac{1}{3} \\ 3\overline{)550} \end{array}$

$\pi = \frac{22}{7}$, $r = 5$ m, $h = 7$ m; $V = ?$

ANSWER: $V = 183\frac{1}{3} \text{ m}^3$

The volume of a pyramid is $\frac{1}{3}$ that of a rectangular solid with the same base and height.

Volume $= \frac{1}{3} \times$ area of the base $\times$ height

$\qquad = \frac{1}{3} \times B \times h$

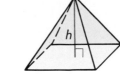

Formula: $V = \frac{1}{3}Bh$

3. Find the volume of a square pyramid if a side of the base is 5 inches and the height is 9 inches.

$V = \frac{1}{3} Bh$

$V = \frac{1}{3} \times 5 \times 5 \times 9$

$V = 75 \text{ in.}^3$

$$\begin{array}{r} 5 \\ \times\ 5 \\ \hline 25 \end{array} \qquad \begin{array}{r} 25 \\ \times\ 9 \\ \hline 225 \end{array}$$

$$\begin{array}{r} 75 \\ 3\overline{)225} \end{array}$$

$h = 9$ in., $B = 25$ in.^{2}; $V = ?$

ANSWER: $V = 75$ in.^{3}

Think About It

1. If the radius of a sphere is 3 inches, the total area (in square units) is equal to the volume (in cubic units). If the radius of the sphere is less than 3 inches, will the area be larger or smaller than the volume?

2. A pyramid has a height of 18 inches and a square base with sides of $2\frac{1}{3}$ feet. Explain how to find the volume in cubic feet.

Practice Exercises

1. Find the volumes of spheres having the following radii.

 a. 30 m   **b.** 10.4 cm   **c.** $6\frac{5}{16}$ in.   **d.** 2 ft 8 in.

2. Find the volumes of spheres having the following diameters.

 a. 26 mm   **b.** 11.3 m   **c.** $12\frac{1}{2}$ in.   **d.** 4 ft 5 in.

3. How many cubic feet of air does a basketball contain if its diameter is 10 in?

4. Find the volumes of cones having the following dimensions.

 a. $r = 8$ cm $h = 23$ cm   **b.** $r = 6.5$ m $h = 9$ m

5. A conical pile of sand is 2.5 m in diameter and 9 m high. How many cubic meters of sand are in the pile?

6. Find the volume of a pyramid when the side of its square base is 40 m and the height is 27 m.

7. What is the volume of a pyramid when its rectangular base is 23 m long and 18 m wide and the height is 35 m?

8. Make a paper model of a square pyramid with a height of 6 in. and a square base with edges of 5 in. What is the volume of the pyramid?

Making Decisions

PLANNING A TRIP

You, your 10-year old brother, and your parents are planning to take a 2-week vacation to visit relatives. You have budgeted $1,500 to spend on transportation for the trip. You live in Boston, Massachusetts. Your relatives live in Dallas, Texas. Dallas is 1,753 driving miles from Boston. You need to decide whether you should drive, fly, or take a bus or train to Dallas. You call a travel agent and record this information on the various forms of transportation.

Forms of Transportation	Travel Time One Way	Best Rate Round Trip
Plane	3 hours	$348—adults and children
Train	2 days	$188—adults $\frac{1}{2}$ fare—children under 12
Bus	$1\frac{1}{2}$ days	$238—adults $\frac{1}{2}$ fare—children under 12
Car	3 days	

Use the following questions to help you decide which form of transportation to use.

1. Your family car averages 21 miles per gallon during highway driving. Assume gasoline will average $1.05 per gallon during your trip.

   a. How many gallons of gasoline will you use during a *round* trip from Boston to Dallas?
   b. How much will the gasoline cost for the round trip?
   c. What other car costs might you have if you drive?

2. What other travel costs might you have if you

   a. fly?
   b. take a train?
   c. take a bus?
   d. drive?

3. Complete a chart like the one below showing the best rate, your estimated other costs, the total cost, and the advantages and disadvantages of using each form of transportation for your family of four.

	Best Rate	Other Costs	Total Cost	Advantages	Disadvantages
Plane	$1392				
Train	$658				
Bus	$833				
Car	$175.35				

4. For which form of transportation would the total cost of travel probably be the least expensive? The most expensive?

5. What factors, other than cost, might influence the form of transportation you choose?

6. Suppose your grandmother decides to go on the trip to Dallas with you. How might this affect your travel plans?

7. Which of these forms of transportation can you use and still stay within your travel budget?

8. Decide how you will travel to Dallas. Be prepared to defend your decision!

In Your Community

With a classmate, choose a place you would like to go for a 2-week vacation. Find out what forms of transportation you could use to get to your destination. Get a road map to find the driving distance. Call a travel agent to find out the costs of the different forms of transportation. How would you choose to travel from your home to your vacation destination?

Career Application

AIR-CONDITIONING AND REFRIGERATION MECHANIC

Many businesses depend on their air conditioning and refrigeration systems. Hospitals need to keep medicines cold. Food companies need to keep food fresh. When air conditioning or refrigeration equipment must be installed or repaired, *air-conditioning and refrigeration mechanics* do the job.

Air-conditioning and refrigeration mechanics read blueprints, install motors and other parts, and connect the equipment to ducts and electrical lines. When a system breaks down, they check it and make repairs.

Air-conditioning and refrigeration mechanics often begin by helping experienced mechanics. Most have taken related courses in high school, vocational school, or junior college.

1. An air-conditioning and refrigeration mechanic needs to wrap some insulation around the perimeter of a rectangular cooling duct. If the duct is 8 inches by 16 inches, how long a piece of insulation is needed?

2. To install the correct size air conditioner, a mechanic needs to know the volume of a room. The room is 10 feet wide, 15 feet long, and 8 feet high. What is the volume?

3. How long a piece of insulation is needed to wrap around a cylindrical duct that is 7 inches in diameter?

4. The cylinder of a compressor has a radius of 1.2 centimeters. What is the area of the opening in the cylinder?

5. A tank for refrigerant is a cylinder 16 inches high. Its radius is 8 inches. What is its volume?

6. A grill has square openings $\frac{3}{4}$ inch on a side. What is the area of each opening?

Chapter 14 Review

Page
498 **1.** Find the perimeter of a rectangle 125 m long and 74 m wide.

Page
499 **2.** Find the perimeter of a square whose side is 7.3 cm.

500 **3.** Find the perimeter of a triangle whose sides measure $2\frac{3}{8}$ in., $4\frac{1}{2}$ in., and $3\frac{5}{8}$ in.

501 **4.** Find the circumference of a circle whose diameter is 42 in.

501 **5.** Find the circumference of a circle whose radius is 20 m.

Find the area.

503 **6.** A rectangle 57 m long and 38 m wide.

504 **7.** A square whose side measures $3\frac{1}{2}$ ft.

505 **8.** A parallelogram with a height of 132 mm and a base of 85 mm.

506 **9.** A triangle with a base of 3.5 m and a height of 6.6 m.

507 **10.** A trapezoid with bases of 130 ft and 74 ft and a height of 82 ft.

508 **11.** A circle with a radius of 22 mm.

12. A circle whose diameter is 0.4 m.

Find the total area of the outside surface.

510 **13.** A rectangular solid 4.3 m long, 3.6 m wide, and 1.5 m high.

511 **14.** A cube whose edge measures 24 mm.

512 **15.** A cylinder with a radius of 21 cm and a height of 32 cm.

512 **16.** A sphere with a diameter of 14 yd.

Find the volume.

515 **17.** A rectangular solid 58 in. by 32 in. by 15 in.

516 **18.** A cube whose edge measures $4\frac{1}{2}$ in.

517 **19.** A cylinder 20 ft high with the diameter of its base 18 ft.

518 **20.** A sphere whose radius is 14 yds.

518 **21.** A cone 33 ft high and a base with radius 7 ft.

518 **22.** A square pyramid 10.3 cm on each side and 12 cm high.

517 **23.** Topsoil is to be placed on a circular flower garden 16 m in diameter to a depth of 0.15 m. How many cubic meters of topsoil must be purchased? (You cannot buy a fraction of a cubic meter.)

503 **24.** The __?__ of any surface is the number of units of square measure contained in the surface.

Chapter 14 Test

Page
498 **1.** Find the perimeter of a rectangle 10.53 cm wide and 12.04 cm long.

Page
498 **2.** How many inches of decorative edging are needed for a tablecloth 54 in. long and 44 in. wide?

499 **3.** What is the perimeter of a square whose side is 47 ft.

500 **4.** Find the perimeter of a triangle with sides of 4 m, 2.3 m, and 3.7 m.

501 **5.** Find the circumference of a circle with a diameter of 6.4 m.

501 **6.** How far will a wheel with a radius of 35 in. travel in one turn?

503 **7.** Find the area of a rectangle whose width is 14 ft and length is 22 ft.

503 **8.** Construction costs based on the area covered are $54 per square foot. What would the cost of construction be for a house 40 ft long and 30 ft wide?

503 **9.** How many 8-inch-square flagstones are needed for a patio 14 ft long and 10 ft wide?

504 **10.** Find the area of a square whose side is $3\frac{3}{8}$ in.

505 **11.** Find the area of a parallelogram 6.2 cm high with a base of 5 cm.

506 **12.** What is the area of a triangle whose base is 14 in. and whose altitude is 28 in?

507 **13.** Find the area of a trapezoid whose height is 26 ft and whose bases are 12 ft and 14 ft.

508 **14.** What is the area of a circular mirror with a radius of 16 cm?

508 **15.** A circle has a diameter of 20 ft, what is its area?

510 **16.** Find the total surface area of a rectangular solid with a length of 9 ft, a width of 11 ft, and a height of 14 ft.

511 **17.** Find the total area of a cube with a side of 2.1 cm.

512 **18.** Find the lateral area of a cylinder 15 in. tall with a 6 in. diameter.

512 **19.** Find the total area for Problem 18.

515 **20.** Find the volume of a rectangular solid, 15.7 cm by 2 cm by 3 cm.

516 **21.** Find the volume of a cube with a side of 3.2 cm.

517 **22.** Find the volume of a cylinder whose base has a diameter of 14 ft and whose height is 23 ft.

518 **23.** Find the volume of a square pyramid whose side is 12 cm and whose height is 12 cm.

518 **24.** Find the volume of a sphere with a radius of 15 yd.

518 **25.** Find the volume of a cone with a radius of 22 mm and a height of 24 mm.

Achievement Test

Page
428 Read, or write in words, each of the following.

1. $\overline{EF}$ **2.** $\overleftrightarrow{MY}$ **3.** $\overrightarrow{LN}$ **4.** $\overleftrightarrow{CR}$ **5.** $\overline{BT}$ **6.** $\overrightarrow{HG}$

Name each of the following and express them symbolically.

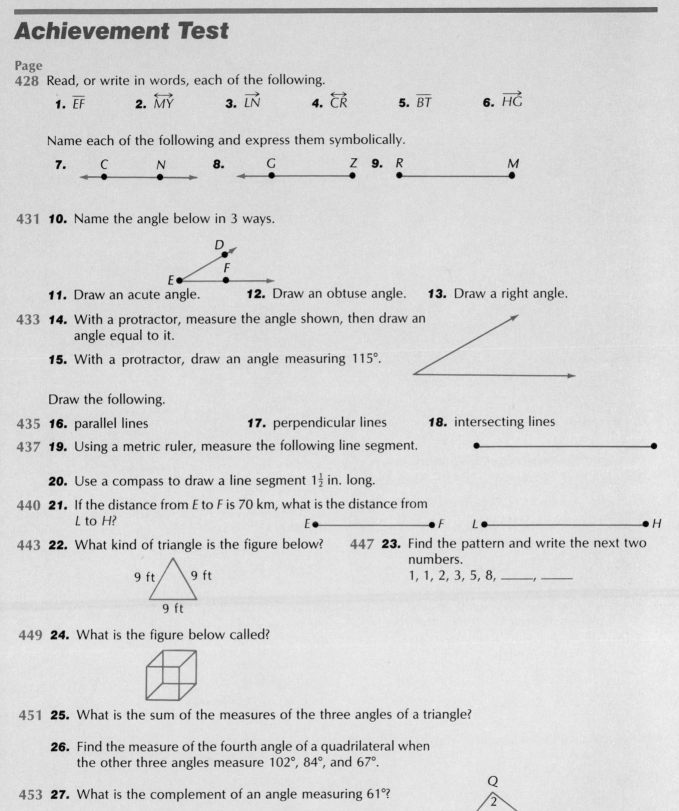

7.   C   N   **8.**   G   Z   **9.**   R   M

431 **10.** Name the angle below in 3 ways.

D   F   E

11. Draw an acute angle.   **12.** Draw an obtuse angle.   **13.** Draw a right angle.

433 **14.** With a protractor, measure the angle shown, then draw an angle equal to it.

15. With a protractor, draw an angle measuring 115°.

Draw the following.

435 **16.** parallel lines   **17.** perpendicular lines   **18.** intersecting lines

437 **19.** Using a metric ruler, measure the following line segment.

20. Use a compass to draw a line segment $1\frac{1}{2}$ in. long.

440 **21.** If the distance from E to F is 70 km, what is the distance from L to H?

E •————————• F   L •————————————• H

443 **22.** What kind of triangle is the figure below?

9 ft   9 ft

9 ft

447 **23.** Find the pattern and write the next two numbers.

1, 1, 2, 3, 5, 8, ____, ____

449 **24.** What is the figure below called?

451 **25.** What is the sum of the measures of the three angles of a triangle?

26. Find the measure of the fourth angle of a quadrilateral when the other three angles measure 102°, 84°, and 67°.

453 **27.** What is the complement of an angle measuring 61°?

28. What is the supplement of an angle measuring 97°?

29. If $m\angle 2 = 47°$ and $m\angle 4 = 121°$, find the measure of $\angle 1$. Of $\angle 3$.

Q   2   1   3   4   S   P   R

30. If m∠2 = 104°, what is the measure of ∠1? ∠3? ∠4?

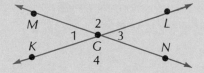

456 31. Parallel lines *AB* and *CD* are cut by transversal *EF*. If m∠8 = 56°, what is the measure of ∠1? ∠2? ∠3? ∠4? ∠5? ∠6? ∠7?

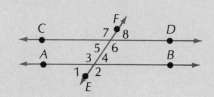

458 32. Find the indicated missing parts.

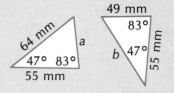

460 33. Find the indicated missing parts.

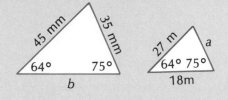

466 34. Make a copy of the following line segment: _____

467 35. Draw with a protractor an angle of 74°. Then construct with a compass an angle equal to it. Check with a protractor.

468 36. Construct a triangle with sides measuring 56 mm, 47 mm, and 63 mm.

37. Construct a triangle with sides measuring 7.2 cm and 5.1 cm and an included angle of 51°.

38. Construct a triangle with angles measuring 39° and 101° and an included side measuring $3\frac{1}{8}$ in.

471 39. Draw any line. Using a compass, construct a perpendicular to this line at a point on the line.

472 40. Draw any line. Using a compass, construct a perpendicular to this line from a point not on the line.

473 41. Moving two adjacent figures at a time, rearrange the top row to look like the bottom row in just 4 moves. The figures you move must stay together and remain in the same order.

□ □ □ □ ○ ○ ○

□ ○ □ ○ □ ○ □

475 42. Draw a line segment 46 mm long. Using a compass, bisect this segment. Check with a ruler.

476 43. Draw an angle measuring 62°. Using a compass, bisect this angle. Check with a protractor.

477 44. Draw any line. Locate a point outside this line. Through this point draw a line parallel to the first line.

482 45. Find the square of 1.6.

483 46. Find $\sqrt{169}$.

526 *Achievement Test*

Page

47. Find $\sqrt{32}$ by the estimate, divide, and average method. *(Page 484)*

48. Find the hypotenuse of a right triangle if the altitude is 63 m and the base is 60 m. *(Page 487)*

49. Find the base of a right triangle if the hypotenuse is 120 cm and the altitude is 72 cm. *(Page 487)*

50. Find the altitude of a right triangle if the hypotenuse is 73 ft and the base is 55 ft. *(Page 487)*

51. Find the height of a TV antenna tower that casts a shadow 6 m long when a nearby light pole, 2.4 m high, casts a shadow 0.4 m long. *(Page 489)*

In right triangle *ABC*, find each of the following.

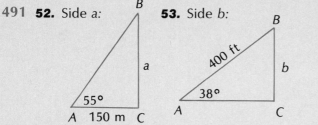

52. Side *a*: *(Page 491)*

53. Side *b*:

54. Side *b*:

55. Find the perimeter of a rectangle 41 m long and 17 m wide. *(Page 498)*

56. Find the perimeter of a square whose side measures 26 cm. *(Page 499)*

57. Find the perimeter of a triangle with sides measuring $4\frac{3}{4}$ in., $3\frac{5}{8}$ in., and $5\frac{11}{16}$ in. *(Page 500)*

Find the circumference of each of the following.

58. A circle whose diameter is 28 mm. *(Page 501)*

59. A circle whose radius is 80 ft.

60. Find the area of a rectangle 51 cm long and 18 cm wide. *(Page 503)*

61. Find the area of a square whose side measures 47 ft. *(Page 504)*

62. Find the area of a parallelogram with an altitude of 24 m and a base of 49 m. *(Page 505)*

63. Find the area of a triangle whose altitude is 39 cm and base is 28 cm. *(Page 506)*

64. Find the area of a trapezoid with bases of 43 mm and 37 mm and a height of 25 mm. *(Page 507)*

Find the area of each of the following. *(Page 508)*

65. A circle whose radius is 9 in.

66. A circle whose diameter is 56 in.

67. Find the total area of the outside surface of a rectangular solid 2.5 m long, 1.7 m wide, and 2 m high. *(Page 510)*

68. Find the total area of the outside surface of a cube whose edge measures $4\frac{1}{2}$ ft. *(Page 511)*

Find the total area of the outside surface of each of the following. *(Page 512)*

69. A cylinder 54 cm in diameter and 75 cm in height.

70. A sphere whose diameter is 49 cm.

71. Find the volume of a rectangular solid 6.9 m long, 1.8 m wide, and 4.7 m high. *(Page 515)*

72. Find the volume of a cube whose edge measures 79 mm. *(Page 516)*

73. Find the volume of a cylinder with a diameter of 70 ft and a height of 46 ft. *(Page 517)*

Find the volume of each of the following. *(Page 518)*

74. A sphere whose diameter is 84 cm.

75. A cone 14 in. in diameter and 10 in. high.

76. A square pyramid 10 m on each side of the base and 15 m high.

Algebra

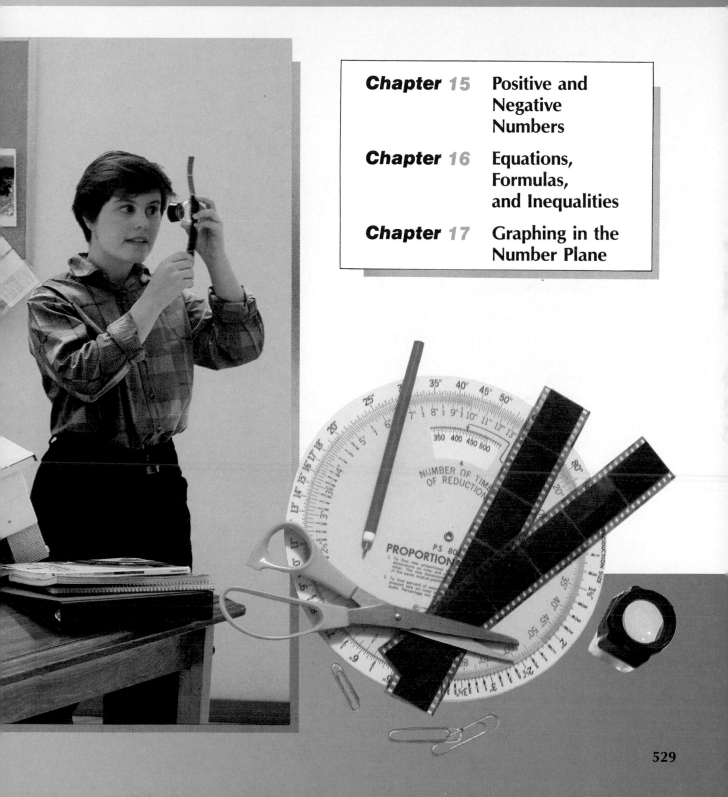

529

Inventory Test

$$-\frac{2}{3} \quad +86 \quad -0.625 \quad -5\frac{1}{2} \quad +1.7 \quad -300 \quad -0.3 \quad +8\frac{3}{4} \quad -72 \quad -\sqrt{64} \quad -\frac{7}{10} \quad +\frac{36}{6}$$

Page

536 **1.** Which of the numbers in the box above name positive numbers?

2. Negative numbers?

3. Integers?

4. Rational numbers?

5. Find the value of $|-63|$.

539 **6.** Draw the graph of -3, -1, 0, and 2 on the number line.

7. Write the coordinates of the following graph.

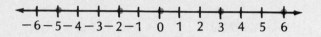

541 **8.** If -5 kg means 5 kg underweight, what does $+3$ kg mean?

9. What is the opposite of -18?

542 Which of the following sentences are true?

10. $-3 > -1$ **11.** $+2 \not< -2$ **12.** $0 < -4$ **13.** $+4 \not> +6$

14. List the following numbers in order from least to greatest.
$+4, -3, +6, -9, +11, -10, -6, +9, 0, +1$

15. Which has the greater opposite number: $+2$ or $+5$?

544 **16.** What is the additive inverse of $+23$?

17. Write symbolically: The opposite of negative ten is positive ten.

545 Add.

18. $(-5) + (+9)$ **19.** $(+6) + (+7)$ **20.** $(-12) + (+8)$

21. $(-9) + (-7)$ **22.** $(-3) + (+3)$

23. $(-7) + (+4) + (-9) + (-11)$

24. Simplify. $-4 + 6 - 3 - 7 + 5$

549 Subtract.

25. $(+6) - (-8)$ **26.** $(+9) - (+11)$ **27.** $(-5) - (-4)$

28. $(-14) - (+5)$ **29.** $(0) - (-4)$ **30.** $(+5) - (-5)$

31. Take 6 from 2.

552 Multiply.

32. $(+9) \times (-7)$ **33.** $(-4) \times (+8)$ **34.** $(-10) \times (-10)$

35. $(+7) \times (+3)$ **36.** $0 \times (-5)$ **37.** $(-1)(-5)(-2)(+7)$

Page

552 **38.** Find the value of $(-2)^4$.

554 Divide.

39. $\frac{-40}{-8}$ **40.** $\frac{-72}{+9}$ **41.** $\frac{+24}{-1}$ **42.** $\frac{+60}{+15}$ **43.** $\frac{-16}{+16}$ **44.** $(-56) \div (-7)$

45. Simplify. $\frac{6(-2) + 5(-3)}{-3(4 - 7)}$

560 Write each of the following as an algebraic expression.

46. The sum of l and w.

47. The difference between ten and four.

48. The quotient of six divided by r.

49. The product of a and nine.

50. Six times the square of the side (s).

562 Write each of the following as an algebraic sentence.

51. Some number y decreased by five is equal to twelve.

52. Each number x increased by two is greater than eight.

53. Ten times each number n is less than or equal to fourteen.

54. Each number s is greater than negative three and less than positive nine.

564 **55.** Express the following as a formula.
Centripetal force (F) equals the product of the weight of the body (w) and the square of the velocity (v) divided by the product of the acceleration of gravity (g) and the radius of the circle (r).

566 Find the value of each of the following.

56. $m + 9n$ when $m = 8$ and $n = 3$.

57. $a + b(a - b)$ when $a = 6$ and $b = 2$.

568 **58.** Find the value of E when $I = 5$, $r = 16$, and $R = 20$.
Formula: $E = Ir + IR$

569 **59.** Which of the equations below have 4 as the solution?

$n - 3 = 7$	$n = 4$	$6n - 7 = 17$
$n + 9 = 13$	$\frac{n}{8} = 2$	$9n + n = 20$

60. Which are equivalent?

61. Which is in simplest form?

571 What operation with what number do you use on both sides of each of the following equations to get an equivalent equation in simplest form? Write this equivalent equation.

62. $7x = 56$ **63.** $m + 8 = 14$

64. $y - 11 = 6$ **65.** $\frac{a}{15} = 9$

Solve and check.

574 **66.** $x + 6 = 13$

575 **67.** $b - 7 = 4$

576 **68.** $-7y = 63$ **69.** $4m = 76$

577 **70.** $\frac{a}{9} = 18$ **71.** $\frac{z}{-3} = 15$

578 **72.** $6n + 17 = 35$ **73.** $8x - 3x = 60$

579 **74.** Find the value of t when $v = 217$, $V = 25$, and $g = 32$ using the formula $v = V + gt$.

581 **75.** Solve by the equation method.
Pedro bought a football at 40% reduction sale, paying $10.65 for it. What was the regular price?

583 Find the solutions of each of the following inequalities when the replacements for the variable are all the real numbers.

76. $8x > 136$ **77.** $n + 3 < 11$ **78.** $t - 5 \not> -7$

79. $4a + 13 \geq 1$ **80.** $6a - 7a < 0$ **81.** $\frac{d}{3} \neq -1$

82. $-7y \not< -21$ **83.** $2n - 5n > -15$ **84.** $\frac{3}{5}x \leq 42$

594 **85.** Write the ordered pair of numbers that has -6 as the first component and 4 as the second component.

595 What are the coordinates of each of the following?

86. Point A **87.** Point B

88. Point C **89.** Point D

90. Point E **91.** Point F

597 On graph paper, draw axes and plot the points with the following coordinates.

92. $G(-5, -2)$ **93.** $H(0, 6)$ **94.** $I(4, -3)$

599 **95.** Draw the graph of the equation: $x + y = 7$.

Chapter 15 Positive and Negative Numbers

Introduction to Algebra

In algebra we continue to use the operational signs $(+, -, \times, \div$ or $\overline{)}\ )$ of arithmetic. The exponent (p. 54), the square root symbol $\sqrt{\ }$, the parentheses ( ), the absolute value | |, and new meanings for the symbols $+$ and $-$ (positive and negative) are also used.

In the lessons in reading, writing, and evaluation of algebraic expressions and sentences (equations, formulas, and inequalities) where the language of algebra is developed, the symbols $=, \neq, <, \nless,$ $>, \ngtr, \leq,$ and $\geq$ are used in addition to the above symbols.

The number system you will use is enlarged by the introduction of negative numbers. Correspondingly the number line is extended to include points for negative integers and rational numbers.

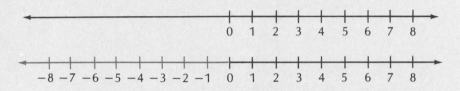

Knowledge of how to solve an equation provides an essential mathematical tool both to solve problems and to deal with the important subject of the formula.

The graphing of the equation and the inequality on the number line and of the equation on the number plane are also included.

PROPERTIES OF THE OPERATIONS

Operations in mathematics have certain characteristics, or properties.

Commutative Property of Addition

The order in adding two numbers does not affect the sum.

Since $6 + 5 = 11$ and $5 + 6 = 11$, then $6 + 5 = 5 + 6$

Note: The operation of subtraction does not have this property.

$$6 - 5 \neq 5 - 6$$

Commutative Property of Multiplication

The order in multiplying two numbers does not affect the product.

Since $6 \times 3 = 18$ and $3 \times 6 = 18$, then $6 \times 3 = 3 \times 6$.

Note: The operation of division does not have this property.

$$\frac{6}{3} \neq \frac{3}{6}$$

Associative Property of Addition

The grouping in adding three numbers does not affect the sum.

Since $5 + (6 + 4) = 15$ and $(5 + 6) + 4 = 15$,

then $5 + (6 + \mathbf{4}) = (5 + 6) + \mathbf{4}$

Note: The operation of subtraction does not have this property.

$$10 - 6 - 2 \neq 10 - (6 - 2)$$

Associative Property of Multiplication

The grouping in multiplying three numbers does not affect the product.

Since $6 \times (3 \times 4) = 72$ and $(6 \times 3) \times 4 = 72$,

then $6 \times (3 \times \mathbf{4}) = (6 \times 3) \times \mathbf{4}$

Note: The operation of division does not have this property.

$$24 \div 4 \div 2 \neq 24 \div (4 \div 2)$$

Distributive Property of Multiplication over Addition

Since $5 \times (7 + 3) = 5(10) = 50$ and $5 \times 7 + 5 \times 3 = 50$,

then $5 \times (7 + 3) = 5 \times 7 + 5 \times 3$

You say multiplication is distributed over addition.

Note: The operation of division is not distributed over addition.

$$12 \div (2 + 4) \neq 12 \div 2 + 12 \div 4$$

The same properties hold for all positive and negative numbers.

15-1 Rational Numbers

Numbers can be classified according to their properties. Whole numbers, integers, and rational numbers are three kinds of numbers.

Vocabulary

Positive numbers are numbers associated with points to the right of 0 on the number line.

Negative numbers are numbers associated with points to the left of 0 on the number line.

The **whole numbers** are the numbers 0, 1, 2, 3, 4, 5, 6, 7, 8, 9, 10, and so on without end.

Integers are all the whole numbers and their opposites.

A **rational number** is a number that can be expressed as a quotient of two integers, with division by 0 excluded.

The **absolute value** (symbol | |) of a number is the value of the number with no sign.

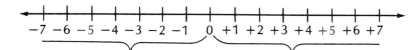

Negative Numbers   Positive Numbers

$+8$, 7, $\frac{3}{5}$, and $+\sqrt{11}$ are positive numbers.

-4, $-\frac{5}{8}$, and $-\sqrt{7}$ are negative numbers.

We read $+8$ as "positive 8." We read -4 as "negative 4." Zero is neither positive nor negative.

Look at the number line. Find $+5$ and -5 They are on opposite sides of 0 but are the same distance from 0. Such numbers are called **opposites** of each other. The opposite of 0 is 0.

The set of numbers consisting of all the whole numbers and their opposites is called the **integers.**

-4, $+6$, 0, -12, and 3 are integers.

Think About It

1. An integer is always a part of what other set of numbers?

2. Name two numbers that have an absolute value of 5.

3. What is the absolute value of 0?

4. What sign does the calculator assume all numbers have? How do you enter -7 on the calculator? What key should you press to change -7 to $+7$?

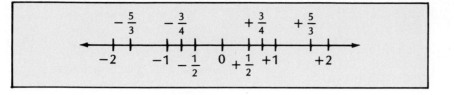

All the fractional numbers and their opposites are **rational numbers.** Integers are also rational numbers, since an integer can be written in fraction form: $-7 = -\frac{7}{1}$ $0 = \frac{0}{3}$

$\frac{1}{2}$, $-\frac{6}{7}$, -7, 0, and $3\frac{1}{2}$ are all rational numbers.

Fractions can be arranged in a definite order on the number line. There is a point corresponding to each fraction. Each has an opposite.

The absolute value of any number, positive or negative, is the value of the number without its sign. $|\ |$ is the symbol for absolute value. We read $|-9|$ as "the absolute value of negative 9." $|-9| = 9$, $|+9| = 9$.

Examples

1. What number is the opposite of -8?

ANSWER: $+8$

Look 8 units to the right of zero on the number line.

2. List all the numbers described by the following.

$-5, -4, -3, \ldots, 6$

The three dots between -3 and 6 mean that this list includes not only -5, -4, -3, and 6 but also all the integers between -3 and 6.

ANSWER: $-5, -4, -3, -2, -1, 0, 1, 2, 3, 4, 5, 6$

3. Is -8 a whole number?

ANSWER: No

Negative numbers do not belong to the set of whole numbers.

Is -8 an integer?

ANSWER: Yes

-8 is the opposite of the whole number, 8

Is -8 a rational number?

ANSWER: Yes

-8 can be written as $\frac{-8}{1}$

4. What is the absolute value of -4?

ANSWER: 4

-4 is 4 units from 0.

Practice Exercises

Read, or write in words, each of the following.

1. a. -6 **b.** $+10$ **c.** -100 **d.** -147

2. a. $+\frac{5}{6}$ **b.** -17.3 **c.** $-6\frac{1}{3}$ **d.** $+0.89$

3. a. $|+11|$ **b.** $|-3|$ **c.** $|-0.09|$ **d.** $\left|5\frac{2}{3}\right|$

What number is the opposite of each of the following?

4. a. $+7$ **b.** -3 **c.** -100 **d.** $+79$

5. a. $+\frac{5}{8}$ **b.** $-\frac{9}{4}$ **c.** 0.07 **d.** -2.75

6. List all the numbers described by the following.

   a. $-11, -10, -9, \ldots , 3$

   b. $-9, -7, -5, \ldots , 9, 11$

7. Write using symbols.

   a. negative twelve   **b.** positive nine

   c. negative one-fifth   **d.** positive nine-hundredths

| -1 | 5 | $+\frac{3}{4}$ | -0.05 | 0 | -17 | 1 | 19 | -6 | 0.18 | $-4\frac{7}{12}$ | $+0.11$ |

8. Which of the numbers in the box above are whole numbers?

9. Which of the numbers in the box above are integers?

10. Which of the numbers in the box above are rational numbers?

11. Write, using the absolute value symbol.

   a. The absolute value of positive three and nine-tenths

   b. the absolute value of negative fourteen-thousandths.

12. Find the absolute value of each of the following.

   a. $|-7|$ **b.** $|0|$ **c.** $|-0.82|$ **d.** $\left|+\frac{9}{4}\right|$

13. In each of the following, first find the absolute value of each number. Then find the answer.

   a. $|-4| + |-8|$ **b.** $\left|-\frac{4}{5}\right| - \left|+\frac{7}{10}\right|$

   c. $|-6.08| \times |-0.07|$ **d.** $\left|3\frac{5}{6}\right| \div \left|-1\frac{7}{12}\right|$

Graphing on a Number Line

All numbers can be represented by points on a number line that indicate their distance from 0.

A number line is endless in both directions. There is one and only one point corresponding to each number and one and only one number corresponding to each point.

Vocabulary

A **number line** is a collection of points corresponding to a set of numbers.

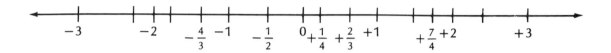

Often a number line is labeled only with the integers. Only a part of it is shown at any one time.

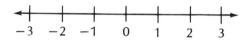

Each point on the number line is called the **graph** of the number to which it corresponds. Each number is called the **coordinate** of the corresponding point on the line. Capital letters are generally used to identify these points.

The **graph of a number** is a point on the number line whose coordinate is the number.

Point *B* is the graph of −2.
−2 is the coordinate of point *B*.

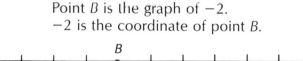

The graph of a set of numbers is made up of the point on the number line whose coordinates are the numbers.

The points *B, C, E,* and *H* are the graph of the numbers −3, −2, 0, and 3.

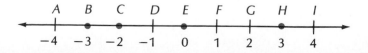

Examples

1. Draw the graph of this set of numbers on a number line: −4, −2, 6, 8

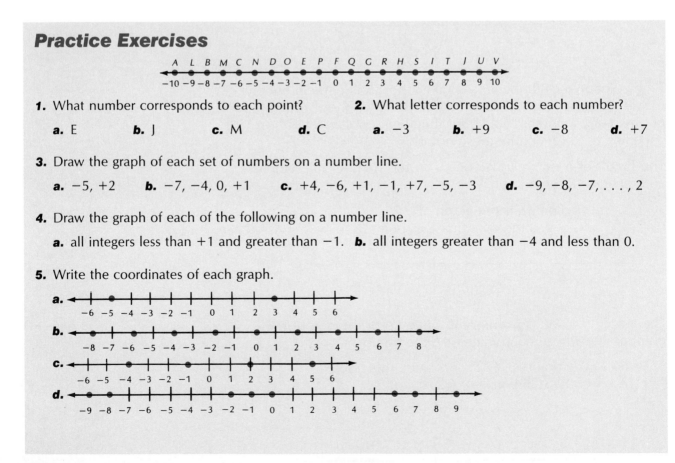

To draw the graph of a set of numbers on a number line, do this.

a. Draw an appropriate number line.

b. Locate the points whose coordinates are the given numbers.

2. Write the coordinates of this graph.

ANSWER: $-4\frac{1}{2}$, -2, -1, 0, $1\frac{1}{2}$, 4, 6

To write the coordinates of a graph, do this.

Write the coordinates of the points marked on the number line.

Think About It

1. If a number is located to the left of 0 on a number line, what type of number is it?

2. How many whole numbers lie between −3 and +3? How many integers lie between −3 and +3? How many rational numbers lie between −3 and +3?

Practice Exercises

1. What number corresponds to each point?

   a. E   **b.** J   **c.** M   **d.** C

2. What letter corresponds to each number?

   a. −3   **b.** +9   **c.** −8   **d.** +7

3. Draw the graph of each set of numbers on a number line.

   a. −5, +2   **b.** −7, −4, 0, +1   **c.** +4, −6, +1, −1, +7, −5, −3   **d.** −9, −8, −7, . . . , 2

4. Draw the graph of each of the following on a number line.

   a. all integers less than +1 and greater than −1.   **b.** all integers greater than −4 and less than 0.

5. Write the coordinates of each graph.

15-3 Opposite Meanings

Positive and negative numbers are used in science, statistics, weather reports, stock reports, sports, and many other fields to express opposite meanings or directions.

Example

If $+1\frac{1}{4}$ points indicates a $1\frac{1}{4}$-point *gain* in a stock price, what does -2 points indicate?

ANSWER: 2-point *loss* in a stock price.

Think About It

1. Describe three real-life situations where opposites are used.

2. Name the opposite of 10 miles southwest.

Practice Exercises

1. If $+250$ m represents 250 m above sea level, what does -60 m represent?

2. If -30 kg represents a downward force of 30 kg, what does $+75$ kg represent?

3. If an increase of 6% in the cost of living is indicated by $+6\%$, how can a 4% decrease in the cost of living be indicated?

4. If 85 degrees west longitude is indicated by $-85°$, how can 54 degrees east longitude be indicated?

5. If $+18$ mi/h indicates a tail wind of 18 mi/h, what does -23 mi/h represent?

6. If a deficiency of 47 mm of rainfall is indicated by -47 mm, how can an excess of 39 mm be indicated?

7. If $-14°$ means 14 degrees below zero, what does $+72°$ mean?

8. If $70 deposited in the bank is represented by $+$70$, how can $25 withdrawn from the bank be represented?

9. If an inventory shortage of 120 items is represented by -120 items, how can 58 items over be represented?

10. If a charge of 15 amperes of electricity is indicated by $+15$ amperes, how can a discharge of 17 amperes of electricity be indicated?

15-4 Comparing Integers

You can compare integers just as you compare whole numbers. You can use the number line to help.

The number line may be drawn either horizontally or vertically.

```
    L   M   N   P   Q   R   S   T   U   V   W
  ◄─●───●───●───●───●───●───●───●───●───●───●──►
   -5  -4  -3  -2  -1   0  +1  +2  +3  +4  +5
```

On the horizontal number line, the number corresponding to the point farther to the right is greater. On the vertical number line, any number corresponding to a point is greater than any number corresponding to a point below it.

Use the symbols >, <, and = to compare integers.

Read > as "is greater than." Read < as "is less than."

```
A ●  +5
B ●  +4
C ●  +3
D ●  +2
E ●  +1
F ●   0
G ●  -1
H ●  -2
I ●  -3
J ●  -4
K ●  -5
```

Example

1. Insert >, <, or = to make a true sentence.

$$-3 \underline{\overset{?}{}} -7$$

```
  ◄─✳─┼─┼─✳─┼─┼─┼─┼─┼─┼─┼─┼─┼─┼─┼─►
   -7-6-5-4-3-2-1  0+1+2+3+4+5+6+7
```

ANSWER: $-3 > -7$

$$-7 \underline{\overset{?}{}} -3$$

ANSWER: $-7 < -3$

$$-3 \underline{\overset{?}{}} -3$$

ANSWER: $-3 = -3$

To compare two integers on the number line, do this.

a. Find the position of each integer on a number line.

b. If the first number is to the right or above the second number, use the symbol >.

c. If the first number is to the left or below the second number, use the symbol <.

d. If the two numbers are represented by the same point, use the symbol =.

Think About It

1. How do 3 and $\frac{12}{4}$ compare?

2. Write two different true sentences using -2 and -1.

3. If $a > b$, describe the positions of a and b in relation to one another on a horizontal number line. Describe their positions on a vertical line.

4. If two numbers are opposites, what is true about their positions on a number line?

Practice Exercises

1. On the horizontal number line, name the point that corresponds to the greater number.

    a. Point T or point M   **b.** Point P or point V
    c. Point Q or point N   **d.** Point U or point L

2. On the vertical number line, name the point that corresponds to the greater number.

    a. Point C or point H   **b.** Point G or point B
    c. Point F or point J   **d.** Point K or point D

3. Which number is greater?

    a. +8 or −8   **b.** −8 or −5   **c.** −7 or 0   **d.** +5 or −10

4. Which number is less?

    a. −8 or +5   **b.** −9 or −10   **c.** 0 or −6   **d.** −4 or −7

5. Which of the following sentences are true?

    a. +3 > −4   **b.** −12 > +10   **c.** −6 < −3   **d.** −9 < +9

6. Rewrite each of the following. Insert between the two numerals the symbol =, <, or > that will make the sentence true.

    a. −8 _____ −1   **b.** +11 _____ −14
    c. +9 _____ −4   **d.** −3 _____ 0

7. List the following numbers in order from the least to the greatest.

    a. +4, −3, +7, −9, −2, +10, −4, 0, −6, +5
    b. −1, +2, 0, −3, +8, −5, −7, +11, −10, +10

8. List the following numbers in order from the greatest to the least.

    a. −3, +8, −7, +4, −8, 0, +3, −12, +9, −1
    b. +5, −6, +1, −1, 0, +10, −9, −11, +8, −4

9. Which number is greater?

    a. The absolute value of −8 or the absolute value of +3
    b. The absolute value of +9 or the absolute value of 9

10. Which has the greater opposite number?

    a. +7 or +4   **b.** −3 or −8   **c.** −1 or +6   **d.** +5 or −10

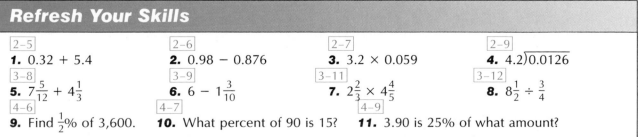

Refresh Your Skills

  2-5
1. 0.32 + 5.4   2-6 **2.** 0.98 − 0.876   2-7 **3.** 3.2 × 0.059   2-9 **4.** 4.2$\overline{)0.0126}$

  3-8
5. $7\frac{5}{12} + 4\frac{1}{3}$ 3-9 **6.** $6 - 1\frac{3}{10}$ 3-11 **7.** $2\frac{2}{3} \times 4\frac{4}{5}$ 3-12 **8.** $8\frac{1}{2} \div \frac{3}{4}$

  4-6
9. Find $\frac{1}{2}$% of 3,600.   4-7 **10.** What percent of 90 is 15?   4-9 **11.** 3.90 is 25% of what amount?

15-5 Additive Inverse of a Number

If the sum of two numbers is zero, each addend is the additive inverse of the other.
−4 is the opposite of or the additive inverse of +4 (or 4).
+4 (or 4) is the opposite of or the additive inverse of −4.

Examples

1. Find the additive inverse of −8.

ANSWER: 8 or (+8)

2. Find the additive inverse of −(+4)

ANSWER: −4

3. Read, or write in words, −(−13).

ANSWER: The opposite of negative thirteen.

Think About It

1. What keys would you press on a calculator to name the additive inverse of 8?

2. What is always true about the opposite of the opposite of a number?

Practice Exercises

1. What number is the additive inverse of each of the following numbers?

   a. −7   **b.** +26   **c.** 0   **d.** $-\frac{5}{8}$

2. What number is the opposite of each of the following numbers?

   a. +10   **b.** −6   **c.** −0.85   **d.** 0

3. Read, or write in words, each of the following.

   a. −9   **b.** −(+2)   **c.** $-\left(-\frac{5}{6}\right)$ **d.** −(15) = −15

4. Write symbolically, using the "opposite" symbol.

   a. The opposite of thirty-six.   **b.** The opposite of negative five.
   c. The opposite of positive fifty-six is negative fifty-six.

5. Find the opposite of the opposite of: −7, +6; $+\frac{3}{8}$; −0.9; $-3\frac{1}{2}$

6. Find the value of each of the following:

   a. −(+9) = ?   **b.** $-\left(3\frac{2}{5}\right)$ = ?   **c.** −(54) = ?   **d.** $-\left(\frac{11}{4}\right)$ = ?

Adding Integers and Rational Numbers

You can use a number line to add integers and rational numbers.

Examples

1. $(+2) + (+3) = ?$

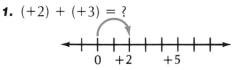

$$0 \quad +2 \qquad +5$$

To add on a number line do this.

a. Start at zero and move to the position of the first number.

b. From that point, move the number of units indicated by the second number. If the second number is positive, move to the right. If the second number is negative, move to the left.

ANSWER: $+5$

---

2. $(-4) + (-3) = ?$

$$-7 \qquad -4 \qquad 0$$

ANSWER: -7

3. $(-6) + (+4) = ?$

$$-6 \qquad -2 \quad 0$$

ANSWER: -2

4. $(-3) + (+7) = ?$

$$-3 \qquad 0 \qquad +4$$

ANSWER: $+4$

You can also add integers and rational numbers algebraically.

5. $(+2) + (+3) = ?$

$|+2| + |+3| = 5$
$\quad +5$

ANSWER: $+5$

To add two positive numbers do this.

a. Find the sum of their absolute values.
b. Write a positive sign before the sum.

---

6. $(-4) + (-3) = ?$

$|-4| + |-3| = 7$
$\quad -7$

ANSWER: -7

To add two negative numbers do this.

a. Find the sum of their absolute values.
b. Write a negative sign before the sum.

7. $(-6) + (+4) = ?$

$|-6| + |+4| = 2$

-2

ANSWER: -2

To add a positive number and a negative number do this.

a. Subtract the smaller absolute value from the greater absolute value.

b. Write the sign of the number with the greater absolute value before the sum.

---

8. $(+2) + (-9) + (+6) + (-3) = ?$

$(+2) + (-9) + (+6) + (-3)$

$(+8) + (-9) + (-3)$

$(+8) + (-12)$

-4

ANSWER: -4

To add three or more numbers do this.

a. Add the positive numbers.

b. Add the negative numbers.

c. Find the sum.

---

9. $6 - 9 + 4 - 3 = ?$

$(+6) + (-9) + (+4) + (-3)$

$(+10) + (-12)$

-2

ANSWER: -2

To simplify a numerical expression do this.

a. Make the given signs part of the numbers. The operation is addition.

b. Find the sum.

Think About It

1. When is the sum of a positive and negative number a negative number? When is the sum a positive number? When is the sum zero?

2. What is the sum of any number and its opposite?

3. Use a calculator to solve the following exercises. List, in order, the keys you pressed.

    a. $(+653) + (-159)$

    b. $(-325) + (-99) + (-5)$

Practice Exercises

Add on the number line.

1. a. $(-4) + (-1)$　　　　　　　　　　**b.** $(+7) + (-3)$
　　c. $(-8) + (+2)$　　　　　　　　　　**d.** $(+2) + (-6)$

Add.

2. a. $(+24) + (+19)$　　　　　　　　　**b.** $(-9) + (-14)$
　　c. $(-27) + (-35)$　　　　　　　　　**d.** $(-83) + (-49)$

3. a. $(+11) + (-3)$　　　　　　　　　　**b.** $(+91) + (-37)$
　　c. $(-9) + (+1)$　　　　　　　　　　**d.** $(-54) + (+39)$

4. a. $(+3) + (-10)$　　　　　　　　　　**b.** $(+20) + (-41)$
　　c. $(-2) + (+15)$　　　　　　　　　　**d.** $(-45) + (+68)$

5. a. $(+6) + (-6)$　　　　　　　　　　　**b.** $(-12) + (+12)$
　　c. $(0) + (+7)$　　　　　　　　　　　**d.** $(-5) + (0)$

6. a. $\left(+\frac{3}{4}\right) + \left(-\frac{1}{4}\right)$　　　　　　　　**b.** $\left(-\frac{7}{8}\right) + \left(-\frac{2}{3}\right)$
　　c. $(+0.4) + (+1.5)$　　　　　　　　**d.** $(-5.28) + (-2.64)$

7. a. $(-3) + (-8) + (-9)$　　　　　　　**b.** $(-3) + (+6) + (-5)$
　　c. $(+5) + (-14) + (+7)$　　　　　　**d.** $(-10) + (-8) + (-2)$

8. a. $(-4) + (+7) + (-8) + (+5)$　　　　**b.** $(-6) + (-9) + (+7) + (-2)$
　　c. $(-12) + (+9) + (-6) + (+9)$　　　**d.** $(+5) + (-3) + (-4) + (-5)$

9. Simplify.

　　a. $-6 + 7$　　　　　　　　　　　　　**b.** $-5 - 3$
　　c. $5 - 6 - 8$　　　　　　　　　　　　**d.** $-2 - 3 + 9 - 1 + 3$

10. Simplify by performing the indicated operations and finding the absolute values.

　　a. $|-4 - 5|$　　　　　　　　　　　　**b.** $|-10| + |+7|$
　　c. $-|14 - 5|$　　　　　　　　　　　　**d.** $-|-(-4) + (-6)|$

11. Solve each problem.

a. The temperature at 9 A.M. was $+15°C$. By noon the temperature had risen 7 degrees. Write the temperature at noon as an integer.

b. A quarterback lost 6 yards on a play. On the next play he completed a pass for a 15-yard gain. Write his total gain or loss as an integer.

Using Mathematics

WIND CHILL FACTOR

On a cold, windy day, have you ever felt a lot colder than the actual temperature would seem to indicate? You were right—because wind does affect how cold we feel. If the still air temperature on Mt. McKinley is 0°F but there is a wind of 40 miles per hour, the wind-chill temperature is −53°F.

People who work outdoors or climb mountains or ski in cold weather must take the wind-chill factor into account.

Here is a table that shows wind-chill factors.

Wind-Chill Chart							
Wind Speed mi/h	Actual Temperature °F						
	40	30	20	10	0	−10	−20	−30
	Wind-Chill Temperature °F						
10	28	16	4	−9	−21	−33	−46	−58
20	18	4	−10	−25	−39	−53	−67	−82
30	13	−2	−18	−33	−48	−63	−79	−94
40	10	−6	−21	−37	−53	−69	−85	−100

Greater wind speeds have little added effect.

Exercises

1. What is the wind-chill temperature if the actual temperature is 10°F and the wind speed is 20 miles per hour?

2. Find the wind-chill air temperature if the actual temperature is 20°F and the wind speed is 30 miles per hour?

3. When the air temperature is 20°F, what is the change in wind-chill air temperature if the wind increases from 20 to 30 miles per hour?

4. Find the wind-chill air temperature if the actual temperature is 40°F and the wind speed is 20 miles per hour.

15-7 Subtracting Integers and Rational Numbers

Subtraction is the inverse of addition.

You can use a number line to subtract rational numbers.

Examples

1. $(+3) - (-2) = ?$

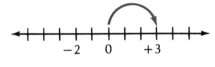

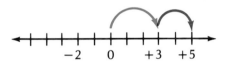

ANSWER: $+5$

To subtract on a number line, do this.

1. Start at zero and move to the position of the first number.

2. From that point, move the number of units indicated by the second number but in the direction opposite to the sign of the number. (Instead of moving 2 units to the left as in addition, move 2 units to the right.)

You can also subtract rational numbers algebraically.

> When we compare
>
> $(+3) - (-2) = +5$ and $(+3) + (+2) = +5$
>
> we see that subtracting 2 from $+3$ gives the same answer as adding $+2$ and $+3$. That is, *subtracting a number* gives the same answer as *adding its opposite.*

2. $(+2) - (+8) = n$
  $(+2) + (-8) = -6$

ANSWER: -6

To subtract a number algebraically, add its opposite.

---

3. $(+2) - (-8) = n$
  $(+2) + (+8) = +10$

ANSWER: $n = +10$

4. $(-2) - (+8) = n$
  $(-2) + (-8) = -10$

ANSWER: $n = -10$

5. $(-2) - (-8) = n$
  $(-2) + (+8) = +6$

ANSWER: $n = +6$

1. What process gives the same result as subtracting a number?

2. Would you choose to use paper and pencil, mental arithmetic, estimation, or a calculator to solve Exercises 7a.–d. below? Why?

3. Use a calculator to solve the following. List, in order, the keys you pressed.
  a. $(152) - (-259)$ **b.** $(-189) - (+576)$

Practice Exercises

Subtract on the number line.

1. a. $(+6) - (+3)$
  c. $(+4) - (-6)$

  b. $(-5) - (-1)$
  d. $(-2) - (+5)$

Subtract.

2. a. $(5) - (7)$
  c. $(+2) - (+14)$

  b. $(6) - (15)$
  d. $(+28) - (+45)$

3. a. $(-6) - (-4)$
  c. $(-7) - (-9)$

  b. $(-12) - (-7)$
  d. $(-51) - (-70)$

4. a. $(+7) - (-2)$
  c. $(-6) - (+19)$

  b. $(+18) - (-27)$
  d. $(-16) - (+25)$

5. a. $(+6) - (+6)$
  c. $(0) - (-1)$

  b. $(-5) - (-5)$
  d. $(+8) - (0)$

6. a. $(+11) - (-11)$
  c. $\left(-\frac{7}{8}\right) - \left(+\frac{3}{4}\right)$

  b. $(-0.6) - (-0.7)$
  d. $\left(-2\frac{1}{2}\right) - \left(-1\frac{2}{3}\right)$

7. a. $(+5) - (+9)$
  c. $(-13) - (+7)$

  b. $(-6) - (-2)$
  d. $(+20) - (-15)$

8. a. $(-3) - (0)$
  c. $(+0.9) - (-1.5)$

  b. $(-14) - (+14)$
  d. $(0) - (-17)$

9. a. $\left(+\frac{1}{2}\right) - \left(-\frac{1}{4}\right)$
  c. $(+0.3) - (+0.6)$

  b. $(-0.7) - (-1.1)$
  d. $\left(-\frac{2}{3}\right) - \left(+\frac{1}{3}\right)$

10. a. $(4 - 3) - (6 - 10)$
  c. $(8 + 7) - (6 - 13)$

  b. $(2 - 6) - (7 - 8)$
  d. $(3 - 1) - (9 - 2)$

11. a. $[(-2) + (-7)] - (-9)$
   c. $[(-5) - (+8)] - (-12)$
   b. $(+4) - [(-5) - (+6)]$
   d. $(-6) - [(-10) + (+3)]$

12. a. From -3 take -7.
   c. Take -9 from $+9$.
   b. Subtract 10 from 3.
   d. From 0 subtract -8.

Simplify by performing the indicated operations and finding the absolute values.

13. a. $|-4| - |-9|$
   b. $|6 - 2| - |-5|$

14. a. $|1 - 3| - |4 - 7|$
   b. $|-3 - 8| - 6$

15. a. $8 - |5 + 9|$
   b. $|(0 - 2) - (4 - 2)|$

16. Solve each problem.

a. The average daily high temperature in Boston in August is 82°F. In January the average daily high temperature is 45° less. Write the average daily high temperature in January as an integer.

b. During the first half of a football game, the Tigers lost 7 yards rushing. During the second half, they gained 63 yards rushing. How much better did they do in the second half than in the first half?

Refresh Your Skills

1–8
1. $7,687 + 80,547 + 3,396$

1–9
2. $140,000 - 69,076$

1–11
3. $6,080 \times 319$

1–13
4. $360\overline{)225,720}$

2–5
5. $0.395 + 0.68 + 0.7$

2–6
6. $7.8 - 0.65$

2–7
7. 0.09×1.02

2–9
8. $15.3\overline{)6.12}$

3–8
9. $\frac{5}{6} + \frac{1}{2} + \frac{4}{5}$

3–9
10. $2\frac{5}{8} - 1\frac{13}{16}$

3–11
11. $6\frac{3}{4} \times 9\frac{1}{3}$

3–12
12. $6 \div 1\frac{1}{3}$

4–6
13. Find 9.2% of 10,000.

4–7
14. What percent of 95 is 95?

4–9
15. 0.10 is 1% of what amount?

Multiplying Integers and Rational Numbers

Multiplication can be thought of as repeated addition.

$(3)(4) = (+3)(+4) = (+4) + (+4) + (+4) = +12$
$(+3)(+4) = +12$

The product of two positive numbers is a positive number.

$(+3)(-4) = (-4) + (-4) + (-4) = -12$
$(+3)(-4) = -12$

The product of a positive number and a negative number is a negative number.

Since $(+3)(-4)$ can be written as $(-4)(+3)$ by using the commutative property of multiplication, then,

The product of a negative number and a positive number is a negative number.

Think About It

1. Does the number of positive factors affect the sign of the product? Why?

2. Does the number of negative factors affect the sign of the product? Why?

To determine what $(-3)(-4)$ equals, observe the following steps:

1. $(-3)(0) = 0$ — The product of any number and zero is zero.

2. But $[(+4) + (-4)] = 0$ — The sum of a number and its inverse is zero.

3. Then $(-3)[(+4) + (-4)] = 0$ — By substituting $[(+4) + (-4)]$ for 0 in step 1.

4. $[(-3)(+4)] + [(-3)(-4)] = 0$ — By the distributive property of multiplication over addition.

5. $-12 + [(-3)(-4)] = 0$ — Since $(-3)(+4) = -12$.

6. Therefore $[(-3)(-4)] = +12$ — In order for $-12 + [(-3)(-4)] = 0$ to be true, the product of $[(-3)(-4)]$ must be the additive inverse of -12, which is $+12$.

$(-3)(-4) = +12$

The product of two negative numbers is a positive number.

Examples

1. $(+5)(+6) = ?$

$|+5| \times |+6| = 30$
$+30$

To multiply two positive numbers, do this.

a. Find the product of their absolute values.
b. Write a positive sign before this product.

ANSWER: $+30$

2. $(-5)(-6) = ?$

$|-5| \times |-6| = 30$
$\qquad +30$

ANSWER: $+30$

To multiply two negative numbers, do this.

a. Find the product of their absolute values.
b. Write a positive sign before this product.

---

3. $(+5)(-6) = ?$

$|+5| \times |-6| = 30$
$\qquad -30$

ANSWER: -30

To multiply one positive number and one negative number, do this.

a. Find the product of their absolute values.
b. Write a negative sign before this product.

---

4. $(-2)(+1)(-4)(-2) = n$

$|-2| \times |+1| \times |-4| \times |-2| = 16$
$\qquad -16$

ANSWER: -16

To multiply more than two factors, do this.

a. Find the product of their absolute values.
b. Write a positive sign before this product when the number of negative factors is even. Write a negative sign if the number of negative factors is odd.

If the product of two numbers is one (1), then each factor is said to be the multiplicative inverse, or the reciprocal, of the other.

5. What is the multiplicative inverse of $\frac{2}{3}$?

ANSWER: $\frac{3}{2}$

6. What is the reciprocal of $-\frac{3}{2}$?

ANSWER: $-\frac{2}{3}$

Practice Exercises

Multiply.

1. a. $(+5)(+9)$ **b.** $(-6)(-7)$ **c.** $(-13)(+5)$ **d.** $(+12)(-8)$

2. a. $(0)(-8)$ **b.** $(-0.4)(-0.2)$ **c.** $(+2.5)(-1.4)$ **d.** $(-3\frac{1}{2})(+2\frac{1}{4})$

3. a. $-8(-7 - 5)$ **b.** $(+6)(-2)(-4)$ **c.** $(-2) \times [(-6)(-7)]$

4. a. $(-3)^4$ **b.** $(-4)^5$ **c.** $(-1)^6$ **d.** $(-1)^5 \cdot (-3)^3$

5. a. $|(-1)(-5)|$ **b.** $|-4| \times |-7|$ **c.** $-10 \times |-6|$

6. a. $|12 - 3| \times |6 - 8|$ **b.** $(-6) \times |(-2)(-9)(-1)|$ **c.** $|-5|^2 \times |(-4)(-11)|$

7. Name the multiplicative inverse of each.

    a. -6 **b.** $-\frac{2}{5}$ **c.** $-\frac{17}{4}$ **d.** $-3\frac{1}{2}$

8. a. Phil lost 4 lb a month for 5 months. Write his total gain or loss as an integer.

    b. Sara saved $25 each month for 6 months. Write her total savings as an integer.

15-9 Dividing Integers and Rational Numbers

Just as subtraction is the inverse of addition, division is the inverse of multiplication.

Since $(+3)(+4) = +12$, then $(+12) \div (+3) = +4$
The quotient of two positive numbers is a positive number.

Similarly,

Since $(-3)(+4) = -12$, then $(-12) \div (-3) = +4$
The quotient of two negative numbers is a positive number.

Since $(+3)(-4) = -12$,   Since $(-3)(-4) = +12$,
then $(-12) \div (+3) = -4$ then $(+12) \div (-3) = -4$

The quotient of a negative number divided by a positive number or a positive number divided by a negative number is a negative number.

Examples

1. $(+8) \div (+4) = ?$
$|+8| \div |+4| = 2$
$\qquad +2$

ANSWER: $+2$

To divide two numbers when both numbers are positive, do this.

a. Find the quotient of their absolute values.

b. Write a positive sign before the quotient.

---

2. $(-8) \div (-4) = ?$
$|-8| \div |-4| = 2$
$\qquad +2$

ANSWER: $+2$

To divide two numbers when both numbers are negative, do this.

a. Find the quotient of their absolute values.

b. Write a positive sign before the quotient.

---

3. $(-8) \div (+4) = ?$
$|-8| \div |+4| = 2$
$\qquad -2$

ANSWER: -2

To divide two numbers when one number is positive and one number is negative, do this.

a. Find the quotient of their absolute values.

b. Write a negative sign before the quotient.

Think About It

1. Name two situations that would result in a negative quotient. Name two situations that would result in a positive quotient.

2. Use a calculator to solve the following. List, in order, the keys you pressed.

$(-108) \div (-9)$

Practice Problems

Divide.

1. a. $\dfrac{+30}{+5}$ **b.** $\dfrac{+54}{+9}$ **c.** $\dfrac{-8}{-2}$ **d.** $\dfrac{-90}{-18}$

2. a. $\dfrac{+56}{-7}$ **b.** $\dfrac{+45}{-5}$ **c.** $\dfrac{-48}{+8}$ **d.** $\dfrac{-80}{+10}$

3. a. $\dfrac{-9}{-9}$ **b.** $\dfrac{-40}{+1}$ **c.** $\dfrac{+7}{-7}$ **d.** $\dfrac{0}{-3}$

4. a. $(+36) \div (-9)$ **b.** $(-20) \div (-4)$
c. $(-42) \div (+7)$ **d.** $(0) \div (-7)$

5. a. $\left(-8\frac{1}{2}\right) \div \left(-\frac{3}{4}\right)$ **b.** $(+0.24) \div (-0.3)$
c. $(+19) \div (-1)$ **d.** $(-100) \div (+25)$

6. a. $+2\overline{)-14}$ **b.** $-8\overline{)-64}$ **c.** $-3\overline{)-27}$

7. a. $-5\overline{)+50}$ **b.** $0.4\overline{)-3.2}$ **c.** $-7\overline{)+21}$

Simplify be performing the indicated operations.

8. a. $\dfrac{8 - 17}{3}$ **b.** $\dfrac{-10 - 4}{-7}$ **c.** $\dfrac{12 - 3(4)}{-2}$

9. a. $\dfrac{8(-1) + 7(-6)}{-5(5)}$ **b.** $\dfrac{(-4)^3 - (-2)^4}{(-10)^2}$ **c.** $\dfrac{6(4 - 7) - 5(6 - 12)}{-3(1 - 5)}$

Find the value of each.

10. a. $\dfrac{|-16|}{8}$ **b.** $\dfrac{|-24|}{|-2|}$ **c.** $\dfrac{|-30|}{|-5|}$

11. a. $\dfrac{|-8| \times |-6|}{|-9| - |-5|}$ **b.** $\dfrac{|(-12)(+7)|}{|8 - 9| - |1 - 5|}$ **c.** $\dfrac{|-3|^3 - |1|^4}{|(-6) + (-7)|}$

12. a. The price of ACE stock changed -20 points in 4 days. Write the average daily change in price as an integer.

b. Mort's Market had a total profit of $650 in 4 weeks. Write the average profit per week as an integer.

Computer Know-How

The computer program below asks for two numbers and computes their *sum* and *difference*.

```
1Ø PRINT "WHAT IS THE FIRST NUMBER";
2Ø INPUT X
3Ø PRINT "WHAT IS THE SECOND NUMBER";
4Ø INPUT Y
5Ø PRINT "THEIR SUM IS:" X + Y
6Ø PRINT "THEIR DIFFERENCE IS:"; X − Y
```

a. Enter the program and then run it to make sure it works.

b. Extend the program so that it *also* computes and prints out the *product* and the *quotient* of X and Y.

Career Application
METEOROLOGIST

Almost everyone is affected by the weather. For farmers and pilots, an unexpected change in the weather can be a disaster. Thousands of workers study weather conditions so they can make accurate predictions. These workers are called *meteorologists*. They record temperatures, wind velocities, air pressure, and humidity. They watch for types of clouds that signal changes in the weather. They use weather balloons to measure conditions in the upper air. Some meteorologists even help with research on air pollution. Others develop better instruments for measuring weather conditions.

Many colleges and technical schools offer two-year programs. Meteorologists learn some of their skills on the job. They may work for government agencies or for private companies such as airlines.

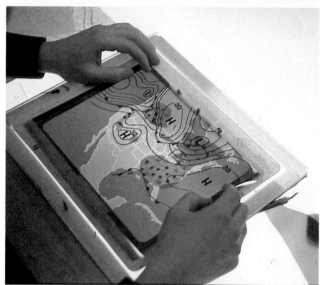

1. On Monday night the lowest temperature was −8°C. On Tuesday night the lowest temperature was −11°C. Which night was colder?

2. A weather balloon at an altitude of 70,000 feet measured the temperature as −67°F. The temperature at ground level was 74°F. How much warmer was it at ground level?

3. A meteorologist measured the air pressure every hour. Each hour the change in pressure was −5 millibars. What was the total change after three hours?

4. The temperature dropped from 0°C to −24°C in six hours. What was the average change per hour?

5. At noon the temperature was 12°F. At 6 P.M. it was 19 degrees lower. What was the temperature at 6 P.M.?

6. The wind velocity changed from 31 mi/h to 23 mi/h. Use a signed number to express the amount of the change.

Chapter 15 Review

Vocabulary

additive inverse p. 544
absolute value p. 536
integer p. 536
multiplicative inverse p. 553

negative number p. 536
number line p. 539
opposites p. 541
positive number p. 536

rational number p. 536
reciprocal p. 553
whole numbers, p. 536

$$+0.117 \quad 0 \quad -2\frac{5}{9} \quad +9\frac{1}{3} \quad -431 \quad -8.1 \quad -\frac{11}{31} \quad +3$$

Page

536

1. Which of the numbers in the box above name negative numbers?

2. Which of the numbers in the box above name integers?

3. Which of the numbers in the box above name rational numbers?

4. Which of the numbers in the box above name whole numbers?

5. Which of the numbers in the box above name negative integers?

6. Find the value of $|-20|$

7. What number is the opposite of -35?

8. On a number line, draw the graph of all integers greater than -4 and less than $+4$.

9. Write the coordinates of this graph.

$$-5 \quad -4 \quad -3 \quad -2 \quad -1 \quad 0 \quad +1 \quad +2 \quad +3 \quad +4 \quad +5$$

541 10. If -87 represents 87 km west, what does $+87$ represent?

542 Which of the following sentences are true?

11. $-11 < 7$

12. $-3 > -8$

13. $-1 > 0$

14. $+1 < -10$

15. List the following numbers in order from least to greatest.
$+7, -2, -5, 0, +4, -6$

544 16. What is the additive inverse of $+1.2$?

17. Find the value of $-(-14)$.

545 Add.

18. $(-7) + (-3)$

19. $(-13) + (+9)$

549 Subtract.

20. $(-3) - (+5)$

21. $(+7) - (-4)$

552 Multiply.

22. $(+10)(-12)$

23. $\left(-7\frac{1}{2}\right)\left(+2\frac{2}{3}\right)$

554 Divide.

24. $\frac{-27}{+9}$

25. $(-9.6) \div (-0.12)$

553 26. Name the multiplicative inverse of $\frac{11}{12}$.

536 27. A number that can be expressed as a quotient of two integers, with division by zero excluded, is a ___?___.

Chapter 15 Test

$$-0.7 \quad +10\tfrac{1}{5} \quad 0 \quad -74 \quad -\tfrac{3}{100} \quad +\tfrac{43}{7} \quad -0.09$$

Page
536   **1.** Which of the numbers in the box above name positive numbers?

2. Which of the numbers in the box above name integers?

3. Which of the numbers in the box above name rational numbers?

4. Find the value of $|-59|$.

5. What number is the opposite of -23?   **6.** What number is the opposite of $-\tfrac{1}{2}$?

539   **7.** On a number line draw the graph of all integers greater than -1 and less than $+5$.

8. Write the coordinates of this graph.

541   **9.** If -4 means 4 meters backward, what does $+4$ mean?

542   Which of the following sentences are true?

10. $-4 > -5$ **11.** $+3 > +7$ **12.** $-3 < -1$

13. List the following numbers in order from least to greatest.
$+6, -2, +5, 0, -15, +1$

544   **14.** What number is the additive inverse of $+34$?

545   Add.

15. $(-6) + (+13)$ **16.** $(-19) + (+11)$ **17.** $(-2.3) + (-8.1)$

18. $(-40) + (+40)$ **19.** $(-6) + (+9) + (-5)$

549   Subtract.

20. $(+5) - (-7)$ **21.** $(+3) - (+5)$ **22.** $\left(+2\tfrac{1}{2}\right) - \left(-3\tfrac{1}{4}\right)$

23. Take 11 from 3   **24.** $(-6) - (-14)$

552   Multiply.

25. $(-6)(+2)$ **26.** $(0)(-7)$ **27.** $\left(-\tfrac{7}{8}\right)\left(-\tfrac{3}{5}\right)$

28. $(-2)(-3)(-4)$ **29.** $(-6)(-6)(-6)$

554   Divide.

30. $\dfrac{-72}{-9}$ **31.** $\dfrac{+36}{-4}$ **32.** $\dfrac{-24}{-0.24}$

33. $(-63) \div (-9)$

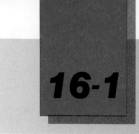

16-1 Algebraic Expressions

A **numerical expression** consists of a single number, or two or more numbers joined by operational symbols.

> 18; 7 + 5; 28 − 14; 63 ÷ 7; and 8 × (3 + 11) are numerical expressions.

A symbol that holds a place open for a number is called a **variable**.

> A; c; b_1; S; and $\triangle$ are variables.

An **algebraic expression** is a mathematical phrase. It expresses a relationship between numbers, variables, and operational symbols.

> b; $a − 5$; and $3y^2 − 7x + 9$ are algebraic expressions.

A number is sometimes called a **constant** because it has a definite value.

In an algebraic expression the multiplication symbol need not be used when the two factors are both variables ($b \times d$ may be expressed as bd) or a number and a variable ($3 \times c$ may be expressed as $3c$). When a number and a variable are expressed as a product, the number always precedes the variable.

Examples

1. Write an algebraic expression for the product of n and 4.

ANSWER: $4n$

2. Write an algebraic expression for ten times the sum of m and 6.

ANSWER: $10(m + 6)$

Think About It

1. Write the following as algebraic expressions.
   a. the sum of the squares of x and y
   b. the square of the sum of x and y

2. For what values of x and y are the algebraic expressions you wrote for Exercises 1a and 1b equivalent?

Practice Exercises

Name the variable in each expression.

1. $14 - x$ **2.** $3y + 36$ **3.** $\frac{z}{5} + 2$ **4.** $b_1 - 70$ **5.** $n(n + 4)$ **6.** $c^3 \times 5$

Write each of the following as an algebraic expression.

7. Six added to four.

8. Twelve times nineteen.

9. From ten subtract two.

10. Fifteen divided by three.

11. The square of nine.

12. The square root of twenty.

13. The sum of eight and three.

14. The difference between six and two.

15. The product of nine and five.

16. The quotient of twelve divided by four.

17. The sum of b and x

18. a times y.

19. The product of g and eight.

20. The difference between c and g.

21. The quotient of d divided by r.

22. Two times the sum of l and w.

23. The cube of r.

24. The square root of s.

25. The product of b and h.

26. Three times the difference between t and nine.

Read, or write in words, each of the following.

27. $5n$ **28.** $19 - 7$ **29.** $y + 25$ **30.** $\frac{m}{x}$

31. cd^2x^3 **32.** $6a - 9b$ **33.** $(m + n)(m - n)$ **34.** $3x^2 - 4xy + 7y^2$

Write each of the following as an algebraic expression.

35. Twice the radius (r).

36. The circumference (C) divided by pi (π).

37. The principal (p) plus the interest (i).

38. 90° decreased by angle B.

39. Three times the side (s).

40. The sum of $m\angle A$, $m\angle B$, and $m\angle C$.

41. The length (l) times the width (w) times the height (h).

42. One half the product of the altitude (a) and base (b).

43. The profit (p) added to the cost (c).

44. Angle B subtracted from 180°.

45. The square of the side (s).

46. The cube of the edge (e).

47. The base (b) multiplied by the height (h) divided by two.

48. The sum of twice the length (l) and twice the width (w).

49. The quotient of the interest (i) for one year divided by the principal (p).

Write an algebraic expression for each of the following.

50. Marilyn has n cents. She spends 95 cents for lunch. How many cents does she have left?

51. Scott is x years old. How old will he be in fifteen years?

52. Steve has d dollars in the bank. If he deposits y dollars, how many dollars will he then have in the bank?

53. How long does it take a train averaging r mi/h to travel m miles?

54. How many notebooks can be bought for x cents if one notebook costs y cents?

An **algebraic sentence** is two or more algebraic expressions joined by equal signs or inequality signs.

Mathematical Sentences
$x - 4 = 9$ "Some number x decreased by 4 is equal to nine."
$8a > 20$ "Eight times each number a is greater than twenty."
$n + 10 < 14$ "Each number n increased by ten is less than fourteen."

An **open sentence** is a mathematical sentence that contains a variable.

An **equation** is an open sentence that has the equal sign (=).

An **inequality** is an open sentence that uses an inequality sign ($<$, $>$, $\leq$, or $\geq$).

Open Sentences	
$x - 4 = 9$ Equation	$n + 10 < 14$ Inequality

A line drawn through symbols like $\neq$, $\not<$, and $\not>$ reverses their meanings.

$x \neq 45$	"Some number x is *not* equal to forty-five."

A **compound sentence** is formed by joining two simple sentences with "or" or "and."

Compound Sentences
$6n \leq 30$ means $6n < 30$ **or** $6n = 30$.
$8 < x < 12$ means $8 < x$ **and** $x < 12$.

Examples

1. Read, or write in words, $b - 6 > 25$.

ANSWER: Some number b decreased by 6 is greater than 25.

2. Write "some number x increased by 12 is less than 15" using symbols.

ANSWER: $x + 12 < 15$

3. Read, or write in words, $4y \geq 12$.

ANSWER: $4y$ is greater than 12 or $4y$ is equal to 12

1. How does an algebraic sentence differ from an algebraic expression?

2. Use the numbers 5 and 9 and the variable x to write two different equations and two different inequalities.

Practice Exercises

Read, or write in words, each of the following.

1. $n < 16$

2. $x > 29$

3. $3y \neq 17$

4. $a - 5 < 53$

5. $7t \not< -14$

6. $9m + 4 = 15$

7. $8c - 9 \not> c + 7$

8. $2n + 6 < 3n - 8$

9. $\frac{b}{5} > 20$

10. $15x - 4 \not> 18 - 7x$

Write each of the following as an algebraic sentence using symbols.

11. Some number x increased by ten is equal to forty-one.

12. Each number n decreased by four is less than eighteen.

13. Four times each number t is greater than twelve.

14. Nine times each number y plus two is not equal to fifty.

15. Each number a increased by five is not less than nine.

16. Each number m divided by seven is not greater than fourteen.

17. Seven times each number x is greater than or equal to thirty.

18. Each number c plus nine is less than or equal to ten.

19. Each number t is less than negative one and greater than negative five.

20. Twelve times each number y is greater than or equal to four and less than or equal to fifteen.

Read, or write in words each of the following.

21. $h \geq -6$

22. $19w \leq 38$

23. $5d + 7 \leq 12$

24. $n - 16 \geq 4n + 9$

25. $18 > x > 0$

26. $4 < b < 27$

27. $12 \geq 7r \geq -6$

28. $2 < a \leq 45$

Formulas

A **formula** is a special kind of equation. It is a frequently used equation expressing the relationship between two or more quantities by using numbers, variables, and symbols of operation. Often quantities are represented by the first letter of a key word.

> $p = 3s$, $E = IR$, $F = 1.8C$ are formulas.

Examples

1. Write as a formula.

The circumference (C) of a circle is equal to pi (π) times the diameter (d).

> To express mathematical and scientific principles as formulas, write numbers, symbols of operation, and variables to show the relationships between the quantities.

ANSWER: $C = \pi d$

---

2. Write as a word statement.

$a = \frac{360°}{n}$ where a = the measure of a central angle of a

regular polygon and n = the number of sides.

> To write a formula as a word statement, write a word rule stating the relationship expressed by the formula.

ANSWER: The measure of a central angle of a regular polygon equals 360° divided by the number of sides.

Think About It

Suppose you wanted to write a formula that involved more than one quantity beginning with the same first letter (such as "costs," "credits," and "charges"). How could you differentiate these quantities in a formula?

Practice Exercises

Express each of the following as a formula.

1. The perimeter of a square (*p*) is equal to four times the length of the side (*s*).

2. The area of a rectangle (*A*) is equal to the length (*l*) multiplied by the width (*w*).

3. The circumference of a circle (*C*) is equal to twice pi (π) times the radius (*r*)

4. The interest (*i*) is equal to the principal (*p*) times the rate of interest per year (*r*) times the time in years (*t*).

5. The radius of a circle (*r*) is equal to the diameter (*d*) divided by two.

6. The total amount (*A*) is equal to the sum of the principal (*p*) and the interest (*i*)

7. The area of a parallelogram (*A*) is equal to the product of the base (*b*) and height (*h*).

8. The volume of a cube (*V*) is equal to the cube of the edge (*e*).

9. The area of a circle (*A*) is equal to pi (π) times the square of the radius (*r*).

10. The area of a triangle (*A*) is equal to one half the product of the altitude (*a*) and base (*b*).

11. The sum of two complementary angles ($\angle A$ and $\angle B$) equals $90°$.

12. The net price (*n*) is equal to the list price (*l*) less the discount (*d*).

13. The volume of a cylinder (*V*) is equal to one fourth pi (π) times the square of a diameter (*d*) times height (*h*)

14. The perimeter of a rectangle (*p*) is equal to twice the sum of the length (*l*) and width (*w*).

15. The rate of commission (*r*) is equal to the commission (*c*) divided by the sales (*s*).

16. The total area of a cylinder (*A*) is equal to twice pi (π) times the radius (*r*) times the sum of the height (*h*) and the radius.

17. The perimeter of an isosceles triangle (*p*) is equal to the base (*b*) added to twice the length of the equal side (*e*).

18. The area of a trapezoid (*A*) is equal to the height (*h*) times the sum of the two parallel bases, b_1 and b_2, divided by two.

Express each formula as a word statement.

19. $p = 3s$ where p = perimeter of an equilateral triangle and s = length of a side of an equilateral triangle.

20. $E = IR$ where E = electromotive force in volts, I = current in amperes, and R = resistance in ohms.

21. $t = \dfrac{d}{r}$ where t = time of travel, d = distance traveled, and r = average rate of speed.

22. $F = 1.8C + 32$ where F = Fahrenheit temperature reading and C = Celsius temperature reading.

23. $d = \dfrac{1}{2}gt^2$ where d = distance a freely falling body drops, g = acceleration of gravity, t = time of falling.

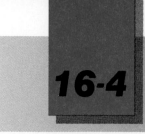

16-4 Evaluating Expressions

To evaluate an expression means to find the value of the expression for given numerical replacement values of the variables. The value of an algebraic expression depends on the numerical values chosen for each variable in the expression. If these values change, the value of the expression usually changes.

Examples

1. Find the value of $2d + x$ when $d = 7$ and $x = 9$.

To evaluate an algebraic expression, follow these steps.

a. Copy the expression. $2d + x$

b. Substitute the given numerical value $2d + x = (2 \cdot 7) + 9$
for each variable.

c. Perform the necessary operations indicated $(2 \cdot 7) + 9 = 14 + 9 = 23$
in the expression.

ANSWER: 23

---

2. Find the value of $\frac{A}{b}$ when $A = 28$ and $b = 4$.

$$\frac{A}{b}$$
$$= \frac{28}{4}$$
$$= 7$$

ANSWER: 7

3. Find the value of $2(l + w)$ when $l = 17$ and $w = 14$

$$2(l + w)$$
$$= 2(17 + 14)$$
$$= 2(31)$$
$$= 62$$

ANSWER: 62

4. Find the value of $c^2 + d^2$ when $c = 8$ and $d = 3$.

$$c^2 + d^2$$
$$= (8)^2 + (3)^2$$
$$= 64 + 9$$
$$= 73$$

ANSWER: 73

Think About It

1. If the value of $4c - d$ is 20, list 3 pairs of possible values for c and d.

2. Use a calculator to evaluate the following expressions where $r = 0.75$, $s = 6.3$, and $t = 2.1$. List, in order, the keys you pressed.

    a. $0.5rs$ **b.** $\frac{17s}{t}$ **c.** $7t - s$

3. Find the value of $2x + 1$ when $x = 2$, when $x = 4$, when $x = 6$, and when $x = 8$. Find the value of $2x + 1$ for three other values of x where x is a positive integer. What is true about all the values you found for $2x + 1$?

Practice Exercises

Find the value of each algebraic expression.

When $a = 12$ and $b = 6$:

1. $a + b$ **2.** $a - b$ **3.** ab **4.** $\frac{a}{b}$

5. a^2 **6.** $7a$ **7.** $4(a - b)$ **8.** $a^2 - b^2$

When $x = 4$ and $y = 3$:

9. $x + 6y$ **10.** $5xy$ **11.** $4x - 8y$ **12.** $x(x - y)$

13. $9\,x^2 y$ **14.** $x^2 - y^2$ **15.** $(x + y)^2$ **16.** $\frac{x + y}{2x - y}$

When $m = -8$, $n = -2$, and $x = 4$:

17. $3mnx$ **18.** $10mn - 7nx$ **19.** $(m + n)(m - n)$

20. $\frac{(n - x)^2}{n^2 - x^2}$ **21.** $m - n(m + n)$ **22.** $5m^2 + 2mx - 3x^2$

23. $4s$ when $s = 6$

24. $2r$ when $r = 12$

25. bh when $b = 10$ and $h = 7$

26. πd when $\pi = 3.14$ and $d = 26$

27. $2\pi r$ when $\pi = \frac{22}{7}$ and $r = 21$

28. lwh when $l = 11$, $w = 7$, and $h = 15$

29. $p + i$ when $p = 500$ and $i = 120$

30. $b + 2e$ when $b = 32$ and $e = 29$

31. $b_1 + b_2$ when $b_1 = 47$ and $b_2 = 35$

32. $2l + 2w$ when $l = 63$ and $w = 54$

33. $90 - B$ when $B = 72$

34. $A - p$ when $A = 214$ and $p = 185$

35. $l - d$ when $l = 67$ and $d = 29$

36. $180 - A$ when $A = 104$

37. $\frac{d}{2}$ when $d = 46$

38. $\frac{ab}{2}$ when $a = 25$ and $b = 30$

39. $\frac{c}{\pi}$ when $c = 12.56$ and $\pi = 3.14$

40. $\frac{BH}{3}$ when $B = 216$ and $h = 17$

41. a^2 when $a = 6$

42. h^2 when $h = 300$

43. πr^2 when $\pi = \frac{22}{7}$ and $r = 28$

44. $0.7854d^2$ when $d = 50$

45. $6s^2$ when $s = 41$

46. e^2 when $e = 19$

47. $\frac{1}{2}ab$ when $a = 72$ and $b = 66$

48. $\frac{1}{3}Bh$ when $B = 135$ and $h = 20$

49. $\frac{1}{4}\pi d^2$ when $\pi = 3.14$ and $d = 55$

50. $\frac{4}{3}\pi r^3$ when $\pi = \frac{22}{7}$ and $r = 42$

51. $a^2 + b^2$ when $a = 11$ and $b = 9$

52. $h^2 - a^2$ when $h = 36$ and $a = 27$

53. $V + gt$ when $V = 150$, $g = 32$, and $t = 7$

54. $p + prt$ when $p = 300$, $r = 0.06$, and $t = 8$

55. $2(l + w)$ when $l = 37$ and $w = 28$

56. $\frac{5}{9}(F - 32)$ when $F = -13$

57. $l(r + R)$ when $l = 8$, $r = 3$, and $R = 10$

58. $(n - 1)d$ when $n = 10$ and $d = 6$

59. $\sqrt{A}$ when $A = 49$

60. $\sqrt{h^2 - a^2}$ when $h = 78$ and $a = 72$

61. The expression $(a + b + c) \div 3$ represents the average of test scores a, b, and c. Marie received scores of 84, 78, and 87 on three math tests. What was her average score?

62. Phil received test scores of 95, 72, 81, 82, and 90 on science quizzes. Write an expression to represent the average of these five scores. Evaluate the expression for Phil's five tests.

16-5 Evaluating Formulas

Formulas are evaluated in the same way as expressions.

Example

Find the value of p when $s = 16$, using the formula $p = 4s$.

To determine the required value when evaluating a formula, follow these steps.

a. Copy the formula and substitute the given values for the variables.

$P = 4s$
$P = 4 \cdot 16$

b. Perform the necessary operations.

$P = 64$

ANSWER: $P = 64$

Think About It

Find a formula in a science book. Identify what each variable, number, and mathematical operation in the formula represents. How might you find values for the variables in the formula?

Practice Exercises

Find the value.

1. d when $r = 85$. Formula: $d = 2r$

2. A when $B = 47$. Formula: $A = 90 - B$

3. A when $b = 32$ and $h = 49$.
Formula: $A = bh$

4. c when $\pi = \frac{22}{7}$ and $d = 84$.
Formula: $c = \pi d$

5. A when $p = 620$ and $i = 53$.
Formula: $A = p + i$

6. P when $l = 73$ and $w = 28$.
Formula: $P = 2l + 2w$

7. A when $p = 150$, $r = 0.08$, and $t = 4$.
Formula: $A = p(1 + rt)$

8. A when $\pi = 3.14$, $r = 48$, and $h = 75$.
Formula: $A = 2\pi rh$

9. A when $s = 50$. Formula: $A = s^2$

10. V when $e = 8$. Formula: $V = e^3$

11. Jon rode his bike at 8 miles per hour for 4 hours. Use the formula $d = rt$ (where d is the distance traveled, r is the rate of travel, and t is the time traveled) to determine how far Jon traveled.

12. Elaine has had 7 hits in her last 20 times at bat. Use the formula $A = \frac{H}{T}$ (where A is the batting average, H is the number of hits, and T is the times at bat) to find Elaine's batting average.

Solving Equations by Replacement

Solving an equation means finding the number represented by the variable that makes the sentence true. Any number that makes a sentence true is called a **solution** of the sentence.

Checking a solution is important. To check whether a number is a solution of a sentence, substitute the number for the variable and simplify. If the resulting sentence is true, the number is a solution.

Equations that have exactly the same solutions are called **equivalent equations.**

> The equations $n + 4 = 20$ and $n = 16$ are equivalent because they both have the same solution, 16.

An equation is in **simplest form** when one member contains only the variable itself and the other member is a constant.

> The equation $n = 16$ is an equation in simplest form.

Examples

1. Is $n = 18$ a solution for $n - 2 = 16$?

To solve an equation by replacement, follow these steps.
a. Copy the equation.

$n - 2 = 16$

b. Substitute the given or chosen number for the variable in the equation.

$18 - 2 = 16$?

c. Simplify. If the resulting sentence is true, the number is a solution. If it is false, the number is not a solution.

$16 = 16$ ✔

ANSWER: $n = 18$ is a solution.

2. Is $n = 12$ a solution for $\frac{n}{6} = 4$?

$$\frac{n}{6} = 4$$
$$\frac{12}{16} = 4?$$
$$2 \neq 4$$

ANSWER: $n = 12$ is not a solution.

Think About It

1. Write an equation with one variable for which there is more than one value for the variable that will make the sentence true.

2. Can you write an equation with one variable for which there is no value for the variable that will make the sentence true?

Practice Exercises

Which value for the variable will make the given sentence true?

1. $x + 4 = 8$ **a.** $x = 12$ **b.** $x = 2$ **c.** $x = -4$ **d.** $x = \frac{1}{2}$ **e.** $x = 4$

2. $n - 2 = 10$ **a.** $n = 8$ **b.** $n = 5$ **c.** $n = 12$ **d.** $n = -8$ **e.** $n = -\frac{1}{5}$

3. $9r = 63$ **a.** $r = 72$ **b.** $r = 54$ **c.** $r = 0$ **d.** $r = 7$ **e.** $r = -6$

4. $\frac{b}{3} = 18$ **a.** $b = 6$ **b.** $b = \frac{1}{6}$ **c.** $b = 15$ **d.** $b = 54$ **e.** $b. = 21$

Which value is a solution of the given equation?

5. $m + 6 = 6$ **a.** $m = 12$ **b.** $m = -12$ **c.** $m = 0$ **d.** $m = 1$ **e.** $m = 6$

6. $8s = 24$ **a.** $s = 16$ **b.** $s = 32$ **c.** $s = -\frac{1}{3}$ **d.** $s = 0$ **e** $s = 3$

7. $d - 9 = 3$ **a.** $d = 6$ **b.** $d = 3$ **c.** $d = -6$ **d.** $d = 12$ **e.** $d = -12$

8. $\frac{a}{8} = 4$ **a.** $a = 2$ **b.** $a = \frac{1}{2}$ **c.** $a = 32$ **d.** $a = 12$ **e.** $a = 4$

9. $12y = 6$ **a.** 6 **b.** 2 **c.** 18 **d.** $\frac{1}{2}$ **e.** -2

10. $n + 7 = 21$ **a.** 3 **b.** 28 **c.** $\frac{1}{3}$ **d.** 14 **e.** -3

11. $\frac{x}{12} = 8$ **a.** 20 **b.** 4 **c.** $\frac{2}{3}$ **d.** 96 **e.** $1\frac{1}{2}$

12. $w - 12 = 12$ **a.** -24 **b.** 0 **c.** 1 **d.** 24 **e.** 144

13. Which of the equations below have 6 as a solution?

14. Which are equivalent equations?

   a. $y - 2 = 8$ **b.** $\frac{y}{3} = 2$ **c.** $9y = 54$ **d.** $y = 6$ **e.** $y + 17 = 23$

15. Which of the equations below have 7 as a solution?

16. Which are equivalent equations?

17. Which one of the equivalent equations is in simplest form?

   a. $9x = 63$ **b.** $x - 4 = 11$ **c.** $x = 7$ **d.** $\frac{x}{7} = 1$ **e.** $x + 15 = 8$

16-7 Properties of Equality

Inverse operations are operations that undo each other. Recall that addition undoes subtraction, subtraction undoes addition, multiplication undoes division, and division undoes multiplication.

$$
\begin{aligned}
(5 + 2) - 2 &= 5 & (8 - 6) + 6 &= 8 \\
(4 \times 3) \div 3 &= 4 & (14 \div 7) \times 7 &= 14
\end{aligned}
$$

The **properties of equality** allow you to use inverse operations to obtain equations in simplest form, which give you the solutions to the equations.

Properties of Equality

I When you subtract the same number from both sides of an equation, the result is an equivalent equation.

II When you add the same number to both sides of an equation, the result is an equivalent equation.

III When you multiply both sides of an equation by the same nonzero number, the result is an equivalent equation.

IV When you divide both sides of an equation by the same nonzero number, the result is an equivalent equation.

These properties of equality tell you that the results are equal when equals are increased, decreased, multiplied, or divided by the same number. Division by zero is excluded.

Examples

1. Solve. $n + 2 = 12$

To solve an equation by using properties of equality, follow these steps.

a. Copy the equation.

$n + 2 = 12$

b. Decide what operation will undo the operation in the equation.

subtraction

c. Choose the number that must be used with the operation.

(2)

d. Operate with this number on both sides of the equation.

$n + 2 - 2 = 12 - 2$

e. Simplify.

$n = 10$

ANSWER: $n = 10$

---

2. Solve. $n - 2 = 12$

$n - 2 = 12$

$(n - 2) + 2 = 12 + 2$

$n = 14$

ANSWER: $n = 14$

3. Solve. $2n = 12$

$2n = 12$

$\frac{2n}{2} = \frac{12}{2}$

$n = 6$

ANSWER: $n = 6$

4. Solve. $\frac{n}{2} = 12$

$\frac{n}{2} = 12$

$2 \cdot \frac{n}{2} = 2 \cdot 12$

$n = 24$

ANSWER: $n = 24$

Think About It

1. Why is it necessary to perform the same operation on both sides of an equation? What would happen if you didn't?

2. How does a balance scale demonstrate the properties of equality?

Practice Exercises

Find the missing number (▓) and, where required, the missing operation (?).

1. $(11 + 8) - ▓ = 11$

2. $(16 + 22) \,?\, ▓ = 16$

3. $(n + 5) \,?\, ▓ = n$

4. $(9 \times 7) \div ▓ = 9$

5. $(54 \times 31) \,?\, ▓ = 54$

6. $(h \times 6) \,?\, ▓ = h$

7. $(26 - 5) + ▓ = 26$

8. $(40 - 18) \,?\, ▓ = 40$

9. $(x - 12) \,?\, ▓ = x$

10. $(3 \times 8) \div ▓ = 8$

11. $(23 \times 43) \,?\, ▓ = 43$

12. $(7m) \,?\, ▓ = m$

13. $(36 \div 4) \times ▓ = 36$

14. $(19 \div 16) \,?\, ▓ = 19$

15. $(y \div 8) \,?\, ▓ = y$

16. $\frac{7}{12} \times ▓ = 7$

17. $\frac{13}{5} \,?\, ▓ = 13$

18. $\frac{n}{4} \,?\, ▓ = n$

19. $(r - 10) \,?\, ▓ = r$

20. $(16n) \,?\, ▓ = n$

21. $(w + 25) \,?\, ▓ = w$

22. $\left(\frac{b}{21}\right) \,?\, ▓ = b$

23. $(11x) \,?\, ▓ = x$

24. $(t - 37) \,?\, ▓ = t$

In each of the following, tell what number you subtract from both sides of the given equation to get an equivalent equation in simplest form. Write this equivalent equation.

25. $n + 3 = 9$

26. $x + 12 = 27$

27. $y + 9 = 33$

28. $r + 24 = 72$

In each of the following, tell what number you add to both sides of the given equation to get an equivalent equation in simplest form. Write this equivalent equation.

29. $w - 4 = 11$ **30.** $s - 19 = 6$ **31.** $h - 28 = 28$ **32.** $x - 15 = 0$

In each of the following, tell what number you divide both sides of the given equation by to get an equivalent equation in simplest form. Write this equivalent equation.

33. $9m = 72$ **34.** $15y = 90$ **35.** $20b = 35$ **36.** $12p = 7$

In each of the following, tell what number you multiply both sides of the given equation by to get an equivalent equation in simplest form. Write this equivalent equation.

37. $\frac{t}{3} = 6$ **38.** $\frac{m}{8} = 2$ **39.** $\frac{x}{14} = 9$ **40.** $\frac{a}{10} = 25$

Tell what operation with what number you use on both sides of the given equation to get an equivalent equation in simplest form. Write this equivalent equation.

41. $x - 5 = 13$ **42.** $m + 11 = 21$ **43.** $\frac{n}{8} = 7$

44. $\frac{t}{6} = 12$ **45.** $8h = 56$ **46.** $4 = \frac{n}{10}$

47. $d - 110 = 40$ **48.** $42 = x - 28$ **49.** $6a = 42$ $a = 7$

50. $12y = 9$ **51.** $b + 9 = 0$ **52.** $s - 31 = 31$

53. $x + 45 = 108$ **54.** $2 = 8n$ **55.** $75 = a + 17$

Write an equation for each problem. Then solve and check.

57. On Monday, Ted hid $75 in a cookie jar. On Tuesday, he added $15 to the cookie jar. On Wednesday, he took $15 from the cookie jar. How much money remains in the cookie jar?

58. Nancy's $10 allowance was cut in half. After a few weeks, her allowance was doubled. How much was her allowance after it was doubled?

Refresh Your Skills

1. 1–8
514 + 96,225 + 698 + 7,649

2. 1–9
631,036 − 59,838

3. 1–11
6,009 × 908

4. 1–13
$1{,}728\overline{)100{,}224}$

5. 2–5
0.58 + 3.69 + 0.08 + 16.93

6. 2–6
0.05 − 0.005

7. 2–7
2.4 × 0.009

8. 2–9
$1.2\overline{)0.852}$

9. 3–8
$2\frac{3}{5} + 1\frac{1}{3}$

10. 3–9
$10\frac{3}{8} - 6\frac{5}{6}$

11. 3–11
$60 \times 3\frac{11}{12}$

12. 3–12
$2\frac{2}{3} \div 4\frac{1}{2}$

13. Find $10\frac{3}{4}\%$ of 800
4–7

14. 16 is what percent of 40?
4–6

15. 100% of what number is 2?
4–8

16-8 Solving Equations by Subtraction

One way of solving equations is to use the properties of equality and inverse operations. It is important to check your answer when you finish. When the operation indicated in the equation is addition, the inverse operation is subtraction.

Example

Solve and check. $n + 4 = 20$

To solve equations involving addition, follow these steps.

a. Copy the equation.

$n + 4 = 20$

b. Subtract from both sides the number that is on the same side of the equation as the variable.

$n + 4 - 4 = 20 - 4$

c. Simplify the equation.

$n = 16$

d. Check your answer by substituting it for the variable in the equation.

Check: Does $16 + 4 = 20$?

$20 = 20$ ✔

ANSWER: $n = 16$

The equations $4 + n = 20$, $20 = n + 4$, and $20 = 4 + n$ are all solved by subtracting 4 from both sides of the equation. The solution, $16 = n$, can be rewritten as $n = 16$.

Practice Exercises

Solve and check.

1. $c + 6 = 14$ **2.** $b + 18 = 31$ **3.** $i + 37 = 63$ **4.** $48 + w = 75$

5. $29 + p = 106$ **6.** $54 + n = 162$ **7.** $67 = a + 28$ **8.** $50 = m + 33$

Write an equation for each problem. Then solve and check.

9. If Maria had 3 more pictures, she would have a total of 20 pictures. How many pictures does Maria have?

10. The bank credited Chou's savings account with $5.50 in interest. If his new balance is $210.25, how much did he have in his account before the interest was credited?

16-9 Solving Equations by Addition

When the operation indicated in the equation is subtraction, the inverse operation is addition.

Example

Solve and check. $n - 4 = 20$.

To solve equations involving subtraction, follow these steps.

a. Copy the equation.

$n - 4 = 20$

b. Add to both sides the number that is on the same side of the equation as the variable.

$n - 4 + 4 = 20 + 4$

c. Simplify the equation.

$n = 24$

d. Check your answer by substituting it for the variable in the given equation.

Check: Does $24 - 4 = 20$?

$20 = 20$ ✔

ANSWER: 24

To solve $20 = n - 4$, you also add 4 to both sides. The solution, $24 = n$, can be written as $n = 24$.

Practice Exercises

Solve and check.

1. $c - 7 = 21$

2. $l - 18 = 5$

3. $A - 42 = 124$

4. $35 = n - 19$

5. $16 = b - 47$

6. $28 = y - 16$

7. $t - 9 = 0$

8. $0 = y - 26$

9. $-5 = r - 5$

10. $52 = m - 52$

11. $18 = n - 1.3$

12. $9 = b - 2\frac{1}{2}$

Write an equation for each problem. Then solve and check.

13. On his way home from school, Bryon spent $8.25 for a new cassette. How much money did Bryon have before this purchase if he now has $14.10?

14. Carla bought some ribbon to decorate a costume. She used $3\frac{1}{2}$ yards for the costume. How much ribbon did Carla buy if she has $2\frac{3}{4}$ yards left?

16-10 Solving Equations by Division

When the operation indicated in the equation is multiplication, the inverse operation is division.

Example

Solve and check. $4n = 20$

To solve equations involving multiplication, follow these steps.

a. Copy the equation.

$4n = 20$

b. Divide both sides by the number by which the variable is multiplied in the given equation.

$$\frac{4n}{4} = \frac{20}{4}$$

c. Simplify the equation.

$n = 5$

d. Check your answer by substituting it for the variable in the given equation.

Check: Does $(4)(5) = 20$?

$20 = 20$ ✔

ANSWER: 5

Think About It

1. Write an equation involving multiplication that has the solution $n = 6$.

2. Use a calculator to solve the following equations. List, in order, the keys you pressed.

a. $2.5x = 5.5$ **b.** $-0.67y = 2.077$

To solve $20 = 4n$, you also divide both sides by 4. The solution, $5 = n$, can be written as $n = 5$.

Practice Exercises

Solve and check.

1. $5a = 45$ **2.** $9c = 54$ **3.** $14n = 42$ **4.** $91 = 7y$

5. $60 = 12x$ **6.** $105 = 15b$ **7.** $2r = 18$ **8.** $48 = 4s$

9. $12m = 300$ **10.** $140 = 15y$ **11.** $10c = -50$ **12.** $10 = -50c$

Write an equation for each problem. Then solve and check.

13. If Robert had three times as many baseball cards as he now has, he would have 102 cards. How many baseball cards does Robert have?

14. Three-fourths of the students plan to go on the field trip to the museum. If 120 students go on the trip, how many students are there in the school?

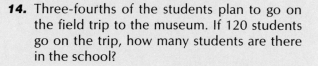

16-11 Solving Equations by Multiplication

When the operation indicated in the equation is division, the inverse operation is multiplication.

Example

Solve and check. $\frac{n}{4} = 20$

To solve equations involving division, follow these steps.

a. Copy the equation.

$\frac{n}{4} = 20$

b. Multiply both sides by the number by which the variable is divided in the given equation.

$4 \cdot \frac{n}{4} = 4 \cdot 20$

c. Simplify the equation.

$n = 80$

d. Check your answer by substituting it for the variable in the given equation.

Check: Does $\frac{80}{4} = 20$?

$20 = 20$ ✔

ANSWER: 80

To solve $20 = \frac{n}{4}$, you still multiply both sides by 4. The solution, $80 = n$, can also be written as $n = 80$.

Practice Exercises

Solve and check.

1. $\frac{d}{2} = 19$ **2.** $\frac{s}{4} = 24$ **3.** $\frac{b}{7} = 3$ **4.** $\frac{n}{5} = 0$

5. $\frac{t}{8} = 8$ **6.** $12 = \frac{b}{3}$ **7.** $45 = \frac{m}{5}$ **8.** $1 = \frac{s}{9}$

9. $-8 = \frac{d}{-11}$ **10.** $\frac{x}{6} = 1$ **11.** $-5 = \frac{z}{7}$ **12.** $\frac{b}{16} = 16$

Write an equation for each problem. Then solve and check.

13. If Billy divided his fish equally into three tanks, he would have six fish in each tank. How many fish does Billy have?

14. Headphones cost $\frac{1}{8}$ the price of a stereo. The headphones cost $23.95. How much does the stereo cost?

16-12 Solving Equations Using Two Operations

The solution of some equations requires more than one operation.

Example

1. Solve and check. $7x + 2 = 72$

To solve equations using two operations, follow these steps.

a. Copy the equation.

$7x + 2 = 72$

b. Simplify the equation by combining like terms (two or more terms with the same variable) where possible.

c. Undo any addition or subtraction by the inverse operation.

$7x + 2 - 2 = 72 - 2$

d. Simplify.

$7x = 70$

e. Undo multiplication or division by the inverse operation.

$\frac{7x}{7} = \frac{70}{7}$

f. Simplify.

$x = 10$

g. Check the answer in the original equation.

Does $7(10) + 2 = 72$?

$70 + 2 = 72$?

$72 = 72$ ✔

ANSWER: $x = 10$

Think About It

Why is it necessary to "undo" the addition/subtraction operation before "undoing" the multiplication/division operation?

Practice Exercises

Solve and check.

1. $2b + 18 = 46$

2. $3 + 5n = 62$

3. $51 = 9y + 6$

4. $120 = 76 + 2w$

5. $22h + 154 = 374$

6. $1.8c + 32 = 113$

7. $7t - 11 = 38$

8. $0 = 10n - 30$

9. $6a - 7 = 29$

10. $37 = 14b - 19$

11. $n + n = 50$

12. $9x + 3x = 84$

13. $p + 0.06p = 689$

14. $7y - y = 18$

15. $a - 0.05a = 760$

16. $6b = 91 - 15$

17. $82 = 5a + 9a$

18. $1.8c + 32 = 50$

19. $l - 0.35l = 195$

20. $\frac{3}{5}b = 21$

16-13 Using Formulas

Sometimes a formula is used to solve a problem. The solution to the problem is found by solving the formula for the unknown quantity.

Example

Find the value of w when $A = 54$ and $l = 9$, using the formula $A = lw$.

To use a formula to solve a problem, follow these steps.

a. Copy the formula.

$A = lw$

b. Substitute the given values for the variables.

$54 = 9w$

c. Perform the necessary operations.

$\frac{54}{9} = \frac{9w}{9}$

d. Solve the resulting equation for the value of the unknown variable.

$6 = w$ or $w = 6$

e. Check in the formula.

Check. $A = lw$ Does $54 = 9(6)$?

$54 = 54$ ✔

ANSWER: $w = 6$

Think About It

1. Suppose you substitute all the given values in a formula and still have two unknown quantities in the equation. Can you solve the equation? Why?

2. Sometimes formulas require numerical values that need to be expressed in specific units, such as feet, or days, or degrees Fahrenheit. What would happen if you used the wrong unit with these numerical values?

Practice Exercises

Find the value.

1. s when $P = 18$. Formula: $P = 3s$

2. A when $C = 42$. Formula: $C = A - 273$

3. d when $C = 15.7$ and $\pi = 3.14$. Formula: $C = \pi d$

4. t when $i = 140$, $p = 400$, and $r = 5\%$. Formula: $i = prt$

Write an equation for each problem. Then solve and check.

5. The area of a square garden is 49 square feet. What is the length of one side of the garden? (Use the formula $A = s^2$ where A is the area of a square and s is the length of a side of the square.)

6. One angle of a triangle measures 40°; another angle measures 50°. What is the measure of the third angle? (Use the formula $a + b + c = 180°$ where a, b, and c are the measures of the three angles of a triangle.)

Using Mathematics

GRAPHING AN EQUATION ON THE NUMBER LINE

Once you have solved an equation, you can list its solution(s) or show them on a graph.

> The graph of an equation in one variable on a number line is the collection of all points on the number line whose coordinates are solutions of the equation.

Example

Draw the graph of $n + 2 = 8$.

To draw the graph of an equation, follow these steps.

a. Solve the given equation.

$$n + 2 = 8$$
$$n = 6$$

b. Locate the point on the number line that corresponds to the solution and indicate the point by a **solid dot**.

Exercises

For each of the following equations, draw an appropriate number line, then graph its solution.

1. $x = 5$ **2.** $b = -3$ **3.** $x + 4 = 7$

4. $a - 5 = 2$ **5.** $8a = 32$ **6.** $3y + 8 = 5$

7. $6n - n = 15$ **8.** $\frac{b}{3} = 2$ **9.** $5b - 6 = 29$

10. $|x| = 3$ **11.** $z = z + 2$ **12.** $5(x + 2) = 5x + 10$

Write a corresponding equation pictured by each graph.

13.

14.

15.

16.

Problem Solving by the Equation Method

Building Problem Solving Skills

Equations may be used to find:

I A fractional part or decimal part or percent of a number.

II What fractional part or decimal part or percent one number is of another.

III A number when a fractional part or decimal part or percent of it is known.

To solve problems using the equation method, follow these steps.

Think About It

Does every fractional part of a number have an equivalent decimal part and an equivalent percent? Why or why not?

1. **READ** the problem carefully to find the facts related to the number you wish to determine.

2. **PLAN**

a. Represent the unknown number by a variable, usually a letter.

b. Form an equation by translating two equal facts, with at least one containing the unknown, into algebraic expressions and writing one expression on each side of the equal sign.

3. **SOLVE** the equation.

4. **CHECK** the answer directly with the facts in the given problem.

Examples

I Find $\frac{3}{4}$ of 20.

$\frac{3}{4} \times 20 = n$

$15 = n$ or $n = 15$

ANSWER: 15

Find 0.75 of 20.

$0.75 \times 20 = n$

$15 = n$ or $n = 15$

ANSWER: 15

Find 75% of 20.

$75\% \times 20 = n$

$0.75 \times 20 = n$

$15 = n$ or $n = 15$

ANSWER: 15

II What fractional part of 20 is 15?

$$n \times 20 = 15$$
$$20n = 15$$
$$\frac{20n}{20} = \frac{15}{20}$$
$$n = \frac{3}{4}$$

ANSWER: $\frac{3}{4}$

What decimal part of 20 is 15?

$$n \times 20 = 15$$
$$20n = 15$$
$$\frac{20n}{20} = \frac{15}{20}$$
$$n = \frac{3}{4} = 0.75$$

ANSWER: 0.75

What percent of 20 is 15?

$$n\% \times 20 = 15$$
$$\frac{n}{100} \times 20 = 15$$
$$\frac{n}{5} = 15$$
$$n = 75$$

ANSWER: 75%

III $\frac{3}{4}$ of what number is 15?

$$\frac{3}{4} \times n = 15$$
$$\frac{3}{4}n = 15$$
$$\frac{4}{3} \cdot \frac{3}{4}n = \frac{4}{3} \cdot 15$$
$$n = 20$$

ANSWER: 20

0.75 of what number is 15?

$$0.75 \times n = 15$$
$$0.75n = 15$$
$$\frac{0.75n}{0.75} = \frac{15}{0.75}$$
$$n = 20$$

ANSWER: 20

75% of what number is 15?

$$75\% \times n = 15$$
$$0.75n = 15$$
$$\frac{0.75n}{0.75} = \frac{15}{0.75}$$
$$n = 20$$

ANSWER: 20

Practice Exercises

Use the equation method to find each required number.

1. a. $\frac{1}{3}$ of 54
b. $\frac{3}{8}$ of 72
c. $\frac{2}{5}$ of 140
d. $\frac{5}{6}$ of 258

2. a. 0.4 of 7
b. 0.75 of 92
c. 0.36 of 250
d. 0.625 of 496

3. a. 8% of 50
b. 60% of 18
c. 25% of 392
d. 140% of 475

4. a. What percent of 50 is 9?
b. 30 is what percent of 48?
c. 28 is what percent of 40?
d. What percent of 300 is 57?

5. a. $\frac{1}{3}$ of what number is 29?
b. $\frac{4}{5}$ of what number is 56?
c. 63 is $\frac{7}{12}$ of what number?
d. $1\frac{3}{4}$ times what number is 84?

6. a. 0.06 of what number is 300?
b. 0.375 of what number is 96?
c. 54 is 0.9 of what number?
d. 1.04 of what number is 364?

7. a. 25% of what number is 18?
b. 8% of what number is 52?
c. 37 is 10% of what number?
d. 477 is 1.06% of what number?

8. a. If the school basketball team won 14 games, or $\frac{2}{3}$ of the games played, how many games were played?

b. José received 0.54 of all votes cast in the election for school president. If he received 405 votes, how many students voted?

c. If 817 students, or 95% of the school enrollment, were promoted to the next grade, how many students were enrolled in the school?

d. There are 18 boys in a certain mathematics class. If this represents $\frac{3}{5}$ of the enrollment, how many girls are in the class?

16-15 Solving Inequalities

To solve an inequality in one variable means to find all the numbers that will make the inequality a true sentence. The solution depends on the replacements allowed for the variable.

- When the replacements for the variable are all the *natural numbers,* the solutions of $n < 4$ are the natural numbers less than 4.

  Therefore, $n = 1, 2,$ or 3.

- When the replacements for the variable are all the *whole numbers,* the solutions of $x > 11$ are the whole numbers greater than 11.

  Therefore, $x = 12, 13, 14, \ldots$

- When the replacements for the variable are all the *rational numbers,* the solutions of $y \neq 6$ are all the rational numbers except 6.

  Therefore, $y =$ every rational number except 6.

- When the replacements for the variable are all the *integers,* the solutions of $b \geq -3$ are -3 and all the integers greater than -3.

  Therefore, $b = -3, -2, -1, 0, 1, 2, \ldots$

- The solutions of $c \leq 5$ do not exist when the replacements for the variable are 6, 7, 8, 9, and 10.

  Therefore, c has no solution.

- When the replacements for the variable are all the *integers,* the solutions of $x \not> 2$ are the same as the solutions of $x \leq 2$: 2 and all the integers less than 2.

  Therefore $x = \ldots, -2, -1, 0, 1, 2.$

- When the replacements for the variable are all the *rational numbers,* the solutions of $n \not< 4$ are the same as the solutions of $n \geq 4$: 4 and all the rational numbers greater than 4.

  Therefore, $n = 4$ or any rational number greater than 4.

Properties of Inequalities

  I The same number can be added to or subtracted from both sides of an inequality without changing the order of the inequality.

II Both sides of an inequality can be multiplied or divided by the same positive number without changing the order of the inequality.

III When both sides of an inequality are multiplied or divided by the same negative number, an inequality of the reverse order results.

Examples

I Solve $n + 2 < 6$ when the replacements for the variable are all the whole numbers.

$$n + 2 < 6$$
$$n + 2 - 2 < 6 - 2$$
$$n < 4$$
$$n = 0, 1, 2, \text{ or } 3$$

ANSWER: 0, 1, 2, or 3

Solve $x - 2 > 3$ when the replacements for the variable are all the rational numbers.

$$x - 2 > 3$$
$$x - 2 + 2 > 3 + 2$$
$$x > 5$$
$$x = \text{all rational numbers}$$
$$\text{greater than 5}$$

ANSWER: All rational numbers greater than 5.

II Solve $3c < 27$ when the replacements for the variable are all the natural numbers.

$$3c < 27$$
$$\frac{3c}{3} < \frac{27}{3}$$
$$c < 9$$
$$c = 1, 2, 3, \ldots, 8$$

ANSWER: 1, 2, 3, . . . , 8

Solve $4y + 7 \geq 7$ when the replacements for the variable are $-2, -1, 0, 1,$ and 2.

$$4y + 7 \geq 7$$
$$4y + 7 - 7 \geq 7 - 7$$
$$4y \geq 0$$
$$y \geq 0$$
$$y = 0, 1, \text{ or } 2$$

ANSWER: 0, 1, or 2

III Solve $-5\,x > 30$ when the replacements for the variable are all the integers.

$$-5x > 30$$
$$\frac{-5x}{-5} < \frac{30}{-5} \quad \textbf{reverse order}$$
$$x < -6$$
$$x = \ldots, -9, -8, -7$$

ANSWER: . . . , −9, −8, −7

Solve $-2b < -6$ when the replacements for the variable are all the integers.

$$-2b < -6$$
$$\frac{-2b}{-2} > \frac{-6}{-2}$$
$$b > 3$$
$$b = 4, 5, 6, \ldots$$

ANSWER: 4, 5, 6, . . .

Think About It

1. Why is it necessary to reverse the inequality sign when you multiply or divide by a negative number? Give an example.

2. Would it be possible to solve an inequality if the variable were in the denominator of a fraction? Why?

Practice Exercises

Find the solutions of each inequality when the replacement values for the variable are all the rational numbers.

I
1. a. $n + 7 > 10$ **b.** $a + 6 > 6$ **c.** $23 + d > 12$

2. a. $c - 5 > 18$ **b.** $x - 3 > -4$ **c.** $r - \frac{1}{2} > 3\frac{1}{2}$

3. a. $b + 1 < 7$ **b.** $g + 9 < 2$ **c.** $m + 0.5 < 2.4$

4. a. $d - 9 < 6$ **b.** $t - 10 < 18$ **c.** $h - 3 < -5$

5. a. $m + 8 \geq 12$ **b.** $f + 17 \geq 0$ **c.** $w + 8 \not< -8$

6. a. $x - 11 \geq 4$ **b.** $k - 3.5 \geq 8.7$ **c.** $a - 7 \not< -4$

7. a. $y + 2 \leq 9$ **b.** $e + \frac{1}{4} \not> \frac{3}{4}$ **c.** $v + 16 \leq 13$

8. a. $r - 4 \leq 10$ **b.** $c - 6 \not> -8$ **c.** $x - 9 \leq 9$

9. a. $s + 9 \neq 17$ **b.** $n + 0.3 \neq 1.6$ **c.** $b + 12 \neq 1$

10. a. $w - 11 \neq 6$ **b.** $y - \frac{7}{8} \neq 1\frac{1}{2}$ **c.** $r - 2 \neq -2$

II
11. a. $7x > 14$ **b.** $9y > -72$ **c.** $8a > 2$

12. a. $\frac{m}{4} > 2$ **b.** $\frac{c}{10} > 7$ **c.** $\frac{r}{9} > -6$

13. a. $6z < 30$ **b.** $12s < -96$ **c.** $16b < 24$

14. a. $\frac{h}{8} < 3$ **b.** $\frac{p}{15} < -1$ **c.** $\frac{w}{7} < 5$

15. a. $10n \geq 50$ **b.** $18a \geq 12$ **c.** $\frac{3}{4}x \not< 24$

16. a. $\frac{b}{30} \geq 0$ **b.** $\frac{r}{4} \geq -6$ **c.** $\frac{z}{5} \not< 3$

17. a. $8x \leq 72$ **b.** $48h \not> 96$ **c.** $0.05n \leq 20$

18. a. $\frac{y}{6} \leq 7$ **b.** $\frac{t}{14} \not> 2$ **c.** $\frac{b}{3} \leq -10$

19. a. $9c \neq 144$ **a.** $\frac{5}{8}x \neq 80$ **c.** $24f \neq 0$

20. a. $\frac{n}{12} \neq 5$ **b.** $\frac{d}{6} \neq -2$ **c.** $\frac{y}{10} \neq 9$

III
21. a. $-5d > 30$ **b.** $-7c > 98$ **c.** $-0.4h > 1.2$

22. a. $-4b > -28$ **b.** $-3m > -40$ **c.** $-\frac{3}{8}z > -15$

23. a. $-9x < 81$ **b.** $-16r < 96$ **c.** $-8a < 36$

24. a. $-2t < -16$ **b.** $-17y < -51$ **c.** $-1.5m < -9$

25. a. $-3y \geq 21$ **b.** $-10r \geq 0$ **c.** $-x \not< 2$

26. a. $-7n \geq -63$ **b.** $-\frac{4}{9}g \geq -72$ **c.** $-4t \not< -17$

27. a. $-9s \leq 180$ **b.** $-3y \not> 0$ **c.** $-18x \leq 90$

28. a. $-6z \leq -216$ **b.** $-7n \not> 63$ **c.** $-\frac{2}{3}y \leq -60$

29. a. $\frac{n}{-3} < 12$ **b.** $\frac{b}{-6} > -2$ **c.** $\frac{y}{-8} \leq 6$

30. a. $\frac{c}{-10} \geq 9$ **b.** $\frac{a}{-3} < -7$ **c.** $\frac{x}{-5} > 8$

Write an inequality to solve the following problems.

31. If Bob has no more than twelve books checked out from the public library, how many books could he have?

32. Roberta's savings account went over $700 when the bank added $10.50 in interest to her account. How much money did Roberta have in the account before the interest was added?

Enrichment

Find the solutions of each inequality when the replacement values for the variable are all the rational numbers.

1. $2x + 7 < 15$

2. $5a - 3 > -28$

3. $6m + 1 \leq 13$

4. $7s - 5 \geq 9$

5. $8n + 3 \neq -61$

6. $3y - 7 < 20$

7. $10b + 5 > 0$

8. $17r - 9 \not< 42$

9. $9t + 6 < -57$

10. $14z - 8 \not> 69$

11. $7w + 15 > 113$

12. $4x - 11 \neq -9$

13. $3x + 2x > 15$

14. $9c - 3c < -72$

15. $5d - 7d \geq 14$

16. $y + y \neq -30$

17. $8m - 3m \leq 0$

18. $6t - 11t > -85$

19. $12b + 13b < -100$

20. $4x - 8x \not< 36$

21. $n - 2n \not> -5$

22. $10r + 3r \neq 65$

23. $12w - 5w > -84$

24. $7a - 15a < 144$

Find the solutions of each of the following inequalities. Select only the solutions that are included in the set of replacements.

25. When replacements for the variable are 0, 1, 2, 3, 4, 5, 6, 7, 8, 9:

a. $x + 5 > 8$ **b.** $n - 6 < 1$ **c.** $2b \leq 6$ **d.** $-3t \not< -12$

26. When replacements for the variable are $-3, -2, -1, 0, 1, 2, 3$:

a. $b - 2 \neq 0$ **b.** $5n + 3 < 13$ **c.** $\frac{d}{6} > -1$ **d.** $2a - 5 \leq -1$

27. When replacements for the variable are all the prime numbers:

a. $12d - 4d < 56$ **b.** $m - 7 \not> 16$ **c.** $6c - c \neq 55$ **d.** $-2n > -8$

28. When the replacements for the variable are all the integers:

a. $d + 5 < -2$ **b.** $-7s \geq 35$ **c.** $\frac{y}{-4} \not> -5$ **d.** $4a - 7 \neq -15$

29. When the replacements are all the even whole numbers:

a. $6b < 45$ **b.** $2a - 5 > 9$ **c.** $8a - 2a \not> 18$ **d.** $\frac{d}{9} \not< 4$

30. When the replacements are all the non-positive integers:

a. $m - 1 > -6$ **b.** $-5x < -20$ **c.** $4t - 9 \not> -13$ **d.** $11y - y \geq -10$

Using Mathematics

GRAPHING AN INEQUALITY ON THE NUMBER LINE

The graph of an inequality in one variable is the collection of all points whose coordinates are solutions of the given inequality.

> The graph of an inequality may be a line, a ray, a line segment, or an interval. An **open dot** indicates that a point is not part of the graph. A **solid dot** indicates that it is part of the graph.

Graph of $x > 2$

Graph of $-3 \leq x < 2$

Example

Draw the graph of $4x - 9 < 3$.

To draw the graph of an inequality, follow these steps.

a. Solve the inequality.

$$4x - 9 < 3$$
$$4x < 12$$
$$x < 3$$

b. Draw the graph.

Exercises

For each of the following inequalities, draw an appropriate number line, then graph its solutions.

1. $a > -1$

2. $a + 5 < 7$

3. $\frac{x}{2} \leq 2$

4. $2x - 1 \neq 1$

5. $-9x > 18$

6. $n + 2n \not> -12$

7. $-7y \not< -7$

8. $2b - 5b > -12$

9. $\frac{m}{3} \geq -1$

Draw the graph for each of the following on a number line.

10. $-3 < x < 4$

11. $-1 \leq x < 3$

12. $2 \geq x \geq -5$

Write an inequality corresponding to each graph.

13.

14.

15.

16.

17.

18.

Making Decisions

HOME FIX-UP

The four outside walls and the two gables on the ends of your house need painting. You decide to get four estimates and to find out how much it would cost to paint the house yourself. You measure and record the outside dimensions of your house.

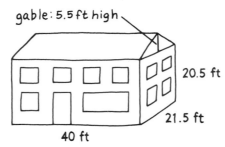

gable: 5.5 ft high

20.5 ft

21.5 ft

40 ft

You estimate that there are about 450 square feet of windows and doors on the outside of the house.

You call four different contractors for estimates and record this information.

Contractor	Estimate	Comments	Discounts/ Warranties
Jones Painting	$5,000	Latex paint, 2 or more coats as needed. Will replace all broken windows and rotting wood. Will sand and powerwash.	$1,000 off if you sign contract within 1 week. 6-year warranty.
Paint-Nu	$2,887	Oil paint, 1 coat. Will replace broken windows and rotting wood for additional charge. Will scrape, powerwash, and prime.	$100 discount for cash.
Rivera Bros.	$3,445	Latex paint, 1 coat. Will replace broken windows and rotting wood for additional charge. Will powerwash and apply 1 coat Peelstop.	1-year warranty
Thom Johnson	$2,000	Latex paint, 1 coat. Will replace all rotting wood. Will scrape and powerwash.	

Use the following questions to help you decide whether to hire one of the contractors or to paint the house yourself. Explain your answers.

1. What is the surface area to be painted?

2. Latex paint costs $21.50 per gallon. One gallon of paint covers 400 square feet. How much will it cost to buy enough paint to apply one coat to the outside of the house?

3. What other costs might you have if you paint the house yourself?

4. Complete a chart like the one below to show the advantages and the disadvantages in hiring each of the contractors and in doing the painting yourself.

5. Your house has several broken windows and rotting boards. How might this affect your decision?

6. Thom is a college student who started painting houses this summer. Rivera Bros. gave you names of people whose houses they had painted this summer. The other two contractors gave you names of people whose houses they had painted in the last 3 years. How might these factors affect your decision?

7. What other factors might affect your decision? (For example, do you have time to paint? Do you have the tools you need? Do you plan to sell your house in the near future?)

8. Decide how you will get your house painted. Be prepared to justify your decision!

9. Contact some contractors in your town to find out how much it would cost to have the outside of your house painted.

Contractor	Cost	Advantages	Disadvantages
Jones Painting			
Paint-Nu			
Rivera Bros.			
Thom Johnson			
"You"			

Career Application

AUTOMOTIVE ENGINEER

Car buyers are always looking for better and more efficient automobiles. *Automotive engineers* are constantly working towards building the better car. They do tests on engines and other equipment to see how well they work. Automotive engineers also design new parts to be safer or more efficient. They may do computer simulations (that is, models) or actually build models.

Automotive engineers work with design, production, and sales specialists. They may specialize in certain systems of an automobile, such as the electrical system or pollution control. Automotive engineers are college graduates. Many of them major in mechanical engineering.

1. An automotive engineer is studying how much a metal part expands when it is heated. The formula is
$$l_H - l_C = \triangle l$$
where l_H is the length of the part when it is hot, l_C is the length when it is cold, and $\triangle l$ is the difference. Solve for l_H if $l_C = 12$ mm and $\triangle l = 2$ mm.

2. An automotive engineer is studying the acceleration of a test car. The formula is
$$v_2 = v_1 + at$$
where v_2 is the final velocity, v_1 is the initial velocity, a is the acceleration, and t is the time. Solve the formula for a.

3. The displacement of an automobile engine is given by the formula
$$D = 0.784\ b^2 sn$$
where b is the bore, s is the stroke, and n is the number of cylinders. Solve the formula for s.

4. An automotive engineer is studying the motion of a car. The formula is
$$v = \frac{d}{t}$$
where v is velocity, d is distance, and t is time. Solve for d if $v = 30$ ft/s and $t = 10$ s.

Chapter 16 Review

Vocabulary

algebraic expression p. 560
algebraic sentence p. 562
compound sentence p. 562
constant p. 560
equation p. 562

equivalent equations p. 569
formula p. 564
inequality p. 562
numerical expression p. 560
open sentence p. 562

properties of equality p. 571
simplest form p. 569
solution p. 569
variable p. 560

Page
560 Write each of the following as an algebraic expression.

1. The difference between z and 12.

2. The product of x and y.

3. The sum of two times b and 5.

4. The square root of the sum of u and v.

562 Write each sentence using symbols.

5. Each number t multiplied by 10 is equal to two hundred.

6. Some number c decreased by nine is equal to sixteen.

564 Express each of the following as a formula.

7. The area (A) of a triangle is equal to one-half the product of the base (b), and the height (h).

8. The sum of the measures of two complementary angles C and D equals 90°.

9. Express the following formula as a word statement. $A = s^2$ where A = area of a square and s = length of a side.

566 Find the value of each algebraic expression.

10. $5n - m^2$ when $n = -1$ and $m = 3$

11. $a^2 + b(b - 4)$ when $a = -4$ and $b = 6$

568 **12.** Find the value of S when $\pi = 3.14$, $r = 3$, and $s = 10$, using the formula $S = \pi r(s + r)$.

569 **13.** Which of the following equations have -6 as the solution?

a. $a - 4 = 2$ **b.** $\frac{1}{2}a = -3$ **c.** $x + (-7) = -13$ **d.** $-2c + 6 = 18$

Solve and check.

574 **14.** $r + 30 = 80$ **575** **15.** $n - \frac{2}{7} = \frac{3}{7}$ **576** **16.** $18y = 360$ **577** **17.** $\frac{1}{5} = 15$

578 **18.** $5a - 3a = -8$ **578** **19.** $3z + 5 = -1$ **579** **20.** Find the value of R when $W = 12.8$ and $I = 8$, using the formula $W = I^2R$.

Solve by the equation method.

581 **21.** If it rained 12 days last month, or $\frac{2}{5}$ of the month, how many days were in the month?

583 **22.** Find the solutions of $24 > 4x$ when the replacements are all positive integers.

569 **23.** Equations that have exactly the same solution are called ____?____.

Chapter 16 Test

560 Write each of the following as an algebraic expression.

1. The difference between p and 15.

2. The product of -3 and x.

3. The sum of a and b.

4. Pi (π) times the diameter (d).

562 Write each sentence using symbols.

5. Each number w decreased by one is greater than forty.

6. Some number a increased by fourteen is equal to fifty-two.

7. Five times each number z is less than or equal to fifteen.

564 **8.** Express this sentence as a formula. The volume (V) of a sphere is four-thirds the product of pi (π) and the cube of the radius (r).

566 Find the value of each algebraic expression.

9. $-3x + y$ when $x = -1$ and $y = 4$

10. $2a^2 - ab + b^2$ when $a = -2$ and $b = 3$

568 **11.** Find the value of P when $r = 2$, $h = 6$, and $\pi = 3.14$, using the formula $P = 2(r + h) + \pi r$

569 **12.** Is $n = 8$ a solution for $4n - 2n = 16$?

Solve and check.

574 **13.** $x + 8 = 4$ **14.** $3 + w = 9$

575 **15.** $b - 10 = 12$ **16.** $8 = d - 4$

576 **17.** $-14y = 28$ **18.** $6n = -4$

577 **19.** $\frac{c}{12} = 6$ **20.** $10 = \frac{1}{2}a$

578 **21.** $7b - 3 = 18$ **22.** $9x + 7x = 80$

579 **23.** Find the value of n when $h = 14.4$ and $d = 3$, using the formula $h = 0.4nd^2$

581 **24.** Solve by the equation method. If $\frac{1}{8}$ of the student body, or 120 students, were absent, how many students are there in the whole school?

583 **25.** Find all the solutions of $d - 7 \geq -2$ when the replacements are all the rational numbers.

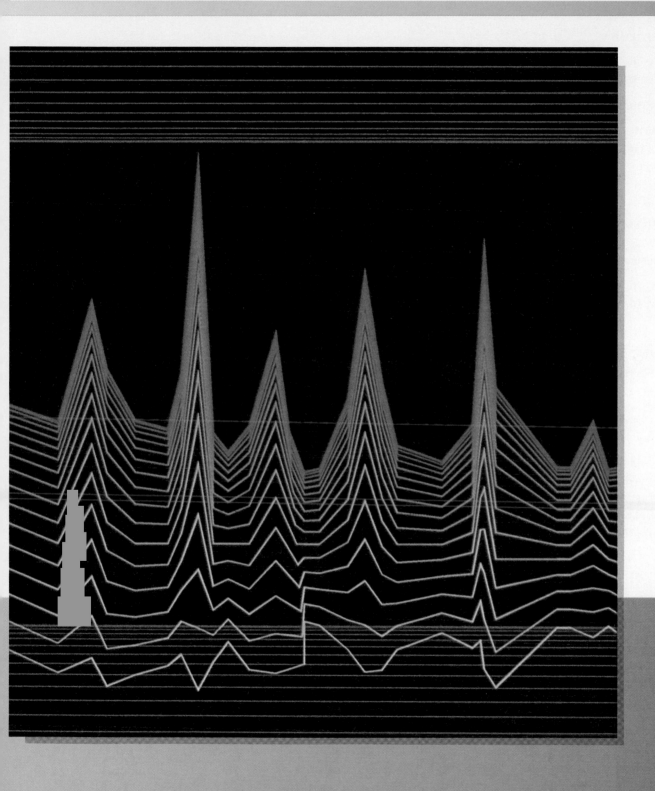

17-1 Ordered Pairs

Example

Write an ordered pair of numbers that has 2 as the first component and 5 as the second component.

To express an ordered pair of numbers, follow these steps.

a. Write the first component, followed by a comma.　　2,

b. Write the second component.　　2,5

c. Enclose them within parentheses.　　(2,5)

first component⌐　　　⌐second component
(2 , 5)

ANSWER: (2,5)

> ### Vocabulary
>
> An **ordered pair of numbers** is two numbers expressed in a definite order so that one number is first (first component) and the other number is second (second component).

Think About It

1. Why is it necessary to separate the numbers in an ordered pair with a comma?

2. Is the ordered pair (3,5) the same as the ordered pair (5,3)? Explain your answer.

Practice Exercises

Write the ordered pair of numbers.

1. 1 as the first component and 3 as the second component.

2. 4 as the first component and 0 as the second component.

3. 7 as the first component and −3 as the second component.

4. −2 as the first component and 5 as the second component.

Write all the ordered pairs of numbers.

5. 3, 4, or 5 as the first component and 4 as the second component.

6. 5 or 6 as the first component and 5 or 6 as the second component.

7. 1, 3, or 5 as the first component and 0 or 1 as the second component.

8. −3, 0, or 3 as the first component and −2, −1, 1, or 2 as the second component.

17-2

Naming Points in the Number Plane

A point in the number plane is identified by an ordered pair of numbers. The x-coordinate of the point is the first component and the y-coordinate of the point is the second component.

Vocabulary

A **number plane** is determined when a vertical number line and a horizontal number line are drawn perpendicular to each other.

The **coordinate axes** are the perpendicular number lines, which are used to locate or plot points in the number plane.

The **x-axis** is the horizontal number line.

The **y-axis** is the vertical number line.

The **x-coordinate** is the number that indicates the horizontal distance, measured parallel to the x-axis, that a point is located to the left or right of the y-axis.

The **y-coordinate** is the number that indicates the vertical distance, measured parallel to the y-axis, that a point is located above or below the x-axis.

The **coordinates** of a point are represented by an ordered pair of numbers in which the x-coordinate is first and y-coordinate is second.

The **origin** is the point where the x-axis and y-axis intersect.

The coordinate axes divide the number plane into four regions, each of which is called a **quadrant**

Example

What are the coordinates of the point on the number plane?

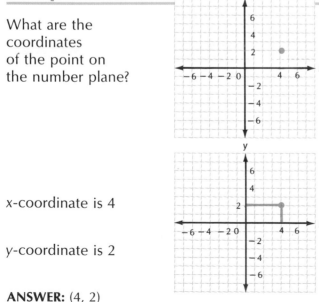

x-coordinate is 4

y-coordinate is 2

ANSWER: (4, 2)

To name a point on a number plane, follow these steps.

a. Draw a perpendicular line from the point to the x-axis. Where the line intersects the x-axis is the x-coordinate of the point.

b. Draw a perpendicular line from the point to the y-axis. Where the line intersects the y-axis is the y-coordinate of the point.

c. Represent the x- and y-coordinates as an ordered pair of numbers in which the x-coordinate is the first component and the y-coordinate is the second component.

In quadrant I, each point has a positive
x-coordinate and a positive y-coordinate.   G(4, 4)

In quadrant II, each point has a negative
x-coordinate and a positive y-coordinate.   D(−5, 1)

In quadrant III, each point has a negative
x-coordinate and a negative y-coordinate.   H(−5, −4)

In quadrant IV, each point has a positive
x-coordinate and a negative y-coordinate.   E(3, −2)

The origin has 0 for both the x-coordinate and
y-coordinate.   I(0, 0)

Any point on the x-axis has 0 for its y-coordinate.   J(4, 0)

Any point on the y-axis has 0 for its x-coordinate.   M(0, −3)

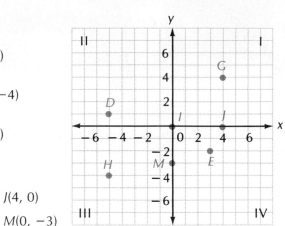

Think About It

1. Why do you think the quadrants are
numbered in the order they are?

2. Why do you need two coordinates to name a
point?

Practice Exercises

Write the coordinates of each indicated point as an ordered pair.

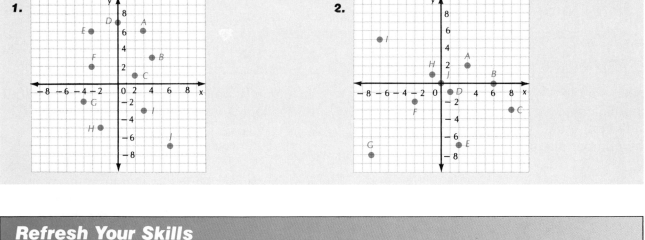

1.

2.

Refresh Your Skills

1. 1–8
  4,726 + 9,867 + 5,388

2. 1–9
  104,003 − 93,985

3. 1–11
  398 × 649

4. 1–13
  5,280)469,920

5. 2–5
  8.62 + 0.97 + 23.88

6. 2–6
  100 − 59.47

7. 2–7
  24 × 61.79

8. 2–8
  15)9

17-3 Plotting Points in the Number Plane

An ordered pair of numbers identifies one point in the number plane.

Examples

1. Plot point K whose coordinates are (2, 3).

x-coordinate is 2.
y-coordinate is 3.

To plot a point in the number plane, follow these steps.

a. Locate the point by using the first component of the given ordered pair as the x-coordinate and the second component as the y-coordinate.

b. Use a dot to indicate the position of the point.

2. Plot point L whose coordinates are (−3,4).

x-coordinate is −3.
y-coordinate is 4.

3. Plot point M whose coordinates are (−2,−6).

x-coordinate is −2.
y-coordinate is −6.

4. Plot point N whose coordinates are (5, 0).

x-coordinate is 5.
y-coordinate is 0.

Think About It

1. The points (4,1), (4,−1), (−4,1) and (−4,−1) name different points. Name three things these points have in common.

2. What can you do if the point you wish to plot has coordinates that are too large or too small for the labels on the graph paper you are using?

Practice Exercises

On graph paper, draw axes and plot the points that have the following coordinates.

1. $A(6, 1)$
2. $B(−8, 4)$
3. $C(5, −3)$
4. $D(−3, −3)$
5. $E(−5, 0)$

6. $F(−3, 7)$
7. $G(4, 9)$
8. $H(0, 5)$
9. $I(1, −6)$
10. $J(−5, −4)$

11. $K(7, 3)$
12. $L(8, −8)$
13. $M(−6, −3)$
14. $N(0, 0)$
15. $O(−3, 2)$

16. $P(0, −1)$
17. $Q(−4, −4)$
18. $R(4, 0)$
19. $S(−7, 6)$
20. $T(6, −9)$

Using Mathematics

LONGITUDE AND LATITUDE

The position of any point on the earth's surface is determined by the intersection of its meridian of longitude and its parallel of latitude. Meridians of longitude are imaginary circles which pass through the North Pole and South Pole. Parallels of latitude are imaginary circles which are parallel to the equator.

The prime meridian from which longitude is calculated is the meridian that passes through Greenwich near London, England. West longitude extends from this prime meridian (0° longitude) westward halfway around the earth to the International Date Line (180° longitude). East longitude extends eastward from the prime meridian to the International Date Line.

The equator is 0° latitude. North latitude is measured north of the equator and south latitude is measured south of the equator. The North Pole is 90° north latitude and the South Pole is 90° south latitude. North latitude is indicated by the letter N, south latitude by S, east longitude by E, and west longitude by W.

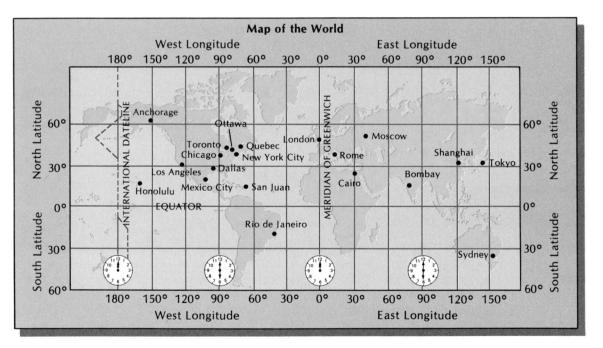

Exercises

Approximate the longitude and latitude for each of the following cities.

1. Sydney　　**2.** London　　**3.** Dallas　　**4.** Los Angeles　**5.** Anchorage　**6.** Chicago

Find the city that is located at each longitude and latitude.

7. 20°N and 100°W　　**8.** 18°N and 66°W　　**9.** 23°S and 43°W　　**10.** 42°N and 12°E

11. 40°N and 74°W　　**12.** 21°N and 158°W　　**13.** 44°N and 80°W　　**14.** 47°N and 71°W

Graphing an Equation in the Number Plane

When you draw the graph of an equation, you make a picture of the solutions to the equation.

Examples

1. Draw the graph of $x + y = 4$.

x				
y				

x	0	2	3	
y				

$$0 + y = 4 \qquad 2 + y = 4 \qquad 3 + y = 4$$
$$y = 4 \qquad\quad y = 2 \qquad\quad y = 1$$

x	0	2	3	
y	4	2	1	

To draw the graph of an equation, follow these steps.

a. Make a table.

b. Select three values for the x-coordinate.

c. Substitute each value of the x-coordinate for the x variable in the equation. Find the corresponding value of the y-coordinate by solving the equation for the y variable.

d. Write the y values in the table.

ANSWER:

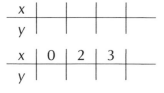

e. Plot the three points in the number plane.

f. Draw a line through the three plotted points.

Sometimes one variable is missing in a given equation. In such cases, no matter what value is substituted for the missing variable, the other variable remains constant.

When the variable for x is missing, the line is parallel to the x-axis. When the variable for y is missing, the line is parallel to the y-axis.

2. Draw the graph of $4y = 12$.

$4y = 12$
$\quad y = 3 \quad$ may be written as:
$\qquad 0x + y = 3$

x	y
0	3
3	3
−4	3

3. Draw the graph of $3x = -15$.

$3x = -15$
$\quad x = -5 \quad$ may be written as:
$\qquad x + 0y = -5$

x	y
−5	0
−5	3
−5	−2

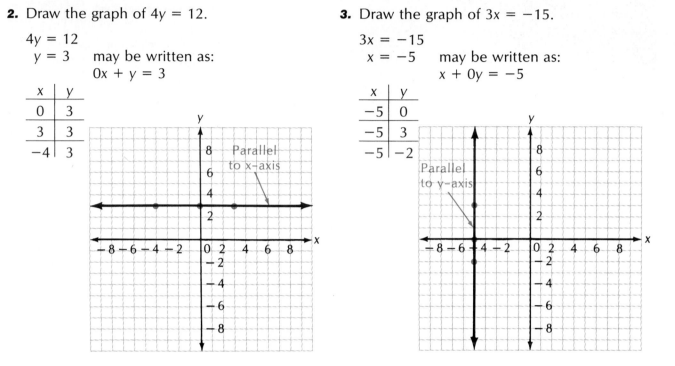

Think About It

1. What do you think is true about the equations of all lines that pass through the origin?

2. Why do you plot three points, rather than two points, to locate a line?

Practice Exercises

Draw the graph of each equation.

1. $y = x + 5$
2. $y = x - 3$
3. $x = y + 1$
4. $x = y - 6$
5. $x + y = 0$

6. $x + y = 2$
7. $x + y = -9$
8. $x - y = 4$
9. $x - y = -2$
10. $y = 5x$

11. $x = 2y$
12. $y = x$
13. $y = -3x$
14. $2y = 4x$
15. $2x = -2y$

16. $-3y = 12x$
17. $4y = 7x$
18. $-12y = -8x$
19. $y = 4$
20. $x = 2$

21. $y = -5$
22. $x = -6$
23. $y = 0$
24. $x = 0$
25. $7y = 21$

26. $12x = -60$
27. $-9y = -18$

Career Application

TEXTILE DESIGNER

C loth, rugs, and needlework kits are available in many patterns. *Textile designers* create and plan these patterns. They use graph paper to draw a design. Each line on the graph paper may represent a thread. The graph paper is an enlarged scale model of the textile. A designer planning a woven fabric marks the paper to show which thread will go over which. A designer planning a needlepoint canvas marks what stitch and what color yarn will be used at each point. Many designers do their work on computers, using special graphics software.

Textile designers need to know about art and graphic design. They should also know the best styles and techniques for different types of textiles. Most textile designers learn their skills in technical schools or colleges.

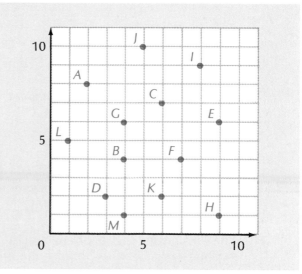

1. A textile designer made the graph shown for a needlepoint pattern. Some points will have blue stitches. What are the coordinates of each point?
 a. A **b.** B
 c. C **d.** D

2. Some points will have red stitches. Name the point for each pair of coordinates.
 a. (6,2) **b.** (1, 5)
 c. (8, 9) **d.** (4, 1)

3. A textile designer makes a pattern using rows of stitches. Graph these equations on the same set of coordinates.
 a. $y = 2x - 1$ **b.** $y = 2x + 2$ **c.** $y = 2x + 5$

Chapter 17 Review

Vocabulary

coordinate p. 595
coordinate axis p. 595
graph of an equation p. 599
number plane p. 595
ordered pair of numbers p. 594
origin p. 599

quadrant p. 595
x-axis p. 595
x-coordinate p. 595
y-axis p. 595
y-coordinate p. 595

Page
594 **1.** Write the ordered pair of numbers that has 0 as the first component and −5 as the second component.

595 What are the coordinates of each point?

2. Point M

3. Point N

4. Point O

5. Point P

6. Point Q

7. Point R

8. Point S

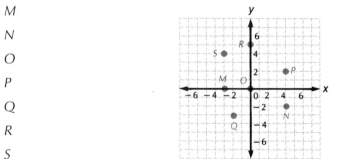

597 On graph paper, draw axes and plot points with the following coordinates.

9. $C(-4, 1)$ **10.** $D(0, -5)$ **11.** $E(-3, -6)$ **12.** $F(5, -1)$

13. $V(4, 2)$ **14.** $W(-6, 0)$ **15.** $X(-5, -5)$ **16.** $Y(0, 6)$

599 Draw the graph of each of the following equations.

17. $y = 2x + 1$ **18.** $x = y - 5$ **19.** $-4y = -2x$ **20.** $x = -1$

599 **21.** Without drawing the graph, determine if the graph of the equation $3x = 5y$ passes through the origin $(0, 0)$.

599 **22.** Without drawing the graph, determine if the graph of the equation $x + y = 11$ is parallel to the y-axis.

599 **23.** Which of the following ordered pairs of numbers are solutions of the equation $x + y = 9$?

a. $(4, -5)$ **b.** $(11, -2)$ **c.** $(2, 7)$ **d.** $(-3, 6)$

599 **24.** Without drawing the graph, determine if $(7, 6)$ and $(-1, -10)$ are coordinates of points on the graph of $2x - y = 8$.

594 **25.** The coordinates of a point are represented by ____?____.

Chapter 17 Test

594   **1.** Write the ordered pair of numbers that has −2 for the first component and 8 for the second component.

595   What are the coordinates of each point?

     2. Point *A*

     3. Point *B*

     4. Point *C*

     5. Point *D*

     6. Point *E*

     7. Point *F*

     8. Point *G*

597   On graph paper, draw axes and plot points with the following coordinates.

     9. $H(4, -5)$ **10.** $I(-2, 0)$ **11.** $J(0, 2)$ **12.** $K(0, 0)$

     13. $L(-3, 3)$ **14.** $M(1, 6)$ **15.** $N(-4, -4)$ **16.** $P(3, 4)$

599   Draw the graph of each of the following equations.

     17. $y = x + 3$ **18.** $x - y = 7$ **19.** $3x + 2y = 0$ **20.** $y = 5$

True or false?

595   **21.** In quadrant III, all the points have negative *x*-coordinates and negative *y*-coordinates.

599   **22.** To draw the graph of the equation $x + y = 10$, you can use the coordinates $(12, -2)$, $(2, 8)$, and $(7, 3)$.

599   **23.** An ordered pair that is a solution of the equation $2x + y = 9$ is $(-2, 5)$.

599   **24.** One of the points on the line forming the graph of the equation $3x - y = 10$ has the coordinates $(4, 2)$.

599   **25.** Without drawing the graph, determine if the graph of the equation $y = 5x$ is a line that is parallel to the *x*-axis.

Achievement Test

$$-41 \quad +6\tfrac{2}{3} \quad -0.084 \quad 195 \quad 0 \quad -\tfrac{5}{8} \quad -48 \quad +6\tfrac{2}{5} \quad -0.9 \quad +\tfrac{20}{5} \quad -\sqrt{144} \quad -\tfrac{2}{3}$$

530 **1.** Which of the numbers in the box above name positive numbers?

2. Negative numbers?

3. Integers?

4. Rational numbers?

5. Find the value of $|-52|$.

539 **6.** Draw a graph of -4, -3, 0, and 3 on a number line.

7. Write the coordinates of the following graph.

541 **8.** If a tailwind of 20 mi/h is indicated by +20 mi/h, how can a headwind of 32 mi/h be indicated?

9. What is the opposite of -20?

542 Which of the following sentences are true?

10. $-3 < -1$ **11.** $-10 > +6$ **12.** $+2 \not< -5$ **13.** $-4 \not> 0$

14. List the following numbers in order from greatest to least.
$-1, +17, -3, 0, +2, -11, -5, +5, +8, -7$

15. Which has the greater opposite number: -3 or $+2$?

544 **16.** What is the additive inverse of $+14$?

17. Write symbolically: The opposite of negative twelve is positive twelve.

545 Add.

18. $(+8) + (-10)$ **19.** $(-4) + (-5)$ **20.** $(-15) + (+7)$

21. $(+9) + (+6)$ **22.** $(-7) + (+7)$ **23.** $(-6) + (+11) + (-7)$

24. $-6 + 8 - 3 - 2 + 7$

549 Subtract.

25. $(-2) - (-9)$ **26.** $-6 - (+6)$ **27.** $+5 - (+8)$

28. $+3 - (-12)$ **29.** $0 - (-7)$ **30.** $(-9) - (-12)$

31. Take 9 from 5.

552 Multiply.

32. $-8 \times (+3)$ **33.** $-9 \times (+7)$ **34.** $+2 \times (+11)$

35. $(+5) \times (-5)$ **36.** $0 \times (-6)$ **37.** $(-3)(+4)(-2)$

38. Find the value of $(-3)^5$.

Divide.

39. $\dfrac{-48}{6}$ **40.** $\dfrac{+32}{-8}$ **41.** $\dfrac{-90}{15}$ **42.** $\dfrac{+35}{+7}$ **43.** $\dfrac{+12}{-12}$

44. $(-54) \div (+9)$ **45.** Simplify: $\dfrac{6(-7+5)-4(-1-1)}{-2(9-13)}$

Write each of the following as an algebraic expression.

46. The product of b and four.   **47.** The quotient of d divided by t.

48. The sum of p and i.   **49.** Pi (π) times the square of the radius (r).

50. The difference between one hundred eighty and A.

Write each of the following sentences symbolically.

51. Some number n increased by seven is equal to thirty-six.

52. Each number x decreased by ten is less than sixteen.

53. Five times each number y is greater than or equal to forty.

54. Each number w is greater than negative nine and less than positive two.

55. Express the following as a formula.
The distance (d) a freely falling body drops is one half the product of the acceleration due to gravity (g) and the square of the time of falling (t).

Find the value of each of the following.

56. $x - 4y$ when $x = 5$ and $y = 2$.

57. $3m^2 + 2mn - n^2$ when $m = 4$ and $n = -2$.

58. Find the value of A when $\pi = \dfrac{22}{7}$, $r = 14$, and $h = 10$, using the formula: $A = 2\pi r(h + r)$.

59. Which of the equations below have 3 as the solution?

$n - 3 = 4$	$n + 6 = 9$	$5n - 8 = 7$
$\dfrac{n}{6} = 2$	$n = 3$	$4n + 2n = 12$

60. Which are equivalent?

61. Which is in simplest form?

What operation with what number do you use on both sides of each of the following equations to get an equivalent equation in simplest form? Also write this equivalent equation.

62. $9x = 72$ **63.** $y + 10 = 15$ **64.** $n - 14 = 3$ **65.** $\dfrac{r}{6} = 12$

Solve and check.

66. $n + 9 = 7$ **67.** $x - 6 = 11$

68. $-12y = 84$ **69.** $8n = 6$

577 **70.** $\frac{c}{14} = 10$

71. $\frac{x}{-5} = 25$

578 **72.** $8b - 19 = 13$

73. $7x + 8x = 75$

579 **74.** Find the value of d when $l = 63$, $a = 7$, and $n = 9$, using the formula $l = a + (n - 1)d$.

581 **75.** Solve by the equation method.
If the school baseball team won 15 games, or $\frac{5}{8}$ of the games played, how many games were lost?

583 Find the solutions of each of the following inequalities when the replacements for the variable are all the real numbers.

76. $5n < 60$ **77.** $b - 4 > 9$ **78.** $b + 2 \not< -3$

79. $3x - 4 \neq -10$ **80.** $-7y \not< 14$ **81.** $\frac{n}{6} \geq -5$

82. $3y - 5y > 0$ **83.** $\frac{2}{3}x < 18$ **84.** $n - 2n \leq -14$

594 **85.** Write the ordered pair of numbers that has 0 for the first component and -7 as the second component.

595 What are the coordinates of each of the following?

86. Point G

87. Point H

88. Point I

89. Point J

90. Point K

91. Point L

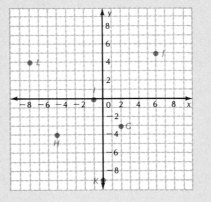

597 On graph paper, draw axes and plot the points with the following coordinates.

92. $M (-4, -6)$ **93.** $N (5, 0)$ **94.** $P (-7, 2)$

599 **95.** Draw the graph of the equation $x - y = 5$.

Table of Trigonometric Values

Angle	Sine	Cosine	Tangent	Angle	Sine	Cosine	Tangent
0°	0.0000	1.0000	0.0000	46°	0.7193	0.6947	1.0355
1°	0.0175	0.9998	0.0175	47°	0.7314	0.6820	1.0724
2°	0.0349	0.9994	0.0349	48°	0.7431	0.6691	1.1106
3°	0.0523	0.9986	0.0524	49°	0.7547	0.6561	1.1504
4°	0.0698	0.9976	0.0699	50°	0.7660	0.6428	1.1918
5°	0.0872	0.9962	0.0875	51°	0.7771	0.6293	1.2349
6°	0.1045	0.9945	0.1051	52°	0.7880	0.6157	1.2799
7°	0.1219	0.9925	0.1228	53°	0.7986	0.6018	1.3270
8°	0.1392	0.9903	0.1405	54°	0.8090	0.5878	1.3764
9°	0.1564	0.9877	0.1584	55°	0.8192	0.5736	1.4281
10°	0.1736	0.9848	0.1763	56°	0.8290	0.5592	1.4826
11°	0.1908	0.9816	0.1944	57°	0.8387	0.5446	1.5399
12°	0.2079	0.9781	0.2126	58°	0.8480	0.5299	1.6003
13°	0.2250	0.9744	0.2309	59°	0.8572	0.5150	1.6643
14°	0.2419	0.9703	0.2493	60°	0.8660	0.5000	1.7321
15°	0.2588	0.9659	0.2679	61°	0.8746	0.4848	1.8040
16°	0.2756	0.9613	0.2867	62°	0.8829	0.4695	1.8807
17°	0.2924	0.9563	0.3057	63°	0.8910	0.4540	1.9626
18°	0.3090	0.9511	0.3249	64°	0.8988	0.4384	2.0503
19°	0.3256	0.9455	0.3443	65°	0.9063	0.4226	2.1445
20°	0.3420	0.9397	0.3640	66°	0.9135	0.4067	2.2460
21°	0.3584	0.9336	0.3839	67°	0.9205	0.3907	2.3559
22°	0.3746	0.9272	0.4040	68°	0.9272	0.3746	2.4751
23°	0.3907	0.9205	0.4245	69°	0.9336	0.3584	2.6051
24°	0.4067	0.9135	0.4452	70°	0.9397	0.3420	2.7475
25°	0.4226	0.9063	0.4663	71°	0.9455	0.3256	2.9042
26°	0.4384	0.8988	0.4877	72°	0.9511	0.3090	3.0777
27°	0.4540	0.8910	0.5095	73°	0.9563	0.2924	3.2709
28°	0.4695	0.8829	0.5317	74°	0.9613	0.2756	3.4874
29°	0.4848	0.8746	0.5543	75°	0.9659	0.2588	3.7321
30°	0.5000	0.8660	0.5774	76°	0.9703	0.2419	4.0108
31°	0.5150	0.8572	0.6009	77°	0.9744	0.2250	4.3315
32°	0.5299	0.8480	0.6249	78°	0.9781	0.2079	4.7046
33°	0.5446	0.8387	0.6494	79°	0.9816	0.1908	5.1446
34°	0.5592	0.8290	0.6745	80°	0.9848	0.1736	5.6713
35°	0.5736	0.8192	0.7002	81°	0.9877	0.1564	6.3138
36°	0.5878	0.8090	0.7265	82°	0.9903	0.1392	7.1154
37°	0.6018	0.7986	0.7536	83°	0.9925	0.1219	8.1443
38°	0.6157	0.7880	0.7813	84°	0.9945	0.1045	9.5144
39°	0.6293	0.7771	0.8098	85°	0.9962	0.0872	11.4301
40°	0.6428	0.7660	0.8391	86°	0.9976	0.0698	14.3007
41°	0.6561	0.7547	0.8693	87°	0.9986	0.0523	19.0811
42°	0.6691	0.7431	0.9004	88°	0.9994	0.0349	28.6363
43°	0.6820	0.7314	0.9325	89°	0.9998	0.0175	57.2900
44°	0.6947	0.7193	0.9657	90°	1.0000	0.0000	
45°	0.7071	0.7071	1.0000				

Glossary

absolute value (p. 536) The absolute value (symbol | |) of any number, positive or negative, is the value of the corresponding arithmetic number with no sign. For example, $|{-5}| = |5| = 5$.

acute angle (p. 431) An angle having a measure more than 0° and less than 90°.

addend (p. 31) Any number that is being added.

additive inverse (p. 544) A number that, when added to a given number, yields a sum of zero.

altitude (p. 444) In polygons, a line segment from any vertex perpendicular to a side or extension of a side opposite the vertex.

angle (p. 431) The figure formed by two rays with a common endpoint.

angle bisector (p. 444) A segment drawn from the vertex of an angle, dividing the angle into two equal angles.

arc (p. 445) Any part of a circle.

area (p. 503) The number of units of square measure contained in a surface.

base (p. 54) A number that is raised to a power. In 7^2, 7 is the base.

BASIC (p. 74) An introductory computer language. BASIC stands for Beginner's All-purpose Symbolic Instruction Code.

chord (p. 445) A segment that has both of its endpoints on a circle.

circumference (p. 501) The distance around a circle; equal to πd where d is the length of the diameter and π is a constant approximately equal to 3.14.

complementary angles (p. 453) Two angles whose sum measures 90°.

complex fraction (p. 191) A fraction in which the numerator or denominator or both are fractions.

composite number (p. 66) Any whole number greater than 1 that can be divided exactly by at least one whole number other than 1 and itself.

congruent (p. 458) Having corresponding sides equal in length and corresponding angles equal in measure; same size and shape. Symbol ≅.

constant (p. 560) A definite value. Compare with *variable*.

cosine (cos) (p. 491) The ratio of the adjacent side of an acute angle in a right triangle to the hypotenuse of the triangle.

decimal (p. 76) A number that contains place values to the right of the ones place to express parts of one, such as tenths, hundredths, and so on. 0.36 and 2.1 are decimals.

denominator (p. 148) In a fraction, the number appearing below the fraction bar.

dependent events (p. 412) Two events are dependent if the occurrence of one of them affects the probability of the occurrence of the other.

diagonal (p. 443) A line segment connecting any two non-adjacent vertices of a polygon.

diameter (p. 445) A chord that passes through the center of a circle.

difference (p. 40) The answer in subtraction.

dividend (p. 58) In a division problem, the dividend is the number that is divided.

divisor (p. 58) In a division problem, the divisor is the number by which you divide.

empirical probability (p. 414) A probability value that is determined through actual observation of a population sample.

equation (p. 562) An open sentence with an equals sign (=) as its verb. For example, $3x = 25$ and $x + 5y = 33$ are equations.

event (p. 405) In probability, an event is a set of outcomes.

exponent (p. 54) A superscript that indicates how many times a base is to be used as a factor. In 4^3, 3 is the exponent.

exterior angle (p. 453) The angle formed by any side of a polygon and the extension of an adjacent side. Also, any of the four outside angles formed when two parallel lines are cut by a transversal.

factor (p. 47) Any of the numbers used in multiplication to form a product.

formula (p. 564) An equation that expresses a known relationship between or among quantities.

fraction (p. 148) A quotient expressing the relation of one or more equal parts to the total number of parts in the whole; a numeral expressing part of a region or part of a group.

frequency (p. 400) In statistics, the number of times that a data value occurs.

frequency distribution (p. 400) The arrangement of statistical data in a table form that shows all data values, a tally count for each value, and the frequency (sum of tally marks) for each value.

frequency polygon (p. 400) A line graph showing the frequency distribution for a given set of statistical data.

greatest common factor (GCF) (p. 69) The greatest whole number that will divide all members of a given set of whole numbers exactly.

histogram (p. 400) A bar graph showing the frequency distribution for a given set of statistical data.

hypotenuse (p. 444) The side opposite the right angle in a right triangle.

improper fraction (p. 151) A fraction in which the denominator is equal to or less than the numerator.

independent events (p. 412) Two events are independent if the occurrence of one of them in no way affects the occurrence of the other.

inequality (p. 562) An open sentence that has an inequality sign ($>$, $<$, $\geq$, $\leq$, $\neq$) as its verb. For example, $7x \geq 14$ and $4x + 3 < 6y$ are inequalities.

integers (p. 536) The set of all natural numbers, their opposites, and zero.

interior angle (p. 454) Any angle formed by two adjacent sides of a polygon.

least common denominator (LCD) (p. 163) The smallest non-zero whole number that is a multiple of all denominators in a given set of fractions.

least common multiple (LCM) (p. 157) The smallest non-zero whole number that is a multiple of all numbers in a given set of numbers.

line (p. 428) A one-dimensional continuum of points that extends indefinitely in opposite directions.

line segment (p. 428) A definite (measurable) part of a line consisting of two endpoints and all the points between.

lowest terms (p. 150) A fraction is in lowest terms when its numerator and denominator have no common factors except 1.

mean (p. 402) The average of a set of values. *Mean* is one of three measures of central tendency. Compare with *median* and *mode*.

median 1. The middle value when a set of values is arranged from smallest to largest. If the number of values is even (there is no single middle value), then the median is the average of the two middle values. Median is one of three measures of central tendency. Compare with **mean and mode** (p. 402). 2. In a triangle, a segment drawn from any vertex to the midpoint of the opposite side (p. 444).

midpoint (p. 475) The point in a line segment that separates the segment into two equal parts.

mixed number (p. 152) A number that has both a whole number part and a fraction part.

mode (p. 402) The data value in a set of data that has the greatest frequency. A set of data can be bimodal (have two modes). Generally, if three or more values have the same greatest frequency, then the data is said to have no mode. *Mode* is one of three measures of central tendency. Compare with *mean* and *median*.

mutually exclusive events (p. 409) Two events are mutually exclusive if when one occurs the other cannot occur.

natural numbers (p. 20) The numbers 1, 2, 3, 4, and so on without end. They are also called the counting numbers.

numerator (p. 148) In a fraction, the number appearing above the fraction bar.

obtuse angle (p. 43) An angle measuring more than 90° but less than 180°.

ordered pair (p. 594) A pair of numbers expressed in a definite order, such as the coordinates of a point on the real number plane.

origin (p. 595) The point of intersection of the horizontal and vertical axes in the coordinate plane, designated by the ordered pair (0, 0).

outcome (p. 405) In probability, one specific result of an experiment. See also *event*.

parallel lines (p. 435) Two lines in the same plane that never meet (have no common points). Symbol ∥.

parallelogram (p. 445) A quadrilateral having two pairs of opposite sides equal and parallel.

percent (p. 224) A measure of parts per hundred.

perfect square (p. 483) A number that has a whole number square root.

perimeter (p. 498) The distance around a polygon, equal to the sum of the lengths of its sides.

perpendicular (p. 435) Two lines intersecting to form a right angle. Symbol ⊥.

perpendicular bisector (p. 475) A line or line segment that passes through another line segment's midpoint at a right angle to the second segment.

pi (π) (p. 501) The ratio of any circle's circumference to its diameter. The value of π is approximately 22/7 or 3.14.

plane (p. 435) An endless two-dimensional flat surface. A minimum of three points is required to define a plane.

point (p. 428) An exact location in space without definable size or shape.

polygon (p. 443) A simple closed curve made up of line segments called sides.

polyhedron (p. 449) A closed geometric figure consisting of four or more polygons and the surfaces that they define, all in different planes.

population (p. 414) In probability and statistics, the total set of all things from which a sample is taken.

power (p. 54) A product that results from multiplying a quantity by itself. For example, 9 is the second power of 3, since $3^2 = 9$.

prime number (p. 66) Any whole number greater than 1 that can be divided exactly only by the number itself and 1.

probability (p. 409) A numerical measure, used to indicate the likelihood that an outcome will occur. Probability is the ratio of favorable outcomes to the total number of outcomes, and is from 0 to 1.

product (p. 47) The answer in multiplication.

proportion (p. 214) An equivalence relation between two ratios. Given the proportion $\frac{a}{b} = \frac{c}{d}$, the extremes are the terms a and d, and the means are the terms b and c. The cross products are the product of the extremes (*ad*) and the product of the means (*bc*), and *ad* = *bc*.

quadrant (p. 595) Any of the four regions defined by the coordinate axes in the real number plane. Quadrants are numbered I through IV counterclockwise from top right.

quotient (p. 58) The answer in division.

radius (p. 445) A segment having one endpoint at the center of a circle and the other endpoint on the circle.

range (p. 402) The difference between the largest and smallest values in a set of data.

rate (p. 211) A ratio that compares two quantities of differing units, such as pay per month or kilometers per hour.

ratio (p. 211) A comparison of two quantities by division. When written as a fraction, the number being compared appears as the numerator, and the number to which it is compared appears as the denominator. Ratios may also be expressed using a colon (:).

rational numbers (p. 536) Numbers that can be expressed as a quotient of two integers, with division by zero excluded.

ray (p. 428) Part of a line that includes one endpoint.

reciprocal (p. 191) Given a number, its reciprocal is that number which when multiplied with the original number yields a product of 1. Also called the multiplicative inverse.

regular polygon (p. 443) A polygon having all sides equal in length and all angles equal in measure.

remainder (p. 58) The number remaining when a dividend does not divide exactly.

right angle (p. 431) An angle having a measure of 90°.

sample (p. 414) In probability and statistics, a selected portion of an entire population, which is assumed to represent the entire population.

scientific notation (p. 132) A method of notation used to express very large and very small numbers as the product of a number between 1 and 10 and a power of 10.

similar (p. 460) Having corresponding sides in equal ratios and corresponding angles equal in measure; same shape, but not necessarily same size. Symbol ~.

simplest form (p. 151) The form of a mixed number when the fraction portion is in lowest terms.

sine (sin) (p. 491) The ratio of the side opposite an acute angle in a right triangle to the hypotenuse of the triangle.

square root (p. 483) The square root of a number is that number which when multiplied by itself yields the given number.

standard form (p. 20) A numeral such as 23,459.

statistics (p. 400) The study of collecting, organizing, analyzing, and interpreting data.

straight angle (p. 431) An angle having a measure of 180°.

sum (p. 31) The answer in addition.

supplementary angles (p. 453) Two angles whose measures added are 180°.

tangent 1. A line exterior to a circle that has only one point in common with the circle (p. 445).   2. (tan) The ratio of the side opposite an acute angle in a right triangle to the adjacent side of the angle (p. 491).

variable (p. 560) A symbol, usually a letter, that holds a place open for a number whose value may vary or be unknown.

vertex (p. 431) The common endpoint of two sides of an angle.

vertical angles (p. 453) The opposite and equal angles formed from the intersection of two straight lines.

volume (p. 515) The number of units of cubic measure contained in a solid.

whole numbers (p. 20) The numbers 0, 1, 2, 3, 4, and so on without end.

Index

Selected Answers

Chapter 1   Related Practice

Lesson 1-1   2a. A ribbon 42 inches long cut into 3 equal pieces.   **3a.** How much did she save?   **4a.** The movie last $1\frac{1}{2}$ hours.   **5a.** No; the number of votes David received.

Lesson 1-2   2a. What was their total; addition   **4a.** How long is each; division   **6a.** Subtraction; sentences will vary
8a. $86 - 49 = x$ **10a.** $24 \times 18 = s$

Lesson 1-3   1a. $20 + 50 = 70$ **2a.** 120   **3a.** $40 + 110 = 150$; $39 + 113$ is greater   **4a.** $900 + 800 = 1,700$
5a. $850 - 400 = 450$ **6a.** $60 \times 60 = 3,600$ **7a.** $210 \div 7 = 30$
8a. $70 + 100 + 90 + 40 + 200 + 400 = 900$

Lesson 1-4   1a. 70

Lesson 1-5   1a. 793,058   **2a.** 61 437   **3a.** Four hundred twenty-six   **4a.** Eight thousand two hundred seventy-eight   **5a.** Six thousand four   **6a.** Twenty-seven thousand four hundred thirty-two   **7a.** One hundred fifty thousand   **8a.** Nine million two hundred fifty thousand   **9a.** Seven million one hundred twenty-two thousand eight hundred forty-three   **10a.** Thirty-two million four hundred twenty-nine thousand seven hundred eighty-four   **11a.** Seven hundred million   **12a.** Five billion   **13a.** One hundred sixty-five million two hundred forty-six thousand three hundred twenty; eighty-two million four hundred forty-one thousand five hundred sixty

Lesson 1-6   1a. 400   **2a.** 4,610   **3a.** 3,900,000
4a. 48,018,007,000   **5a.** 4,950,000,000,000   **6a.** 14 600
7a. 2,175,597 km²

Lesson 1-7   1a. 30   **2a.** 30   **3a.** 200   **4a.** 600   **5a.** 9,000
6a. 9,000   **7a.** 90,000   **8a.** 60,000   **9a.** 300,000
11a. 5,000,000   **13a.** 6,000,000,000   **15a.** 6,000,000,000,000

Lesson 1-8   1a. 59   **2a.** 67   **3a.** 142   **4a.** 20   **5a.** 158   **6a.** 19
7a. 247   **8a.** 20   **9a.** 94   **10a.** 688   **11a.** 9,898   **12a.** 78,799
13a. 939   **14a.** 8,769   **15a.** 89,986   **16a.** 1,596   **17a.** 8,783
18a. 79,906   **19a.** 1,525   **20a.** 9,302   **21a.** 2,098   **22a.** 32
23a. 616   **24a.** 2,000   **25a.** 53   **26a.** 11,553 spectators

Lesson 1-9   1a. 55   **2a.** 15   **3a.** 24   **4a.** 17   **5a.** 30   **6a.** 3
7a. 24   **8a.** 336   **9a.** 512   **10a.** 484   **11a.** 210   **12a.** 217
13a. 7   **14a.** 227   **15a.** 6,223   **16a.** 3,217   **17a.** 3,737
18a. 6,168   **19a.** 2,349   **20a.** 4,815   **21a.** 12,241   **22a.** 33,427
23a. 22,166   **24a.** 35,895   **25a.** 68,668   **26a.** 13,955
27a. 65,539   **28a.** 331,722   **29a.** 5,524,025   **30a.** 219
31a. 401   **32a.** 148   **33a.** 23   **34a.** 251   **35a.** 6,000   **36a.** 87
37a. 619 students

Lesson 1-10   1a. 64   **2a.** 189   **3a.** 92   **4a.** 288   **5a.** 963
6a. 868   **7a.** 752   **8a.** 4,286   **9a.** 96,956   **10a.** 160
11a. 2,100   **12a.** 208   **13a.** 1,620   **14a.** 4,008   **15a.** 42,240
16a. 21,280   **17a.** 168   **18a.** 104   **19a.** 1,152   **20a.** 320
21a. 279   **22a.** 130 miles

Lesson 1-11   1a. 276   **2a.** 1,296   **3a.** 3,312   **4a.** 86,373
5a. 2,235,672   **6a.** 7,488   **7a.** 46,656   **8a.** 643,566
9a. 7,382,928   **10a.** 3,049,212   **11a.** 21,332,160
12a. 393,478,722   **13a.** 3,100   **14a.** 1,770   **15a.** 13,026
16a. 373,474   **17a.** 348,783   **18a.** 32,800   **19a.** 820,615
20a. 1,234,240   **21a.** 72,384   **22a.** 3,230,700   **23a.** 512
24a. 952   **25a.** 2,100   **26a.** 2,136   **27a.** 5,184 envelopes

Lesson 1-12   1a. 31   **2a.** 17   **3a.** 231   **4a.** 141   **5a.** 236
6a. 41   **7a.** 72   **8a.** 3,224   **9a.** 1,651   **10a.** 2,497   **11a.** 749
12a. 23,292   **13a.** 9,122   **14a.** 210   **15a.** 304   **16a.** 105

17a. 3,001   **18a.** 2,004   **19a.** 2 R4   **20a.** 10 R5   **21a.** 3 R8
22a. 19 R2   **23a.** 60 R5   **24a.** 146   **25a.** 233   **26a.** 100
27a. 23   **28a.** 64 tins

Lesson 1-13   1a. 5   **2a.** 7   **3a.** 28   **4a.** 76   **5a.** 123   **6a.** 548
7a. 2,415   **8a.** 5,219   **9a.** 189   **10a.** 705   **11a.** 220
12a. 2,004   **13a.** 862 R25   **14a.** 172 R4   **15a.** 7   **16a.** 52
17a. 48   **18a.** 629   **19a.** 754   **20a.** 37   **21a.** 301   **22a.** 320
23a. 426   **24a.** 528 R113   **25a.** 467 R80   **26a.** 8   **27a.** 42
28a. 467   **29a.** 5,485   **30a.** 654   **31a.** 17   **32a.** 39   **33a.** 60
34a. 3   **35a.** About 9 min

Lesson 1-14   1a. yes   **2a.** 3   **3a.** 1, 2, 3, 4, 6, 8, 12, 16, 24, 48
4a. 1, 2, 4   **5a.** 1, 2, 3, 4, 6, 12   **6a.** 2   **7a.** 34

Chapter 2   Related Practice

Lesson 2-1   1a. eight tenths   **2a.** three hundredths
3a. twenty-four hundredths   **4a.** one and six tenths
5a. two and fifty-one hundredths   **6a.** five thousandths
7a. twenty-four thousandths   **8a.** eight hundred thirty-two thousandths   **9a.** six and five thousandths   **10a.** seven ten-thousandths   **11a.** six hundred-thousandths   **12a.** one millionth   **13a.** eight point zero zero two five; eight and twenty-five ten-thousandths   **14a.** zero point nine three; ninety-three hundredths   **15a.** thirty-six point eight nine; thirty-six and eighty-nine hundredths   **16a.** zero point twelve; twelve hundredths

Lesson 2-2   1a. 0.4   **2a.** 6.5   **3a.** 0.08   **4a.** 0.36   **5a.** 6.04
6a. 0.003   **7a.** 0.069   **8a.** 0.274   **9a.** 2.017   **10a.** 0.0008
11a. 0.00003   **12a.** 0.000006   **13a.** 500.0058   **14a.** 61.2
15a. 0.45 oz

Lesson 2-3   1a. 0.3   **2a.** 0.1   **3a.** 0.52   **4a.** 0.32   **5a.** 0.215
6a. 0.545   **7a.** 0.2359   **8a.** 0.3414   **9a.** 0.000088
10a. 0.00008   **11a.** 0.000003   **12a.** 0.000005   **13a.** $.27
14a. $64   **15a.** 6   **16a.** 10   **17a.** $600   **18a.** $2.00   **19a.** 1.15

Lesson 2-4   1a. 0.3   **2a.** 0.89   **3a.** 0.1, 0.01, 0.001, 0.0001
4a. True   **5a.** 0.060   **6a.** Mrs. Goldstein

Lesson 2-5   1a. 0.8   **2a.** 1.6   **3a.** 0.07   **4a.** 0.15   **5a.** 0.94
6a. 1.32   **7a.** 1.00   **8a.** 5.9   **9a.** 14.3   **10a.** 12.13   **11a.** 5.41
12a. 10.85   **13a.** 6.18   **14a.** 4.6   **15a.** 8.009   **16a.** 34.83
17a. 391.77   **18a.** 2.69   **19a.** $2.78   **20a.** $17.24   **21a.** 8.7
22a. 10.38   **23a.** $1.49   **24a.** 15.462   **25a.** 7   **26a.** 1.5
27a. $141.03

Lesson 2-6   1a. 0.4   **2a.** 0.26   **3a.** 0.18   **4a.** 0.07   **5a.** 0.09
6a. 0.3   **7a.** 4.2   **8a.** 2.8   **9a.** 5.0   **10a.** 0.621   **11a.** 4.11
12a. 12.3   **13a.** 0.0863   **14a.** 0.10709   **15a.** 0.3007
16a. 0.4600 or 0.46   **17a.** 0.0021   **18a.** 0.8   **19a.** 2.247
20a. 0.5   **21a.** 0.326   **22a.** 0.7   **23a.** 4.4   **24a.** $2.25   **25a.** 0.5
26a. 0.37   **27a.** 0.07   **28a.** 3.36   **29a.** 7.5   **30a.** $.69
31a. 0.45   **32a.** 15   **33a.** 0.6   **34a.** $22.51

Lesson 2-7   1a. 0.4   **2a.** 2.07   **3a.** 1.235   **4a.** 7.6176   **5a.** 10
6a. 0.24   **7a.** 0.09   **8a.** 0.018   **9a.** 0.009   **10a.** 52.2   **11a.** 0.72
12a. 0.06   **13a.** 0.102   **14a.** 0.048   **15a.** 0.2072   **16a.** 0.0134
17a. 0.0012   **18a.** 0.09568   **19a.** 0.00004   **20a.** 0.247632
21a. 0.000048   **22a.** 3.5   **23a.** 0.06   **24a.** $1.68   **25a.** $14.91
26a. $3.56   **27a.** $525   **28a.** 15   **29a.** $4   **30a.** 4.8
31a. $187.25

Lesson 2-9 **1a.** 19 **2a.** 2.8 **3a.** 0.42 **4a.** 0.841 **5a.** 0.002
6a. 24.3 **7a.** 35 **8a.** 30 **9a.** 5 **10a.** 0.563 **11a.** 0.05
12a. 97 **13a.** 17 **14a.** 11.6 **15a.** 234.77 **16a.** 0.992
17a. 0.03 **18a.** 57 **19a.** 1,900 **20a.** 720 **21a.** 300
22a. 181.481 **23a.** $15.50 **24a.** 25 **25a.** 30 **26a.** 109
27a. 1,004.7 **28a.** 14.91 **29a.** 0.09 **30a.** 3.2 **31a.** 2,000
32a. 270 **33a.** 87,700 **34a.** 3,000 **35a.** 4,714.286
36a. 6,000 **37a.** 200 **38a.** 5 notebooks

Lesson 2-10 **1a.** 0.7 **2a.** 0.9 **3a.** 0.07 **4a.** 0.75 **5a.** 0.375
6a. 0.14 **7a.** 0.5 **8a.** 0.83 **9a.** 2.5 **10a.** 0.25 **11a.** 0.05
12a. 0.2 **13a.** 0.444 **14a.** 0.29 **15a.** 0.16

Lesson 2-11 **1a.** 40 **2a.** 50 **3a.** 700 **4a.** 4,000 **5a.** 37.5
6a. 200 **7a.** 400 **8a.** $20 **9a.** 200

Lesson 2-12 **1a.** 50 **2a.** 400 **3a.** 3 **4a.** 2.6 **5a.** 0.3 **6a.** 58
7a. 700 **8a.** 2,000 **9a.** 42 **10a.** 20 **11a.** 72.1 **12a.** 854
13a. 3,000 **14a.** 50,000 **15a.** 0.657 **16a.** 350 **17a.** 265.3
18a. 6,582 **19a.** 952,000,000 **20a.** 83,000,000,000
21a. $5,000

Lesson 2-13 **1a.** 2 **2a.** 3.4 **3a.** 0.5 **4a.** 0.09 **5a.** 0.35
6a. 2 **7a.** 3.82 **8a.** 0.59 **9a.** 0.04 **10a.** 0.0021 **11a.** 0.295
12a. 8 **13a.** 3.725 **14a.** 0.628 **15a.** 0.085 **16a.** 0.000925
17a. 0.2849 **18a.** 617 **19a.** 39 **20a.** 8 **21a.** $1,284.00

Lesson 2-14 **2.** 128 different burgers (2^7) **4.** 52 people
($2 \times 25 + 2$) **6.** 204 squares

Lesson 2-15 **1a.** 83,000,000 **2a.** 6,800,000 **3a.** 17,450,000
4a. 8,562,000 **5a.** $4,500,000 **6a.** 78,000,000,000
7a. 9,400,000,000 **8a.** 12,060,000,000 **9a.** 7,568,000,000
10a. $8,300,000,000 **11a.** 4,000,000,000,000
12a. 3,900,000,000,000 **13a.** 5,920,000,000,000
14a. 8,227,000,000,000 **15a.** $9,600,000,000,000 **16a.** 9,000
17a. 8,700 **18a.** $19,000 **19a.** 3,800 **20a.** 850
21a. 93,000,000

Lesson 2-16 **1a.** 9 hundred **2a.** 13.4 hundred **3a.** 25.68
hundred **4a.** 6 thousand **5a.** 8.5 thousand **6a.** 7.13
thousand **7a.** 8 million **8a.** 7.3 million **9a.** 11.47 million
10a. 9 billion **11a.** 7.5 billion **12a.** 16.82 billion **13a.** 51.006
billion **14a.** 6 trillion **15a.** 4.7 trillion **16a.** 32.08 trillion
17a. 17.526 trillion **18a.** $63 thousand **19a.** $84 million
20a. $58 billion **21a.** $6.5 trillion **22a.** $25 billion

Chapter 3 Related Practice

Lesson 3-1 **1a.** $\frac{1}{2}$ **2a.** $\frac{1}{2}$ **3a.** $\frac{4}{5}$ **4a.** $\frac{5}{9}$ **5a.** $\frac{4}{7}$ **6a.** $\frac{5}{6}$ **7a.** $\frac{3}{10}$
8a. $\frac{4}{9}$

Lesson 3-2 **1a.** 1 **2a.** $1\frac{2}{5}$ **3a.** $1\frac{1}{2}$ **4a.** $4\frac{1}{2}$ **5a.** $3\frac{1}{2}$ **6a.** 3
7a. $4\frac{2}{3}$ **8a.** $7\frac{3}{5}$ **9a.** $4\frac{1}{2}$ **10a.** $40\frac{5}{6}$ ft

Lesson 3-3 **1a.** $\frac{4}{8}$ **2a.** $\frac{8}{12}$ **3a.** $\frac{70}{100}$ **4a.** $\frac{8}{64}$ **5a.** $\frac{20}{10}$ **6a.** $\frac{103}{412}$;
The library ordered 103 children's books.

Lesson 3-4 **1a.** $\frac{2}{4}, \frac{3}{6}, \frac{4}{8}, \frac{5}{10}, \ldots$ **2a.** $\frac{1}{6}, \frac{2}{12}, \frac{3}{18}, \frac{4}{24}, \ldots$ **3a.** $\frac{3}{8} = \frac{3}{8}$;
yes **4a.** $576 \neq 672$; no **5a.** No

Lesson 3-5 **1a.** 42, 51 **2a.** 0, 5, 10, 15, 20, 25, ... **3a.** 0, 15,
30 **4a.** 20, 40, 60, 80, ... **5a.** 24 **6a.** 12 **7a.** 48 **8a.** Rows 5,
10, 15, 20, ... were flowers

Lesson 3-6 **1a.** $\frac{1}{3}$ **2a.** $\frac{1}{6}$ **3a.** False **4a.** False **5a.** $\frac{1}{6}, \frac{1}{4}, \frac{1}{2}$
6a. 9 **7a.** Suzanne

Lesson 3-7 **1a.** 4, $\frac{2}{4}, \frac{1}{4}$ **2a.** 12, $\frac{4}{12}, \frac{3}{12}$ **3a.** 12, $\frac{3}{12}, \frac{2}{12}$ **4a.** 12,
$\frac{6}{12}, \frac{4}{12}, \frac{3}{12}$ **5a.** $\frac{5}{9} = \frac{25}{45}, \frac{1}{5} = \frac{9}{45}, \frac{25}{45} > \frac{9}{45}$

Lesson 3-8 **1a.** $\frac{4}{5}$ **2a.** $\frac{1}{2}$ **3a.** 1 **4a.** $1\frac{1}{3}$ **5a.** $1\frac{2}{3}$ **6a.** $\frac{3}{4}$
7a. $1\frac{5}{12}$ **8a.** $\frac{2}{3}$ **9a.** 1 **10a.** $8\frac{7}{8}$ **11a.** $13\frac{3}{5}$ **12a.** $7\frac{3}{4}$ **13a.** $13\frac{3}{4}$
14a. 12 **15a.** $4\frac{1}{3}$ **16a.** $13\frac{1}{2}$ **17a.** $6\frac{2}{5}$ **18a.** $10\frac{11}{24}$ **19a.** $6\frac{1}{2}$

20a. $15\frac{1}{4}$ **21a.** $11\frac{1}{2}$ **22a.** $7\frac{11}{16}$ **23a.** 4 **24a.** $10\frac{7}{8}$ **25a.** $\frac{3}{8}$
26a. $3\frac{5}{8}$ **27a.** $1\frac{1}{4}$ **28a.** $1\frac{1}{2}$ **29a.** $4\frac{5}{8}$ yd

Lesson 3-9 **1a.** $\frac{1}{5}$ **2a.** $\frac{1}{2}$ **3a.** $\frac{11}{16}$ **4a.** $\frac{1}{6}$ **5a.** $\frac{1}{12}$ **6a.** $4\frac{2}{5}$
7a. $2\frac{1}{4}$ **8a.** $4\frac{3}{5}$ **9a.** $5\frac{1}{3}$ **10a.** 1 **11a.** $2\frac{1}{2}$ **12a.** $6\frac{3}{4}$ **13a.** $4\frac{5}{8}$
14a. $2\frac{1}{2}$ **15a.** $3\frac{2}{15}$ **16a.** $3\frac{7}{12}$ **17a.** $2\frac{3}{4}$ **18a.** $5\frac{5}{6}$ **19a.** $7\frac{17}{20}$
20a. $\frac{1}{4}$ **21a.** $4\frac{3}{4}$ **22a.** $\frac{2}{5}$ **23a.** $4\frac{1}{2}$ **24a.** $\frac{9}{16}$ **25a.** $\frac{1}{24}$ **26a.** $6\frac{1}{2}$ lb

Lesson 3-10 **1a.** $\frac{6}{5}$ **2a.** $\frac{13}{3}$ **3a.** $\frac{41}{5}$ **4a.** $\frac{39}{7}$ **5a.** 19 times

Lesson 3-11 **1a.** $\frac{1}{8}$ **2a.** $\frac{3}{10}$ **3a.** $\frac{3}{5}$ **4a.** $\frac{5}{16}$ **5a.** $\frac{1}{6}$ **6a.** $\frac{2}{3}$ **7a.** $\frac{5}{6}$
8a. 5 **9a.** 6 **10a.** $1\frac{1}{2}$ **11a.** $2\frac{1}{4}$ **12a.** $2\frac{1}{3}$ **13a.** 3 **14a.** 10
15a. $\frac{3}{4}$ **16a.** $5\frac{1}{4}$ **17a.** $2\frac{2}{3}$ **18a.** 34 **19a.** $8\frac{1}{2}$ **20a.** $8\frac{4}{5}$ **21a.** 42
22a. $25\frac{1}{2}$ **23a.** $12\frac{2}{3}$ **24a.** 2 **25a.** $1\frac{7}{8}$ **26a.** $1\frac{3}{5}$ **27a.** $1\frac{3}{32}$
28a. 2 **29a.** $2\frac{4}{5}$ **30a.** $3\frac{3}{16}$ **31a.** $\frac{1}{12}$ **32a.** $11\frac{55}{64}$ **33a.** 126
34a. 258 **35a.** $29\frac{1}{4}$ **36a.** 12 **37a.** 1,760 **38a.** $10 **39a.** $82
40a. $1\frac{1}{2}$ **41a.** 1 **42a.** $2.32 **43a.** $.20 **44a.** 100 ft

Lesson 3-12 **1a.** $\frac{3}{4}$ **2a.** $2\frac{1}{2}$ **3a.** $\frac{1}{3}$ **4a.** 2 **5a.** $1\frac{1}{7}$ **6a.** $\frac{1}{6}$
7a. $\frac{3}{40}$ **8a.** 18 **9a.** 20 **10a.** $6\frac{2}{3}$ **11a.** $\frac{1}{2}$ **12a.** $1\frac{1}{2}$ **13a.** $\frac{13}{16}$
14a. 4 **15a.** $3\frac{1}{3}$ **16a.** $\frac{3}{4}$ **17a.** $3\frac{3}{5}$ **18a.** $2\frac{2}{5}$ **19a.** $2\frac{2}{5}$ **20a.** $7\frac{1}{12}$
21a. $\frac{1}{20}$ **22a.** $\frac{8}{15}$ **23a.** 5 **24a.** $2\frac{2}{5}$ **25a.** $\frac{2}{3}$ **26a.** $\frac{21}{40}$ **27a.** 1
28a. $7\frac{1}{2}$ **29a.** $\frac{3}{4}$ **30a.** $\frac{32}{45}$ **31a.** 640 km/h

Lesson 3-13 **1a.** $\frac{1}{4}$ **2a.** $\frac{7}{25}$ **3a.** $\frac{1}{2}$ **4a.** $\frac{1}{8}$ **5a.** $\frac{3}{5}$ **6a.** $1\frac{1}{8}$ **7a.** 2
8a. 1 **9a.** 1,000 **10a.** 12 **11a.** 8 **12a.** 72 **13a.** Meyer
fielded the ball cleanly $\frac{23}{25}$ of the time **14a.** 1,062 students

Lesson 3-14 **2.** 17, 18 **4.** 8 min **6.** $1\frac{1}{2}$ mi **8.** 35, 36

Lesson 3-15 **1a.** 0.1 **2a.** 0.8 **3a.** 0.39 **4a.** 0.75 **5a.** $0.37\frac{1}{2}$
6a. $0.33\frac{1}{3}$ **7a.** 1.15 **8a.** $0.33\frac{1}{3}$ **9a.** $0.57\frac{1}{7}$ **10a.** 0.5 or 0.50
11a. $0.37\frac{1}{2}$ **12a.** $0.77\frac{7}{9}$ **13a.** 1.5 or 1.50 **14a.** 1.6 or 1.60
15a. $1.55\frac{5}{9}$ **16a.** $2.37\frac{1}{2}$ **17a.** 0.571 **18a.** 0.9514 **19a.** Matt

Lesson 3-16 **1a.** $\frac{3}{5}$ **2a.** $\frac{3}{4}$ **3a.** $\frac{1}{50}$ **4a.** $\frac{2}{5}$ **5a.** $\frac{7}{8}$ **6a.** $1\frac{1}{5}$
7a. $1\frac{1}{4}$ **8a.** $1\frac{1}{3}$ **9a.** $\frac{1}{8}$ **10a.** $\frac{9}{250}$ **11a.** $1\frac{3}{8}$ **12a.** $\frac{5}{16}$ **13a.** $\frac{1}{400}$
14a. $3\frac{9}{16}$ **15a.** $3\frac{673}{1000}$ oz

Lesson 3-17 **1a.** How much did Jill spend? **2a.** $52 \div 13 = 4$;
Mimi buys 4 pairs of sneakers a year. **3a.** $1,799.82

Lesson 3-18 **1a.** $\frac{8}{15}$ **2a.** 2:9 **3a.** $\frac{5}{8}$ **4a.** $\frac{1}{3}$ **5a.** $\frac{11}{2}$ **6a.** $\frac{8}{1}$
7a. $\frac{3}{2}$ **8a.** $\frac{n}{7}$ **9a.** $\frac{4}{5}$ **10a.** $\frac{7}{2}$ **11a.** 20 or $\frac{20}{1}$ **12a.** $\frac{1}{4}$ **13a.** $\frac{88}{1}$ or
88 **15b.** $\frac{32}{4}$ or $\frac{8}{1}$ or 8

Lesson 3-19 **1a.** $\frac{26}{13} = \frac{6}{3}$ **2a.** $\frac{x}{30} = \frac{9}{54}$ **4a.** $t = 24$ **5a.** $a = 5$
6a. $n = 144$ **7a.** $x = 63$ **8a.** $y = 4$ **9a.** 56 **10a.** 15 items

Chapter 4 Related Practice

Lesson 4-1 **1a.** 5% **2a.** 47% **3a.** 140 **4a.** 23%, 0.23, $\frac{23}{100}$
5a. $\frac{18}{100}$ or 18:100 **6a.** 43%

Lesson 4-2 **1a.** 0.06 **2a.** 0.16 **3a.** 0.4 **4a.** 1.34 **5a.** 1
6a. 0.375 or $0.37\frac{1}{2}$ **7a.** 0.045 or $0.04\frac{1}{2}$ **8a.** 0.005 or $0.00\frac{1}{2}$
9a. 0.035 **10a.** 0.0125 **11a.** 0.26375 **12a.** 0.007 **13a.** 0.07

Lesson 4-3 **1a.** 1% **2a.** 28% **3a.** 60% **4a.** 139% **5a.** 120%
6a. $12\frac{1}{2}$% **7a.** $1\frac{1}{2}$% **8a.** $10\frac{1}{2}$% **9a.** $137\frac{1}{2}$% **10a.** 87.5%
11a. 26.25% **12a.** 25% **13a.** 124.5% **14a.** 100% **15a.** $\frac{1}{4}$%
16a. 0.5% **17a.** 250%

Lesson 4-4 **1a.** $\frac{1}{2}$ **2a.** $\frac{1}{3}$ **3a.** $\frac{3}{50}$ **4a.** $1\frac{1}{10}$ **5a.** $\frac{5}{6}$

Lesson 4-5 **1a.** 25% **2a.** $83\frac{1}{3}$% **3a.** 7% **4a.** 54% **5a.** 9%
6a. $66\frac{2}{3}$% **7a.** $71\frac{3}{7}$% **8a.** $55\frac{5}{9}$% **9a.** 100% **10a.** 175%
11a. $166\frac{2}{3}$% **12a.** 150% **13a.** $262\frac{1}{5}$% **14a.** $266\frac{2}{3}$%
15a. 25%

Lesson 4-6 **1a.** 12.48 **2a.** 0.36 **3a.** 0.412 **4a.** 160 **5a.** 31 **6a.** 100 **7a.** 54 **8a.** 72 **9a.** 0.75 **10a.** 0.24 **11a.** 0.9 **12a.** 0.016 **13a.** $.23 **14a.** 9.80 **15a.** 15 words

Lesson 4-7 **1a.** 60% **2a.** 25% **3a.** $83\frac{1}{3}$% **4a.** $33\frac{1}{3}$% **5a.** $42\frac{6}{7}$% **6a.** $28\frac{4}{7}$% **7a.** 100% **8a.** 400% **9a.** 120% **10a.** $133\frac{1}{3}$% **11a.** 75% **12a.** 75% **13a.** 25% **14a.** 15% **15a.** 35% **16a.** 3% **17a.** $37\frac{1}{2}$%

Lesson 4-8 **1a.** 25% **2a.** 50% **3a.** 37.5% **4a.** 10% **5a.** $33\frac{1}{3}$% **6a.** 25.83% **7a.** 20%

Lesson 4-9 **1a.** 200 **2a.** 50 **3a.** 15 **4a.** 234 **5a.** 500 **6a.** 100 **7a.** 59 **8a.** 45 **9a.** 78 **10a.** 800 **11a.** 400 **12a.** 18 **13a.** 500 **14a.** 2,000 **15a.** $200 **16a.** $50,000

Lesson 4-10 **1a.** Helen's average; Miguel's average. **2a.** Gloria's team; 8 points

Chapter 5 Practice Exercises
Lesson 5-1 **1a.** $164 **2a.** $193.20 **3a.** $247.65 **5a.** $4.75 **6a.** $4,420 **7a.** $8,400 **8a.** $8,970 **9a.** $180 **10a.** $675 **11a.** $216 **12a.** $180 per week **15a.** $105.70 **19a.** $560
Lesson 5-2 **2a.** $59.64; $366.36 **7a.** $2,300 **8a.** $56.52; $684.52
Lesson 5-3 **1a.** $4.14 **2a.** $3.30 **3a.** $208 **4a.** $502.32
Lesson 5-4 **1a.** $4.29 **2a.** $28.24
Lesson 5-5 **1a.** $1.70 **2a.** $2.04 **3a.** $4.55 **4a.** $15.75
Lesson 5-6 **2a.** $87.28 **3a.** $84.22
Lesson 5-7 **2.** $2.09
Lesson 5-8 **2.** 6 **4.** $3,929.04 **6.** 12

Chapter 6 Practice Exercises
Lesson 6-1 **1a.** $.36 **2a.** $3.58 **3a.** $.75 **4a.** $.63 **5a.** $.80 **6a.** $.42 **7a.** $7.80
Lesson 6-2 **1a.** $.95 **2a.** 3¢ **3a.** $1.49 **4a.** 86¢ **5a.** 29.7¢ **6a.** 90¢ **9a.** $.29 **10a.** 5 for 49¢ **11a.** 10¢ **12a.** 0.916¢ **13a.** a half-gallon at $1.12 **14a.** 6¢ **15a.** 23.8¢ **16a.** 62.5¢ **17a.** 2.26-kg bag at 95¢ **18a.** $.45
Lesson 6-3 **1a.** Tables $69.30; Lamps $38.15; Sofas $297.50; Bedroom sets $629.30; Bookcases $50.93 nearest cent; Mirrors $26.59 nearest cent **5a.** $2.10 **6a.** 25% **8a.** $40 **12a.** $49.41; $133.59 **14a.** $115.20 **17a.** $56
Lesson 6-4 **1a.** $1.00 **2a.** $42.10 **3a.** $88.56
Lesson 6-5 **1a.** 8352; 8417 **2a.** $53.14 **3a.** 6529; 6774 **4a.** $7.35 **5a.** 8.6¢ **6a.** 25,240 gal **7a.** $47.83
Lesson 6-7 **3a.** $9.54 **4a.** $10
Lesson 6-8 **1a.** $294.20 **3a.** $5,000
Lesson 6-9 **1a.** $.007 **2a.** 0.9¢ **3a.** 80 mills **4a.** 4¢ per $1 **5a.** 50 mills per $1 **6a.** $8 per $100 **7a.** $45 per $1,000 **8a.** 4% **9a.** $464 **10a.** $592 **11a.** $520 **12a.** $200 **13a.** $150
Lesson 6-10 **1a.** $385.42 **2a.** $68,125 **3a.** $622 **5a.** $5,200
Lesson 6-11 **2.** 8 **6.** 3:31 P.M. **10.** $50

Chapter 7 Practice Exercises
Lesson 7-1 **2.** Deposit: $406.75 **4.** Balance: $1,249.47
Lesson 7-2 **1a.** $30 **2a.** $14 **3a.** $45 **4a.** $8 **5a.** $4 **6a.** $5.95 **7a.** $45; $345 **8a.** 12%
Lesson 7-3 **1a.** $45 **2a.** $1,980 **3a.** $349.84
Lesson 7-4 **1a.** $451.14 **2a.** $563.92 **3a.** $12,775.40 **4a.** $179,204.40
Lesson 7-5 **1a.** $2,185.25; $785.25 **2a.** $6,755.51; $4,455.51 **3a.** $443.37; $93.37 **4a.** $530.60 **5a.** $82.43 **6a.** $788.77 **7a.** $31.71 **8a.** $81.21 **9a.** $5.62 **10a.** $627.50
Lesson 7-6 **2.** $844.99 **4.** $1,612.50

Chapter 8 Practice Exercises
Lesson 8-1 **1a.** kilometer **2a.** 2 cm **3a.** 4 cm **4a.** 7 cm **5a.** 66 mm **6a.** 80 mm **7a.** 14; 1; 4 **8a.** 6.5 cm **9a.** 10 **10a.** 1,000 **11a.** 7,583 m **12a.** 340 mm **13a.** 47 cm **14a.** 4,390 dm **15a.** 780 m **16a.** 8.1 km **17a.** 0.7; 7; 70; 7,000; 70,000; 700,000 **18a.** 10 m **20.** 84.5 mm; 8.45 km; 8,450,000 cm
Lesson 8-2 **1a.** milligram **2a.** 10 **4a.** 90 mg **5a.** 3.9 cg **6a.** 53 g **7a.** 7.864 kg **8a.** 3 t **9a.** 5; 500,000; 5,000,000 **10a.** 56 **11a.** $1.49
Lesson 8-3 **1a.** kiloliter **2a.** 10 **3a.** 100 L **4a.** 70 mL **5a.** 8,500 cL **6a.** 310 dL **7a.** 0.9 L **8a.** 6.582 kL **9a.** 74 **10a.** 5 liters **11a.** 65 mL
Lesson 8-4 **1a.** square centimeters **3.** 1,000,000 mm² **5.** 10,000 m² **7a.** 600 **8a.** 3,700 **9a.** 8.19 **10a.** 45,000,000 **11a.** 80 **12a.** 7.5 **13a.** 5
Lesson 8-5 **1a.** cubic meter **5a.** 8,000 **6a.** 27,000,000 **7a.** 63,000 **8a.** 17 **14a.** 5 kg **15a.** 7 **16a.** 6 **17a.** 13 **18a.** 0.8 **19a.** 7 **20a.** 79 **21a.** 11 **22a.** 6,000 **23a.** 15 **24a.** 7 **25a.** 2.6 **26a.** 800 dm³

Chapter 9 Practice Exercises
Lesson 9-1 **1a.** $1\frac{3}{4}$ in. **3a.** 96 **4a.** 18 **5a.** 12,320 **6a.** 3 **7a.** $\frac{1}{3}$ **8a.** $\frac{5}{6}$ **9a.** $\frac{1}{4}$ **10a.** 13 ft 11 in. **11a.** 5 yd 2 in. **12a.** 20 ft 8 in. **13a.** 3 ft 2 in. **14a.** 7 yd
Lesson 9-2 **1a.** a **2a.** c **3a.** b **5a.** 96 **6a.** 17.500 **7a.** 3 **8a.** 3 **9a.** $\frac{3}{4}$ **10a.** $\frac{3}{5}$ **11a.** $\frac{1}{4}$ **12a.** 4 lb 14 oz **13a.** 3 lb 9 oz **14a.** 18 lb 12 oz **15a.** 2 lb 3 oz
Lesson 9-3 **2a.** 144 **3a.** 6 **4a.** 32 **5a.** 5 **6a.** $\frac{7}{16}$ **7a.** $\frac{1}{2}$ **8a.** $\frac{3}{4}$ **9a.** 8 gal 3 qt **10a.** 3 gal 1 pt **11a.** 4 gal 1 qt
Lesson 9-4 **3a.** 2,304 **4a.** 11 **5a.** 5 **6a.** 1.84 **7a.** 17 **8a.** $\frac{8}{9}$ **9a.** $\frac{3}{8}$ **10a.** 0.275
Lesson 9-5 **2.** $2\frac{1}{2}$ ft
Lesson 9-6 **2a.** 65,664 **3a.** 1,080 **4a.** 5 **5a.** $\frac{1}{4}$ **6a.** $\frac{1}{3}$
Lesson 9-7 **3a.** 1,440 **4a.** 45 **5a.** 720 **6a.** 45 **7a.** 30 **8a.** 156 **9a.** 25 **10a.** $\frac{3}{4}$ **11a.** $\frac{2}{3}$ **12a.** $\frac{1}{6}$ **13a.** $\frac{2}{7}$ **14a.** $\frac{3}{4}$ **15a.** 23 h **16a.** 9 yr 9 mo **17a.** 15 yr 9 mo **18a.** 1 da 16 h 20 min **19a.** 5 min **20a.** 16 h **22a.** 61
Lesson 9-8 **2.** 6:25 P.M.; 8:17 P.M. **6.** 235; 12:40 P.M.; CST; meal.

Chapter 10 Practice Exercises
Lesson 10-1 **1a.** 400 mi; 100 mi
Lesson 10-2 **1a.** $5,000; $2,500
Lesson 10-3 **1a.** $\frac{1}{4}$; $\frac{1}{5}$
Lesson 10-5 **1a.** 7 **2a.** 6 **3a.** 15 **4a.** 21 **5a.** Mean 46; Median 47; Mode 47; Range 10
Lesson 10-6 **1a.** (1), (2), (3), (4), (5) **3a.** 3 × 2 = 6 **4a.** 3 × 2 = 6; (W,B), (W,G), (B,B), (B,G), (G,B), (G,G)
Lesson 10-7 **1a.** $\frac{4}{9}$ **2a.** $\frac{26}{42}$ or $\frac{13}{21}$ **3a.** $\frac{2}{12}$ or $\frac{1}{6}$ **4a.** $\frac{34}{40}$ or $\frac{17}{20}$
Lesson 10-8 **1a.** $P = \frac{1}{40}$ **2a.** $P = \frac{5}{48}$ **3a.** $P = \frac{2}{21}$ **4a.** $P = \frac{1}{56}$
Lesson 10-9 **1a.** 2.3% **2a.** 126

Chapter 11 Practice Exercises
Lesson 11-1 **1a.** line AG or GA, $\overleftrightarrow{AG}$, $\overleftrightarrow{GA}$ **2a.** line segment EF or FE, $\overline{EF}$, $\overline{FE}$ **3a.** line ST or TS, $\overleftrightarrow{ST}$, $\overleftrightarrow{TS}$ **4a.** ray AB, $\overrightarrow{AB}$ **6a.** line CN **9a.** $\overline{FG}$, $\overline{GH}$, $\overline{HF}$
Lesson 11-2 **3a.** ∠4 **5a.** 360°

Lesson 11-3 1a. 80° **3a.** 150° **4a.** 180°
Lesson 11-4 2. intersecting lines **4.** perpendicular line and ray
11a. $\overrightarrow{CE}$ **12a.** D **14.** No
Lesson 11-5 2a. $2\frac{3}{8}$ in. **5a.** m$\overline{BR}$ = 44 mm; m$\overline{AD}$ = 3.5 cm;
m$\overline{ST}$ = 42 mm; m$\overline{NH}$ = 35 mm
Lesson 11-6 1a. 180 km **2a.** 240 mi **3a.** 12 ft **4a.** 2,400 m
or 2.4 km **5a.** 3 in. **6a.** 6 in. **7a.** 3 cm **8a.** 4 mm
10. 1:192 **12.** 1:576 **13a.** 3 cm **14a.** 3 in. **17a.** 72 mi
18a. 60 km **20.** 130 km
Lesson 11-7 1a. plane ABC **2a.** pentagon **3a.** obtuse
5a. three
Lesson 11-8 2. 36; 81; 400; 10,000
Lesson 11-9 1a. rectangular solid **2a.** 6 **3a.** 5
Lesson 11-10 1a. 64° **2.** 62° **4a.** 42° **5a.** 133° **6.** 65°; 115°;
115° **8a.** 900°
Lesson 11-11 2a. 69° **3a.** 64° **4a.** 35°, 55°, 35° **7a.** 136°
8a. 148° **11a.** 68° **12a.** 110° **14.** 127°
Lesson 11-12 1a. transversal **2.** equal; $\angle 6$; $\angle 4$
4. supplementary **6.** vertical or opposite **8a.** 130° **10a.** Yes
Lesson 11-13 2. $\triangle GHI \cong \triangle JKL$, SAS **4a.** 67°

Chapter 12 Practice Exercises
Lesson 12-3 4a. 2 in., $2\frac{3}{4}$ in., $1\frac{1}{8}$ in. **5a.** 20 mm, 35 mm,
29 mm **8.** Yes **12.** Yes; yes
Lesson 12-6 2. 24 **4a.** 8

Chapter 13 Practice Exercises
Lesson 13-1 1a. 16 **2a.** 0.04 **3a.** $\frac{1}{4}$ **4a.** 36 **5a.** 49 in.²
6a. 25 **7a.** 26
Lesson 13-2 1a. 2 **2a.** 20 **3a.** $\frac{3}{5}$
Lesson 13-3 1a. 5.385 **2a.** 2.449
Lesson 13-4 1a. 15 m **2a.** 12 mm **3a.** 7 cm **4a.** 10 cm
6. 127.3 ft
Lesson 13-5 2a. $\overline{AC}$ **3a.** $\overline{EB}$
Lesson 13-6 2a. 0.9816 **4a.** 21° **6a.** 26° **7a.** 171.56 cm
8a. 392.5 **9a.** 64.95 m **10a.** 21.1 ft **11a.** 69.47 m
12a. 45.1 cm **13a.** 5,724.33 m

Chapter 14 Practice Exercises
Lesson 14-1 2a. 224 ft **4.** 7 ft
Lesson 14-2 2a. 21,120 ft **4.** $104.07
Lesson 14-3 2a. 189 cm **4.** 592 m
Lesson 14-4 1a. 14 cm **2a.** 19 m **3a.** 109.9 cm
4a. 43.96 mm **6a.** 56 m **8.** 21.98 ft **10.** 44 ft
Lesson 14-5 2a. 391 mm²
Lesson 14-6 2a. 100 cm²
Lesson 14-7 2a. 364 in.²
Lesson 14-8 2a. 108 cm²
Lesson 14-9 2a. 56 cm²
Lesson 14-10 2a. 26,002.34 mm² **4a.** 283.385 yd²
6. 1,962.5 ft²
Lesson 14-11 1a. 2,050 cm² **2.** 800 ft²; 4 gal; $58.76
Lesson 14-12 1a. 8,214 cm² **2.** e = 7 in.
Lesson 14-13 1a. 125.6 m² **3a.** 836 m² **6a.** 1,808.64 m²
7a. 803.04 mm²
Lesson 14-14 2a. 144 mm³
Lesson 14-15 2a. 19,683 cm³
Lesson 14-16 2. 1,256 ft³ **4.** second cylinder; $9\frac{3}{7}$ cm³ more
6. 12 lb $\frac{1}{4}$ oz
Lesson 14-17 1a. 113,142$\frac{6}{7}$ m³ **2a.** 9,206.48 mm³
4a. 1,540.69 cm³ **6.** 14,400 m³

Chapter 15 Practice Exercises
Lesson 15-1 1a. negative six **2a.** positive five-sixths **3a.** The
absolute value of positive eleven **4a.** -7 **5a.** $-\frac{5}{8}$ **6a.** -11,
-10, -9, -8, -7, -6, -5, -4, -3, -1, 0, 1, 2, 3 **7a.** -12
8. 5, 0, 1, 19 **11a.** $|+3\frac{9}{10}|$ or $|+3.9|$ **12a.** 7 **13a.** 4 + 8 = 12
Lesson 15-2 1a. -2 **2a.** 0 **5a.** -5, 3
Lesson 15-3 2. an upward force of 75 kg **4.** $+54°$
6. $+39$ mm **8.** $-\$25$ **10.** -17 amperes
Lesson 15-4 1a. Point T **2a.** Point C **3a.** $+8$ **4a.** -8 **5a.** T
6a. $<$ **8a.** $+9, +8, +4, +3, 0, -1, -3, -7, -8, -12$ **10a.** $+4$
Lesson 15-5 1a. $+7$ **2a.** -10 **3a.** The opposite of nine
4a. $-(36)$ **6a.** -9
Lesson 15-6 1a. -5 **2a.** $+43$ **3a.** $+8$ **4a.** -7 **5a.** 0
6a. $+\frac{1}{2}$ **7a.** -20 **8a.** 0 **9a.** $+1$ **10a.** 9
Lesson 15-7 1a. $+3$ **2a.** -2 **3a.** -2 **4a.** $+9$ **5a.** 0
6a. $+22$ **7a.** -4 **8a.** -3 **9a.** $\frac{3}{4}$ **10a.** $+5$ **11a.** 0 **12a.** $+4$
14. -1 **16a.** $+37°F$
Lesson 15-8 1a. $+45$ **2a.** 0 **3a.** $+96$ **4a.** $+81$ **5a.** 5
6a. 18 **7a.** $-\frac{1}{6}$ **8a.** -20 lb
Lesson 15-9 1a. $+6$ **2a.** -8 **3a.** 1 **4a.** -4 **5a.** $11\frac{1}{3}$
6a. -7 **7a.** -10 **8a.** -3 **9a.** $+2$ **10a.** 2 **11a.** 12
12a. -5 points

Chapter 16 Practice Exercises
Lesson 16-1 2. y **4.** b_1 **8.** 12 × 19 or 12 · 19 **12.** $\sqrt{20}$
16. $\frac{12}{4}$ or 12 ÷ 4 **20.** $c - g$ **24.** $\sqrt{s}$ **28.** Nineteen minus
seven **32.** Six times a minus 9 times b **36.** $C \div \pi$ or $\frac{C}{\pi}$
40. $A + B + C$ **44.** $180° - B$ **48.** e^3 **52.** $d + y$
Lesson 16-2 2. Each number x is greater than twenty-nine.
6. Nine times each number m plus four is equal to fifteen.
10. Fifteen times each number x minus four is not greater than
eighteen minus seven times the number x. **14.** $9y + 2 \neq 50$
18. $c + 9 \leq 10$ **22.** Nineteen times each number w is less
than or equal to thirty-eight. **26.** Each number b is greater
than four and less than twenty-seven.
Lesson 16-3 2. $A = lw$ **6.** $A = p + i$ **10.** $A = \frac{1}{2}ab$
14. $P = 2(l + w)$ **18.** $A = h\frac{b_1 + b_2}{2}$ **22.** Fahrenheit temperature
reading equals one and eight-tenths times the Celsius
temperature reading plus 32 degrees.
Lesson 16-4 2. 6 **6.** 84 **10.** 60 **14.** 7 **18.** 216 **22.** 208
26. 81.64 **30.** 90 **34.** 29 **38.** 375 **42.** 90,000 **46.** 6,859
50. 310,464 **54.** 444 **58.** 54 **62.** $\frac{a + b + c + d + e}{5}$; 84
Lesson 16-5 2. 43 **6.** 202 **10.** 512
Lesson 16-6 2. c **6.** e **10.** d **14.** b, c, d, e
Lesson 16-7 2. -22 **6.** $+6$ **10.** 3 **14.** × 6 **18.** × 4
22. × 21 **26.** 12; $x = 15$ **30.** 19; $s = 25$ **34.** 15; $y = 6$
38. 8; $m = 16$ **42.** subtract 11; $m = 10$ **46.** multiply by 10;
$n = 40$ **50.** divide by 12; $y = \frac{3}{4}$ **54.** divide by 8; $n = \frac{1}{4}$
58. $n = \frac{10}{2} \times 2$; $n = \$10$
Lesson 16-8 2. $b = 13$ **6.** $n = 108$
10. $n + \$5.50 = \$210.25$; $n = \$204.75$
Lesson 16-9 2. $l = 23$ **6.** $y = 24$ **10.** $m = 104$
14. $n - 3\frac{1}{2} = 2\frac{3}{4}$; $n = 6\frac{1}{4}$ yd
Lesson 16-10 2. $c = 6$ **6.** $b = 7$ **10.** $h = 1$ **14.** $y = 9\frac{1}{3}$
18. $0.75s = 120$; $s = 160$
Lesson 16-11 2. $s = 96$ **6.** $b = 36$ **10.** $x = 6$

14. $\frac{n}{8} = \$23.95;\ n = \$191.60$

Lesson 16-12 2. $n = 11\frac{4}{5}$ **6.** $c = 445$ **10.** $b = 4$ **14.** $y = 3$
18. $c = 10$

Lesson 16-13 2. 28 **6.** $50° + 40° + c = 180°;\ 90° + c = 180°;$
$c = 90°$

Lesson 16-14 1a. 18 **2a.** 2.8 **3a.** 4 **4a.** 18% **5a.** 87
6a. 5,000 **7a.** 72 **8a.** 21 games

Lesson 16-15 2a. $c > 23$ **4a.** $d < 15$ **6a.** $x \geq 15$ **8a.** $r \leq 14$
10a. $w \neq 17$ **12a.** $m > 8$ **14a.** $h \leq 24$ **16a.** $b \geq 0$
18a. $y \leq 42$ **20a.** $n \neq 60$ **22a.** $b < 7$ **24a.** $t > 8$ **26a.** $n \leq 9$
28a. $z \geq 36$ **30a.** $c \leq -90$ **32.** $S > \$689.50$

Chapter 17 Practice Exercises
Lesson 17-1 2. (4,0) **6.** (5,5); (5,6); (6,5); (6,6)

Photo Credits

Pages 2, 3 Ken O'Donoghue. **Page 5,** Ellis Herwig/The Picture Cube. **Page 7,** Stuart Cohen/Stock Boston. **Page 12,** David Woo/Stock Boston. **Pages 15, 19,** Susan Van Etten. **Page 23,** Larry Dale Gordon/The Image Bank. **Pages 24, 29, 35, 36,** Susan Van Etten. **Page 43,** Jose Carillo/Stock Boston. **Page 44,** Clyde H. Smith/The Stock Shop. **Page 45,** Dr. J. Muller/Uniphoto. **Pages 46, 53, 55,** Susan Van Etten. **Page 65,** Jeff Jackman/The Stock Shop. **Page 66,** Terry McCoy/The Picture Cube. **Page 70,** Ellis Herwig/Taurus Photos. **Page 75,** Steve Weinrebe/Stock Boston. **Pages 77, 80,** Tom Tracy/The Stock Shop. **Page 83,** David Dempster. **Page 88,** Own Franken/Stock Boston. **Page 102,** David Dempster. **Page 108,** Don Klumpp/The Image Bank. **Page 109,** Peter Chapman. **Page 110,** Susan Van Etten. **Page 115,** Charles Marden Fitch/Taurus Photos. **Pages 119, 126,** Susan Van Etten. **Page 127,** Henry Reis/The Stock Market. **Page 133,** Susan Van Etten. **Page 135,** Paul Conklin. **Page 141,** Al Satterwhite/The Image Bank. **Page 142,** top, Jim Pickerell/FPG. **Page 142,** bottom, Bryce Flynn/Stock Boston. **Page 147,** Owen Franken/Stock Boston. **Page 153,** Susan Van Etten. **Page 170,** Peter Chapman. **Page 177,** Susan Van Etten. **Page 187,** Scott Berner/The Stock Shop. **Page 188,** Paul Conklin. **Page 195,** Donald Dietz/Stock Boston. **Page 198,** Susan Van Etten. **Page 200,** Martin M. Rotker/Taurus Photos. **Pages 206, 207,** Susan Van Etten. **Page 209,** David Dempster. **Pages 216–217,** Lionel Atwill/The Stock Shop. **Page 217,** top, Richard Pasley/Stock Boston. **Page 218,** top, Bob Daemmrich/Uniphoto. **Page 218,** bottom, Michael Philip Manheim/Southern Light. **Pages 223, 224,** Susan Van Etten. **Page 227,** Gary Gladston/The Image Bank. **Page 231,** Susan Van Etten. **Pages 233, 242,** David Dempster. **Page 243,** Murray Alcosser/The Image Bank. **Page 248,** Paul Mozell/Stock Boston. **Page 251,** Daniel MacDonald/The Stock Shop. **Page 255,** Paul Conklin. **Page 260,** Freda Leinwand/Monkmeyer Press Photo Service. **Page 262,** insert, David Dempster. **Page 262,** Charles Gupton/Southern Light. **Page 262,** bottom, Jacques Chenet/Woodfin Camp and Associates. **Pages 268–269,** Bill Leatherman. **Page 271,** Eric Roth/The Picture Cube. **Page 283,** Fred Leinwand/Monkmeyer Press Photo Service. **Page 292,** top, David Witbeck/The Picture Cube. **Page 292,** bottom, Murray Alcosser/The Image Bank. **Page 295,** Charles Anderson/Monkmeyer Press Photo Service.

Page 296, Dick Luria/FPG. **Page 298,** Susan Van Etten. **Page 304,** Raoul Hackel/Stock Boston. **Page 306,** Susan Van Etten. **Page 314,** Jon Reis/The Stock Shop. **Page 318,** David Dempster. **Page 320,** D. Cody/FPG. **Page 320,** bottom, Daniel Brody/Stock Boston. **Page 323,** Richard Laird/The Stock Shop. **Page 337,** Ken Kamanisy/The Picture Cube. **Page 339,** Dave Schaefer. **Pages 341, 342, 343,** David Dempster. **Page 344,** top, Ellis Herwig/The Picture Cube. **Page 344,** bottom, Dick Luria/The Stock Shop. **Pages 348–349,** Bill Leatherman. **Page 351,** Julie Houck/Uniphoto. **Page 353,** Thomas Zimmermann/FPG. **Pages 363, 366,** Paul Conklin. **Page 368,** Billy E. Barnes/Southern Light. **Page 371,** Frank Siteman/Stock Boston. **Page 373,** R. Lee/Uniphoto. **Page 375,** Susan Van Etten. **Page 381,** Steve Satushek/The Image Bank. **Page 383,** Robert Rathe/FPG. **Page 390,** top, Mimi Forsyth/Monkmeyer Press Photo Service. **Pages 390,** bottom, 410, Susan Van Etten. **Page 414,** Bill Gallery/Stock Boston. **Page 418,** inset, David Dempster. **Page 418,** Eric Kroll/Taurus Photos. **Pages 422–423,** Ken O'Donoghue. **Page 427,** K. Michalek/Taurus Photos. **Page 429,** Chuck Fishman/Woodfin Camp and Associates. **Page 435,** Harald Sund/The Image Bank. **Page 443,** John Coletti/Stock Boston. **Page 462,** Paul Conklin. **Page 465,** Don Klumpp. **Page 473,** Stacy Pick/Uniphoto. **Page 478,** top, Berle Cherney/Uniphoto. **Page 478,** bottom, Alec Duncan/Taurus Photos. **Pages 481,** Mike J. Howell/The Picture Cube. **Page 494,** Franz Kraus/The Picture Cube. **Page 497,** Gabriel Covian/The Image Bank. **Pages 513, 520–521,** Susan Van Etten. **Page 522,,** Gary Gladstone/The Image Bank. **Page 522,** bottom, Freda Leinwand/Monkmeyer Press Photo Service. **Pages 528–529,** Bill Leatherman. **Page 533,** Comstock. **Page 534,** Richard Slade/Uniphoto. **Page 546,** Susan Van Etten. **Page 548,** Richard Pasley/Stock Boston. **Page 556,** John Bowden/Uniphoto. **Page 556,** bottom, Daniel Brody/Stock Boston. **Page 559,** Comstock. **Page 569,** Rogers/Monkmeyer Press Photo Service. **Page 583,** Susan Van Etten. **Pages 588–589,** Tim Bieber/The Image Bank. **Page 590,** top, Michael L. Abramson/Woodfin Camp and Associates. **Page 590,** bottom, Joan Menschenfreund/Taurus Photos. **Page 590,** insert, David Dempster. **Page 593,** Michael Rochipp/The Image Bank. **Page 601,** top, Mikki Ansin/Taurus Photos.